STRUCTURAL CRASHWORTHINESS AND FAILURE

STRUCTURAL CRASHWORTHINESS AND FAILURE

Edited by

NORMAN JONES
Department of Mechanical Engineering
The University of Liverpool
Liverpool, UK

and

TOMASZ WIERZBICKI
Department of Ocean Engineering
Massachusetts Institute of Technology
Cambridge, Massachusetts, USA

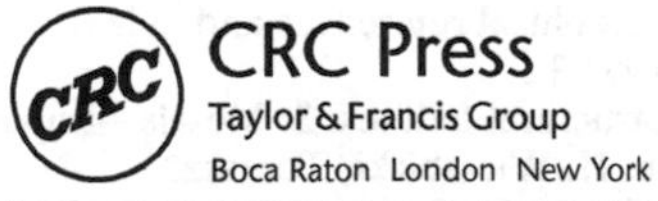

CRC Press is an imprint of the
Taylor & Francis Group, an **informa** business
A TAYLOR & FRANCIS BOOK

CRC Press
Taylor & Francis Group
6000 Broken Sound Parkway NW, Suite 300
Boca Raton, FL 33487-2742

First issued in paperback 2019

ISBN-13: 978-1-85166-969-1 (hbk)
ISBN-13: 978-0-367-86474-3 (pbk)

British Library Cataloguing in Publication Data

Structural Crashworthiness and Failure
I. Jones, Norman II. Wierzbicki, Tomasz
624.1

Library of Congress CIP Data

Structural crashworthiness and failure/edited by Norman Jones and Tomasz Wierzbicki.
p. cm.
Includes bibliographical references and index.
ISBN 1-85166-969-8
1. Impact. 2. Structural failures. 3. Metals—Impact testing.
I. Jones, Norman. II. Wierzbicki, Tomasz.
TA654.2.S77 1993
624.1'71—dc20 92-37840
CIP

Visit the Taylor & Francis Web site at
http://www.taylorandfrancis.com

and the CRC Press Web site at
http://www.crcpress.com

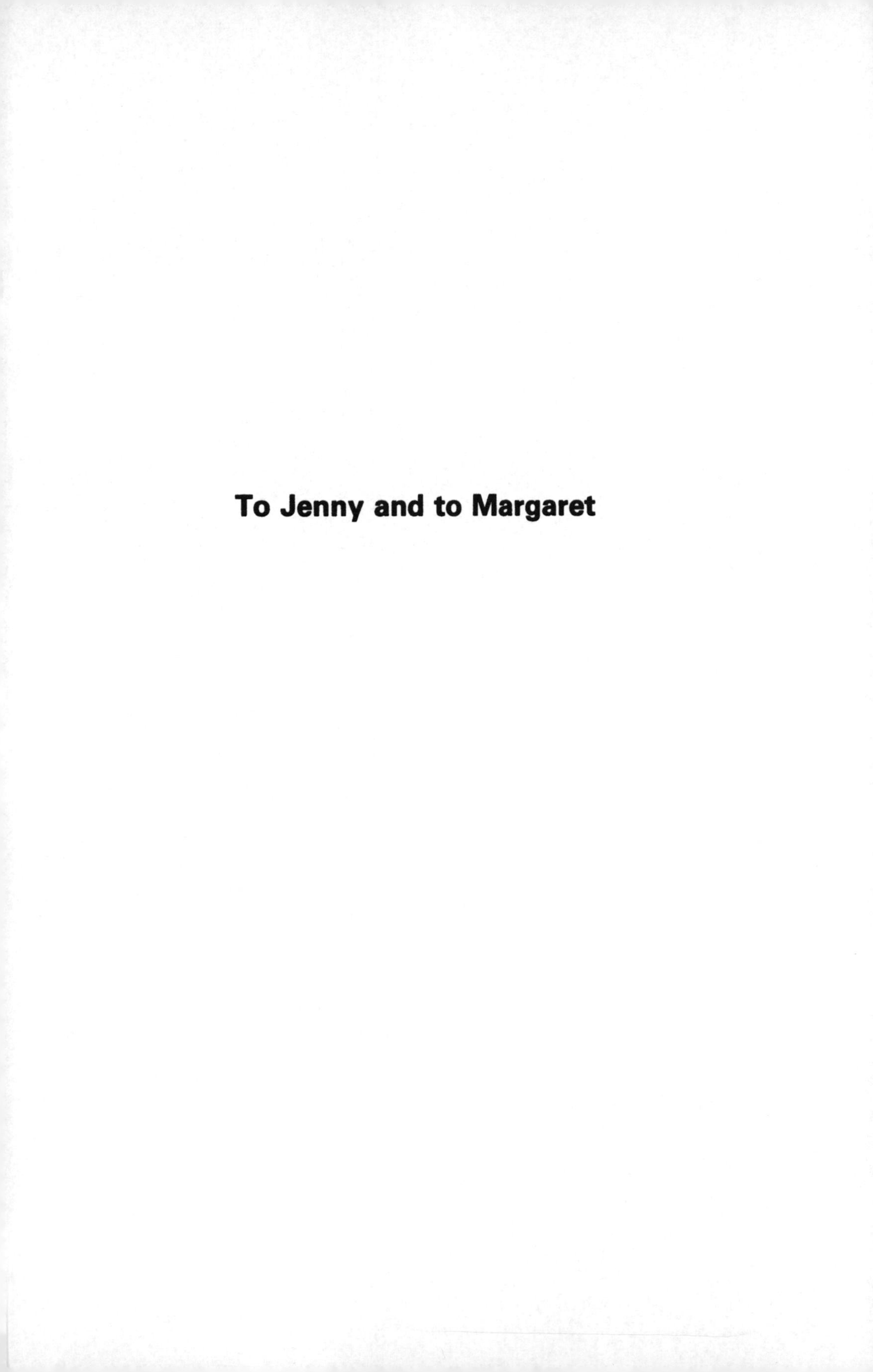

To Jenny and to Margaret

PREFACE

This book contains 12 chapters that were presented as invited lectures at the Third International Symposium on Structural Crashworthiness held at the University of Liverpool, Liverpool, UK, 14–16 April 1993.

This Symposium was a logical continuation and extension of two previous Symposia held in 1983 and 1988, and was organised in order to examine recent developments in structural crashworthiness and failure. Particular emphasis has been given in this book to the failure predictions for ductile metal structures under large dynamic loads and to the behaviour of composite and cellular structures.

The First International Symposium of Structural Crashworthiness was organised by the present editors in 1983, at the University of Liverpool. The proceedings were published in the book *Structural Crashworthiness*[1] and in the special issues of two international journals.[2,3] It was the principal objective of the first meeting to summarize the advances in the structural crashworthiness of a wide range of transportation modes (aircraft, automobiles, buses, helicopters, ships, trains and offshore platforms) and to review the developments in the non-linear large-deflection response of structural members and energy absorbing systems. This meeting emphasized the ductile behaviour of structures and components.

The Second International Symposium on Structural Crashworthiness was organised five years later at MIT, with the proceedings published in the book *Structural Failure*[4] and in special issues of two international journals.[5,6] This meeting emphasized various aspects of structural failure that limit integrity and energy absorbing capacity in many practical situations.

The topics discussed in the first and second symposia are fundamental to the areas of structural crashworthiness and energy absorbing systems as well as for safety calculations and hazard assessments in many industries. For example, a study of the dynamic inelastic behaviour of a side plating on a ship involved in a collision is also of interest for the hazard assessment of an offshore platform deck struck by a dropped object. Similarly, the dynamic transverse indentation of a cylindrical shell or tube is relevant to the structural crashworthiness of an offshore platform brace or leg struck by a ship or the hazard assessment of a land-based gas pipeline struck by a dropped object during maintenance operations, for example. Many other examples can be found from a wide spectrum of industries. These are highly non-linear dynamic structural problems with complex geometries, for which it is essential to integrate the material behaviour into the structural response.

The chapters in this book are now introduced briefly.

Chapters 1–4 focus on the failure of ductile metal structures, and Chapter 5 examines a related topic on the perforation of metal tubes. Bammann *et al.* in Chapter 1 develop an internal state variable type of criterion specifically for numerical applications, which can be used to predict the failure of ductile metal structures subjected to large dynamic loads. These approaches are becoming increasingly possible for designers because of the increase in computing capability per unit cost and other developments over the past 10 years since the First Symposium, though they are not yet quite cheap enough and easy enough to use in many non-military related industries. Homes *et al.* in Chapter 2 use a simpler local damage model and assume that failure occurs when the equivalent plastic strain reaches a critical failure strain that is a function of the stress triaxiality.

An energy density theory is discussed in Chapter 3 for ductile beams subjected to large dynamic loads. This criterion is developed for use with theoretical and numerical formulations based on generalized stresses. Ferron and Zeghloul review in Chapter 4 the topics of strain localization and fracture in metal sheets and thin-walled structures, mainly from a metal-forming viewpoint, but the ideas should have value for future studies on structural failure. Stronge in Chapter 5 reports on the indentation and perforation of metal tubes struck by a missile. This is an area of increasing interest to designers because of the requirement to assess the safety of many engineering systems from a wide range of hazards.

Chapters 6–8 focus on the behaviour of non-metal structures subjected to various large dynamic loads. Kindervater and Georgi in Chapter 6 review the experimental literature on the energy absorption and structural resistance of high-performance structures made from composite materials, mainly from an aircraft crashworthiness perspective, but their treatment contains important information for many other areas. Haug and de Rouvray in Chapter 7 study the numerical simulation and prediction of the crashworthy behaviour of structures and components made from fibre-reinforced composite materials. Particular emphasis is placed on the collision protection of automobiles, but again the results are relevant to many other industries. Reid *et al.* in Chapter 8 review recent studies on the dynamic response of cellular structures and cellular materials. This article concentrates on the basic mechanics and points out the importance of microbuckling and micro-inertia of the individual cells.

Yu in Chapter 9 explores the importance of material elasticity for the behaviour of structures that are subjected to dynamic loads producing large plastic strains.

McGregor *et al.* in Chapter 10 examine the impact performance of aluminium structures, mainly from the viewpoint of the crashworthiness of automobile structures. The principal objective of this study is to provide data for the design of aluminium structures, and, for example, the crushing and bending behaviour of various thin-walled open cross-sections is examined and compared with the behaviour of similar steel components.

Carney in Chapter 11 discusses the energy dissipation characteristics of impact attenuation systems such as guardrail end protection and crash cushions used for motorway safety applications in the United States. The issues for the further development of these protection systems are ventilated, and some suggestions made for further progress.

Finally, Chapter 12 returns to the failure of structural members, but this time for the crushing, cutting and tearing of ships scantlings associated with grounding damage. Analytical methods of analysis are developed and simplified, and compared with the results of small-scale experiments.

The contents of this book and the companion proceedings published in special issues of the *International Journal of Impact Engineering*[7] and the *International Journal of Mechanical Sciences*[8] are relevant to a wide range of safety calculations, hazard assessments and energy absorbing systems throughout engineering, as well as for developments in the field of structural crashworthiness. The public is becoming increasingly aware of safety and environmental issues as a result of several recent accidents,[9,10] and very often dynamic loads play an important role in disasters. Further progress towards the optimization of engineering structures and components, which leads to lighter structures and vehicles, again increases the importance of dynamic loads and structural crashworthiness, and the requirement to estimate the actual failure loads of various systems. Thus it is likely that the field of structural crashworthiness and failure will continue to grow in importance. It is evident from this book that much has already been achieved, but also that much more understanding of the basic mechanics is still required, as well as the generation of design methods and design data.

The editors hope that this book will help to attract research workers into this challenging field and stimulate them to tackle the associated complex non-linear problems from theoretical, experimental and numerical vicwpoints. The prize at the end of the day is a deeper understanding of non-linear structural mechanics as well as the achievement of a cleaner environment and the protection of the public by contributing to a safer world with fewer major accidents.

The editors would like to take this opportunity to thank all of the authors and reviewers for their assistance in the preparation of this book. Thanks are also due to Mrs M. White of the Department of Mechanical Engineering at the University of Liverpool.

Norman Jones
Liverpool, England

Tomasz Wierzbicki
Cambridge, Masschusetts

September 1992

REFERENCES

1. Jones, N. & Wierzbicki, T., eds, *Structural Crashworthiness.* Butterworth, Guildford, UK, 1983.
2. *Int. J. Impact Engng,* **1,** No. 3 (1983).
3. *Int. J. of Mech. Sci.,* **25,** Nos 9/10 (1983).
4. Wierzbicki, T. & Jones, N., *Structural Failure.* Wiley, New York, 1989.
5. *Int. J. Impact Engng,* **7,** No. 2 (1988).
6. *Int. J. Mech. Sci.,* **30,** Nos 3/4 (1988).
7. *Int. J. Impact Engng,* **13,** No. 2 (1993).
8. *Int. J. Mech. Sci.,* **35,** No. 4 (1993).
9. Department of Transport, *Investigation into the Clapham Junction Railway Accident,* HMSO, London, 1989.
10. *Tanker Spills—Prevention by Design.* National Research Council, National Academy of Sciences, Washington, DC, 1991.

CONTENTS

LIST OF CONTRIBUTORS

D. J. Bammann
Sandia National Laboratories, Livermore, California 94550, USA

J. F. Carney, III
Department of Civil Engineering, Vanderbilt University, Nashville, Tennessee 37235, USA

M. L. Chiesa
Sandia National Laboratories, Livermore, California 94550, USA

G. Ferron
LPMM, ISGMP, URA CNRS 1215, Ile du Saulcy, 57045 Metz Cedex 01, France

H. Georgi
Institute for Structures and Design, DLR, Pfaffenwaldring 38–40, 7000 Stuttgart 80, Germany

J. H. Giovanola
SRI International, Menlo Park, California 94025, USA

E. Haug
Research and Development, Engineering System International, SA, 20 rue Saarinen, Silic 270, 94578 Rungis Cedex, France

B. S. Holmes
SRI International, Menlo Park, California 94025, USA

M. F. HORSTEMEYER
Sandia National Laboratories, Livermore, California 94550, USA

N. JONES
Department of Mechanical Engineering, The University of Liverpool, PO Box 147, Liverpool L69 3BX, UK

C. M. KINDERVATER
Institute for Structures and Design, DLR, Pfaffenwaldring 38–40, 7000 Stuttgart 80, Germany

S. W. KIRKPATRICK
SRI International, Menlo Park, California 94025, USA

I. J. MCGREGOR
Alcan International Limited, Banbury Laboratory, Southam Road, Banbury, Oxfordshire OX16 7SP, UK

D. J. MEADOWS
Alcan Automotive Structures (UK) Limited, Southam Road, Banbury, Oxfordshire OX16 7SA, UK

C. PENG
Department of Mechanical Engineering, UMIST, PO Box 88, Manchester M60 1QD, UK

T. Y. REDDY
Department of Mechanical Engineering, UMIST, PO Box 88, Manchester M60 1QD, UK

S. R. REID
Department of Mechanical Engineering, UMIST, PO Box 88, Manchester M60 1QD, UK

A. DE ROUVRAY
Research and Development, Engineering System International, SA, 20 rue Saarinen, Silic 270, 94578 Rungis Cedex, France

C. E. SCOTT
Alcan Automotive Structures (UK) Limited, Southam Road, Banbury, Oxfordshire OX16 7SA, UK

L. SEAMAN
SRI International, Menlo Park, California 94025, USA

A. D. SEEDS
Alcan Automotive Structures (UK) Limited, Southam Road, Banbury, Oxfordshire OX16 7SA, UK

W. Q. SHEN
School of Systems Engineering, University of Portsmouth, Anglesea Building, Anglesea Road, Portsmouth, PO1 3DJ, UK

J. W. SIMONS
SRI International, Menlo Park, California 94025, USA

W. J. STRONGE
Department of Engineering, University of Cambridge, Trumpington Street, Cambridge CB2 1PZ, UK

P. THOMAS
Department of Ocean Engineering, Massachusetts Institute of Technology, Cambridge, Massachusetts 02139, USA

L. I. WEINGARTEN
Sandia National Laboratories, Livermore, California 94550, USA

T. WIERZBICKI
Department of Ocean Engineering, Massachusetts Institute of Technology, Cambridge, Massachusetts 02139, USA

T. X. YU
Department of Mechanics, Peking University, Beijing 100871, China
Present address: Department of Mechanical Engineering, UMIST, PO Box 88, Manchester M60 1QD, UK

A. ZEGHLOUL
LPMM, ISGMP, URA CNRS 1215, Ile du Saulcy, 57045 Metz Cedex 01, France

Chapter 1

FAILURE IN DUCTILE MATERIALS USING FINITE ELEMENT METHODS

D. J. Bammann, M. L. Chiesa, M. F. Horstemeyer & L. I. Weingarten
Sandia National Laboratories, Livermore, California 94550, USA

ABSTRACT

A constitutive model is developed that is capable of predicting deformation and failure in ductile materials. The parameters for the model can be determined by simple tension and compression tests on the material of interest. This plasticity/failure model is implemented into explicit Lagrangian finite element codes. Examples are presented showing that the model can predict ductile failure mechanisms, as benchmarked by carefully planned experiments duplicating real world problems.

1 INTRODUCTION

Sandia National Laboratories has developed a plasticity/failure model that has been implemented into finite element codes to predict failure in ductile metals. The formulation is based on basic theoretical aspects that introduce internal state variables to track plasticity and damage. The model has been used to analyze different types of loading conditions and has proved general enough to solve useful engineering problems that were previously difficult to solve.

A ductile failure criterion has been implemented into the DYNA2D[1] and DYNA3D[2] finite element codes and used to analyze various quasi-static and dynamic loading conditions. These two- and three-dimensional, Lagrangian finite element codes were developed by Lawrence Livermore National Laboratory and are widely used by engineers throughout the world. The damage model has also been implemented in PRONTO2D,[3] PRONTO3D,[4] and JAC3D.[5] The plasticity-specific portion has been implemented into ABAQUS-implicit[6] and NIKE2D.[7]

Until recently, the primary method of predicting failure in ductile materials used simplistic criteria, like the degree of deformation, maximum effective plastic strain, or maximum principal stress. We show in this chapter that such methods are inadequate since the strain at failure is highly dependent on many factors: strain path, strain rate and temperature history, pressure, and effective stress. Only a model that includes all of the pertinent factors like the one presented in this chapter is capable of predicting failure in complex stress states. Some other models are limited to predicting failure in stress states similar to those used to fit the model parameters. The verification and usefulness of the Sandia damage model is demonstrated by showing the solutions to several complex problems and comparing with test results. Current applications of the model include designing penetration-resistant barriers, predicting damage in submarine materials (noted for their high strength and ductility), and design of accident-resistant containers used to transport hazardous materials and nuclear weapons. This chapter covers the theoretical development of the model plus numerical applications that have been verified by carefully designed experiments.

2 THEORY DEVELOPMENT

Over the past two decades, several constitutive models have been developed to describe the damage and ensuing failure of materials. The models of interest in this chapter are of the internal state variable type. In these models, internal variables are introduced to describe the state of the material. Therefore knowledge of the strain rate, temperature, and the current value of the internal variables is sufficient to describe the state of the material. It is well established that the plastic strain is a poor choice for a state variable, both for the deviatoric plasticity and for the damage in the material (i.e. strain to failure). One of the first internal state variable models to be implemented into a finite element code was the Gurson model.[8] This is based on the growth of spherical voids as predicted by Rice and Tracy.[9] The model initially considered the growth of voids in a viscous material and was extended to rigid–plastic, then hardening materials. Finally a tensor variable was added to describe the Bauschinger effect upon reverse loading. In each of these models, care was taken to maintain the form of the yield surface and the evolution of the voids as predicted by Rice and Tracy for a viscous material. As a result, the form of the deviatoric plasticity part of the model was constrained to maintain consistency with the Rice–Tracy equations, which resulted in some inaccuracies in the model's plasticity predictions. Consequently the addition of two arbitrary damage parameters was required to improve the failure predictions.

In the Sandia damage model, we describe the deviatoric plasticity independently of the voids, and then couple the effect of the voids on the plastic flow and the degradation of the elastic moduli. The model in this chapter accounts for the deviatoric deformation resulting from the presence of dislocations and dilatational deformation and ensuing failure from the growth of voids. The kinematics of the model are based upon the multiplicative decomposition of the deformation gradient into elastic, deviatoric plastic, and dilatational plastic parts. The constitutive model is formulated with respect to the natural (stress-free) configuration defined by the plastic deformation. This results in a structure whose current configuration stress variables are convected with the elastic spin. To model the deviatoric plastic flow, both a scalar and a tensor internal variable are introduced to describe the effects of dislocations in cell walls and cell interiors respectively. The evolution equations for these variables are motivated from dislocation mechanics and are in a hardening-minus-recovery format. The use of state variables and the proposed evolution equations enables the prediction of strain rate history and temperature history effects. These effects can be quite large and cannot be modelled by equation-of-state models that assume that stress is a unique function of the total strain, strain rate, and temperature, independent of the loading path. The temperature dependence of the hardening and recovery functions results in the prediction of thermal softening during adiabatic temperature rises. Additional load-dependent softening is possible due to the anisotropy associated with the evolution of the tensor variable.

The effects of damage due to the growth of voids is included through the introduction of a scalar internal variable. This variable tends to degrade the elastic moduli of the material as well as to concentrate the stress in the deviatoric flow rule, thereby increasing the plastic flow and allowing another mechanism of softening. When the damage reaches a critical value, failure occurs. The Cocks–Ashby model of the growth of a spherical void in a rate-dependent plastic material is used to describe the evolution of the damage.[10] This equation introduces only two new parameters, since it is strongly dependent upon the deviatoric plasticity model.

2.1 Constitutive Assumptions

The kinematics associated with the model are discussed thoroughly in Refs 11–13. Based upon a multiplicative decomposition of the deformation gradient into elastic and plastic parts, and assuming linear isotropic elasticity with respect to the natural configuration associated with this decomposition, the assumption of linear elasticity can be written (tensors are denoted by **boldface**),

$$\mathring{\boldsymbol{\sigma}} = \lambda(1-\mathscr{D})\,\mathrm{tr}\,(\mathbf{D}^{\mathrm{e}})\mathbf{1} + 2\mu(1-\mathscr{D})\mathbf{D}^{\mathrm{e}} - \frac{\dot{\mathscr{D}}}{1-\mathscr{D}}\boldsymbol{\sigma} \tag{1}$$

where λ is the Lamé constant, μ is the shear modulus, and the Cauchy stress $\boldsymbol{\sigma}$ is convected with the elastic spin $\mathbf{W}^{\mathrm{e}}$ as

$$\mathring{\boldsymbol{\sigma}} = \dot{\boldsymbol{\sigma}} - \mathbf{W}^{\mathrm{e}}\boldsymbol{\sigma} + \boldsymbol{\sigma}\mathbf{W}^{\mathrm{e}} \tag{2}$$

where, in general, for any tensor variable $\mathbf{x}$, $\mathring{\mathbf{x}}$ represents the convective derivative. It should be noted that the rigid body rotation is included in the elastic spin as discussed in Ref. 12. This is equivalent to writing the constitutive model with respect to a set of directors whose direction is defined by the plastic deformation.[14,15] The damage or porosity is represented by the variable $\mathscr{D}$. Note that the damage enters the elastic assumption in a manner that tends to degrade the elastic moduli. Decomposing both the skew symmetric and symmetric parts of the velocity gradient into elastic and plastic parts, we write for the elastic stretching $\mathbf{D}^{\mathrm{e}}$ and the elastic spin $\mathbf{W}^{\mathrm{e}}$

$$\mathbf{D}^{\mathrm{e}} = \mathbf{D} - \mathbf{D}^{\mathrm{p}} - \mathbf{D}^{\mathrm{V}} - \mathbf{D}^{\mathrm{th}}, \qquad \mathbf{W}^{\mathrm{e}} = \mathbf{W} - \mathbf{W}^{\mathrm{p}} \tag{3}$$

Within this structure, it is necessary to prescribe an equation for the plastic spin $\mathbf{W}^{\mathrm{p}}$ in addition to the normally prescribed flow rule for $\mathbf{D}^{\mathrm{p}}$ and the stretching due to the thermal expansion $\mathbf{D}^{\mathrm{th}}$. As proposed in Refs 15–18, we assume a deviatoric flow rule of the form

$$\mathbf{D}^{\mathrm{p}} = \begin{cases} f(\theta)\sinh\left[\dfrac{|\boldsymbol{\xi}| - \kappa - Y(\theta)(1-\mathscr{D})}{V(\theta)(1-\mathscr{D})}\right]\dfrac{\boldsymbol{\xi}}{|\boldsymbol{\xi}|} \\ \qquad \text{for } |\boldsymbol{\xi}| - \kappa - Y(\theta)(1-\mathscr{D}) \geqslant 0 \\ 0 \quad \text{for } |\boldsymbol{\xi}| - \kappa - Y(\theta)(1-\mathscr{D}) < 0 \end{cases} \tag{4}$$

where θ is the temperature, κ is a scalar hardening variable, $\boldsymbol{\xi}$ is the difference between the deviatoric Cauchy stress $\boldsymbol{\sigma}'$ and the tensor hardening variable $\boldsymbol{\alpha}$,

$$\boldsymbol{\xi} = \boldsymbol{\sigma}' - \tfrac{2}{3}\boldsymbol{\alpha} \tag{5}$$

and $f(\theta)$, $Y(\theta)$, and $V(\theta)$ are scalar functions. Whereas the damage entered the assumption of elasticity in a manner that tended to degrade the elastic moduli, in eqn (4) it tends to concentrate the stress, thereby enhancing plastic flow. From the kinematics, the dilatational plastic flow rule is of the form

$$\mathbf{D}^{\mathrm{v}} = \frac{\dot{\mathscr{D}}}{1-\mathscr{D}}\mathbf{1} \tag{6}$$

Assuming isotropic thermal expansion and introducing the expansion coefficient $\hat{A}$, the thermal stretching can be written as

$$\mathbf{D}^{\mathrm{th}} = \hat{A}\dot{\theta}\mathbf{1} \tag{7}$$

The evolution of the plasticity internal variables $\boldsymbol{\alpha}$ and κ are presceibed in a hardening-minus-recovery format:

$$\mathring{\boldsymbol{\alpha}} = h(\theta)\mathbf{D}^{\mathrm{p}} - [r_{\mathrm{d}}(\theta)\bar{d} + r_{\mathrm{s}}(\theta)]\bar{\alpha}\boldsymbol{\alpha} \tag{8}$$

$$\dot{\kappa} = H(\theta)\,|\mathbf{D}^{\mathrm{p}}| - [R_{\mathrm{d}}(\theta)\bar{d} + R_{\mathrm{s}}(\theta)]\kappa^2 \tag{9}$$

where h and H are the hardening moduli, $\bar{d} = \sqrt{\tfrac{2}{3}}\,|\boldsymbol{D}^{p}|$, $\bar{\alpha} = \sqrt{\tfrac{2}{3}}\,|\boldsymbol{\alpha}|$, $r_{\mathrm{s}}(\theta)$ and $R_{\mathrm{s}}(\theta)$ are scalar functions describing the diffusion-controlled 'static' or 'thermal' recovery and $r_{\mathrm{d}}(\theta)$ and $R_{\mathrm{d}}(\theta)$ are the functions describing dynamic recovery.

The model introduces nine functions to describe the inelastic response. They can be grouped into three basic types: those associated with the initial yield, the hardening functions, and the recovery functions. The temperature dependences of the yield functions are assumed to be of the forms

$$V(\theta) = C_1 \exp\left(-\frac{C_2}{\theta}\right) \tag{10}$$

$$Y(\theta) = C_3 \exp\left(\frac{C_4}{\theta}\right) \tag{11}$$

$$f(\theta) = C_5 \exp\left(-\frac{C_6}{\theta}\right) \tag{12}$$

$Y(\theta)$ decreases rapidly with temperature, while $f(\theta)$ and $V(\theta)$ exhibit much weaker temperature dependence in most cases. The temperature dependences of the hardening functions $H(\theta)$ and $h(\theta)$ should in general be proportional to the temperature dependence of the shear modulus, as described in Ref. 19. Since the temperature of the shear modulus is not readily available for many materials, it is generally simpler to assume an Arrhenius temperature dependence for these hardening functions. Since the recovery functions, in particular for the static recovery, increase rapidly with temperature, the temperature dependences for these functions are also assumed to be of the form

$$r_d(\theta) = C_7 \exp\left(-\frac{C_8}{\theta}\right) \tag{13}$$

$$r_s(\theta) = C_9 \exp\left(-\frac{C_{10}}{\theta}\right) \tag{14}$$

$$R_d(\theta) = C_{11} \exp\left(-\frac{C_{12}}{\theta}\right) \tag{15}$$

$$R_d(\theta) = C_{13} \exp\left(-\frac{C_{14}}{\theta}\right) \tag{16}$$

In the interest of simplicity, we have chosen the same functional form for the evolution of both internal variables. Future research will better reflect the underlying dislocation processes in the internal variables.

The evolution of the damage parameter is motivated from the solution of Cocks–Ashby for the growth of spherical voids in a rate-dependent plastic material and is given by

$$\dot{\mathscr{D}} = \chi\left[\frac{1}{(1-\mathscr{D})^{\bar{m}}} - (1-\mathscr{D})\right]|\mathbf{D}^p| \tag{17}$$

where $\bar{m}$ is a void growth constant and χ is the stress triaxiality factor, given by

$$\chi = \sinh\left[\frac{2(2\bar{m}-1)p}{(2\bar{m}+1)\bar{\sigma}}\right] \tag{18}$$

Note the strong dependence of the evolution of the damage upon the ratio of the tensile pressure p to the effective stress $\bar{\sigma}$.

There are several choices for the form of $\mathbf{W}^p$. If we assume that $\mathbf{W}^p = 0$, we recover the Jaummann stress rate, which, when coupled with a Prager kinematic hardening assumption,[20] results in the prediction of an oscillatory shear stress response in simple shear. Alternatively we can choose

$$\mathbf{W}^p = \mathbf{R}^T[\dot{\mathbf{U}}\mathbf{U}^{-1} - \mathbf{U}^{-1}\dot{\mathbf{U}}]\mathbf{R} \tag{19}$$

which recovers the Green–Naghdi rate of Cauchy stress and has been shown to be equivalent to Mandel's isoclinic state.[11] $\mathbf{U}$ is the right stretch tensor and $\mathbf{R}$ is a proper orthogonal rotation tensor. The model employing this rate allows a reasonable prediction of directional softening for some materials, but in general underpredicts the softening and does not accurately predict the axial stresses that occur in the torsion of a fixed end, thin walled tube. A microstructurally motivated assumption has been evaluated that uses a plastic spin of the form

$$\mathbf{W}^{p} = \left(-\frac{1}{\alpha_n}\right)(\boldsymbol{\alpha}\mathbf{D}^{p} - \mathbf{D}^{p}\boldsymbol{\alpha}) \tag{20}$$

where the evolution of the scalar variable α_n is

$$\dot{\alpha}_n = r_s(\theta)|\boldsymbol{\alpha}|\ \alpha_n \tag{21}$$

This choice for $\mathbf{W}^p$ allows greater flexibility in the prediction of evolving anisotropy. However, more research is necessary to relate this parameter to the underlying physical mechanisms and to develop experimental techniques to better quantify it. For efficiency and ease of implementation, the Jaumman rate has been used for the examples shown in this chapter.

The final equation necessary to complete our description of high-strain-rate deformation is one that allows us to compute the temperature change during the deformation. For problems involving high strain rates, we assume a non-conducting temperature change and follow an empirical assumption that 90% of the plastic work is dissipated as heat. Hence

$$\dot{\theta} = \frac{0{\cdot}9}{\rho C_v}(\boldsymbol{\sigma} \, . \, \mathbf{D}^{p}) \tag{22}$$

where ρ is the material density and C_v the specific heat.

2.2 Uniaxial Stress

The inelastic behavior of the model is governed by the yield functions $f(\theta)$, $Y(\theta)$, and $V(\theta)$; the hardening functions $h(\theta)$ and $H(\theta)$; and the recovery functions $r_s(\theta)$, $R_s(\theta)$, $r_d(\theta)$, and $R_d(\theta)$. To develop a better understanding of the role of these nine functions in the model, we first consider the case of uniaxial compression in which σ is the only nonvanishing component of Cauchy stress and where damage can be neglected. Let α, ε, and ε_p be the normal components along the stress axis of the tensors $\boldsymbol{\alpha}$, $\boldsymbol{\varepsilon}$, and $\boldsymbol{\varepsilon}_p$, respectively. In this case, the flow rule, elasticity relation, and evolution equation for α and κ can be written as

$$\dot{\varepsilon}_p = f(\theta)\sinh\left[\frac{|\sigma - \alpha| - \kappa - Y(\theta)}{V(\theta)}\right] \tag{23}$$

$$\mathring{\sigma} = E(\dot{\varepsilon} - \dot{\varepsilon}_p) \tag{24}$$

and

$$\dot{\alpha} = h(\theta)\dot{\varepsilon}_p - [r_s(\theta) + r_d(\theta)\dot{\varepsilon}_p]\alpha^2 \operatorname{sgn}(\alpha) \tag{25}$$

$$\dot{\kappa} = H(\theta)\dot{\varepsilon}_p - [R_s(\theta) + R_d(\theta)\dot{\varepsilon}_p]\kappa^2 \tag{26}$$

The dependence on the porosity has been neglected in this case, since the initial porosity is negligible, and voids do not grow in compression. Note that for the case of uniaxial stress, all of the factors involving the multiplier $\frac{2}{3}$ disappear. To facilitate the determination of the parameters, we seek an approximate solution to this system of equations. We begin by inverting the flow rule and substituting the total strain rate for the plastic strain rate, resulting in an expression similar to a yield surface. This is a reasonable approximation, since the total strain rate is nearly equal to the plastic strain rate (at least after the first few percent strain) for a constant-strain-rate tensile test. Then

$$\phi = \sigma - \alpha - \kappa - \beta(\dot{\varepsilon}, \theta) \tag{27}$$

where β is given by

$$\beta(\theta, \dot{\varepsilon}) = Y(\theta) + V(\theta) \sinh^{-1}\left[\frac{\dot{\varepsilon}}{f(\theta)}\right] \tag{28}$$

The function $\beta(\theta, \dot{\varepsilon})$ represents the initial rate- and temperature-dependent yield stress of the material.

The rate-independent limit of the yield stress is given by $Y(\theta)$, which dominates β when $\dot{\varepsilon}$ is small. The function $f(\theta)$ determines the strain rate at which the yield behavior exhibits a transition from being essentially rate-independent (low rates) to being rate-dependent (high rates). The value of the function $V(\theta)$ determines the magnitude of the rate dependence at the higher strain rates, as shown in Fig. 1. The function $V(\theta)$ can therefore be directly related to the rate sensitivity parameter m,

$$m = \frac{\partial \ln \sigma}{\partial \ln \dot{\varepsilon}} \tag{29}$$

as

$$V(\theta) = \sigma m \tag{30}$$

Since $V(\theta)$ describes the rate sensitivity of the yield stress in our model, we can write

$$V(\theta) = Y(\theta)m \tag{31}$$

In lieu of experimental data, $V(\theta)$ can be approximated from handbook values of the rate sensitivity parameter m and the yield stress.

Based upon experimental data of the yield stress at different strain rates and temperatues, all of the parameters in the function β can be determined with a nonlinear least squares analysis utilizing the data at

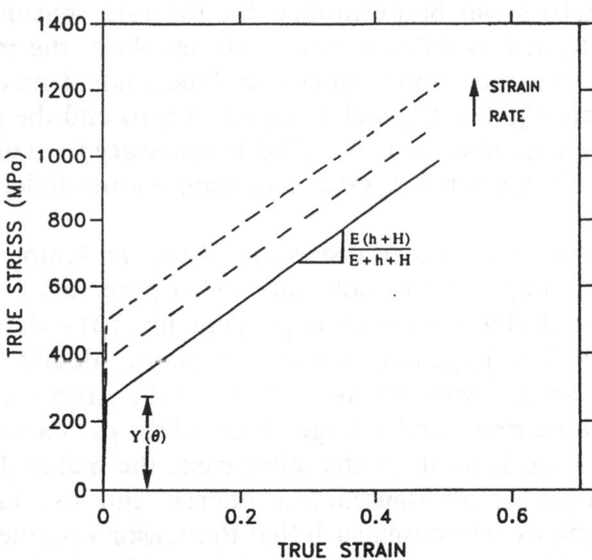

Fig. 1. Rate dependence of the yield stress is obtained by choosing appropriate values for $V(\theta)$ and $f(\theta)$. This, along with the assumptions of zero recovery ($R_d = R_s = r_d = r_s = 0$), results in rate-dependent bilinear hardening.

all available temperatures simultaneously. In order to perform this analysis, it is often convenient to determine the values of $Y(\theta)$, $V(\theta)$, and $f(\theta)$ at two different temperatures and then use these values to determine initial 'guesses' for the parameters $C_1, \ldots, C_6$ when employing all available temperature data (interpolating the model between extreme temperatures is always important, rather than attempting to extrapolate outside the regime of experimental data). In fact, if a nonlinear least squares program is not available, these guesses can be sufficient to determine useful parameters for the model. As an example, consider the function $Y(\theta)$. If the yield stress is known at two extreme temperatures, the constants C_3 and C_4 in eqn (11) can be determined. Assume that two values of the yield stress, such as Y_1 and Y_2, are known from low-strain-rate tensile tests at temperatures θ_1 and θ_2. Then it easily follows that

$$C_4 = \frac{\theta_1 \theta_2}{\theta_2 - \theta_1} \ln\left(\frac{Y_1}{Y_2}\right) \tag{32}$$

and

$$\begin{aligned} C_3 &= Y_1 \exp\left(-\frac{C_4}{\theta_1}\right) \\ &= Y_1 \exp\left[-\frac{\theta_2}{\theta_2 - \theta_1} \ln\left(\frac{Y_1}{Y_2}\right)\right] \end{aligned} \tag{33}$$

A similar analysis can be performed for the rate sensitivity functions $V(\theta)$ and $f(\theta)$ at two different temperatures. Also, the model reduces quite simply to a rate- and temperature-independent model by setting the parameters C_1, C_2, C_4, and C_6 equal to zero and the parameter C_5 equal to a finite number such as 1 (this is necessary to avoid dividing by zero). Then C_3 describes a rate- and temperature-independent yield stress.

Before we begin a detailed discussion of the hardening and recovery functions, it is important to note that the degree to which a material hardens isotropically is strongly dependent upon the definition of the yield stress.[21] For large-deformation problems, details of the short transients associated with the initial 'knee' of the stress–strain curve are generally unimportant, and a large strain offset or 'back-extrapolated' definition of yield is used. In the latter case, the material hardening is mostly isotropic (small Bauschinger effect), and the hardening and recovery functions are chosen such that the tensor variable $\boldsymbol{\alpha}$ represents the directional hardening of the material associated with the development of texture at the microscopic level. Conversely, for small-strain problems, the short transient associated with the initial 'knee' of the stress–strain curve is critical. In these cases, a small-strain definition of the yield stress is used, and the parameters related to the variable α are chosen to describe this small-strain behavior, resulting in a significant amount of kinematic hardening (large Bauschinger effect).

It is instructive at this point to illustrate how the model can be used to describe material behavior ranging from perfect plasticity to the complex behavior described above, simply by appropriate choices of the various parameters. Simultaneously, we will gain a better understanding of the parameter fitting process in the absence of nonlinear least squares programs, or in the case of limited experimental data.

Consider the situation in which $Y(\theta)$ approaches infinity. Here the effective stress will always be less than $Y(\theta)$ and the resulting description is linear elastic. If, however, we choose a more realistic value for $Y(\theta)$, the model simulates elastic–perfectly plastic response. $Y(\theta)$ now represents the rate-independent yield stress of the material.

Next consider the case of a tensile or compressive test conducted at a constant true strain rate $\dot{\varepsilon}$ that is small enough to approximate isothermal conditions. If we initially neglect the effects of recovery ($R_d = r_d = R_s = r_s = 0$), the model is equivalent to bilinear plasticity as seen in Fig. 1. Note that the slope of the stress–strain curve in the plastic regime is approximately given as the sum of the hardening functions h and H. Owing to the manner in which the model is implemented in the finite element codes, the slope of the stress–strain

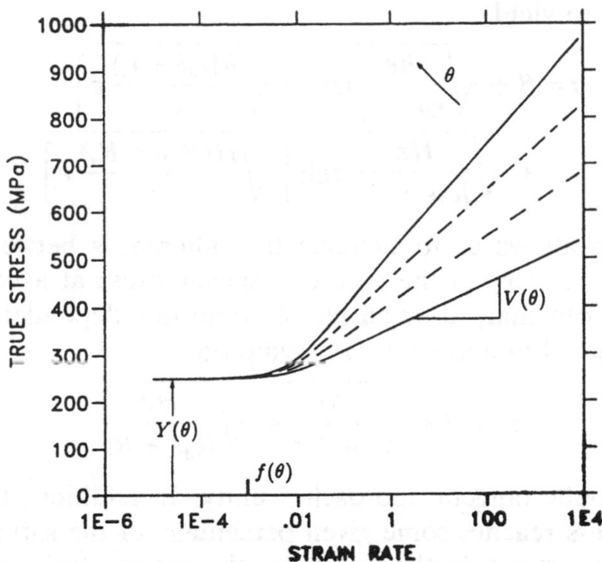

Fig. 2. The rate sensitivity function $V(\theta)$ is the slope of the stress versus logarithmic strain rate curve. In general, $V(\theta)$ increases slightly with increasing temperature. Increasing $f(\theta)$ increases the strain rate at which transition from rate-independent to rate-dependent yield occurs.

curve is described more precisely by

$$\frac{d\sigma}{d\varepsilon} = \frac{E(H+h)}{E+H+h} \tag{34}$$

where E is Young's modulus. In many cases, E is two orders of magnitude larger than $H+h$, and, within experimental accuracy, the slope is just $H+h$. This is particularly true in the case of finite deformation, where the short-term transient is neglected. The relationship between H and h is easily determined from the observed Bauschinger effect in a reverse loading test. The rate dependence of the yield stress is obtained by choosing appropriate values for $V(\theta)$ and $f(\theta)$ as discussed previously and depicted in Figs 1 and 2. It is important to note that this is the strain-rate dependence of the yield stress and not the rate dependence observed at larger strain measures resulting from the operation of different recovery mechanisms.

The manner in which the strain-rate dependence of recovery enters the model is best illustrated by considering all of the parameters in eqns (23)–(26). Utilizing our previous assumptions of constant strain rate and isothermal uniaxial stress together with the assumption that shortly after the yield the plastic strain rate can be reasonably approximated by the total strain rate, the system of constitutive equations (23)–(26) can be integrated to yield

$$\sigma = \beta + \sqrt{\frac{h\dot{\varepsilon}}{r_d\dot{\varepsilon} + r_s}}\tanh\left[\sqrt{\frac{h(r_d\dot{\varepsilon} + r_s)}{\dot{\varepsilon}}}\,\varepsilon\right] + \sqrt{\frac{H\dot{\varepsilon}}{R_d\dot{\varepsilon} + R_s}}\tanh\left[\sqrt{\frac{H(R_d\dot{\varepsilon} + R_s)}{\dot{\varepsilon}}}\,\varepsilon\right] \quad (35)$$

This equation allows us to examine the differences between dynamic and thermal recovery in the case of uniaxial stress at a constant true strain rate. Note that, after an initial strain-rate-dependent yield, the stress is predicted to approach the asymptote

$$\sigma_\infty = \beta + \sqrt{\frac{h\dot{\varepsilon}}{r_d\dot{\varepsilon} + r_s}} + \sqrt{\frac{H\dot{\varepsilon}}{R_d\dot{\varepsilon} + R_s}} \quad (36)$$

as the hyperbolic tangent approaches unity. In addition, the strain at which the stress reaches some given percentage of the saturation stress (the saturation stress is the point on the stress–strain curve where hardening vanishes) is proportional to

$$\varepsilon_{ss}^{\alpha} \propto \sqrt{\frac{\dot{\varepsilon}}{h(r_d\dot{\varepsilon} + r_s)}} \quad (37)$$

for the case in which $\boldsymbol{\alpha}$ represents the longer transient and therefore reaches saturation last, or

$$\varepsilon_{ss}^{\kappa} \propto \sqrt{\frac{\dot{\varepsilon}}{H(R_d\dot{\varepsilon} + R_s)}} \quad (38)$$

if κ approaches its asymptotic value at the larger strain.

Consider now the nature of σ_∞ at various strain rates. For very low strain rates, or at high temperatures when R_s is of the same order as R_d, the static recovery term dominates the denominator of the radical in eqn (36), and σ_α becomes

$$\sigma_\infty(\dot{\varepsilon} \to 0) = \beta + \sqrt{\frac{h\dot{\varepsilon}}{r_s}} + \sqrt{\frac{H\dot{\varepsilon}}{R_s}} \quad (39)$$

The strong rate dependence of recovery leads to predicting responses ranging from perfect plasticity to bilinear hardening over a few orders of magnitude of strain rate, due to the strong rate dependence of the saturation strain. For this case, the saturation strain is given from eqns (37) and (38) as

$$\varepsilon_{ss}^{\alpha} \propto \sqrt{\frac{h\dot{\varepsilon}}{h + r_s}} \quad \text{or} \quad \varepsilon_{ss}^{\kappa} \propto \sqrt{\frac{h\dot{\varepsilon}}{H + R_s}} \quad (40)$$

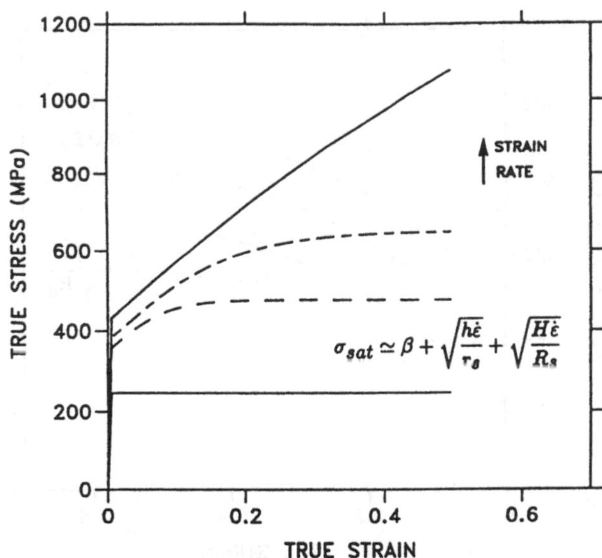

Fig. 3. Predicted stress response at low strain rates or in the case that $R_d = 0$, illustrating the transition from nearly perfect plasticity to nearly bilinear hardening with increasing strain rate.

Figure 3 pictorially illustrates the transition from nearly perfect plasticity to nearly bilinear hardening with increasing strain rate.

At high strain rates, or at low temperatures when R_d is much larger than R_s, the dynamic recovery dominates the denominator in the radical in eqn (36), so that

$$\sigma_\infty(\dot{\varepsilon} \to 0) = \beta + \sqrt{\frac{h}{r_d}} + \sqrt{\frac{H}{R_d}} \tag{41}$$

In this situation, the strain-rate dependence of σ_∞ is solely due to the rate dependence of the initial yield function β. In addition, the strain-rate dependence of the saturation strain also vanishes, so that the asymptotes $\sigma_\infty(\dot{\varepsilon})$ are approached at the same strain for all large $\dot{\varepsilon}$, as illustrated in Fig. 4.

For this case, the saturation strains are given by eqns (37) and (38) as

$$\varepsilon_{ss}^{\alpha} \propto \sqrt{\frac{1}{hr_d}} \quad \text{or} \quad \varepsilon_{ss}^{\kappa} \propto \sqrt{\frac{1}{HR_d}} \tag{42}$$

Unfortunately this loading condition is difficult to achieve, since the heat generated by the plastic deformation prevents the desired isothermal condition at the large strain rates involved.

We are now able to explore in greater detail our earlier statements concerning the choice of α as representing a short-term or a long-term

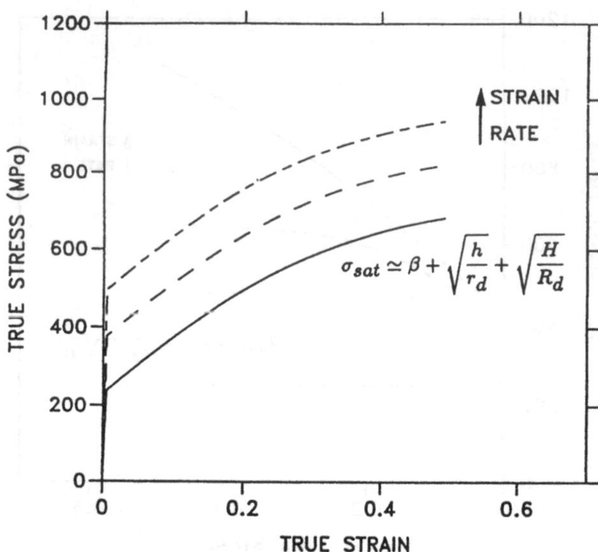

Fig. 4. Predicted stress response at high strain rates or when $R_s = 0$. The rate dependence in this case is due solely to the rate dependence of the yield stress.

transient. This discussion is facilitated by choosing either the static or dynamic recovery variables to be zero. Therefore, for simplicity, assume $R_s = r_s = 0$. For finite-deformation problems in which the short-term transient (i.e. the 'knee' of the stress–strain curve) is neglected, the stress response of many polycrystalline metals can be described as follows. The initial post-yield slope of the stress–strain curve is usually relatively steep, about two orders of magnitude less than the elastic slope. (This is termed stage II hardening by material scientists.) This is followed by a gradual decrease of the slope with increasing strain, and in many materials eventual saturation of the stress (stage III hardening). To model this situation, the hardening function h is chosen to be smaller than the function H, so that the variable κ will dominate the initial hardening behavior. In order to approximate the shape of the stress response, the recovery function R_d will generally have a value such that the saturation of κ occurs at an intermediate strain. At this point, any additional hardening must be modelled by the variable $\boldsymbol{\alpha}$. Owing to the relatively small hardening at this strain, the recovery function R_d can usually be chosen such that a large saturation strain is modelled, as illustrated in Fig. 5. The solid line depicts the stress–strain response due solely to the variable κ. The dashed line shows the predicted response due to the addition of the variable $\boldsymbol{\alpha}$. This allows a more accurate prediction of the stress to larger strain and shifts the strain at which saturation occurs to a more realistic

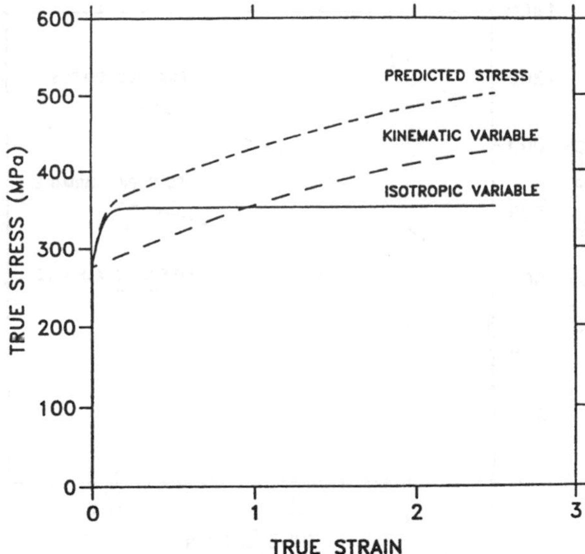

Fig. 5. In this figure, $\boldsymbol{\alpha}$ is used to model large-strain directional hardening. Note that the isotropic hardening function reaches steady state first, and the hardening and recovery of $\boldsymbol{\alpha}$ predict continuous hardening to a larger saturation strain.

value. In addition, the continued evolution of the tensor variable $\boldsymbol{\alpha}$ allows a prediction of anisotropic hardening at large strains due to changes in loading paths as discussed in Ref. 15.

For small-strain problems, in which modelling the short-term transient is important, the function h will be much larger than H (which will still model the same portion of the stress–strain curve as in the long-transient case, and will therefore have similar values) and will describe the initial hardening during the first few percent strain. Figure 6 shows that the variable $\boldsymbol{\alpha}$ will reach a steady state at a much smaller strain, and its saturation value $\sqrt{h/r_d}$ will be much smaller than $\sqrt{H/R}$, the steady state value of the variable κ. The dashed line is the initial rapid hardening due to the variable $\boldsymbol{\alpha}$ and its early saturation, while the solid line results from the continued hardening and eventual saturation of the variable κ. If an accurate description of both the long- and short-term transients are required, an additional variable could be added.

The examples thus far have assumed isothermal conditions. In many applications, deformation occurs rapidly enough that the effects of conduction are negligible. In this situation, the temperature is assumed to increase with the inelastic work as defined by eqn (20). An example of model prediction for the case of isothermal and adiabatic conditions is given in Fig. 7. This figure illustrates the significant thermal softening that occurs at large strains.

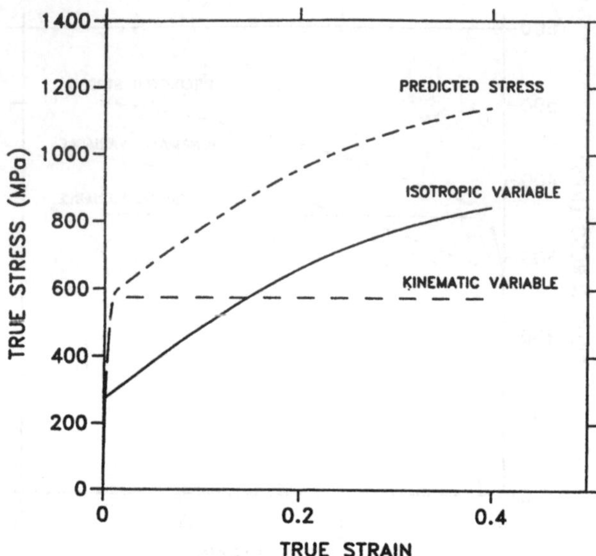

Fig. 6. Here, **α** models the short transient. Note that it reaches a steady state quickly, approximating the 'knee' in the transition from elastic to plastic response.

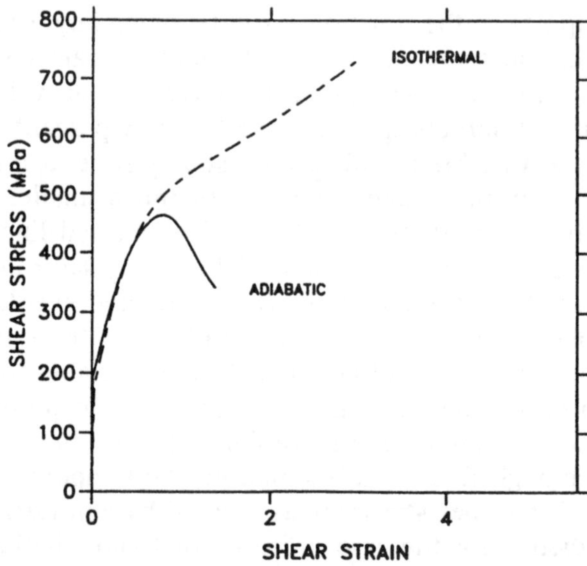

Fig. 7. Comparison of the differences in the prediction of isothermal versus adiabatic response.

3 NUMERICAL IMPLEMENTATION

The damage model has been implemented into and tested in the explicit finite element codes DYNA2D, DYNA3D, PRONTO2D, and PRONTO3D, and the conjugate gradient code JAC3D. The plasticity model (without damage) has been implemented in NIKE2D and the implicit version of ABAQUS. The implementation is based upon the formulation of a numerical flow rule at each time step derived from a numerical consistency condition. It is instructive to consider the general structure of the codes before discussing the details of the implementation. The solution in finite element codes is based on an operator split. This means that equilibrium is enforced assuming that the state of stress is constant over the time step, resulting in the determination of a velocity gradient (or strain increment for implicit codes). The stress is then updated at the end of the time step, assuming the velocity gradient is constant. When implementing a damage or temperature change associated with this model, it is convenient to introduce a second operator split. This is accomplished by assuming that damage (and/or temperature) is also constant when updating the stress. The damage is then updated for the next time step using the new values of stress, velocity gradient, and state variables. This results in a simpler and more efficient implementation, and is consistent with the operator split between the constitutive equations and the equilibrium equations. For explicit codes, in which the time step is extremely small (based on the wave speed through the smallest element), this operator split is almost always insignificant. For implicit codes, where the time step may be several orders of magnitude greater, this may cause inaccuracies. A detailed study of the inaccuracies introduced by these operator splits is in progress.

For maximum efficiency, the radial return method as proposed by Krieg[22] is employed to integrate the stresses. The implementation of the damage model into an explicit code is described as follows.

Step 1. The temperature- and rate-dependent functions in eqns (10)–(16) are evaluated using θ_N and $\mathbf{D}_{N+1/2}$, where the N subscript denotes the value at time step N, and $N+1$ denotes the value at time step $N+1$.

Step 2. Trial stresses and state variables are calculated assuming elasticity using eqns (1), (8), and (9):

$$\boldsymbol{\sigma}^{\mathrm{TR}} = \lambda(1-\mathscr{D}N)\,\mathrm{tr}\,(\mathbf{D}_{N+1/2})\,\mathbf{1} + 2\mu(1-\mathscr{D}_N)\mathbf{D}_{N+1/2} - \frac{\dot{\mathscr{D}}_N}{1-\mathscr{D}_N}\boldsymbol{\sigma}_N. \quad (43)$$

$$p_{N+1} = \tfrac{1}{3}\operatorname{tr}(\boldsymbol{\sigma}^{\mathrm{TR}}) \tag{44}$$

$$\boldsymbol{\sigma}'^{\mathrm{TR}} = \boldsymbol{\sigma}^{\mathrm{TR}} - p_{N+1}\mathbf{1} \tag{45}$$

$$\boldsymbol{\alpha}^{\mathrm{TR}} = \boldsymbol{\alpha}_N - (r_s + r_d\bar{d})\bar{\alpha}\boldsymbol{\alpha} \tag{46}$$

$$\kappa^{\mathrm{TR}} = \kappa_N - (R_s + R_d\bar{d})\kappa^2 \tag{47}$$

Step 3. The elastic assumption is now checked by substitution into the yield condition:

$$\phi^{\mathrm{TR}} = |\boldsymbol{\sigma}^{\mathrm{TR}} - \tfrac{2}{3}\boldsymbol{\alpha}^{\mathrm{TR}}| - \sqrt{\tfrac{2}{3}}\,(\kappa^{\mathrm{TR}} + \beta)(1 - \mathcal{D}_N) \tag{48}$$

Step 4. If $\phi^{\mathrm{TR}} \leqslant 0$ then the step is elastic, stresses at step $N+1$ are set to trial values, and temperature and damage are not changed; if $\phi^{\mathrm{TR}} > 0$ then plastic deformation has occurred and the trial stresses must be returned to the new yield surface:

$$\boldsymbol{\sigma}'_{N+1} = \boldsymbol{\sigma}'^{\mathrm{TR}} - \int 2\mu\mathbf{D}^{\mathrm{p}}\,\mathrm{d}t \tag{49}$$

$$\boldsymbol{\alpha}_{N+1} = \boldsymbol{\alpha}^{\mathrm{TR}} + \int h\mathbf{D}^{\mathrm{p}}\,\mathrm{d}t \tag{50}$$

$$\kappa_{N+1} = \kappa^{\mathrm{TR}} + \int H\bar{d}\,\mathrm{d}t \tag{51}$$

Step 5. Using the radial return method, the assumption is then made that the plastic strain rate is constant over the time step and in the direction of the effective stress:

$$\boldsymbol{\xi} = \boldsymbol{\sigma}'^{\mathrm{TR}} - \tfrac{2}{3}\boldsymbol{\alpha}^{\mathrm{TR}} \tag{52}$$

$$\mathbf{n} = \frac{\boldsymbol{\xi}}{|\boldsymbol{\xi}|} \tag{53}$$

$$\int \mathbf{D}^{\mathrm{p}}\,\mathrm{d}t = \Delta\gamma\,\mathbf{n} \tag{54}$$

Step 6. Substituting the expression for the plastic strain increment yields

$$\boldsymbol{\sigma}'_{N+1} = \boldsymbol{\sigma}'^{\mathrm{TR}} - 2\mu\,\Delta\gamma\,\mathbf{n} \tag{55}$$

$$\boldsymbol{\alpha}_{N+1} = \boldsymbol{\alpha}^{\mathrm{TR}} + h\,\Delta\gamma\,\mathbf{n} \tag{56}$$

$$\kappa_{N+1} = \kappa^{\mathrm{TR}} + H\sqrt{\tfrac{2}{3}}\,\Delta\gamma \tag{57}$$

Step 7. These expressions are then used in the consistency condition ($\phi = 0$) to solve for $\Delta\gamma$. (Note that for kinematic hardening this is not a straightforward solution.) For the assumptions made here, this results in a linear algebraic equation. It is then a simple matter to find the stress and state variables at $t = N+1$ by substituting the value for $\Delta\gamma$ into eqns (55)–(57):

$$\Delta\gamma = \frac{\phi^{TR}}{2\mu + \frac{2}{3}(h+H)} \tag{58}$$

Step 8. Temperature is then updated according to eqn (22), and the damage is updated assuming stresses were constant over the time step. The equation for rate of void growth, eqn (17), can be very stiff numerically, and integration using an Euler forward difference can generate numerical errors. Equation (17), however, can be integrated analytically for constant values of p, $\bar{\sigma}$, and $\mathbf{D}_p$. It should be noted that for 32-bit computers the 'exact' solution can contain numerical roundoff errors, so the Euler method must be used. The exact solution is used on 64-bit machines or 32-bit machines with double precision.

The exact solution is

$$\chi = \sinh\left[\frac{2(2\bar{m}-1)p}{(2\bar{m}+1)\bar{\sigma}}\right] \tag{59}$$

$$\mathcal{D}_{N+1} = 1 - \{1 + [(1-\mathcal{D}_N)^{1+\bar{m}} - 1]\exp[\bar{d}^p\chi(1+\bar{m})\,\Delta t]\}^{1/(1+\bar{m})} \tag{60}$$

The Euler method solution is

$$\mathcal{D}_{N+1} = \mathcal{D}_N + \Delta t\,\dot{\mathcal{D}}_{N+1} \tag{61}$$

In the DYNA implementation, damage is limited to a maximum value of 0·99 to prevent numerical problems. At this value, the stiffness and yield of the material are 1% of their original values. This results in a very soft element incapable of carrying significant loads. Large element distortions may sometimes occur, since the element is not removed from the calculations, which may cause large time-step reductions or code failure. In PRONTO, the DEATH option may be used to completely remove failed elements from further calculations. After an element fails, the applied load is redistributed to surrounding elements, which can then subsequently fail. In this manner, a diffuse crack (one-element wide) is then propagated through the structure. It should be noted that only the first element failure is accurately modelled, since the stress concentration due to the failed element (and formation of a sharp crack) is not precisely described. Implementation of a dynamic automated rezoner would be required to more accurately model this effect. However, in most dynamic cases, reasonable results can be obtained with the present formulation.

Using more accurate assumptions in the implementation results in more complex solutions for $\Delta\gamma$ and thus less computational efficiency. For example, using the plastic strain rate $|\mathbf{D}^p|$ in eqns (46) and (47) results in a quadratic solution for $\Delta\gamma$. Using the plastic strain rate in the expression for β, from eqn (28), results in a highly nonlinear equation for $\Delta\gamma$ that must be solved iteratively. For most large-deformation problems, the more accurate solutions are unnecessary. This is discussed in more detail in Ref. 13.

The model has been implemented into shell elements in the DYNA3D code. The implementation follows that of Hallquist[2] and Whirley.[23] Use of shell elements for thin structures results in significantly less computer time due to larger time step sizes and reduced number of elements. Solid elements, however, must be used for problems in which the thickness stresses are important. More details of the shell implementation can also be found in Ref. 13.

Integration of the model into implicit codes requires significantly more sophistication due to the larger time steps and associated numerical inaccuracies. The requirement of an accurate material stiffness matrix ($\mathbf{c} = d\boldsymbol{\sigma}/d\boldsymbol{\varepsilon}$) further increases the difficulty. The implementation into the implicit version of the ABAQUS code, using a Simo–Taylor stiffness matrix[24] will be discussed in a forthcoming paper.

4 MATERIALS PROPERTY TESTS

As previously described, the equations of the model can be integrated analytically for the case of isothermal, constant-strain-rate, uniaxial stress, eqn (35). The deviatoric plasticity parameters can then be determined utilizing a nonlinear regression analysis of tension or compression test data at constant strain rate.

Consider the response of 304L stainless steel from 20 to 900°C shown in Fig. 8.[25] At 20°C, the rate dependence is due almost entirely to the rate dependence of the yield stress. With increasing temperature, the static recovery term R_s, which is quite small at 20°C, increases, resulting in a large rate dependence of the saturation (peak) stress. In Fig. 9 the predicted response in torsion of a thin-walled tube is compared with the data of Semiatin & Holbrook,[26] utilizing the parameters determined from the tension tests depicted in Fig. 8. These results were computed with the assumption that the state of stress and deformation in the thin-walled tube are uniform, and therefore the constitutive equations

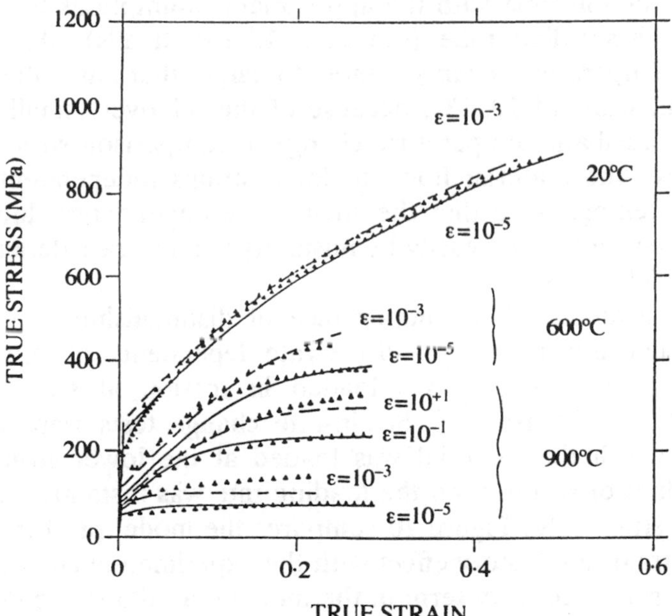

Fig. 8. Parameter fit for 304*L* SS in uniaxial tension. Note the strong effects of dynamic recovery at 20°C and static recovery at 900°C. Model calculations are shown by lines and data by markers.

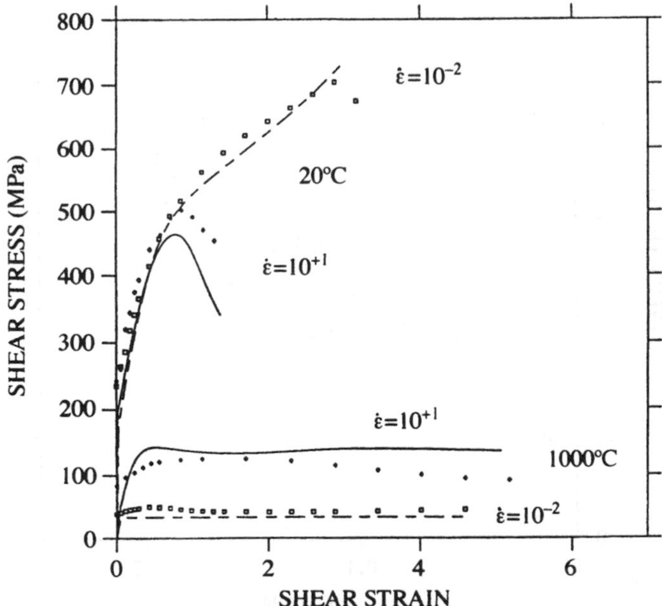

Fig. 9. Comparison of model and data for torsion of thin-walled tubes of 304*L* SS at 20 and 1000°C for strain rates of 0·01 and 10·0 s^{-1}. Note the thermal softening at high rate and low temperature. Calculations are shown by lines and data by markers.

were integrated in time with the appropriate assumptions for the stress state in a thin-walled tube (see Ref. 12 for details). At 20°C, the adiabatic temperature change leads to large thermal softening and localization, while at 1000°C, because of the relatively small contribution of the adiabatic temperature change in comparison with the initial temperature, the material flows to large strains independently of the thermal softening. Note that the strong rate dependence due to static (thermal) recovery could easily be misinterpreted as rate dependence of the initial yield stress.

Another example of the importance of distinguishing between the rate dependence of recovery and the rate dependence of initial yield is seen in Fig. 10. Copper was loaded in torsion at strain rates of $2 \times 10^{-4}\,s^{-1}$ and $3 \times 10^{2}\,s^{-1}$.[27] Strain-rate change tests were then conducted in which the material was loaded at the lower strain rate to various values of strain; then the loading rate was instantly switched to the higher strain rate. Figure 10 compares the model prediction of this apparent strain-rate history effect with the experimental curves. It is the thermal or static recovery term in the model that allows the description of this type of response. It is important to note that this type of response cannot be predicted by empirical models in which the stress is

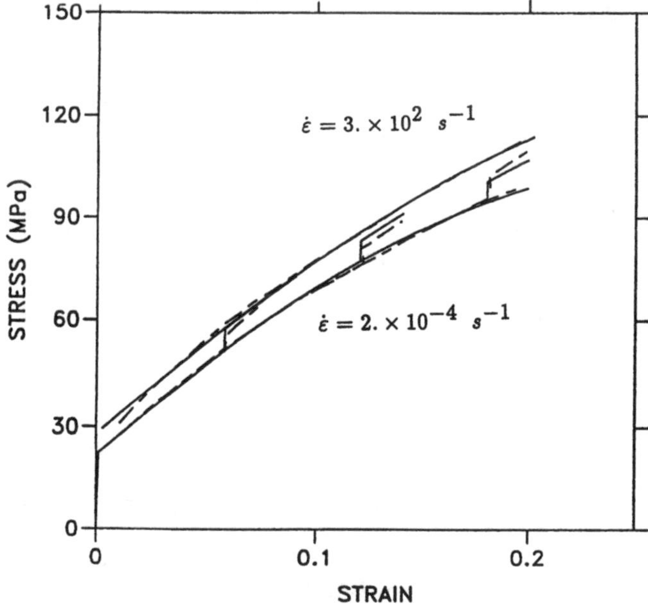

Fig. 10. Strain-rate change tests for copper, loaded in compression, illustrating the model's ability to predict an apparent strain-rate history effect. Solid lines are analysis and dashed lines are data.

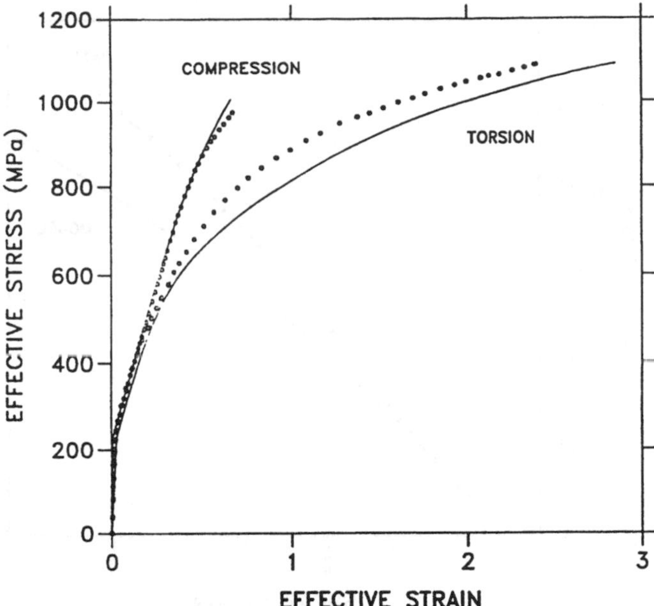

Fig. 11. Comparison of stress response of model (solid line) and data (dots) for free-end torsion of thin-walled tubes of 304*L* SS and uniaxial compression at 20°C.

assumed to be a function of the current values of strain, strain rate, and temperature.

Figures 11 and 12 illustrate the anisotropic hardening that occurs in torsion in comparison to compression. The experiments are compression of a solid cylinder[28] and the free-end torsion of a thin-walled tube.[29] The difference between the response of the two loading cases results from the textural evolution at the microscopic level. The model's ability to predict this response results from the evolution of the tensor variable $\boldsymbol{\alpha}$. Figure 12 illustrates the axial strain that occurs during the torsion of the tube. Note that a significant amount of axial strain is generated. If a fixed-end tube is considered, the stress state is changed significantly, since the axial stress is of the same order of magnitude as the shear stress. This is important in situations such as damage evolution, where an accurate determination of the three-dimensional stress state is essential.

As another example consider the parameter determination for aluminum 6061-T6 at various temperatures. The plasticity parameters are determined as before. Figure 13 illustrates how well the model fits the test data for a large strain compression test, while Fig. 14 shows the temperature-dependent fit of the model versus uniaxial tension test data. The damage parameters are then determined utilizing axisym-

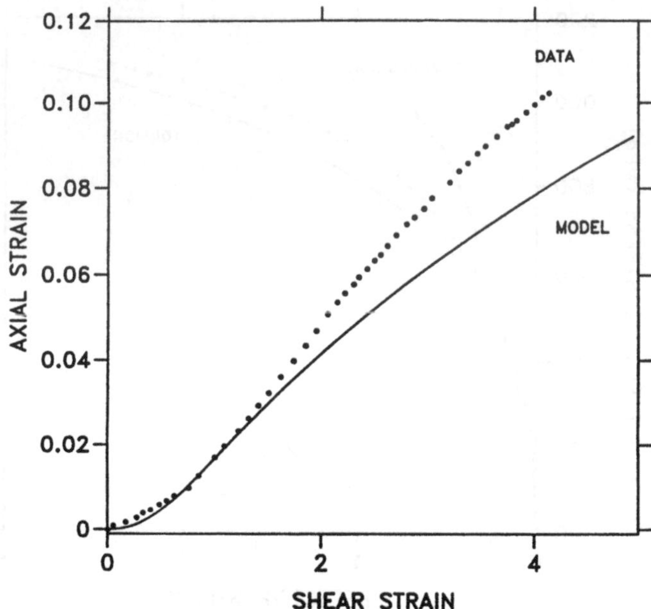

Fig. 12. Axial strain generated during the torsion of the 304*L* SS tube in Fig. 11.

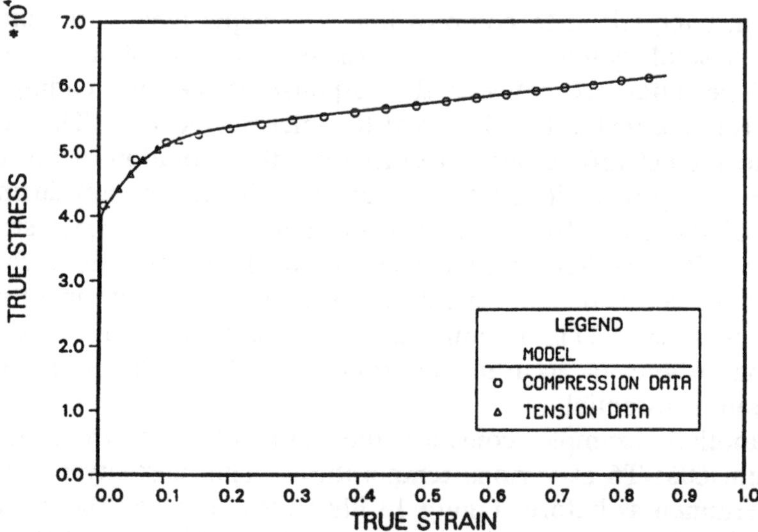

Fig. 13. Comparison of the model versus test for tension and compression tests of Al 6061-T6 conducted at 20°C. The compression data were used to determine the model parameters.

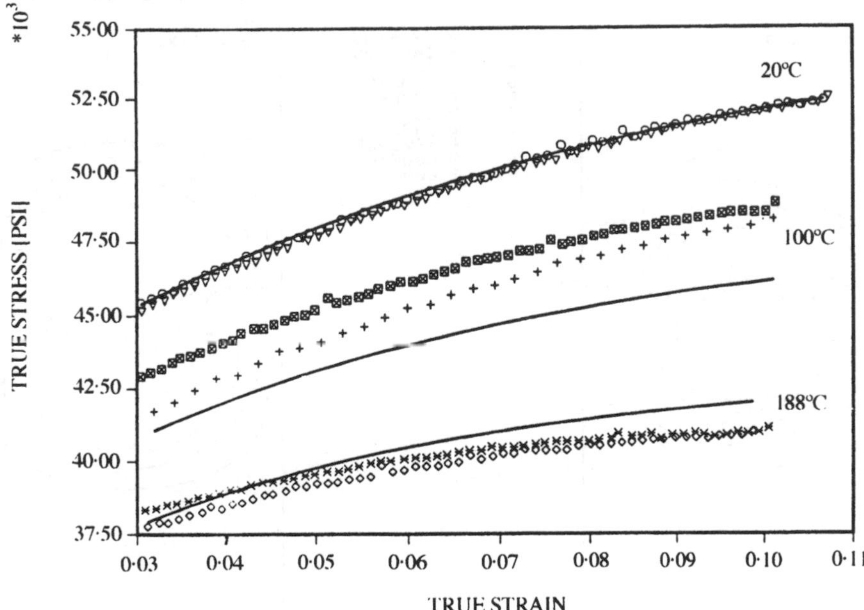

Fig. 14. Comparison of temperature-dependent model prediction versus tension tests for Al 6061-T6. The tests were conducted at 20, 100, and 188°C.

Fig. 15. Finite element meshes of the axisymmetric notched tensile specimens of varying radii of curvature. (Gauge diameter 0·25 (6·35 mm), with notch radii from left to right of 1·0, 2·0, 4·0, and 9·9 mm.)

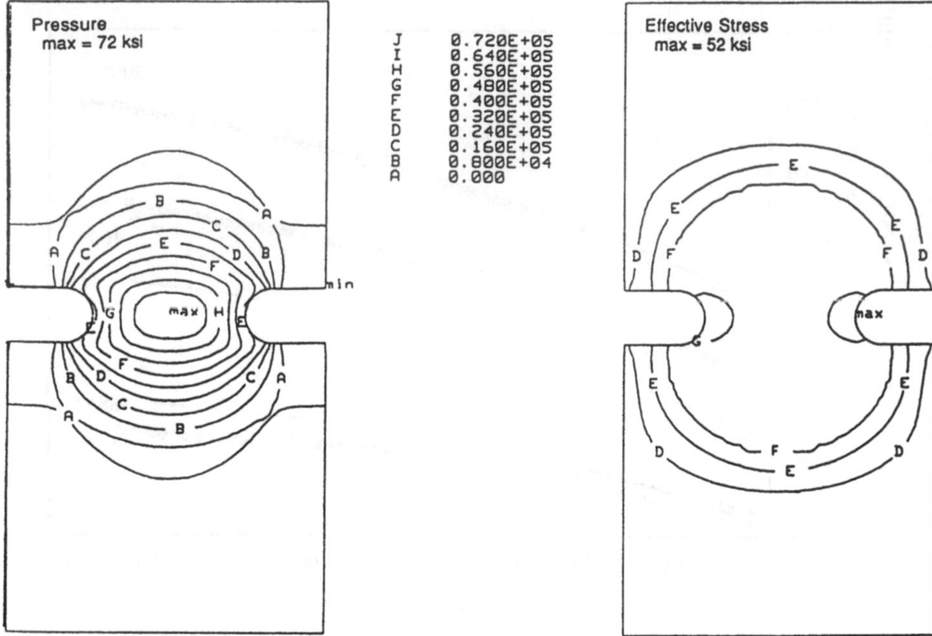

Fig. 16. A comparison of the tensile pressure with the von Mises effective stress in a notched tensile specimen. The ratio of pressure to effective stress at the center of the specimen is 1·65. (Contours are in psi; 1 MPa = 145 psi)

metric notched tensile specimens (Fig. 15). The determination of these parameters requires a trial and error finite element analysis. The notch specimens are useful because the radius of curvature of the notch strongly influences the evolution of damage, due to the significant tensile pressure that develops in the specimen as seen in Fig. 16. The

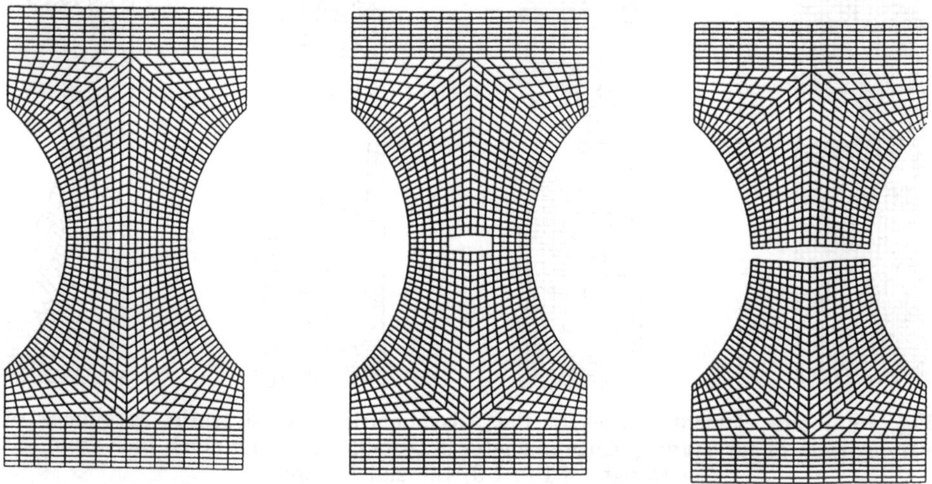

Fig. 17. Progressive failure of a notched tensile specimen.

Table 1
Strain to Failure: Test versus Prediction

Radius (mm)	*Number of tests*	*Test average*	*Calculation*
9·9	3	0·043	0·044
4·0	5	0·021	0·023
2·0	5	0·014	0·015
1·0	5	0·011	0·013

material parameter for void growth is typically determined from one of the notched tests, and the remaining three tests are then used as verification of the model. The progressive failure of a notched specimen is shown in Fig. 17. Two elements appear to fail simultaneously in the axial direction, since only half the specimen was actually analyzed. The predicted effective plastic strain at failure ranged from 4·6% (for a notch radius of 1 mm) to 32% (for a notch radius of 9·9 mm). A comparison of the predicted strain to failure with the test data is given in Table 1. This strain is the elongation over a 1 in gauge length at first observed material failure. The model accurately predicts the strain to failure over the entire range of radii tested.

5 APPLICATIONS

The damage model has been instrumental in accurately predicting deformations and damage in structures under dynamic loads for a variety of problems. The applications mentioned in this section include the study of responses to impinging penetrators and explosively generated blast loads. These analyses have been performed for the design of penetration-resistant structures and for the response of engineering structures that contain high-strength and ductile materials.

5.1 Penetrators

Penetration is an often occurring event that causes large, nonlinear deformation, sometimes coupled with failure of the target material. Penetration studies that exercised the damage model included flat, spherical, and cylindrical targets. The model was very accurate for predicting deformation and failure in metals. These analyses also showed the inaccuracies of using effective plastic strain as a failure criteria.

Disk Targets

Experiments were performed in which 2·24 in (5·72 cm) diameter, 6061-T6 aluminum disks that were 0·125 in (3·2 mm) thick were impacted with a hardened steel rod. The disks were held in a manner that simulated free boundary conditions.[30] The plasticity and damage parameters were defined from tensile, compressive, and notched tests. These tests were described in the previous section.

Initial failure, in the form of visible cracks on the side of the disk opposite the bar, occurred at an impact velocity between 3120 and 3290 in/s (79 and 84 m/s). A finite element model of the experiment using the DYNA2D analysis code predicted initial failure to occur at impact velocities between 3300 and 3500 in/s (84 and 89 m/s). In the experiments, the disk failed completely through its thickness at an impact velocity between 3610 and 4180 in/s (92 and 106 m/s). The analysis indicated that an impact velocity between 4000 and 4200 in/s (102 and 107 m/s) would be required to fully cut a plug from the disk. Figures 18 and 19 illustrate the excellent comparison between predicted failure and experimental results.

Figures 20 and 21 show the effective plastic strain and damage contours from a DYNA2D analysis for a bar velocity of 2500 in/s (64 m/s), a velocity much less than that which initiated failure. Note that the bar hit the disk and rebounded by this time. The predicted maximum effective plastic strain is approximately 25%, nearly twice the 15% failure elongation for a uniaxial tension test. This illustrates the inappropriateness of using uniaxial tensile strain to predict failure. Also note that the damage variable is much more localized than the plastic strain and that the maximum values of damage and plastic strain occur at the boundary of the far side of the disk.

Figures 22 and 23 show the plastic strain and damage contours for an impact velocity of 3300 in/s (84 m/s). The peak value of the damage variable was still on the back surface of the disk, and no failure occurred. However, the peak value of the plastic strain was no longer located at the boundary, but was found in the interior of the disk. The maximum plastic strain in the figure is 42%, much larger than the uniaxial tensile failure strain for this material.

Figure 24 shows the contours of plastic strain just before failure occurred in an element on the back of the disk. The initial impact velocity used here was 3500 in/s (89 m/s), which was high enough to cause two elements through the disk thickness to fail. Figure 24 was produced with the damage model 'active'. Figure 25 was produced with the damage model 'inactive'. The sets of contours are quite similar, and the maximum strains are very close, 45·1% and 45·8%. Figures 26 and

3500 in/s

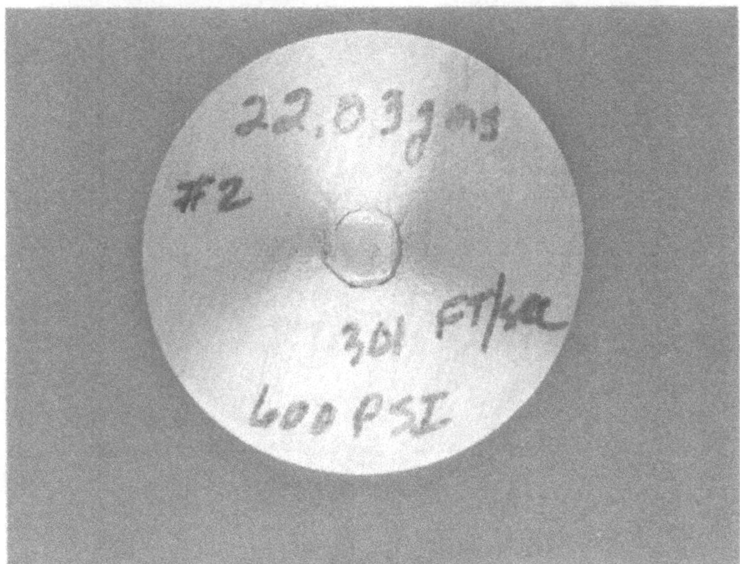

3610 in/s

Fig. 18. Comparison of test and analysis showing that failure has formed on the convex side of disk.

27 show the plastic strain contours later in time. The distributions of the effective plastic strains are very different between the analysis that includes a damage model and the one that does not.

Cylindrical/Spherical Targets

Another example of the use of the model was a study involving a series of finite element analyses that were performed to study penetrations of

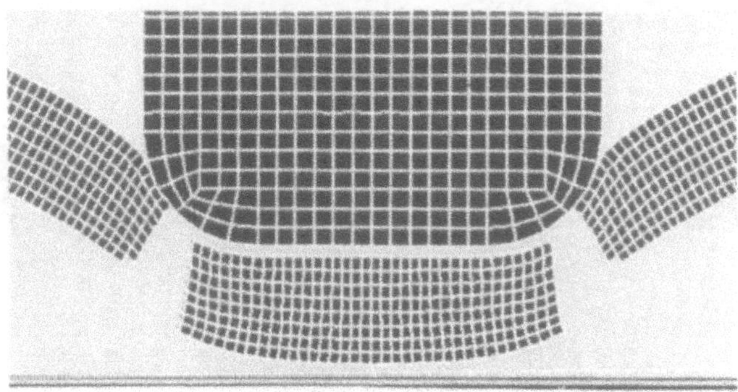

4200 in/s

4180 in/s

Fig. 19. Comparison of test and analysis showing that a crack has formed through the full thickness of the disk.

medium velocity projectiles (~200 m/s) impacting thin spherical and cylindrical shells.[31] The material used for these cylinders and spheres was 6061-T6 aluminum. The penetrator's velocity was varied to determine the onset of target failure. DYNA2D calculations were performed to determine the amount of damage caused by varying the penetrator's velocity. The results obtained from this study provided

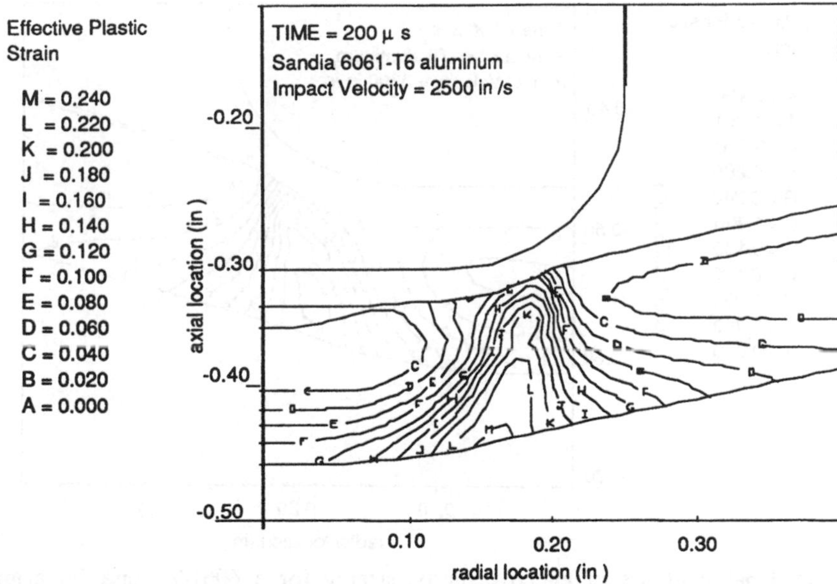

Fig. 20. Final contours of effective plastic strain for a 6061-T6 disk impacted at 2500 in/s (64 m/s), where no failure occurs.

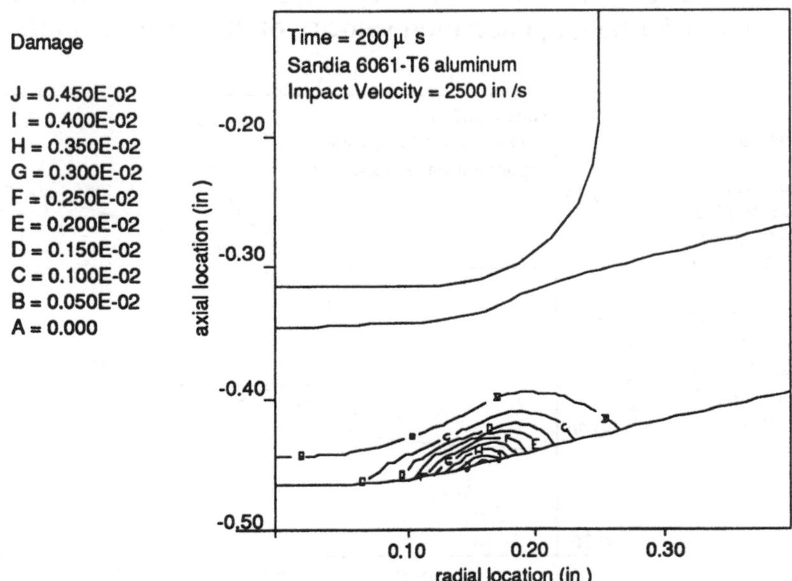

Fig. 21. Final contours of damage for a 6061-T6 disk impacted at 2500 in/s (64 m/s), where no failure occurs.

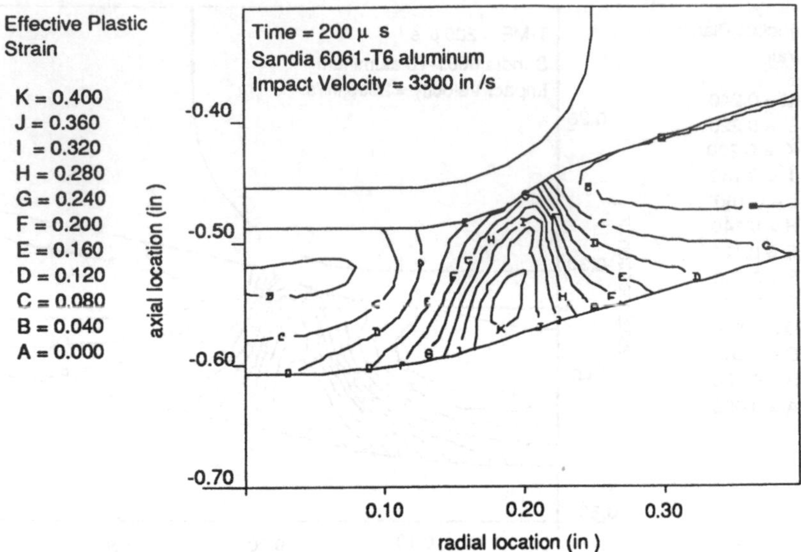

Fig. 22. Final contours of effective plastic straing for a 6061-T6 disk impacted at 3300 in/s (84 m/s), where no failure occurs.

designers with critical data required to optimize the design; for example, the spherical barrier needed a greater penetrator velocity than the corresponding cylindrical barrier to cause failure. The spherical barrier had a penetration velocity of 7656 in/s (194 m/s), and the cylindrical barrier had a penetration velocity of 4606 in/s (117 m/s).

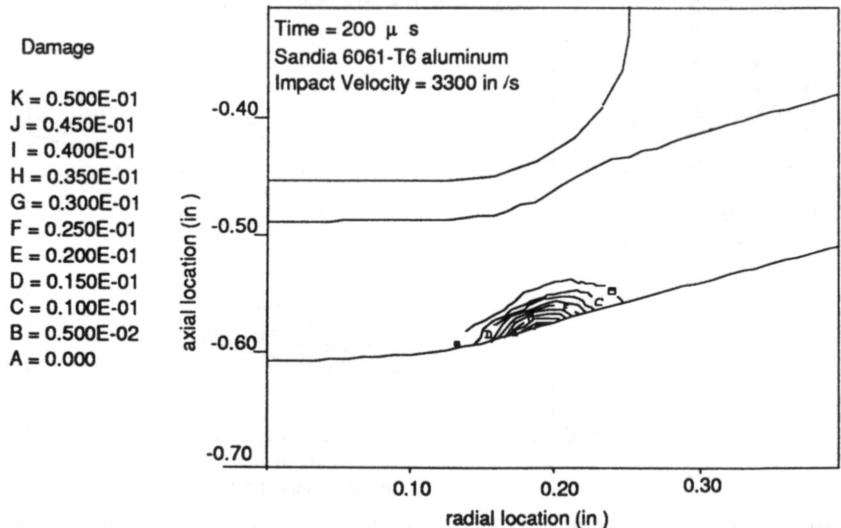

Fig. 23. Final contours of damage for a 6061-T6 disk impacted at 3300 in/s (84 m/s), where no failure occurs.

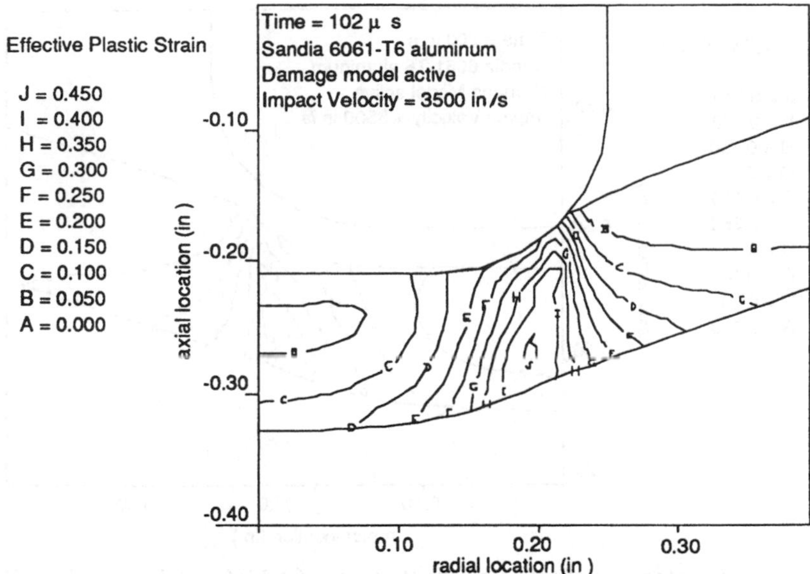

Fig. 24. Contours of effective plastic strain prior to initial failure in a 6061-T6 disk impacted at 3500 in/s (89 m/s) with the damage model active.

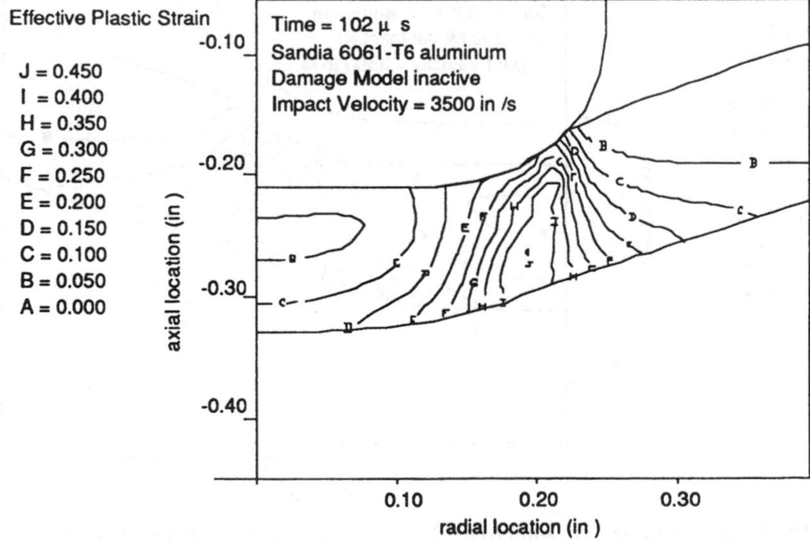

Fig. 25. Contours of effective plastic strain prior to initial failure in a 6061-T6 disk impacted at 3500 in/s (89 m/s) with the damage model inactive.

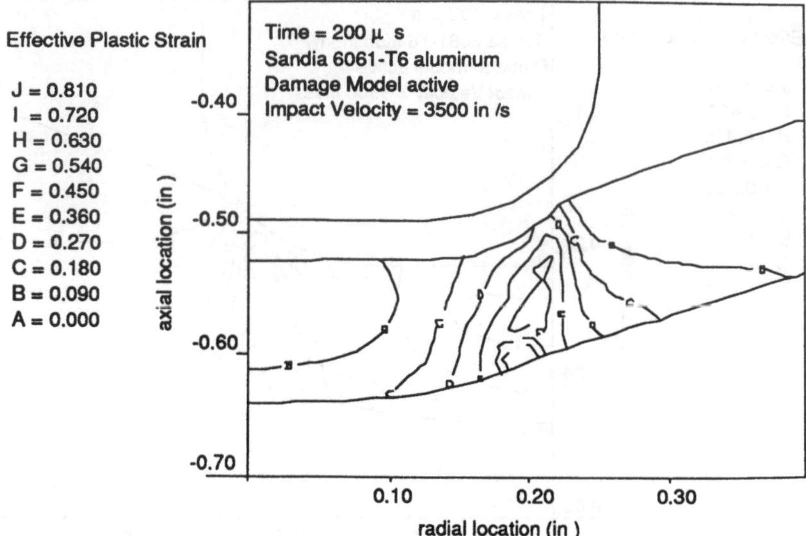

Fig. 26. Final contours of effective plastic strain in a 6061-T6 disk impacted at 3500 in/s (89 m/s) with the damage model active.

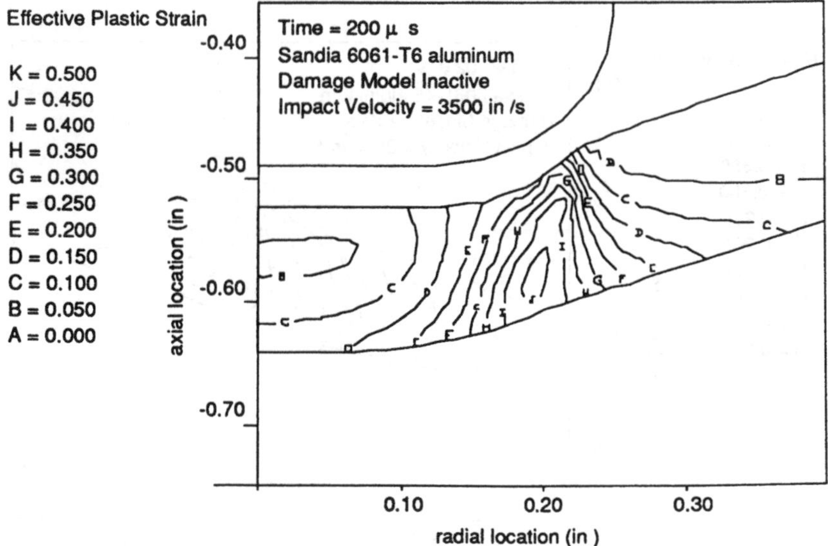

Fig. 27. Final contrours of effective plastic strain in a 6061-T6 disk impacted at 3500 in/s (89 m/s) with the damage model inactive.

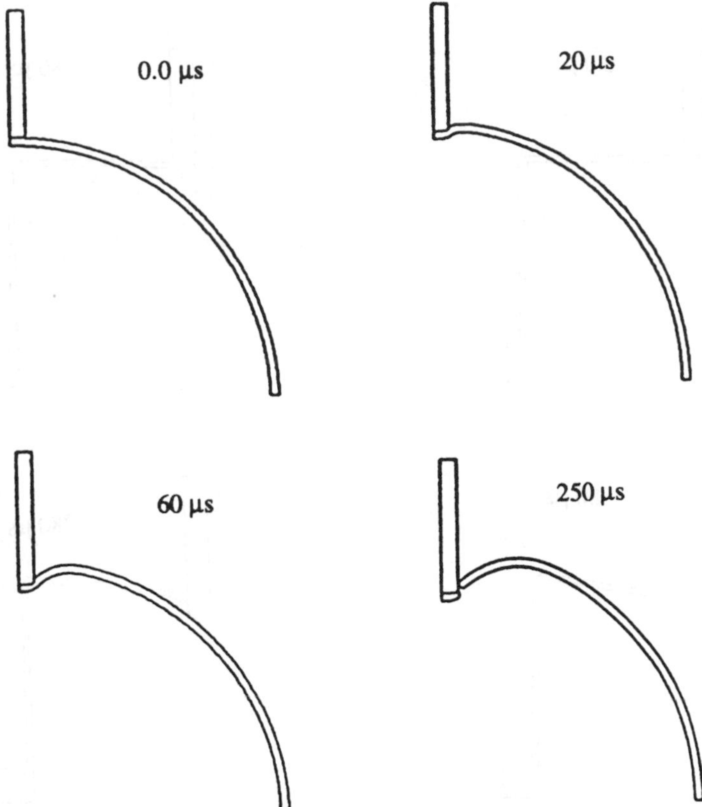

Fig. 28. Spherical barrier response to a penetrator with a velocity of 7656 in/s (194 m/s).

Figures 28 and 29 show the progression of failure for the spherical and cylindrical barrier at different times. Figure 30 shows the localization of the damage in the spherical target near the onset of failure for a penetrator velocity of 7656 in/s (194 m/s). This snapshot in time occurred 120 μs after the first contact. Final penetration occurred 225 μs after initial impact.

5.2 Focused Blasts on Plates

In this section, we show how well the model predicts damage for steel plates from focused blasts. Studying blast effects has been an on-going research topic for many military applications. Understanding how this environment causes large target deformation and target damage is crucial to many design and lethality/vulnerability studies. Under these loading conditions, the model accurately predicted plate deformations, tear initiation sites, tear lengths, tear paths, and damage area.[32,33]

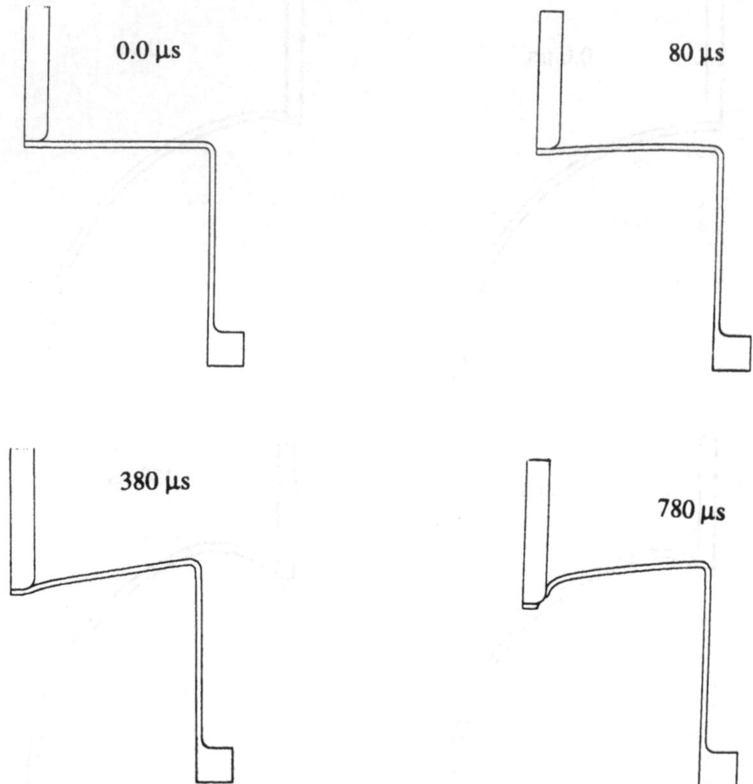

Fig. 29. Cylinder with end-cap barrier response to a penetrator with a velocity of 4606 in/s (117 m/s).

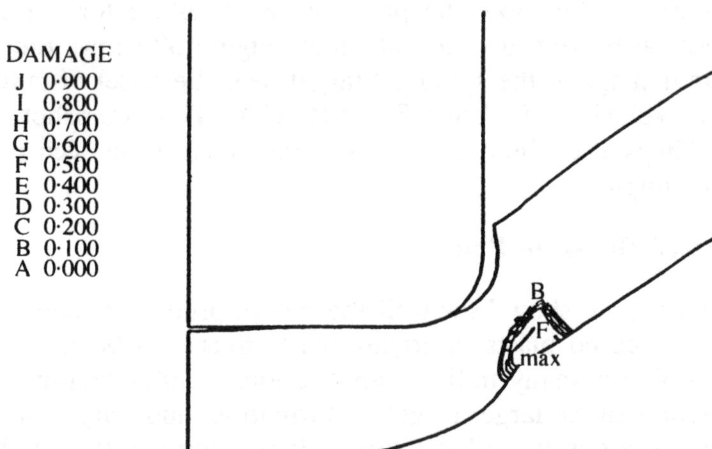

Fig. 30. Final contours of damage for a spherical barrier impacted at 7656 in/s (194 m/s), where failure occurs.

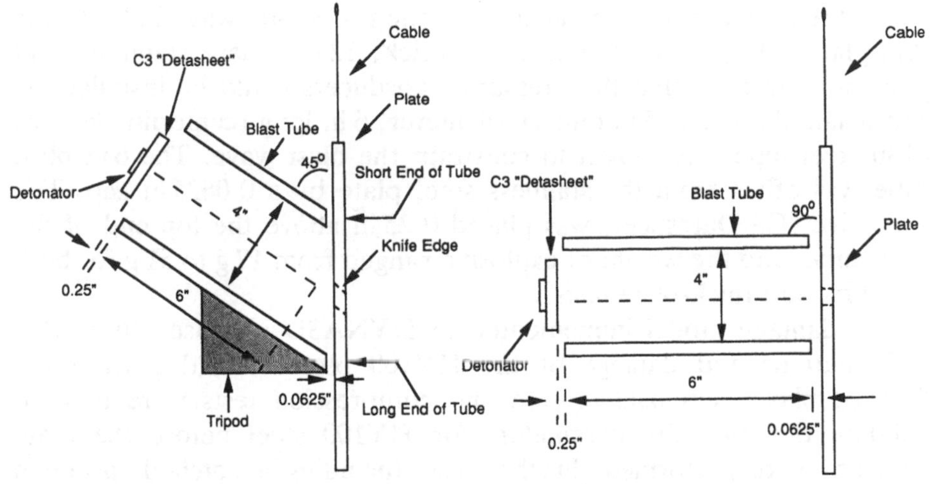

Fig. 31. Schematic of set-up for the 45° and 90° thin-plate experiments.

Experiments were performed in which blasts were constrained by a steel tube that was inclined at an orthogonal (90°) and an oblique angle (45°) as shown in Fig. 31. The blasts were focused onto different steel plates and included several weights of explosive. For the 90° experiments, some target plates had a pre-existing 1 in circular hole, causing a stress concentration. For the 45° experiments, the 1 in hole was drilled in the center of the plate at a 45° angle, so that two types of stress concentrations were induced: one from the elliptical hole and one from the slanted knife edge through the thickness of the plate.

Two types of tests were performed; (1) pressure measurement tests using a 'thick' stainless steel plate were used to estimate the loading functions for the finite element analyses; and (2) tests using 'thin' HY100 and HY130 steel plates were used for comparison with the finite element analyses for damage. For brevity's sake, we will only discuss the HY100 results, although our predictions were equally as good for HY130.

The thin steel plates were 18 in square, and the thickness was selected based on two-dimensional axisymmetric finite element calculations as a value that would allow failure with small amounts of explosive and at pressure levels that could be measured with available transducers (Kistler 607). The pressure transducers were rated to 100 ksi. From these considerations, a plate thickness of 0·0625 in was chosen for the damage tests.

For the pressure measurement tests, the test set-up was similar to the thin-plate set-up, except that a 2 in thick, 12 in square stainless steel plate was used so that the pressure transducers could be installed. A 4 in inside diameter, 5 in outside diameter, 6 in long (centerline length) 4130 steel pipe was chosen to constrain the blast wave. The 6 in blast tube was offset from the stainless steel plate by a 0·0625 in gap. The explosive (C3 Datasheet) was placed 0·25 in above the top end of the blast tube, and the weight of explosive ranged from 17 g to 51 g for both the thin- and thick-plate tests.

The damage model implemented in DYNA3D was used to predict deformations and damage of the HY100 plates resulting from the focused blasts. Uniaxial tension and compression tests were used to validate the plasticity parameters for HY100 steel before the plate analyses were performed. Further tests including a notched specimen with a radius of 0·020 in (0·5 mm) were used to fit damage parameters. Also, a 0·156 in (40 mm) notched specimen was used to validate the damage parameters for the HY100 steel. This independent test gave confidence in using the HY100 model. This HY100 model was then used in the blast analyses.

Orthogonal Interactions

The calculations and experiments showed that the plates without pre-existing holes typically exhibited a plug type failure, where a central disk of the plate was removed. The lack of the pre-existing hole allowed higher radial stresses to develop, which caused localization of the strain and subsequent plug-type failure. After this plug formed, radial tearing sometimes occurred. This plug failure was different than that occurring in plates with pre-existing holes. For the plates that had a pre-existing hole, failure occurred nearly simultaneously at several sites around the circular hole perimeter. Hoop stresses growing large enough to induce localization of the hoop strain caused these failures. Once the tearing began, the sharp cracks acted as stress concentrators to make tear growth the easiest way to relieve the plate stresses. As a result, petals formed as shown in Fig. 32.

The loading functions for the finite element model were based on pressure measuements from 38 g of explosive. Exponentially decaying pressure histories were specified that gave the same specific impulses as obtained from the pressure measurements. Peak measured pressures were about 50–100 ksi, with a decay constant of about 20 μs. The pressure and time constant selected had little effect on the results, since the load is impulsive with regard to the plate response. For instance, a pressure and time constant pair of 50 ksi/20 μs produced the same deformation as a 100 ksi/10 μs pair.

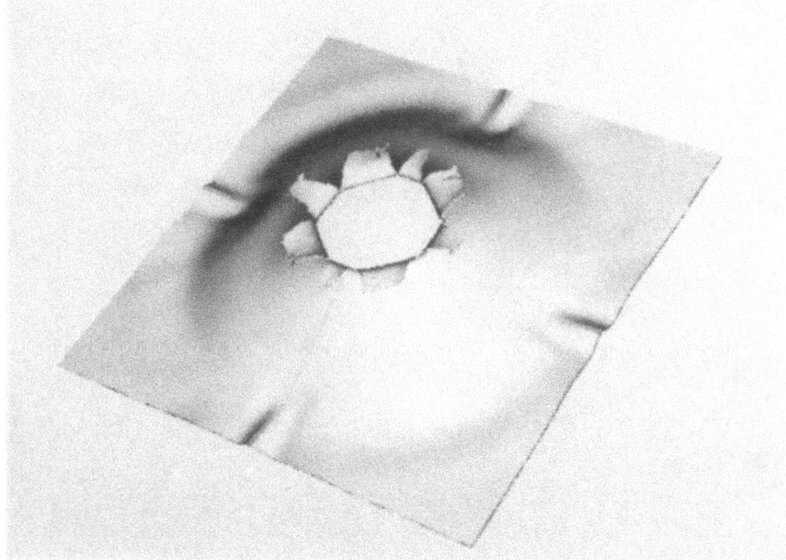

Analysis

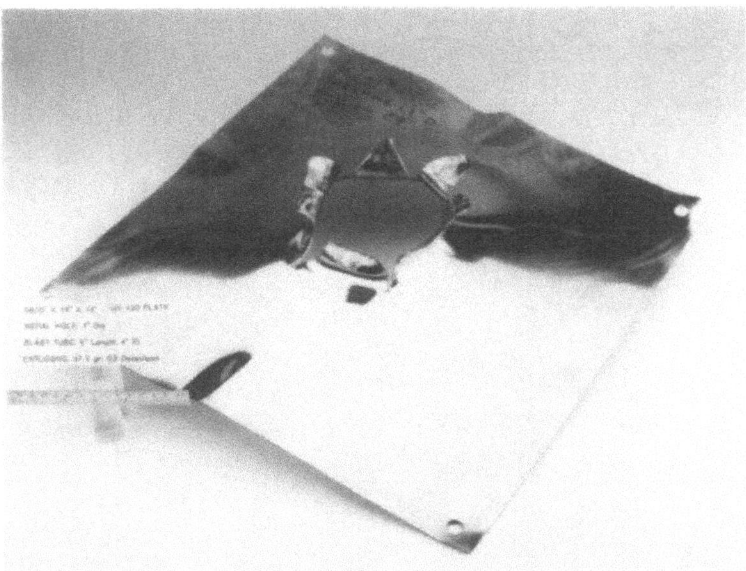

Experiment

Fig. 32. Comparison of experiment and DYNA3D numerical result for 38·5 g of explosive for an orthogonal blast. The plate included a pre-existing hole.

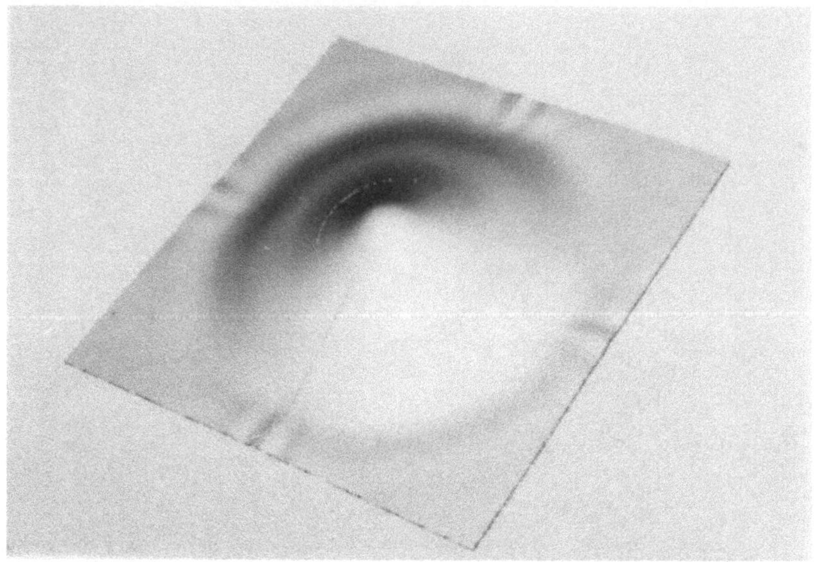

Analysis

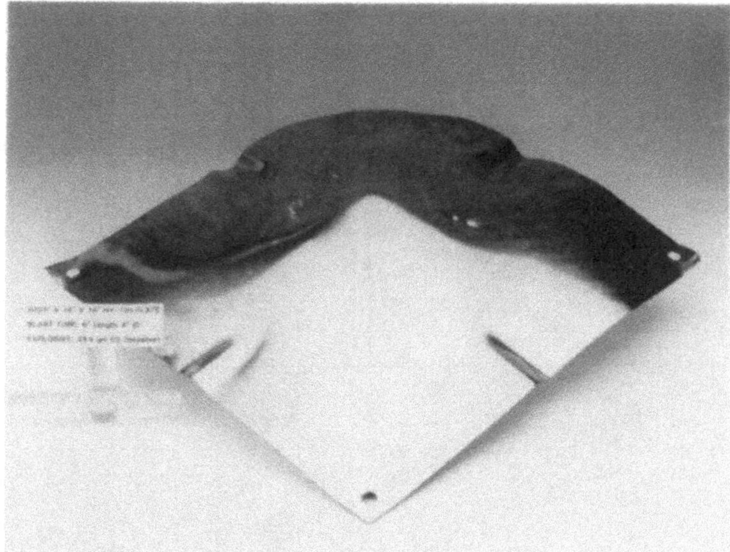

Experiment

Fig. 33. Comparison of experimental and DYNA3D numerical deformations resulting from 38·5 g of explosive on a plate without the pre-existing hole.

The finite element analyses assumed the same radial distribution for the specific impulses that were measured in the 38 g pressure measurement experiment. For instance, an analysis of a 16 g test used the same shape curve as a 38 g analysis, but the magnitude of the curve was decreased appropriately by adjusting the pressures and time constants of the pressure–time histories. Again, the actual pressure history is not critical, as long as the specific impulse is accurate and the rate of loading is of the correct magnitude.

Figure 33 shows a comparison of the numerical result and the experimental result for 38·5 g of explosive. The plate did not include a pre-existing 1 in hole. The predicted final displacement of the plate center was 2·5 in (6·35 cm), while the measured displacement was approximately 2·75 in (6·985 cm). The center of the plate thinned from its original 0·0625 in (1·5875 mm) to about 0·05 in (1·27 mm), a 20% reduction in thickness. A 1·75 in (4·445 cm) diameter ring of localized straining, concentric with the center of the plate, was visible near the center of the bulge.

Note that the numerical analyses showed buckling at the midpoints of each edge, as observed in the experiments. The buckles did not begin to form until fairly late in the analysis, at approximately 400 μs. The analyses also showed a 'ring' of plastic deformation moving from the center out to the edges of the plate. This ring did not reach the edge of the plate until after 550 μs.

Plates with hole diameters of 0·125, 0·5, and 1 in were also numerically analyzed. The goal was to compare deformation predictions with test results for the 1 in diameter holes and to numerically examine the effect of pre-existing hole size with the amount of explosive required to initiate target failure.

The analyses failure predictions for the weights of explosives tested (approximately 16 g and 38 g) against plates with 1 in diameter pre-existing holes agreed well with the experimental results. No failure was predicted for 16 g of explosive, while 38 g of explosive caused failure. A quarter-plate analysis using the 38 g of explosive load predicted the formation of eight 'petals' (for a whole plate), as shown in Fig. 32. The petals are approximately 2 in long, agreeing well with the experiment. In the experiment, seven large petals were formed, and the original 1 in diameter pre-existing hole opened up to approximately 5 in in diameter. The analysis also predicted a 5 in diameter hole.

Numerical analyses of HY100 steel plates indicated that, before tears formed, localization occurred at approximately 24 locations around the hole circumference. As the deformation progressed, only 8 of these localizations formed into long tears.

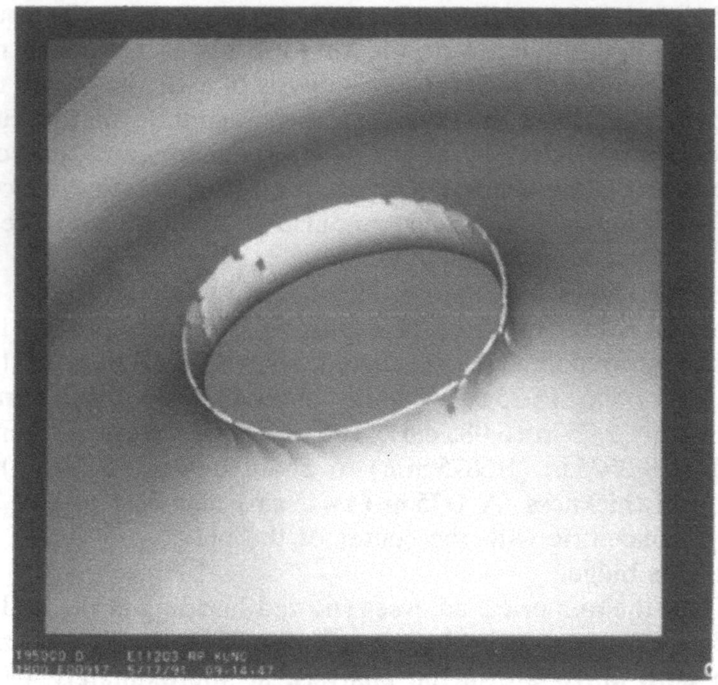

Analysis

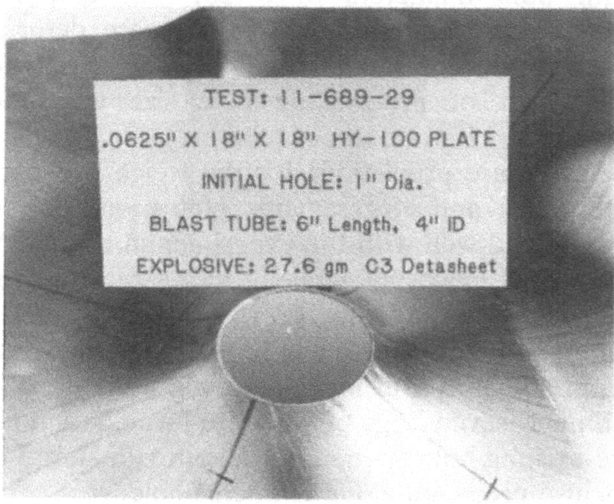

Experiment

Fig. 34. Multiple localization sites in the plates with the pre-existing 1 in hole for the orthogonal blast.

This localization at many sites was observed in experiment as well as in the analysis, as shown with the HY100 plate in Fig. 34. The ability of this numerical technique to capture this detail of the response provides additional confidence in the methodology.

No perturbation in either the material or the geometry was intentionally introduced into any of the finite element models. Numerical differences created during the generation of the mesh by numerical roundoff errors or by edge effects may be enough to allow for deformation to proceed asymmetrically. An analysis of a rectangular plate (24 in by 18 in) with the 38 g load produced the same deformation and tearing around the hole, indicating that the number and location of tears is not a function of the boundaries; however, the rectangular plates buckled only on the long edges.

The predicted time at which failure occurs in the plate without a pre-existing hole agrees well with the experimental results. The analysis predicted plate failure between 90 and 100 μs for a 51 g load, while high-speed photography of the experiment indicated that failure occurred between 105 and 120 μs.

Oblique Interactions

Like the orthogonal analyses, the blast loads for the oblique interactions were based on the pressure measurements from the thick-plate tests. Each element on the top surface was loaded by a linear interpolation scheme using the closest data points from the pressure measurement tests.

Overall, comparisons of numerical and experimental results for the plate deformations, tear lengths, tear initiation sites, and tear propagation paths showed good agreement. The DYNA3D calculations, as did the experiments, distinguished between the different failure mechanisms that arose from the different explosive amounts. Furthermore, both the analyses and experiments showed that an oblique blast caused more deformation and damage than an orthogonal blast. Figure 35 shows the comparison between the 90° and 45° blasts for 38 g of explosive.

Figure 36 is a schematic showing the tear initiation sites and fracture patterns of the plate from the experiment for the 45° shot with 38 g of explosive. Point A shows the site where onset of failure occurred. Post-test analysis showed that the failure initiation sites could be determined from large thinning areas, and the propagation paths could be determined by 45° edges indicative of running cracks. Not much thinning was observed in the areas of the running cracks. High-speed video also confirmed the location of the failure initiation sites and

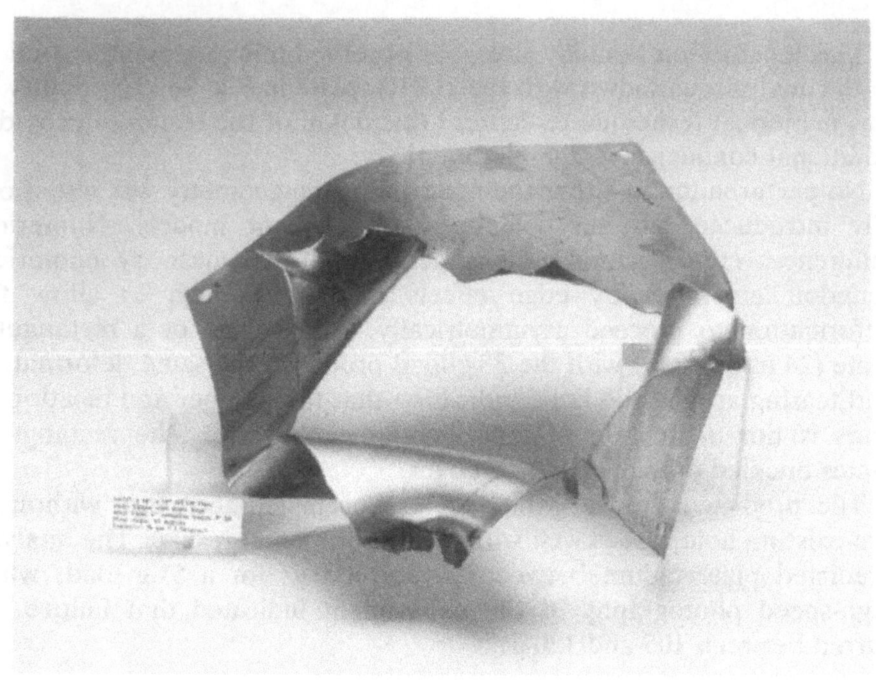

45 Degree Blast Angle

90 Degree Blast Angle

Fig. 35. Photographs comparing the damage from 90° and 45° blast orientations for 38 g of explosive using plates with pre-existing holes.

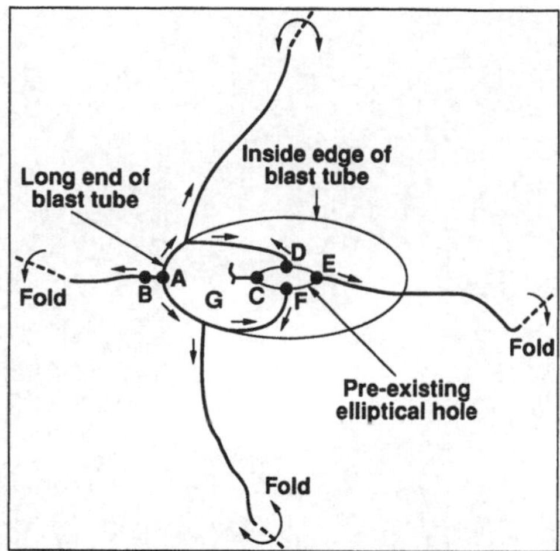

Fig. 36. Damage pattern for 38 g of explosive experiment oriented at a blast angle of 45° based on post-test examination of the plate.

propagation paths by taking frames of the deformation and damage every 10 μs.

Figure 37 shows the DYNA3D result at 130 and 170 μs after detonation for 38 g of explosive. Points A, B, C, and E correspond to the same labels in the schematic shown in Fig. 36. Figures 36 and 37 show how the tears initiated and propagated. The failure model predicted that the tear initiated near the stagnation point of the blast wave, which is what post-test examination of the experiment also showed. The blast pressures increased considerably because the blast wave stagnated at Point A. This stagnation increases the pressure in the local area at a fairly rapid rate. Point B (Fig. 37) shows a second tear initiation site that started at 170 μs after detonation. Point C (Fig. 37) is a third tear initiation site starting at the edge of the major axis of the elliptical hole near the long end of the tube.

Dimpling (Point E) can also be observed from Fig. 37 near the edge of the major axis of the elliptical hole. Other tear initiation sites, denoted by Points C, D, and F in Fig. 36, originated after the plugging started. The tear initiation sites and direction in which the tear propagated in the experiment agreed with the numerical result. The experimental result from the 38 g, 45° angle shot shown in Fig. 38 demonstrated that the tears propagated toward the outside edges of the plate. The failure initiation sites and propagation paths were similar even if the plate did not have a pre-existing hole for 38 g of explosive. The similar damage area is illustrated by Fig. 38.

The plates with and without a pre-existing hole for the 38 g load both

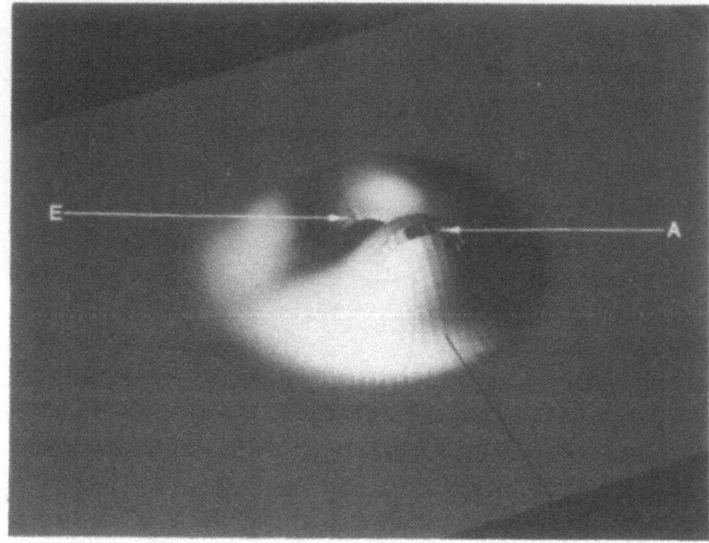

130 μs

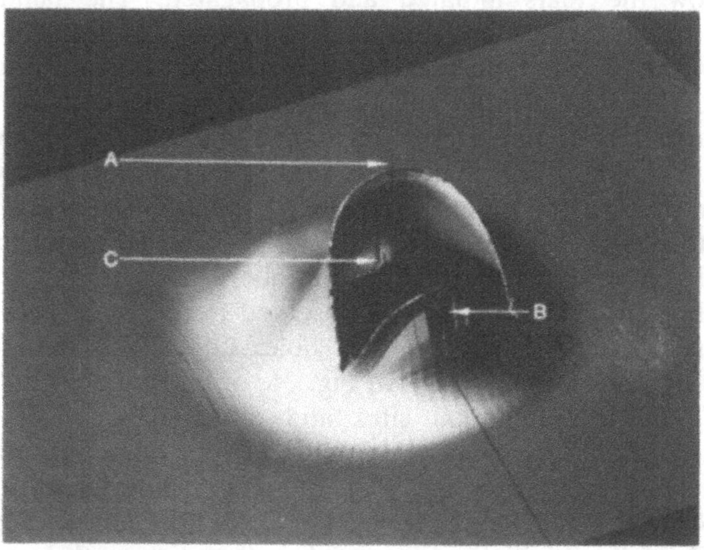

170 μs

Fig. 37. Numerical result of 38 g of explosive simulation at 130 and 170 μs after detonation with a blast angle of 45°.

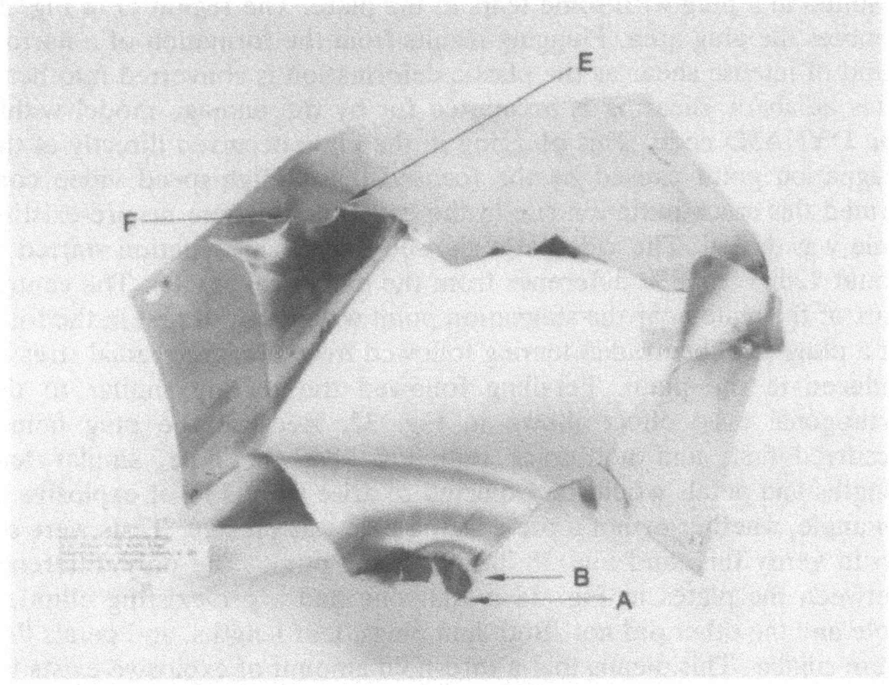

Plate with a pre-existing hole

Plate without a pre-existing hole

Fig. 38. Photographs comparing the damage of plates with and without the pre-existing elliptical hole oriented at a 45° blast orientation with 38 g of explosive.

resulted in a plug with some tears in the plate. The region G in Fig. 36 denotes the plug area. Plugging results from the formation of a narrow band of intense shear as the plastic deformation is converted into heat. This adiabatic shearing is accounted for by the damage model within the DYNA3D code. This plugging in the plate occurred directly at the stagnation point caused by the focused blast. High-speed video confirmed this crack initiation site in the experiment where no pre-existing hole was drilled. The video also showed that crack initiation started at about 120 μs (an 8% difference from the numerical result). The central part of the plate near the stagnation point was removed first in the form of a plug, and then radial tearing followed from the large radial stresses induced in the plate. Petalling followed the tearing similar to the orthogonal blast effect shown in Fig. 32. Because the plug failure occurred first, and not crack initiation near the hole, similar tear lengths and petals would be expected to arise from 38 g of explosive at an angle, whether or not a pre-existing hole was present. Tests were set up to verify this, and Fig. 38 illustrates the point. The only difference between the plates in Fig. 38 is that one had a pre-existing elliptical hole and the other did not. Both had plugs, tear lengths, and petals that were similar. This means that a threshold amount of explosive exists for a blast angle where pre-existing stress concentrations in the plate will not result in a larger damage area in the plate. This results because plugging near the stagnagion point occurred first, thus creating the failure pattern.

Figure 39 shows the DYNA3D result at 500 μs after detonation for 17 g of explosive compared with the experimental result. The numerical and experimental results agreed in that tearing started at the major axis of the elliptical hole near the stagnation point (Point A), and they also agreed where the dent (Point B) occurred resulting from high pressures at the stagnation point. The maximum deformation in the plate was 2·75 in in the experiment, while the DYNA3D calculation predicted 2·66 in (a 4% difference). The buckling at the edge of the plate is not shown by the calculation in Fig. 39, because the problem was not run to 700 μs. The tear length was not as long in the DYNA3D calculation for several reasons. First, the DYNA3D input came from the thick-plate experiment, which had 16·9 g of explosive, while the thin plate experiment had 17·6 g (4% more). Second, if the problem time had been extended, the inertia in the plate may have extended the predicted tear length.

The analyses and experiments using 17 g of explosive showed that the main failure mechanism was tearing initiated at the elliptical hole edge near the long end of the tube. As discussed earlier, the analysis and

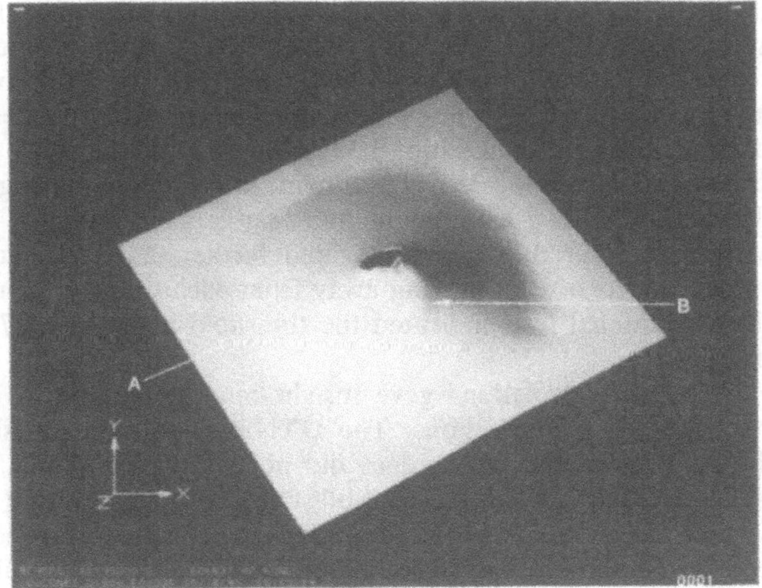

Analysis

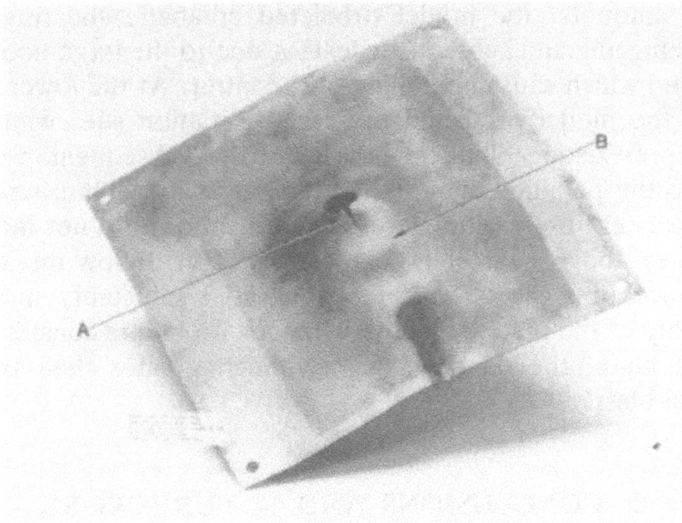

Experiment

Fig. 39. Comparison of experimental and numerical results for 17 g of explosive oriented at 45°

experiments using 38 g of explosive showed that initial tearing did not start at the elliptical hole but near the edge of the long end of the tube. Experiments were performed with 27 g and 32 g of explosives to converge on the threshold where the failure mechanism changes. The 32 g experiment had tear initiation at the long end of the tube, so the response was similar to the 38 g experiment, although less damage area was observed. The 27 g experiment had tear initiation similar to the 17 g experiment, but the tear propagated further, as suspected, and then turned into two tears moving away from each other in form of a 'T'. These additional tests indicated the threshold is between 27 g and 32 g of explosive.

The DYNA3D calculations gave insight into the timing of events from the blast/plate interactions. The DYNA3D calculations showed that the buckling at the plate edges did not start until about 500 μs. DYNA3D also showed for the 38 g blast that plugging started about 130 μs after detonation, where the high-speed video showed that plugging started about 120 μs in the experiment. DYNA3D calculations also agreed with video in that the deformation near the short end of the tube occurred between 40 and 50 μs after detonation and about 80 μs at the long end of the tube.

DYNA3D, which employed the damage model, performed well in predicting failure initiation and propagation for ductile steel under oblique focused blasts with different explosive amounts. At the higher explosive amounts, the model predicted adiabatic shearing and the resulting plugging and subsequent tearing due to the large hoop stresses in the plate, which caused petalling of the latter. At the lower explosive amounts, the model predicted the crack initiation site, which started near the pre-existing elliptical hole, and the subsequent propagation that led to the petalling. The finite element analyses and experimentation showed that pre-existing stress concentrations did not increase the damage area above a threshold explosive weight. Below this threshold, the presence of a stress concentration did significantly increase the vulnerability of the plates. Finally, when a 45° oblique constrained blast impacts a plate, the damage will be much greater than that of an orthogonal blast.

6 CONCLUSIONS AND FUTURE WORK

A constitutive model has been described in this chapter that is capable of predicting deformation and failure in ductile materials. Starting from basic kinematics of continuum mechanics, scalar and tensor internal variables have been introduced to more accurately describe the deviatoric plastic flow that enables the theory to predict damage. The rate of damage growth at a given location is based on values of the effective stress, mean stress (pressure), plastic strain rate, and current damage level. It is important to note that many types of responses that can be predicted by this model cannot be calculated by empirical equation of state models in which the stress is assumed to be a function of the current values of strain, strain rate, and temperature. These models are incapable of predicting any temperature, strain rate, or load path histories, and yet history dependences can be extremely important in the solution of many structural problems. The model is currently being modified to account for the initial anisotropy in a material, and to improve the description of evolving anisotropy. These improvements will result in a yield surface that changes shape and has several regions where the radius of curvature is quite sharp. This is critical in describing localizations that may occur in many situations. A more accurate dependence of the deviatoric plastic flow on pressure is also being included in the model. In addition to following naturally from the micromechanics of void growth, this dependence is necessary in the prediction of some aspects of sheet metal forming. We plan to extend the techniques developed to include anisotropic and composite materials. In addition, extensions to account for pressure dependence on the yield surface are being studied.

The parameters that are introduced by this constitutive model can be determined by simple tension and compression tests that can be performed in most material laboratories. The necessary equipment should include those that can subject specimens to high strain rates and temperatures. Presently, material properties have been determined for several aluminums and steels. In order to use the damage model on a wide variety of applications, the material database must be extended to contain the lower-strength aluminums and steels that are used by mechanical and civil engineers for many structural shapes.

The numerical implementation in the DYNA finite element codes were completed without a significant effect on computational speed. As noted before, the technique can be easily implemented in other explicit dynamic codes. The model is not just restricted to Lagrangian codes. The implementation into the Eulerian code, CTH,[34] is currently being evaluated.

A theoretical effort or its numerical implementation cannot be duly recognized unless it is applied to problems for which we have no previous solution or only cost-prohibitive solutions. Examples presented in this chapter have shown that this damage model can predict ductile failure mechanisms, as benchmarked by carefully planned experiments duplicating real world problems. Now that there is a verified computational tool for predicting these failure mechanisms, significant cost savings are available to the engineer who uses this tool and the industry he or she serves. Use of this model can significantly reduce, and in some cases eliminate, costly tests in which the many parameters in product design and manufacturing are varied. Often, especially when product failure is being determined, these types of tests can be both expensive and/or prone to hazardous effects. Examples include the post-failure response of structures due to severe events such as earthquakes, hazardous weather, or blasts. More mundane, but offering cost savings to society, is the optimization of material properties for applications such as metal forming or welding.

ACKNOWLEDGEMENTS

The authors would like to acknowledge several of their colleagues at Sandia National Laboratories. Though it was not practical, they all could be co-authors of this chapter. These contributors influenced extensions of early versions of the damage model by benchmarking their analyses with experiments. These people, listed in alphabetical order, are D. B. Dawson, J. J. Dike, W. A. Kawahara, A. McDonald, V. D. Revelli, D. Shah, and K. V. Trinh.

This work was performed at Sandia National Laboratories and was supported by the US Department of Energy under Contract DE-AC04-76DP00789.

REFERENCES

1. Hallquist, J. O., User's manual for DYNA2D—an explicit two-dimensional hydrodynamic finite element code with interactive rezoning and graphical display. Lawrence Livermore National Laboratory Report UCID-18756 Rev. 3, March 1988.
2. Hallquist, J. O., DYNA3D user's manual (nonlinear dynamic analysis of structures in three dimensions). Lawrence Livermore National Laboratory Report UCID-19592 Rev. 4, April 1988.
3. Taylor, L. M. & Flanagan, D. P., PRONTO2D—a two-dimensional transient solid dynamics program. Sandia National Laboratories Report SAND86-0594, March 1987.
4. Taylor, L. M. & Flanagan, D. P., PRONTO3D—a three-dimensional transient solid dynamics program. Sandia National Laboratories Report SAND87-1912, March 1989.

5. Biffle, J. H., JAC3D—a three-dimensional finite element computer program for nonlinear quasi-static response of solids with the conjugate gradient method. Sandia National Laboratories Report SAND87-1305, September 1990.
6. Hibbit, Karlsson, and Sorensen, Inc., ABAQUS Theory and Users Manual, Version 4.8, 1989, Providence, RI.
7. Hallquist, J. O., NIKE2D—a vectorized implicit, finite deformation finite element code for analyzing the static and dynamic response of 2-D solids with interactive rezoning and graphics. Lawrence Livermore National Laboratory Report UCID-19677 Rev. 1, December 1986.
8. Gurson, A. L., Continuum theory of ductile rupture by void nucleation and growth: Part I—yield criteria and flow rules for porous ductile media. *J. Engng Mater. Technol.* (Jan. 1977) 2015.
9. Rice, J. R. & Tracy, D. M., On the void ductile enlargement of voids in triaxial stress fields. *J. Mech. Phys. Solids,* **17** (1969) 201–17.
10. Cocks, C. F. & Ashby, M. G., Intergranular fracture during power-law creep under multiaxial stresses. *Metal Sci.* (1980) Aug.–Sept. 395–402.
11. Bammann, D. J. & Johnson, G. C., On the kinematics of finite-deformation plasticity. *Acta Mech.*, **70** (1987) 1–13.
12. Bammann, D. J. & Aifantis, E. C., A model for finite-deformation plasticity. *Acta. Mech.*, **69** (1987) 97–117.
13. Bammann, D. J., Johnson, G. C. & Chiesa, M. L., A strain rate dependent flow surface model of plasticity. Sandia National Laboratories Report SAND90-8227.
14. Bammann, D. J., An internal variable model of viscoplasticity. *Int. J. Engng Sci.* **22** (1984) 1041–53.
15. Bammann, D. J., An internal variable model of elastic-viscoplasticity. In *The Mechanics of Dislocations,* ed. E. C. Aifantis & J. P. Hirth. American Society of Metals, Metals Park, OH, 1985, pp. 203–12.
16. Bammann, D. J., Modelling the large strain–high temperature response of metals. In *Modeling and Control of Casting and Welding Processes IV,* ed. A. F. Giamei & G. J. Abbaschian. TMS Publications, Warrendale, PA, 1988, 329–38.
17. Bammann, D. J. & Aifantis, E. C., A damage model for ductile metals. *Nucl. Engng Des.*, **116** (1989) 355–62.
18. Bammann, D. J. & Johnson, G. C., Development of a strain rate sensitive plasticity model. In *Engineering Mechanics in Civil Engineering,* ed. A. P. Boresi & K. P. Chong. ASCE, New York, 1984, pp. 454–7.
19. Bammann, D. J., Modeling temperature and strain rate dependent large deformations of metals. *Appl. Mech. Rev.*, **43** (1990) S312–19.
20. Johnson, G. C. & Bammann, D. J., A discussion of stress rates in finite deformation problems. *Int. J. Solids Structures,* **20** (1984) 737–57.
21. Stout, M. G., Helling, D. E., Martin, T. L. & Canova, G. R., Multiaxial yield behavior of 1100 aluminum following various magnitudes of pre-strain. *Int. J. Plasticity,* **1** (1985) 163–74.
22. Krieg, R. D. & Krieg, D. B., Accuracies of numerical solution methods for the elastic–perfectly plastic model. *ASME. J. Pressure Vessel Technol.* **99** (1977) 510–15.
23. Whirley, R. G., Hallquist, J. O. & Goudreau, G. L., An assessment of numerical algorithms for plane stress elastoplasticity on supercomputers. *Engng Comp.* **6** (1989) 116–26.

24. Simo, J. C. & Taylor, R. L., Consistent tangent operators for rate-independent elastoplasticity. *Comp. Meth. Appl. Mech. Engng*, **48** (1985) 101–81.
25. Gilbertson, N. D., Fuchs, E. A. & Kawahara, W. A., Unpublished data, Sandia National Laboratories, Livermore, CA.
26. Semiatin, S. L. & Holbrook, J. H., Isothermal plastic flow behavior of 304L stainless steel. Battelle Laboratory Report SN4165-P092-9342, 1982.
27. Sensey, P. E., Duffy, J. & Hawley, R. H., Experiments on strain rate history and temperature effects during the plastic deformation of close packed metals. *J. Appl. Mech.*, **45** (1984) 60–6.
28. Kawahara, W. A., Effects of specimen design in large-strain compression. *Experimental Techniques* (March/April 1990) 58–60.
29. Lipkin, J., Chiesa, M. L. & Bammann, D. J., Thermal softening of 304L stainless steel: experimental results and numerical simulations. In *Impact Loading and Dynamic Behaviour of Materials*, ed. C. Y. Chiem, H.-D. Kunze & L. W. Meyer. DGM Informationsgesell schaft Verlag, 1989, p. 687.
30. Bammann, D. J., Chiesa, M. L., McDonald, A., Kawahara, W. A., Dike, J. J. & Revelli, V. D., Prediction of ductile failure in metal structures. In *Failure Criteria and Analysis in Dynamic Response*, ed. H. E. Lindberg. ASME AMD, Vol. 107, November 1990, pp. 7–12.
31. Trinh, K. V. & Gruda, J. D., Failure resistance of thin shells against projectile penetration. Presented at ASME 1991 Pressure Vessel and Piping Conference, June 1991.
32. Dike, J. J. & Weingarten, L. I., Failure in steel plates subjected to focused blasts: analysis/experiment comparisons. Sandia National Laboratories Report SAND91-8222, August 1991.
33. Horstemeyer, M. F., Damage of HY100 steel plates from oblique constrained blast waves. Presented at ASME 1992 Applied Mechanics Conference, April 1992.
34. Bell, R. L., Elrick, M. G., Hertel, E. S., Kerley, G. I., McGlaun, J. M., Rottler, J. S., Silling, S. A., Thompson, S. L. & Zeigler, F. J., CTH user's manual. Sandia National Laboratories Report, Version 1.023, May 1991.

Chapter 2

MODELING THE PROCESS OF FAILURE IN STRUCTURES

BAYARD S. HOLMES, STEVEN W. KIRKPATRICK, JEFFREY W. SIMONS, JACQUES H. GIOVANOLA & LYNN SEAMAN

SRI International, Menlo Park, California 94025, USA

ABSTRACT

In this chapter, we explore the use of finite element codes to model the processes of structural failure. Of particular concern is predicting what happens after failure initiation. Examples used as illustrations are the calculation of postbuckling deformation in thin shells, the accumulation of fatigue damage in pavement, and the direction of crack propagation through a welded joint. In the first example, the solution of the dynamic buckling of a thin shell is used to illustrate the use of changes in code input to reflect random imperfections in structure geometry, load or material properties. In a second example, a new constitutive model is developed and implemented into a finite element code to solve problems in the cracking of airport pavement. Finally, the fracture of welded T-joints is studied using both code material model modifications and techniques to change finite element geometry during the calculation.

NOTATION

A	Critical number of cycles for pavement damage (cycles)
B	Material constant used in Paris' fatigue equation (cm)
D	Normalized damage parameter
$\mathrm{d}\varepsilon^{\mathrm{p}}_{\mathrm{eq}}$	An increment in plastic strain
$\mathrm{d}N$	Differential increase in the number of cycles during fatigue loading
$\mathrm{d}R$	Differential increase in the radius of a microcrack (cm)
$\mathrm{d}\tau$	Differential increase in the amount of damage (dimensionless)
E	Young's modulus (MPa)

f	Material constant with the units of stress in the equation for pavement damage (MPa)
f_c'	Compressive strength of concrete (MPa)
f_t'	Tensile strength of concrete (MPa)
g	Growth coefficient used in the proposed fatigue equation ($cm^{1-n/2}/MPa^n$)
J	Number of microcracks per unit volume
k	Subscript denoting any plane of the multiple-plane model
K_I	Stress intensity factor ($MPa\,cm^{1/2}$)
K_{Ic}	Fracture toughness ($MPa\,cm^{1/2}$)
n	Material constant appearing as an exponent in the fatigue equations
N	Number of loading cycles
R	Microcrack radius (cm)
R_m	Critical microcrack radius (cm)
R_{MIC}	Linear dimension of microdamage process zone
R_0	Initial microcrack radius (cm)
T_F	Material constant determining the volume of a fragment formed by a set of microcracks (dimensionless)
V_{MIC}	Characteristic microdamage process volume
$[\Delta\varepsilon]$	Tensor of the total imposed strain increment
$[\Delta\varepsilon^p]$	Tensor of the plastic strain increment
ε	Strain component
ε_{ck}	Strain representing the crack opening on the kth plane
ε_k	Strain imposed on the kth plane
σ	Stress (MPa)
σ_{cr}	Critical stress for crack growth (MPa)
σ_k	Stress acting normal to the kth plane (MPa)
$[\sigma^N]$	Stress tensor computed by taking all the imposed strain as elastic (MPa)
$[\sigma_0]$	Stress factor from the previous time increment (MPa)
τ_{1k}, τ_{2k}	Shear stresses on the kth plane (MPa)
ν	Poisson's ratio (dimensionless)

1 INTRODUCTION

Increasingly, the objective of finite element (FE) calculations is to predict the behavior of a structure beyond the threshold of failure. Structural failures usually occur as a result of localized yielding and fracture, sometimes in combination with global or local buckling. Three reasons why the prediction of failure processes using FE codes is difficult are (1) the failure process is influenced by random and undefined imperfections in shape or material properties, (2) the material models available for analysis do not accurately describe the physical damage processes involved, or (3) the response algorithms commonly available cannot correctly model failure processes such as crack propagation.

In general, predicting the initiation of failure is not as difficult as predicting the process and final resolution of the failure. The processes leading to failure initiation, even when geometric and material nonlinearities are present, are better understood than the failure process. For example, constitutive models that can accurately describe nonlinear material behavior in standard cases of elastic–plastic response are commonly found in finite element codes. However, predicting the progression of failure after initiation is made difficult by strongly nonlinear processes such as dramatic changes in material properties and structural geometry, or by redistribution of the loads resulting from damage or other often unanticipated features of the failure process.

Good finite element solutions can be obtained when the response variables (e.g. stresses and strains) in the problem vary smoothly and a sufficient number of finite elements are used to discretize the structure. Problems can arise in the analysis of failure when the response variables vary sharply; for example, high-deformation gradients caused by local buckling, localization of strains in shear bands, or, in the case of fracture, discontinuous variation of stresses and strains across the crack. These responses may then require many elements for spatial resolution or very small time steps for temporal resolution, increasing the demand on computer resources. However, today solutions to these problems are often obtainable on engineering workstations. For the analyses of failure described in this chapter, solutions were obtained by using the DYNA3D finite element code[1] on SUN SPARCstation2 computers.

The solutions to problems involving failure usually require capabilities beyond those normally available in finite element codes, such as a description of material or geometric imperfections that govern local failure and buckling, constitutive models for materials that include calculation of damage, and solution algorithms that modify the finite element mesh to simulate crack propagation.

In the remainder of this paper, we describe finite element calculations for three different failure processes: (1) dynamic buckling of thin shells: (2) fatigue damage to airport pavements; and (3) crack propagation through a weld. These calculations will be used to illustrate specific points regarding the prediction of failure processes. To perform these calculations, we used a modified version of DYNA3D, an explicit nonlinear, three-dimensional finite element code for solid and structural mechanics. DYNA3D is under continuous development at Lawrence Livermore National Laboratory.

DYNA3D is organized in a modular fashion, with each module performing a specific task. To describe the modifications we made to DYNA3D, we first outline the solution procedure used in this code. This procedure is common to most explicit finite element codes and is similar to those described by other authors (see e.g. Ref. 2). Figure 1 shows a simplified flow chart outlining the tasks performed by DYNA3D. We will use this outline to describe where modifications to

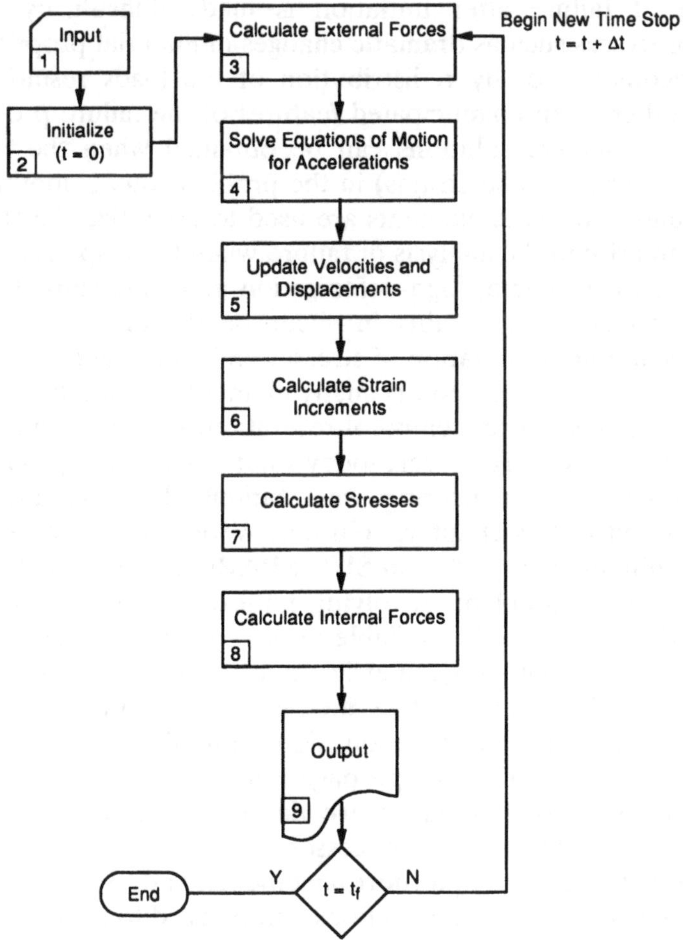

Fig. 1. Simplified flowchart for DYNA3D.

the code were made in the three failure analyses. The specific tasks shown in Fig. 1 can be summarized as follows.

Task 1. *Input*—Data are input to describe the geometry, materials, boundary conditions, loading, and solution control parameters.

Task 2. *Initialize*—Initial values are loaded into the structural and material data arrays.

Task 3. *Calculate nodal forces*—Forces at the nodes are calculated as the difference between internal and external forces. External loading may include pressure, concentrated loads, and gravity. Internal forces are calculated from element stresses.

Task 4. *Solve equations of motion for accelerations*—At each node, the acceleration is calculated from the force divided by the nodal mass. For explicit codes such as DYNA3D, no global stiffness matrix coupling the degrees of freedom is needed.

Task 5. *Update velocities and displacements*—Nodal velocities and displacements are calculated and updated by using the nodal acceleration and the current time step.

Task 6. *Calculate strain increment*—From the nodal velocities, strain rates are calculated from which strain increments can be calculated.

Task 7. *Calculate element stresses*—Element stresses are calculated in the material constitutive models from the strain increments.

Task 8. *Calculate internal forces*—Nodal forces are calculated by integrating the element stresses.

Task 9. *Output and check for problem termination*—Output results if specified. If the calculation time is less than the specified termination time, return to Step 3.

The changes that we use here to solve failure process problems are made in the input (Task 1), in the way in which element stresses are calculated (in Task 7) and in the way in which elements are connected together (these changes are made in Task 4). Of these changes, the simplest type of modifications are changes to code input. For example, the buckling of a rod or shell is often very sensitive to imperfections in its initial shape or eccentricities in load. The use of standard mesh generation techniques using available preprocessors generates near-perfect structure and load geometries that can result in erroneous problem solutions. However, simple modifications to the preprocessor can be used to generate random variations in the geometry and loading that are needed for accurate solutions.

Although most buckling problems can be solved with changes to the input, the analyses of other failure processes may require constitutive models with capabilities beyond those commonly found in finite element codes. For example, the response of airport pavements produces damage in the paving materials over many load cycles and several years. Thus a constitutive model that describes the physical damage processes involved in fatigue response is needed to address the problem of calculating the lifetimes of these pavements.

To describe the fatigue response of pavements, we developed an elastic, viscoplastic constitutive model, called SRI-PAVEMENT, that calculates stress in pavement as a function of stress history, strain, and time. This model goes into the code in Task 7. As described above, DYNA3D sends the constitutive model a series of strain increments and the current values of stress, and SRI-PAVEMENT returns the new calculated stresses. As part of the stress calculation, SRI-PAVEMENT calculates fatigue damage and cracking on discrete planes and stores these values as internal state variables, which then modify the material properties for subsequent time steps.

The modeling of other failure processes involves changes in the solution algorithm in addition to the development of constitutive models. To describe the propagation of a crack through a weld, we implemented a constitutive model in Task 7 that calculates damage in a metal on the basis of a local damage measure, which is a function of the plastic strain and the triaxiality of the stress state. We then used this measure of damage in Task 4 to affect the connectivity of element nodes. In the region of expected damage, each element has its own set of nodes. While the material is undamaged, the nodes of adjacent elements are constrained to move together. When the material is fully damaged, the nodal constraints are removed and the nodes can move independently. The release of nodes is used to model the physical processes of cracking.

In the following sections, we address in more detail the techniques of introducing changes in input, modifying constitutive behavior, and changing mesh connectivity. We will use as examples the analysis of shell buckling, fatigue in airport pavements and the dynamic fracture of steel weldments. We believe that methods used to obtain solutions to these examples are representative of methods that can be applied successfully to a wide variety of failure processes.

2 EXAMPLE 1—SHELL BUCKLING

Modifications to the input to the finite element code are necessary for the solution of problems where random imperfections in load, geometry, or material properties dominate the response of the structure. Imperfections are important when failure is dominated by unstable processes such as the buckling of columns or the necking of thin sheets. For example, the buckling of thin shell structures is often dominated by minute geometric imperfections that are not readily visible in the unloaded structure.

To illustrate the importance of including an accurate description of the initial imperfections for calculating buckling processes in thin shell structures, we use an example of dynamic buckling experiment and calculations shown in Fig. 2. In the experiment shown in Fig. 2(a), a 30·25 cm diameter aluminum shell with a wall thickness of 0·635 mm was impulsively loaded with an external blast loading on one side. The inward motion of the loaded side of the shell caused large compressive circumferential stresses and the resultant buckling seen in Fig. 2(a). The irregular nature of the buckles is seen in Fig. 2(a) (right), where the shell has been cut in half at the middle.

Three different FE calculations were used to illustrate the effect of subtle changes in the problem input. In each calculation, the thin shell was represented by Hughes–Liu shell elements as implemented in DYNA3D. The aluminum constitutive behavior was modeled using an elastic–plastic model with a von Mises yield function and combined linear kinematic/isotropic hardening.

In the first calculation, the initial mesh was generated using the INGRID preprocessor of DYNA3D. Thus the shell represented by the calculation was nearly perfectly round, with the only imperfections in shape being generated by numerical round-off errors. Figure 2(b) shows the final predicted shape for this calculation. Note that the shell shows little evidence of buckling (Fig. 2(b), right) and has a flat profile (Fig. 2(b), left) rather than the dished profile of the experiment. The relatively inaccurate calculation shown in Fig. 2(b) results from an unrealistically smooth initial FE mesh. Because buckling load and the final shape are functions of the initial shape of the shell, this type of calculation usually grossly overestimates the critical impulsive load for the shell and underestimates the final plastic strain in the shell.

In solutions to static buckling problems, the absence of imperfections causes large errors in critical load and sometimes errors in buckling mode. To overcome these problems, some authors have introduced a

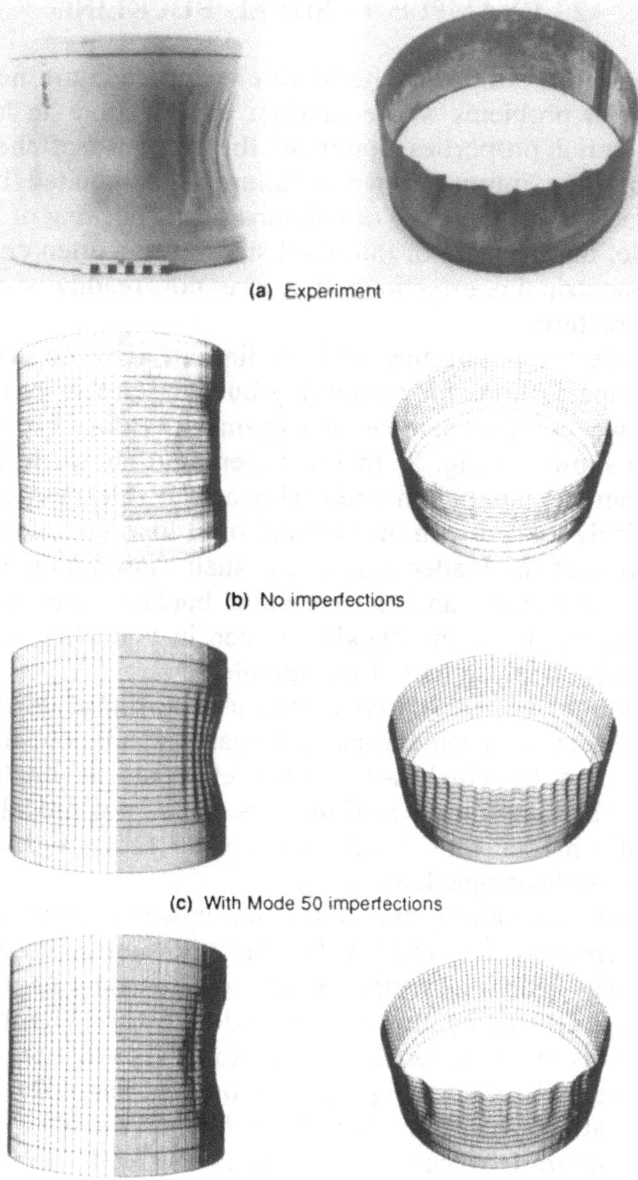

Fig. 2. Comparison of experiment with analysis using different initial imperfection distributions (calculations shown at 600 μs).

single-mode imperfection to trigger buckling at lower loads (see e.g. Refs 3, 4). This approach often leads to errors as illustrated in the calculation shown in Fig. 2(c). In this calculation, we introduced a single-mode imperfection of mode 50 into the initial mesh, forming a series of 50 ridges in the initial shape of the shell. This mode number was chosen because the preferred buckling mode of the shell under impulsive loads is mode 50. The amplitude of the imperfection was arbitrarily taken to be 1% of the shell thickness or 0·000 04 times the shell radius, so that the initial shape appears undistorted to the eye. As seen in Fig. 2(c), the calculated deformed shape exhibits a pronounced pattern of 50 regular buckles, but has a profile similar to that seen in the experiment. Further calculations have shown that the amount of deformation is a function of the arbitrary initial imperfection amplitude and that the final shape and pattern of plastic strains are not similar to those observed in the experiment.

A much preferred method of introducing the effect of imperfections into the finite element mesh is to use the actual initial shell geometry (from measurements) or an approximate random shape based on experience with other shells. For physical shells, the imperfect shape contains all modes of imperfections, usually with decreasing amplitude at increasing mode numbers. In the final calculation, shown in Fig. 2(d), the measured shape of the shell was input into the finite element mesh along with an estimate of the imperfections in the impulsive radial load (using a random number generator). This calculation compares very well with the experiment showing both the deformed shape and the strains measured in the shell experiment.[5,6]

This series of calculations illustrates how subtle changes in finite element code input can affect the response in problems where instabilities are important. This is a simple reflection of the physical processes involved, where small perturbations tend to grow rapidly. Thus, the code input must include these imperfections in order to obtain accurate results. For a more detailed discussion on the inclusion of imperfections in the numerical solution of shell buckling problems, see Refs 5–8.

3 EXAMPLE 2—FATIGUE OF AIRPORT PAVEMENTS

Damage to airport pavements is caused by many passes of airplanes over the pavement. The damage is typically widespread, consisting of patterns of cracks spread over a large area. The cracking is often along certain orientations related to the direction of traffic. The loads and stresses developed in the pavement with each pass are typically small compared with the strength of the pavement materials. However, after many passes, damage to the pavement is noticeable in the form of cracks at the surface. This form of damage clearly changes material properties: cracked material is less strong and less stiff than intact material. Furthermore, because the cracks are in specific directions, the properties of the damaged material are strongly anisotropic: the pavement is weaker in some orientations than in others.

Most constitutive models available in finite element codes are elastic–plastic and thus respond elastically to stresses below the yield stress. The stresses exerted by the passing of a plane would produce no damage in these materials at these loads, and damage, when it occurs (i.e. plasticity), is isotropic.

Our approach to analyzing fatigue damage in airport pavements was to develop a new constitutive model for the fatigue damage of pavement and to implement the model into DYNA3D.[9] As shown in Fig. 1, the task of the constitutive model in Task 7 is to use increments of strain and time step to calculate the stresses. When stresses are functions of history, as in fatigue problems, the constitutive model must also keep track of these internal variables that define the state of the material.

Fatigue cracking can occur by many mechanisms, depending on the material and the stress state.[10] During a cycle of loading, beginning with the tensile opening of the crack, the material near the crack tip may be stretched inelastically. When the crack is then closed again by the overall loading, this stretched material does not recompress to its original position. Hence a small amount of damage occurs at each loading cycle, depending on the range of stress during the cycle. Both Portland cement concrete and asphalt concrete exhibit fatigue cracking under repeated loading. When fatigue cracking occurs it begins by the initiation of small flaws of the order of the aggregate dimensions in the cement matrix or the aggregate or at the interface between the aggregate and the cement. At each loading cycle, these cracks grow by a small amount. After thousands (or millions) of loading cycles, the cracks have become so large that they are noticeable and require repair.

Experimental observations of pavement damage have been made by passing several airplanes or trucks over a pavement section and observing the development of cracking. Thus ‘damage’ here refers to the presence of large cracks that have grown from the bottom of the pavement slab to the top. Observations of this type are then collected from a range of slab thicknesses (different thicknesses respond at different stress levels under a given wheel load) and represented by an equation of the form (see, e.g. Refs 11–14):

$$N = A\left(\frac{f}{\sigma}\right)^n \tag{1}$$

where N is the number of cycles to damage, σ is the peak tensile stress at the base of the slab during a loading cycle, and A, f, and n are constants. This equation defines a curve corresponding to some arbitrary damage level.

3.1 Fatigue Models

The most commonly used fatigue model is that proposed by Paris and co-workers.[15,16] This model gives the rate of crack growth per loading cycle N as a function of the applied stress σ normal to the crack with radius R:

$$\frac{dR}{dN} = B\left(\frac{\Delta K_I}{K_{Ic}}\right)^n \tag{2}$$

where B and n are material constants, K_{Ic} is the fracture toughness, and ΔK_I is the change in the stress intensity factor during a loading cycle. According to linear elastic fracture mechanics, K_I and K_{Ic} are related to the stress on a penny-shaped crack by

$$K_I = 2\sigma\sqrt{\frac{R}{\pi}}, \quad K_{Ic} = 2\sigma_{cr}\sqrt{\frac{R}{\pi}} \tag{3}$$

where R is the crack radius and σ_{cr} is the normal tensile stress just sufficient to cause crack growth. This form for crack growth in eqn (3) represents fatigue data in many metals and other materials. In this fatigue model, we assume that the loading is sinusoidal and that the stress ranges between σ_{max} and σ_{min}. The change in stress intensity factors is the difference between the K_I values associated with these two stresses. Although the stresses reached during the loading are well below those that would be critical for linear elastic fracture mechanics, the K_I and K_{Ic} concepts can be used to evaluate the severity of the stress level reached.

The fatigue damage is presumed to exist in the material in the form of microcracks with a distribution of sizes, locations, and orientations. To simplify the mechanics of the computer simulation, we have selected a discrete number of orientations (nine), designated by the spherical coordinates θ and ϕ on which to track damage. Thus, for the plane in the kth orientation, crack damage is the number of cracks per unit volume (J_k).

The level of damage at any time can now be represented by the quantity τ, defined as

$$\tau = T_F \sum_k J_k R_k^3 \tag{4}$$

where T_F is a constant of the order of 1, J_k is the number of cracks per unit volume, and R_k is the crack radius; k is summed over all the planes. Conveniently, other work on fracture has already shown that, in this definition, τ is proportional to the fragmented fraction of the solid.[17] Thus we define T_F so that $\tau = 1$ means full fragmentation. On this scale, $\tau = 0{\cdot}001$ is barely observable damage and $\tau = 0{\cdot}1$ is serious damage.

Further framework for this formulation is also readily available from the literature. When a tensile loading is applied, the cracks open in proportion to the stress normal to their plane. The opening displacement is given by

$$\delta_k = \frac{4(1-\nu^2)}{\pi E} R_k \sigma_k \quad \text{(half crack opening)} \tag{5}$$

where E and ν are Young's modulus and Poisson's ratio respectively, and σ_k is the stress normal to the plane of the crack. The crack opening strain ε_{ck} is defined as the sum of the crack volumes for all the cracks:

$$\varepsilon_{ck} = \tfrac{4}{3}\pi J_k R_k^2 \delta_k = \frac{16(1-\nu^2)}{3E} J_k R_k^3 \sigma_k \tag{6}$$

Under tensile loading, the applied strain $\Delta\varepsilon_k$ is taken partially by the opening of the crack ($\Delta\varepsilon_{ck}$) and partially by the elastic strain ($\Delta\varepsilon_{sk}$) of the solid, intact material. That is,

$$\Delta\varepsilon_k = \Delta\varepsilon_{sk} + \Delta\varepsilon_{ck} \tag{7}$$

Finally, the crack opening strain for J cracks per unit volume is derived by using the volume of a single crack in eqn (6):

$$\Delta\varepsilon_{ck} = \frac{16(1-\nu^2)}{3E} J_k (R_{2k}^3 \sigma_2 - R_{1k}^3 \sigma_1) \tag{8}$$

Here we account for the growth of the crack radius from R_{1k} to R_{2k} during the time interval.

Fatigue Crack Growth

Equations (5)–(8) provide a handy framework for relating crack dimensions to the additional strain associated with fracture damage.

Now we need to predict the fatigue crack growth process. The cracks grow while the normal stress is tensile yet below the critical stress level for rapid growth. We assume that inelastic strains are occurring at the crack tip at all stress levels. During repeated loadings, these inelastic strains lead to gradual growth of the crack during each loading cycle. A general discussion of such mechanisms in metals and ceramics has been given in Ref. 10. The growth expression must be appropriate for the finite element situation in which the strain (or stress) is imposed in small increments, so that even a single loading cycle might be imposed in 10 or more steps. Hence, the expression should contain a term related to the stress increment size $\Delta\sigma$ and should be independent of whether loading or unloading is occurring. The growth also increases in relation to the stress level σ to some power. In addition, we want an expression that can be integrated over a loading cycle and that will match Paris' equation. Therefore we propose the following form for the increase in crack radius R in the constitutive fatigue model:

$$dR = \frac{nB}{2}\left(\frac{2}{K_{Ic}}\sqrt{\frac{R}{\pi}}\right)^n \sigma^{n-1}\,|d\sigma| \tag{9a}$$

$$= gR^{n/2}\sigma^{n-1}\,|d\sigma| \tag{9b}$$

$$= \frac{nB}{2}\frac{\sigma^{n-1}}{\sigma_{cr}^n}\,|d\sigma| \tag{9c}$$

where g is a growth-rate coefficient with complex units, n and B are factors from Paris' equation, and σ_{cr} is the critical stress from eqn (3).

Equation (9) is suitable for incorporation into a constitutive model for use in a finite element code. Then, at each cycle, the cracks in tensile orientations grow by a small amount proportional to the loading increment. Growth occurs during both loading and unloading.

3.3 Integration to Form Paris' Equation

Having built the framework required to implement the fatigue model into a finite element code, we need to show that the model is equivalent to Paris' law. Let us now integrate eqn (9) over half a loading cycle from a stress of zero to σ_{max}. We can use any of the three forms in eqn (9) for this integration. If we integrate eqn (9a) or (9b), we obtain a complicated function of R that must be approximated to obtain the Paris equation (2). If we integrate eqn (9c), keeping σ_{cr} constant during the interval under the assumption that the change in R is small, we obtain an expression for ΔR, the crack extension over half a loading cycle:

$$\Delta R = \frac{B}{2}\left(\frac{\sigma_{\mathrm{max}}}{\sigma_{\mathrm{cr}}}\right)^n \tag{10}$$

We double this result to account for the unloading half of the cycle. Because of the similarity of the two expressions in eqn (3), we can equate the stress and K factors for a given radius:

$$\frac{\sigma}{\sigma_{\mathrm{cr}}} = \frac{K_{\mathrm{I}}}{K_{\mathrm{Ic}}} \tag{11}$$

The result of integrating our fatigue equation over a loading cycle from zero stress to the maximum and back to zero is then Paris' equation.

3.4 Integration to Form the Pavement Equation

Further work is needed to complete the implementation of the model. The next step is to relate eqn (2) to the pavement damage equation, (1); to do this, we replace K_{I} in eqn (2) with its expression from eqn (3). Then

$$\frac{\mathrm{d}R}{\mathrm{d}N} = \frac{B}{K_{\mathrm{Ic}}^n}\left(\frac{2}{\sqrt{\pi}}\right)^n \sigma^n R^{n/2} \tag{12}$$

or

$$\frac{\mathrm{d}R}{R^{n/2}} = \frac{B}{K_{\mathrm{Ic}}^n}\left(\frac{2}{\sqrt{\pi}}\right)^n \sigma^n\, \mathrm{d}N \tag{13}$$

Here σ is a constant, the maximum stress during a loading cycle. We want to integrate eqn (13) from $N = 1$ to the cycle at which the cracks grow catastrophically. This limit occurs when the crack radius reaches R_{m}, which is given by either $K_{\mathrm{I}} = K_{\mathrm{Ic}}$ (eqn (3)),

$$R_{\mathrm{m1}} = \frac{\pi}{4}\left(\frac{K_{\mathrm{Ic}}}{\sigma}\right)^2 \tag{14}$$

or, when τ reaches 1 in eqn (4),

$$R_{\mathrm{m2}} = (T_{\mathrm{F}}J)^{-1/3} \tag{15}$$

Then we integrate eqn (13) from the initial crack size R_0 to R_{m} and obtain

$$N = \left(\frac{\sqrt{\pi}}{2}\right)^n \frac{K_{\mathrm{Ic}}^n}{B\sigma^n(1 - \frac{1}{2}n)}\left(R_{\mathrm{m}}^{1-n/2} - R_0^{1-n/2}\right) \tag{16}$$

By comparing eqn (16) with eqn (1), we see that

$$A = \left(\frac{\sqrt{\pi}}{2}\right)^n \frac{K_{\mathrm{Ic}}^n}{B(1 - \frac{1}{2}n)f^n}\left(R_{\mathrm{m}}^{1-n/2} - R_0^{1-b/2}\right) \tag{17}$$

and that we were correct in using n as the exponent in both equations. Thus, when we know K_{Ic}, R_0, and J, we can determine n and B in eqn (2) from the pavement damage relation (1).

3.5 Incorporation into a Multiplane Plasticity Model

Completion of a full constitutive relation for DYNA3D was accomplished by combining a standard thermodynamic Mie–Grüneisen relation for pressure, a multiplane plasticity model for deviator stresses, and the constitutive fatigue model. The new model is called SRI-PAVEMENT. The multiplane feature was used so that fatigue crack growth could occur in several specific orientations. The arrangement of the planes in the multiplane model allows a distribution of planes that are nearly uniform in orientation. As the damage develops on some of the planes, the behavior becomes distinctly anisotropic because the stiffness reduces in directions normal to the crack faces.

The multiplane plasticity model was constructed to allow Tresca-like plastic flow in planes with specific orientations. These planes are illustrated in Fig. 3 in their usual orientations; nine planes are oriented to produce approximately isotropic behavior. Three planes are normal to the three coordinate directions, and six are at 45° between these directions. The planes have orientations but no specific locations within a computational element, so there is no direct geometric interaction among them. All plastic flow is assumed to occur on these planes, not homogeneously throughout the material as in a Mises plasticity model, for example.

Model Algorithm

During the stress computation process for an element, we first compute the stresses elastically, and then allow for stress relaxation through

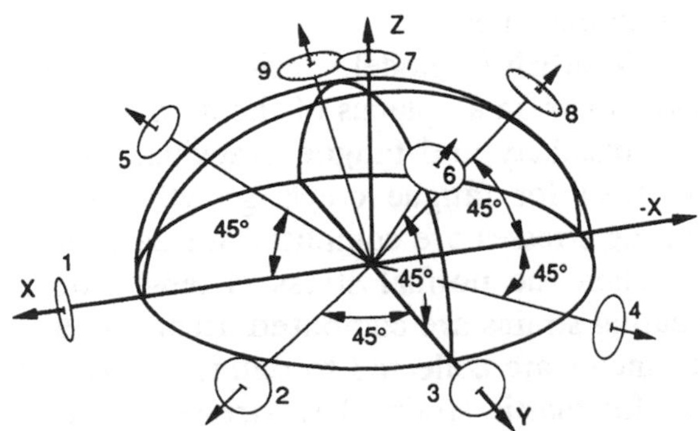

Fig. 3. Relative locations of the initial orientations of the fatigue damage planes with respect to the coordinate directions.

plastic flow on planes where the shear stress exceeds the current yield strength. The following computational steps are undertaken:

Step 1. From the imposed strain increments $\Delta\varepsilon$ and the old stress state σ_0, we compute the new stress state σ^N, treating the strain increments as elastic. Hence this is a standard linear elastic procedure.

Step 2. We transform the new stresses to normal (σ_k) and shear stresses (τ_{1k} and τ_{2k}) on each of the nine planes.

Step 3. The current yield strength is determined on each plane. A nonlinear, tabular work-hardening and thermal softening process is provided.

Step 4. We compute the plastic shear strains on each plane, using a stress-relaxation algorithm.

Step 5. From the known plastic strains on each plane, we compute the plastic strain increment tensor $[\Delta\varepsilon^p]$.

Step 6. We recompute the stress tensor $[\sigma]$, taking the elastic strain increments as the imposed strain increments minus the plastic strain increments, $[\Delta\varepsilon] - [\Delta\varepsilon^p]$.

The foregoing procedure would be a complete solution in the standard isotropic treatment of stress relaxation. However, in the current context, many planes may absorb plastic strain independently. These multiple planes interact in such a way that they may absorb too much plastic strain (so the elastic strain is misrepresented) or too little. To provide for proper interaction among the planes, we proceed through the foregoing steps many times and allow for only a small amount of stress relaxation at any step. The iteration and convergence procedure used here is designed to provide for high fidelity to the intended stress–strain relations and to permit relatively rapid solution even for large strain increments.

In the SRI-PAVEMENT model, fatigue cracks and the fatigue cracking processes occur on all planes of the model. For the model with a combination of plasticity and fatigue cracking, we first provide for plastic flow and then for fatigue crack growth. At each time step, normal and shearing stresses are computed for each plane (as noted in Step 2 above). When the normal stress is tensile on a plane, crack opening and shearing strains are computed according to eqn (6). These crack strain increments are collected to define a crack strain tensor, as in Step 5 above, for plastic strain. The stresses are recomputed as in Step 6 above. Then crack growth computations are made for each plane by following eqn (9).

Combination of SRI-PAVEMENT with DYNA3D

SRI-PAVEMENT was designed as a standard material model for use in DYNA3D; however, it imposed some special requirements. The biggest change was the increase in the possible number of history variables stored for each element; SRI-PAVEMENT needs 65 history variables to track the cracking processes for each element, compared with the previous maximum of 20 history variables for models available in DYNA3D.

3.6 Simulations of Aircraft Rolling over Pavement

We simulated the rolling of a pair of wheels configured as the landing gear of a Boeing 727 aircraft over a concrete pavement, as shown in Fig. 4. The problem was laid out with the wheels on the right-hand side of the pavement slab and moving toward the left at a speed of 100 mph (54 m/s). In the initial configuration, the wheels were off the right-hand edge of the pavement. The calculation was performed to a time of 0·15 s, by which time the landing gear had traveled 22 ft (6·7 m). Although our calculational model allows the load to be applied via pressurized tires, in these calculations the load was applied by specifying for each tire a load footprint that moves along the pavement surface at the speed of the landing gear. The loading footprint for each wheel was a uniform pressure of 200 psi (1·38 MPa) over a rectangular area 16 in × 12 in (40 cm × 30 cm) wide. At the leading and trailing edges of the wheel, the pressure was linearly decreased to zero over a

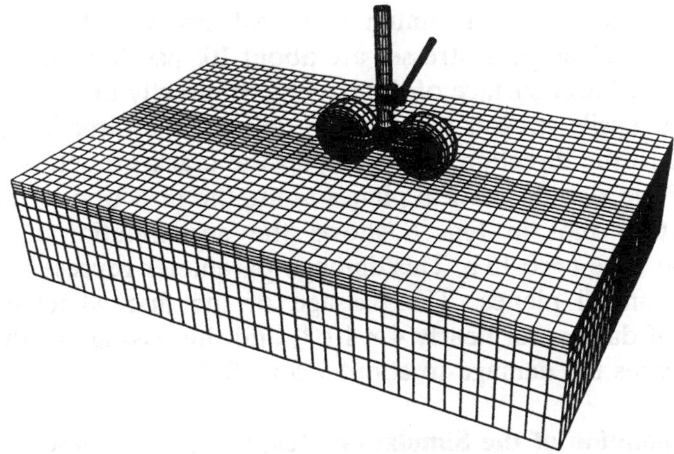

Fig. 4. Finite element layout of an aircraft landing gear on a 12 in thick Portland cement concrete pavement, provided by the preprocessor with DYNA3D.

Table 1
Properties of Portland Cement Concrete

Property	*Symbol*	*Value*	*Units*
Young'modulus	E	30·3	GPa
Poisson's ratio	n	0·15	—
Compressive strength	f_c'	20·0	MPa
Tensile strength	f_t'	2·8	MPa
Fracture toughness	K_{Ic}	2·0	MPa cm$^{1/2}$
Initial flaw radius	R_0	0·4	cm
Paris' growth coefficient	B	$1{\cdot}5 \times 10^{-6}$	cm/cycle
SRI growth coefficient	g	$4{\cdot}04 \times 10^{-7}$	cm$^{1-n/2}$/MPa^n
Crack density	J	0·1	number/cm^3
Crack growth exponent	n	3·0	—
Fragmentation coefficient	T_F	4·0	—

2 in (5 cm) length, for a total load of 86 kips (383 kN) for the two wheels. By assuming a vertical plane of symmetry through the center of the landing gear, we analyzed half the configuration shown in Fig. 4.

The runway was modeled as a single slab of concrete pavement over subgrade. The dimensions of the slab are 25 ft × 45 ft (7·6 m × 13·7 m) with a thickness of 24 in (61 cm). The pavement properties, listed in Table 1, were all obtained from data on concrete. The subgrade was modeled as an elastic layer 6 ft (1·8 m) thick with an elastic modulus of $4{\cdot}4 \times 10^5$ psi (3·0 Gpa) and a Poisson's ratio of 0·15. We assumed that the midplane of the pavement was pinned in the vertical direction and that the vertical surfaces of the subgrade were transmitting boundaries. The interface between the pavement and the subgrade was allowed to slip without friction.

The contours of the maximum principal stresses at $t = 0{\cdot}15$ s are shown in Fig. 5. The peak stresses are about 105 psi (0·72 MPa) and are located on the lower surface of the pavement directly under the wheels. The highest tensile stresses were computed in the z-direction (horizontal, normal to the direction of travel), so the crack growth is largest at the base of the slab on vertical planes along the direction of motion.

The contour plot in Fig. 6 shows the accumulation of cracking damage τ at time $t = 0{\cdot}15$ s. This damage parameter represents the sum of damage on all planes. The passage of the first wheel causes an increment of damage of nearly 4×10^{-8}, and the passage of the second wheel increases the damage to about $7{\cdot}5 \times 10^{-8}$.

3.7 Extrapolation of the Simulation Results to a Lifetime

We may recast the lifetime prediction in eqn (16) to a convenient form when we know dR/dN. Both laboratory tests and finite element

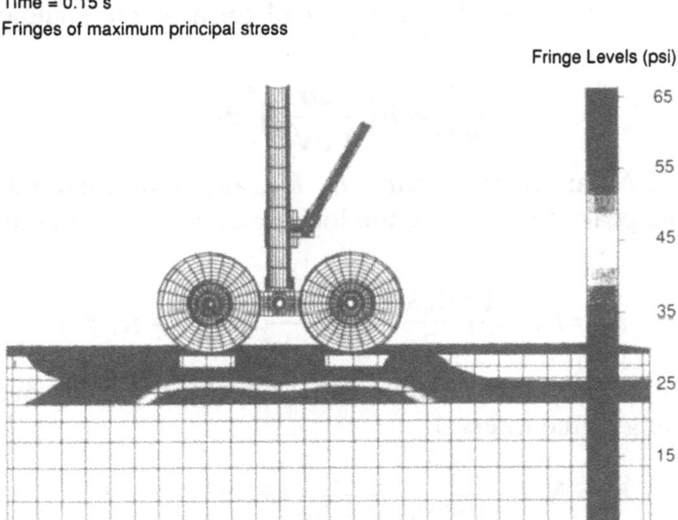

Fig. 5. Horizontal tensile stress contours under the landing gear at 0·15 s.

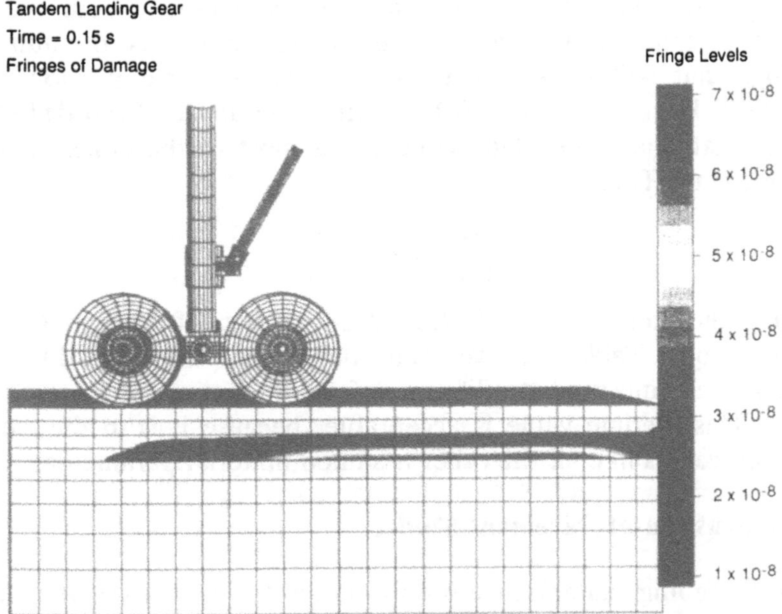

Fig. 6. Contours of fatigue damage on a vertical section of the concrete slab at 0·15 s.

calculations could naturally lead to values of $\mathrm{d}R/\mathrm{d}N$. For this derivation, we start with eqn (13), a modified form of the standard fatigue model:

$$\frac{\mathrm{d}R}{R^{n/2}} = B\left(\frac{2\sigma}{K_{\mathrm{IC}}\sqrt{\pi}}\right)^n \mathrm{d}N \tag{18}$$

Only R and N vary here, because B, K_{Ic}, and n are material constants and σ is the peak stress during the loading cycle. We integrate eqn (18) to obtain

$$N_2 = N_1 + \frac{1}{B}\left(\frac{K_{\mathrm{Ic}}\sqrt{\pi}}{2\sigma}\right)^n \frac{1}{1-\frac{1}{2}n}(R_2^{1-n/2} - R_1^{1-n/2}) \tag{19}$$

Next we replace B with its value from eqn (2), using the quantities at the beginning of the interval:

$$\begin{aligned} N_2 &= N_1 + \frac{R_1^{n/2}}{(\mathrm{d}R/\mathrm{d}N)_1}\frac{1}{1-\frac{1}{2}n}(R_2^{1-n/2} - R_1^{1-n/2}) \\ &= N_1 + \frac{1}{(1-\frac{1}{2}n)(\mathrm{d}R/\mathrm{d}N)_1}\left[R_2\left(\frac{R_1}{R_2}\right)^{n/2} - R_1\right] \end{aligned} \tag{20}$$

With this equation, we can now predict the number of loading cycles, N_2, required to reach a crack radius of R_2. The only information required is the material parameter n, the crack radius R_1, and the rate $(\mathrm{d}R/\mathrm{d}N)_1$.

Using the value of $7{\cdot}5 \times 10^{-8}$ for $\mathrm{d}\tau/\mathrm{d}N$ obtained from the finite element calculation, we can estimate fatigue life; that is, the number of loadings that will cause complete fracture of some portion of the pavement. Here we use eqn (4) to determine $\mathrm{d}R/\mathrm{d}N$ from $\mathrm{d}\tau/\mathrm{d}N$. We assume that one crack plane contributes most of the crack growth in each element. Then

$$\frac{\mathrm{d}\tau}{\mathrm{d}N} = 3T_{\mathrm{F}}JR^2\frac{\mathrm{d}R}{\mathrm{d}N} \tag{21}$$

Using the values of T_{F}, J, $R_{\mathrm{m}} = R_{\mathrm{m}2} = 2{\cdot}92$ cm from eqn (15), and $R = R_0$ from Table 1, we can determine that $\mathrm{d}R/\mathrm{d}N = 3{\cdot}9 \times 10^{-7}$ cm/cycle and predict a lifetime of 1·3 million cycles from eqn (20) above. This lifetime value is acceptable, because it is in the range of observed values like all the other assumed material parameters.

3.8 Discussion of Pavement Model

In the foregoing example, we used a model for fatigue crack growth to further illustrate the concepts of damage process prediction. This model is based on the presence of flaws that open during periods of tensile stress. The fatigue model was incorporated into a multiplane plasticity model to provide for the strong anisotropy that develops during cracking. This combined model, in a computer subroutine, was incorporated into the three-dimensional finite element code DYNA3D at Task 7 in Fig. 1. The action of an aircraft landing gear rolling over a pavement was simulated, and we then made a lifetime prediction based on fatigue cracking in the base of the pavement. An examination of the process of developing the constitutive model reveals its complexity, but also demonstrates the ability to model complex path-dependent behavior. Clearly this path dependence may lead to accumulated errors unless care is taken in formulating and implementing the model. Extensive validation of the model, by comparing calculated results with those from experiments, is also required.

4 EXAMPLE 3—DUCTILE FRACTURE USING LOCAL DAMAGE MODELS

In this section, we describe an approach to predicting the dynamic ductile fracture of a welded steel T-joint.[18] The features of the dynamic structural response in this problem, up to the point of fracture initiation, can be modeled by using finite element techniques that are common in existing nonlinear finite element codes such as DYNA3D. However, the processes of damage accumulation, fracture initiation, and fracture propagation require additional constitutive models and algorithms not commonly found in existing codes. Modifications were made to DYNA3D at the level of the constitutive model (Task 7 in Fig. 1) to introduce a measure of damage accumulation. In addition, the mesh was constructed to allow fracture propagation through changes introduced at Task 4.

In ductile metals, the damage process may be associated with the nucleation and growth of voids near the crack tip. In general, the nucleation and growth of voids are a function of the stresses and strains as well as of other parameters such as strain rate and temperature. In calculations where the characteristic dimensions of the voids are much smaller than the size of the elements used, damage in the elements is treated as a continuous function tracked as a state parameter within the constitutive model.

Finite element modeling of fracture propagation can be accomplished by using either smeared or discrete fracture representation.[19] In the smeared fracture representation, the fracture propagates through the mesh, and the crack opening displacements are accounted for within the element displacement field. This approach can be accomplished in the constitutive model by reducing the strength of the fractured elements so that the resistance to tensile opening is eliminated. This smeared fracture representation has advantages: the fracture can choose an arbitrary path through the mesh,[20] and no additional algorithms are required to allow separation along interfaces. However, the smeared approach may require additional complexity in the constitutive model, and can lead to computational difficulties such as stress locking.[19]

Discrete fracture representation models crack propagation and opening by separating element regions along an interface. The elements on either side of the interface have redundant nodes and degrees of freedom that are constrained to move together until the fracture criterion is exceeded. The discrete fracture approach has commonly been used in applications where the fracture path is known *a priori* from experimental results or from the problem symmetry.[21] Modeling of a region through which the fracture can choose an arbitrary path has seldom been done because of the computational requirements of redundant degrees of freedom required for fracture separation and the mesh resolution desired in an expanded fracture region. However, with advances in both software and hardware for finite element analysis, this approach has become feasible for many problems. An example by Horstemeyer[22] is the analysis of the fracture of steel plates from blast loading using the discrete shell element failure options in DYNA3D. Other approaches include methods that use rezoning as crack extension occurs;[23] however, the extension of these methods to three-dimensional geometries is very difficult.

In this example, a simple local fracture criterion is used to predict the fracture behavior of steel T-weldments of different sizes and geometries, fabricated by different welding processes. The purpose of the analysis was to understand the effects of weld process and experimental scale on fracture behavior. An issue is the ability to predict the path of failure through the structure.

Implementing the fracture model required modification or addition of two separate algorithms: (1) a local damage model in the constitutive model (Task 7 in Fig. 1), which calculates damage as a function of plastic strain increments and the stress state; combined with (2) a fracture algorithm, which allows discrete separation of the elements along an arbitrary fracture path within a region of the mesh (Task 4 in Fig. 1). Using this approach, we can calculate not only fracture initiation but also the path of fracture propagation through the structure.

4.1 Experiments

T-shaped weldments were produced at two scales by welding two stiffners normal to a slotted base plate (Fig. 7a). Each stiffener was bolted to a steel anchor, and the specimen plate was bolted to a steel die, as shown in Fig. 7(b). The center of the specimen plate was loaded with strips of sheet explosive backed by two blocks of tamping material (polymethylmethacrylate, PMMA) over an area bounded by the two slots but not extending all the way to each stiffener. With this test arrangement, only the center portion of the specimen plate is significantly deformed during the experiments, while the portion sup-

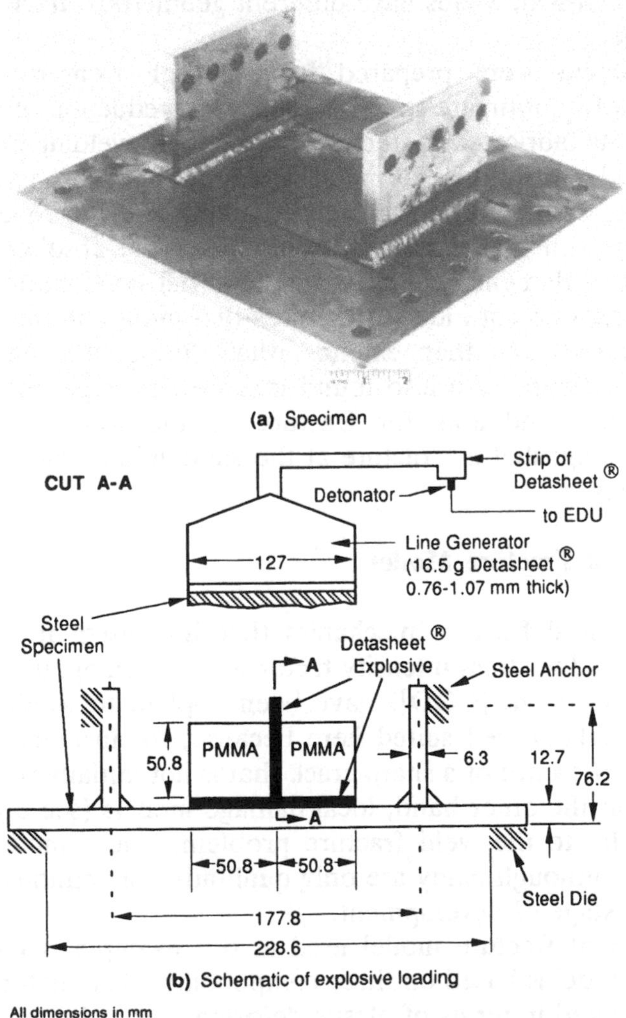

Fig. 7. Dynamic fracture experiment for T-shaped weldments.

ported by the die acts as a reaction frame, inducing membrane stresses in the plate.

Two sizes of specimen, large-scale and small-scale, were tested with the small scale half the size of the large scale. Both scale sizes were smaller than many actual structures of interest, so the computational methods need to be developed to extrapolate the experimental results to full scale. In replica-scaled experiments such as these, the effects of strain rate, gravity, and fracture generally do not scale. Fracture processes do not follow replica scaling because microfailure processes involve characteristic material lengths that remain constant for all specimen sizes.[24,25] Furthermore, because the cooling rates and number of passes can never be exactly reproduced in welded scale models, large- and small-scale welds have different geometries, microstructures, and properties.

The specimens were prepared from a high-strength steel (yield strength 930 MPa, ultimate strength 1000 MPa, reduction of area 70%). The weldments fabricated by the gas tungsten arc welding process were undermatched, with a weld metal yield strength of approximately 690 MPa. The welding process produced a heat-affected zone (HAZ) of approximately constant width, independent of the absolute weldment size. For the weldment geometry and material combinations that we studied, the experiment induces fractures that initiate in the plate HAZ of the weldments and then extend either through the base plate or through the stiffener. We also found that (within experimental uncertainty) loading conditions for fracture of the weldments followed geometric scaling; that is, fracture at the same initial velocities in both weldment sizes.

4.2 Weldment Fracture Model

Although classical fracture mechanics theories based on single parameters (such as the stress intensity factor *K*, the *J*-integral, or the crack opening displacement [COD]) have been applied to weldments, they are not particularly well suited here because, for instance, they must postulate the presence of a sharp crack that is not initially present in the weldment. On the other hand, local damage models (see e.g. Ref. 26) apply naturally to the weld fracture problem. These models are still quite simple, although many are only qualitative or semiquantitative at their current stage of development.

The weldment fracture model used in this example is illustrated in Fig. 8. The model is based on three components: (1) a material damage model formulated in terms of plastic deformations weighted by a stress

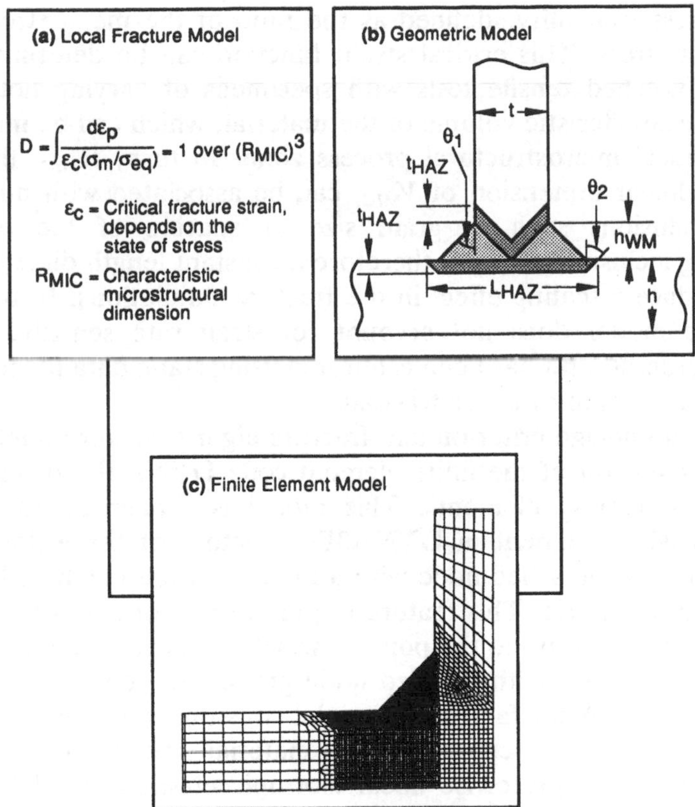

Fig. 8. Elements of the weldment fracture model.

state function; (2) a geometric and strength model of the weldment, based on metallographic observations and hardness measurements; and (3) a finite element formulation that implements these models with an algorithm permitting independent degrees of freedom for nodes on either side of the calculated fracture path. As implemented, the fracture model is the simplest form of a ductile fracture criterion.[27] It assumes that failure at a material location occurs when the damage within a surrounding characteristic volume V_{MIC} exceeds a critical value.

Mathematically, the damage failure criterion can be written in the form

$$D = \int \frac{\mathrm{d}\varepsilon^{\mathrm{p}}_{\mathrm{eq}}}{\varepsilon_{\mathrm{c}}(\sigma_{\mathrm{mean}}/\sigma_{\mathrm{eq}})} = 1 \quad \text{over } V_{\text{MIC}} \approx R^3_{\text{MIC}} \tag{22}$$

where D is the normalized damage parameter, $\mathrm{d}\varepsilon^{\mathrm{p}}_{\mathrm{eq}}$ is an increment in plastic strain, and $\varepsilon_{\mathrm{c}}(\sigma_{\mathrm{mean}}/\sigma_{\mathrm{eq}})$ is the critical failure strain as a function of the stress triaxiality, defined as the ratio of the mean stress to the equivalent stress. This critical strain function can be determined by a series of notched tensile tests with specimens of varying notch radii. V_{MIC} is a characteristic volume of the material, which can be interpreted as the critical microstructural process zone. In turn, R_{MIC}, the representative linear dimension of V_{MIC}, can be associated with microstructural dimensions such as grain size or spacing of the microvoid nucleating inclusions. R_{MIC} is therefore a constant length dimension that will introduce a scaling effect in the fracture simulations. In its current form, the model does not account for strain-rate sensitivity of the fracture process, and has been calibrated using static data for the failure strain as a function of stress triaxiality.

The local damage criterion and fracture algorithm were implemented into SRI's version of the finite element code DYNA3D for eight-node hexahedron (brick) elements. This model is similar to an available model for shell element in DYNA3D. Fracture of the weldment was simulated by using a tied node with a failure feature combined with the local damage criteria. This feature required that node groups have tied degrees of freedom up to the point at which the average damage for the elements associated with the tied node group exceeded a critical value (1 in this case). After failure, the nodal constraints are removed which allows elements to separate. The material model used in these calculations was a piecewise linear isotropic plasticity model. Integration of the damage function, eqn (1), occurred within the material constitutive model subroutine.

Figure 9 shows the results of a calculation that simulates a specimen subjected to an initial velocity sufficient to cause complete fracture compared with the experimental result for a small-scale specimen. The calculated deformed shape of the specimen agreed well with that produced in the experiments. Most importantly, the crack path across the plate was also faithfully reproduced in the simulation. Furthermore, the calculated deflection history was in good agreement with the experimental measurement, and, by using a lower initial velocity, we were able to simulate the arrest of the crack in the base plate. The good agreement between simulations and experiments partially validates the use of the fracture model for performing qualitative analyses of the weldment fracture, as described below. Additional validation and development are needed before truly quantitative analyses can be performed.

4.3 Application to the Optimization of Weld Design

Analyses using the simple local fracture model can be used to optimize the design of weldments for specific engineering applications, and to

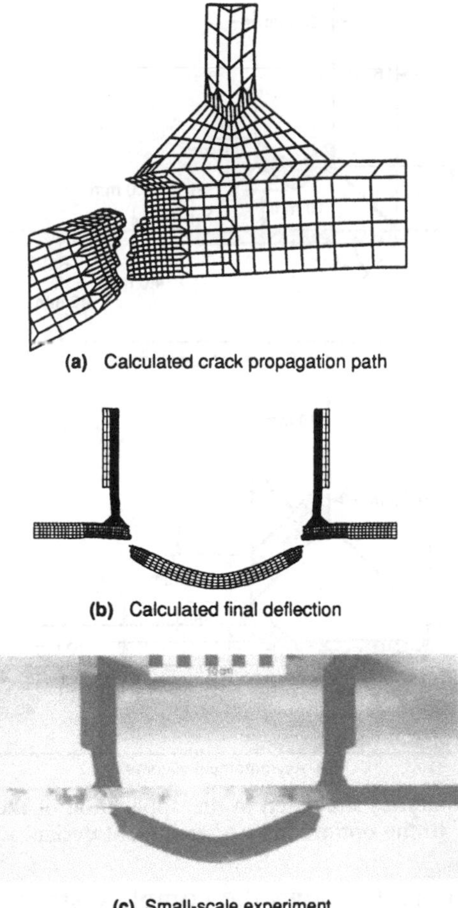

(a) Calculated crack propagation path

(b) Calculated final deflection

(c) Small-scale experiment

Fig. 9. Dynamic weld fracture experiment and calculation.

investigate the effects of weldment geometry, strength, and fracture properties on fracture behavior. Figure 10 illustrates two weldment geometries considered, one with two weld metal beads deposited symmetrically on either side of the stiffener (Fig. 10a), and the other with only one weld metal bead deposited asymmetrically on one side of the stiffner (Fig. 10b). The weld dimensions used for the calculations are also indicated in this figure.

In these calculations, we assumed that the material properties of the base metal, weld metal, and the HAZ are different from each other but do not vary within individual regions. Properties were selected to represent material combinations encountered in actual weldments. Table 2 summarizes the two combinations of material properties, termed 'high-strength weldment' and 'low-strength weldment', used in

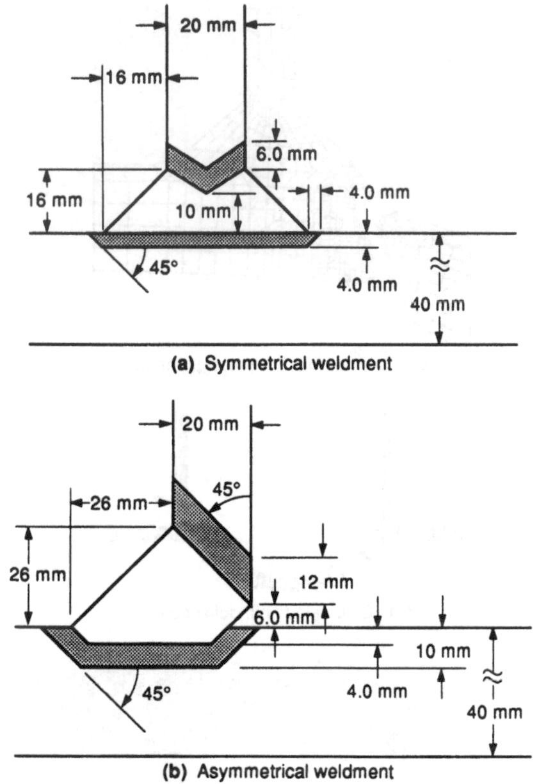

Fig. 10. Weldment geometries simulated in the application of the local fracture model to the optimization of weldment design.

the simulations. Figure 11 shows the complete stress–strain curves and the failure envelopes (plastic strain at fracture, ε_c, as a function of stress triaxiality). The low-strength weldment had a lower flow stress than the high-strength weldment, but higher fracture resistance, except for the HAZ.

Table 2
Material Properties for Weldment Design Simulations

Material	*High-strength weldment*		*Low-strength weldment*	
	Flow stress (MPa)	*Strain shift of failure envelope*	*Flow stress (MPa)*	*Strain shift of failure envelope*
Plate base metal	930	0	800	0·325
Stiffener base metal	930	0	620	0·325
Weld metal	744	0	800	0·325
HAZ	1 116	−0·08	930	−0·08

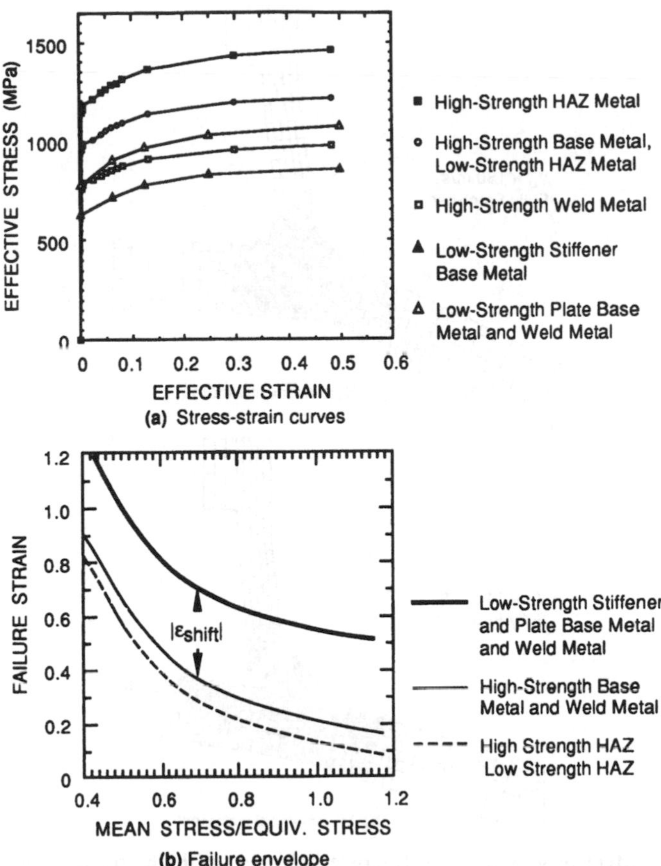

Fig. 11. Material properties used in the application of the local failure model to the optimization of weldment design.

The different weldments studied in the simulations were part of the overall specimen configuration illustrated in Figs 7 and 9 and were loaded dynamically by imparting an initial velocity V_0 to the center of the base plate. Figure 12 illustrates the effect of geometry on the fracture behavior of high-strength weldments. The difference in geometry leads to the initiation and propagation of fracture at different locations and to significant differences in the initial velocity required to produce complete fracture. The asymmetrical weldment is less resistant to fracture than the symmetrical weldment (V_0 for full fracture of 140 versus 180 m/s), and fracture initiates and propagates entirely in the HAZ of the stiffener. In contrast, in the symmetrical weldment, fracture initiates in the HAZ of the plate and then extends through the plate base metal, as observed in the experiments. We also found that changes in the fracture location and resistance can be induced by

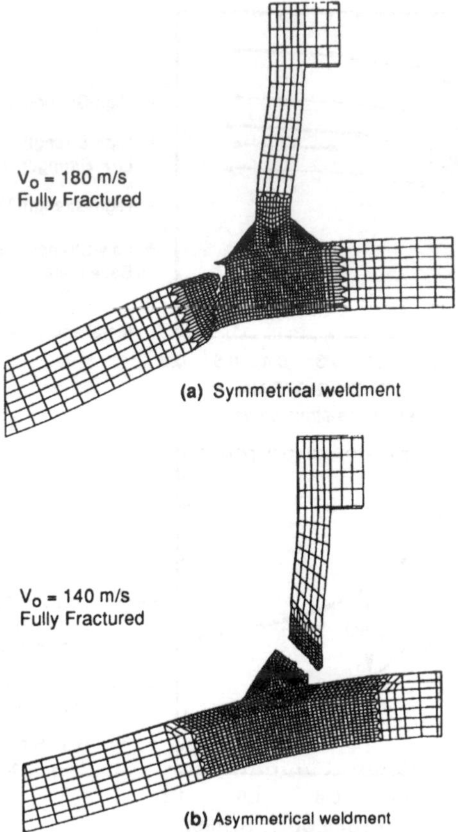

Fig. 12. Effect of geometry on fracture behavior of high-strength weldment.

differences in the material combination of the weldment (not illustrated here). In actual structures, we have observed the change in fracture location with weldment strength predicted by the simulations.

4.4 Estimation of Scaling Effect on Weldment Fracture with the Local Fracture Model

One important application of any fracture theory is to predict fracture in large structures on the basis of experimental data obtained with small laboratory specimens. Fracture processes do not follow geometric scaling laws, because microfailure processes involve characteristic material lengths that do not scale with specimen size.[24,25] Thus, in any computational fracture model, the ability to predict a size effect is important. However, unless rate effects, such as strain-rate sensitivity in the constitutive model, are incorporated in the computational model, the computational results will follow geometric scaling. Thus we need to introduce a characteristic material dimension into the fracture model to introduce scale-size effects.

The ability of the local damage model approach to directly address the effects of scale is particularly important in the weldment fracture problem. Two scaling effects were considered important in the weldment fracture response: (1) the effect associated with the fracture process itself (fracture scaling effect) and (2) the effect associated with the constant thickness of the HAZ (geometric scaling effect). Scaling effects that might be due to differences in the material properties of the weldment regions of different sizes (metallurgical scaling effect) may also be important, but were not considered here. Scaling effects caused by gravity and strain rate are considered negligible in this problem.

To assess the effect of specimen size, we performed separate calculations simulating the effects of the fracture scaling and the geometric scaling; then their combined influence was estimated. The numerical predictions were evaluated by comparing them with the experimental results. Three specimen sizes, termed baseline-scale geometry, double-scale geometry, and one-half-scale geometry, were used in the calculations. Table 3 gives the dimensions of these three configurations, which are based on observations from actual weldments.

The local damage approach is adapted to the study of these scaling problems by introducing length parameters in the otherwise dimensionless finite element meshes. Different specimen scales are simulated by changing the characteristic length R_{MIC} of the material (to introduce a fracture scaling effect) or by changing the dimensions of the HAZ relative to the plate thickness (to introduce a geometric scaling effect), or both.

Table 3
Scaled Weld Specimen Geometries

	Double-Scale geometry	*Baseline-scale geometry*	*One-half-scale geometry*
Stiffener spacing L_S	15·0 cm	7·50 cm	3·75 cm
Stiffener height H_S	6·0 cm	3·0 cm	1·5 cm
Stiffener thickness t	3·0 cm	1·5 cm	0·75 cm
Base plate thickness h	2·0 cm	1·0 cm	0·5 cm
Weld metal height h_{WM}	15·0 mm	7·50 mm	3·75 mm
HAZ thickness t_{HAZ}	2·5 mm	2·5 mm	2·5 mm
HAZ length L_{HAZ}	3·50 cm	2·0 cm	1·25 cm
Weld metal angle 1, q_1	45°	45°	45°
HAZ angle 2, q_2	45°	45°	45°
R_{MIC}	625 μm	625 μm	625 μm

Different methods can be used to introduce the material characteristic length parameter R_{MIC} in the simulation. One approach is to maintain the element size as a constant and vary the size of the specimen. This method, called mesh scaling, is equivalent to changing the mesh resolution so that the number of elements along any given direction in the fracture zone is doubled when the size of the specimen is doubled. However, this method of introducing characteristic length will also introduce mesh refinement effects.

The fracture damage we are trying to describe is typically the growth and linking together of voids that occur in small volume ahead of the crack tip where large gradients in stress and plastic strain occur. When the mesh scaling method is used, the larger-scale calculations will have a greater mesh refinement, and can therefore resolve higher stresses and strains locally in the damage region at the crack tip. Thus this approach will predict a reduction in strength with increasing scale size that is unrealistically large.[18]

The above observations illustrate that the local fracture model is mesh-size-dependent. However, mesh-size effects do not eliminate the usefulness of current fracture models for qualitative or even quantitative analyses. The mesh dependence does, however, require that the model be validated against a known result for new applications, and care must be exercised by the analyst to maintain consistent mesh resolutions in calculations for which comparisons are being made.

The approach used in this study to vary R_{MIC} in the simulation is to maintain a constant mesh resolution in the fracture zone, independent of the scale, but to impose a constraint that fracture occurs only when the failure criterion is met over a region of the mesh that increases with decreasing specimen size. The area (volume) over which damage is averaged for three scale sizes is illustrated in Fig. 13. As the scale size is reduced by a factor of 2, the damage zone radius increases by a corresponding factor of 2 to maintain a constant, absolute size of the damage region. This method, called the damage zone-scaling method, maintains finite element mesh refinement and hence the accuracy with which we calculate the stress and strain gradients in the fracture region.

To complete the investigation of scaling effects, we performed a series of calculations for various initial loading velocities to bracket the initial velocity at which complete fracture occurs. Because close bracketing requires many computations, an estimate of the initial velocity (or energy) required to produce full fracture was made by using the internal energy required by a configuration that fully fractures (equal to the total energy minus the residual kinetic energy of the fractured segment of the base plate) and back-calculating the initial

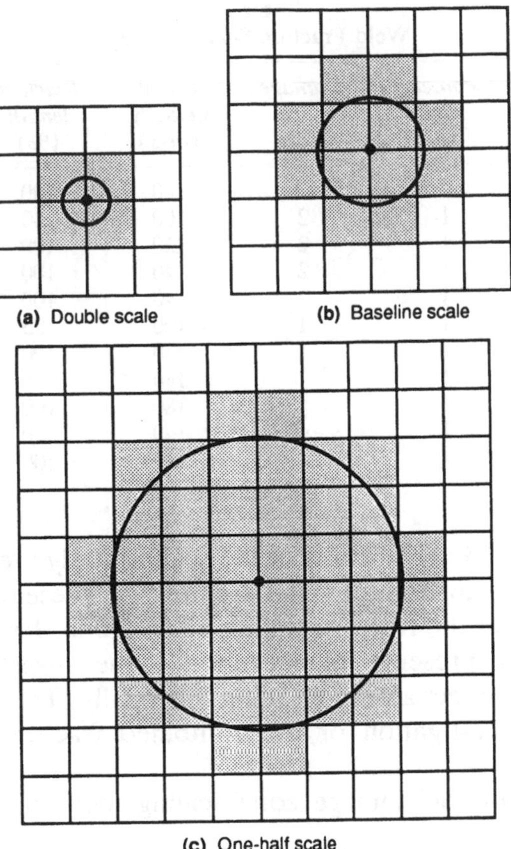

Fig. 13. Domains of integration of the damage function for various specimen sizes (damage zone scaling method).

velocity corresponding to that energy. Because the specimen deformation and the fracture process can change at different velocities, this method gives only an approximation of the true fracture energy, but we estimate that it is accurate to within 5–10%.

Table 4 summarizes all the calculations investigating the scaling effect on the dynamic fracture of weldments, and lists (1) the geometry scale corresponding to the weld geometries in Table 3, (2) the damage zone scale corresponding to the zone size in Fig. 13, (3) the initial velocity applied to the base plate, (4) the relative length of the fracture as a percentage of the base plate thickness, and (5) the residual velocity of the fracture base plate section.

The calculations listed in Table 4 were performed modeling the fracture scaling effect by using the damage zone scaling method

Table 4
Weld Fracture Scaling Study

Calculation number	*Geometry scale*	*Damage zone scale*	*Initial velocity (m/s)*	*Fracture length (%)*	*Residual velocity (m/s)*
1	2	2	140	100	65
2	1	2	100	40	0
3	1	2	120	100	35
4	1	2	140	100	70
5	$\frac{1}{2}$	2	140	100	80
6	1	1	100	25	0
7	1	1	140	100	45
8	1	$\frac{1}{2}$	140	0	0
9	1	$\frac{1}{2}$	180	100	10
10	$\frac{1}{2}$	$\frac{1}{2}$	140	10	0
11	$\frac{1}{2}$	$\frac{1}{2}$	180	100	60

illustrated in Fig. 13. The results of the scaling study are summarized in Table 5. Reading horizontally, the second row presents the results of the investigation of the fracture scaling effect alone. Reading vertically, the first column presents the results of the investigation of the geometric scaling effect alone. Reading diagonally, Table 5 presents the results of the investigation of the combined fracture and geometry scaling effects.

With the use of the damage zone scaling method, the simulations

Table 5
Summary of Results of Weldment Fracture Scaling Study Using Damage Zone Scaling Method

	Double-scale damage zone calculation	*Baseline-scale damage zone calculation*	*One-half-scale damage zone calculation*	
Double-scale HAZ geometry	$V_f = 124$ m/s (10)[a]			
Baseline-scale HAZ geometry	$V_f = 118$ m/s (12, 13)	$v_f = 133$ m/s (16)	$V_f = 176$ m/s (18)	Fracture scaling effect
One-half-scale HAZ geometry	$V_f = 113$ m/s (14)		$V_f = 169$ m/s (20)	
	Geometric scaling effect			Combined fracture and geometric scaling effects

[a] The numbers in parentheses correspond to the calculation numbers in Table 4 from which the fracture energy and fracture velocity were determined.

predict fracture velocities from approximately 118 m/s at double scale to 133 and 176 m/s at baseline and one-half scale, respectively. The fracture scaling effect is stronger than expected for the one-half-scale specimen. Potentially this increase results because the damage zone used may have been unrealistically large (equivalent to having specimen sizes much smaller than those tested, in which the scale effect might actually be large). In addition, metallurgical changes that may occur with different scale sizes were not considered. These results suggest that the damage zone scaling method may be appropriate for simulating the fracture scaling effect; however, further validation on a problem of simpler geometry is needed.

The geometric scaling effect on the dynamic fracture of the weldment results in fracture velocities from approximately 124 m/s at double-scale to 118 and 113 m/s at baseline and one-half-scale geometries respectively. The weld geometric scaling effect produces lower fracture velocities for smaller-scale weldment specimens. We anticipated this trend, because the relatively brittle HAZ is a proportionately larger fraction of the base plate cross section in the smaller-scale weldments. However, the magnitude of this effect is small.

The fracture and geometric scaling effects tend to offset each other, and result in fracture velocities varying from approximately 124 m/s at double-scale to 133 and 169 m/s at baseline and one-half-scale geometries respectively. Again the magnitude of the combined scaling effect appears to be stronger than expected for the one-half-scale specimen by extrapolation of our experimental results.

4.5 Discussion of Fracture Model

The results obtained demonstrate that this model can qualitatively predict the crack path. If the selection of fracture parameters and mesh refinement is based on experimental results, the model can also be used to obtain some quantitative data about structural strength and failure loads. The model provides the opportunity to define better weldment design practices, develop more precise characterization tests for weldments, and improve weldment models. Further improvements to this approach are also possible. For example, more sophisticated and better calibrated and verified damage/local fracture models are already available and should provide better quantitative fracture predictions (see e.g. Refs 28–30).

Further work on local damage models should investigate their ability to predict size effects on fracture and should include further comparison with experimental observations. This ability represents a crucial test for the usefulness of the models, because one main purpose of any fracture theory is to make fracture predictions for large structures on the basis of experimental data obtained with small laboratory specimens.

The last problem is difficult because different phenomena may contribute to the size effect, depending on the stage of the fracture process. Moreover, different microfailure modes will induce different size effects.

In our attempt to assess size effects by using a simple local fracture model, we observed that a consistent numerical formulation of the fracture problem must be selected for all sizes; that is, the number of elements in the fracture region should be kept constant to retain the same resolution for the calculated stress, strain, and damage. As stated previously by other authors, the numerical implementation, which must be performed keeping in mind both the physics of the fracture process and the need for reasonable computation times, becomes an integral part of the model. The approach taken here to select a constant mesh geometry (therefore scaling the element size) and to integrate the damage over a process zone of constant absolute dimensions should eliminate any finite element mesh dependence of the scaling effect predictions.

Thus future development of this type of fracture model should be validated by experiments with simpler laboratory specimens and loading configurations, such as statically loaded notched and precracked fracture mechanics specimens, using materials that do not contain pronounced heterogeneities as in the case of weldments. In addition, comparisons between the computational methods applied here and other models that incorporate softening of the material properties as the damage evolves may provide useful insight into the mechanisms controlling the effects of size in ductile fracture.

5 CONCLUSIONS

As discussed earlier, finite element codes can be made to provide good solutions to structures problems, at least until failure is imminent, but these solutions sometimes break down as failure starts or progresses. The process of failure often involves changes in constitutive or structural response outside the normal capabilities provided by most finite element codes. In this chapter, we have used three examples to illustrate procedures for solving problems involving failure.

The easiest type of failure processes to model are those such as buckling or collapse that do not cause local material damage or fracture. Solutions to these problems require only that a realistic description of the random variations in material properties or geometry be included in the mesh geometry to initiate the buckling process. The nonlinear finite element code provides the necessary framework for calculating the large displacements and material plasticity needed to determine postfailure response.

Many failure process problems are dominated by material damage or fracture processes that produce changes in constitutive behavior or structural integrity not easily described by standard finite element models. For these applications, calculation of the failure processes requires additional computational algorithms, the most common of which are constitutive models that include a description of the physical damage processes.

This approach was illustrated in the example of calculating fatigue damage in airport pavements. The airport pavement damage problem is characterized by loadings that are small compared with static pavement failure loads and by the many load cycles that result in crack formation on discrete planes. The need here was to produce a new constitutive model that included a fatigue damage model on discrete planes within the material. Using this approach, we were able to calculate the fatigue damage in the pavement produced by a single load cycle from the weight of a passing aircraft and then use the damage to estimate the lifetime of the pavement.

The final example given here was the calculation of the dynamic ductile fracture of a steel weldment. To calculate the failure processes in this example, we included a simple damage model in the constitutive model that calculates local (element) damage as a function of the stress state and plastic strain increments. Structural failure was then modeled by tying local damage to an interface routine that constrained the relative motion of nodes on either side of the crack until the damage criterion was exceeded.

These examples illustrate that successful calculation of failure processes is possible with finite element techniques. However, successful calculation of failure often requires the implementation of new computational algorithms in the finite element code for modeling the physical damage processes involved.

REFERENCES

1. Hallquist, J. O. & Whirley, R. G., DYNA3D user's manual (nonlinear dynamic analysis of structures in three dimensions). Report UCID-19592,

University of California, Lawrence Livermore National Laboratory, Livermore, 1989.
2. Owen, D. R. J. & Hinton, E., *Finite Elements in Plasticity: Theory and Practice.* Pineridge Press, Swansea, UK.
3. Liu, W. K. & Lam, D., Numerical analysis of diamond buckles. In *Finite Elements in Analysis and Design,* **4** (1989) 291–302.
4. Kennedy, J. M. & Belytschko, T. B., Buckling and postbuckling behavior of the ACS support columns. *Nucl. Engng Des.,* **75** (1982) 323–42.
5. Kirkpatrick, S. W. & Holmes, B. S., Effect of initial imperfections on dynamic buckling of shells. *ASCE J. Engng Mech.,* **115** (1989) 1075–93.
6. Kirkpatrick, S. W. & Holmes, B. S., Static and dynamic buckling of thin cylindrical shells. In *Computational Aspects of Contact, Impact, and Penetration,* ed. R. F. Kulak and L. E. Schwer. Argonne National Laboratory, Argonne, IL, 1992, pp. 77–98.
7. Kirkpatrick, S. W. & Holmes, B. S., Axial buckling of a thin cylindrical shell: Experiments and calculations. *Computational Experiments,* PVP, Vol. 176. ASME Publications, New York, 1989, pp. 67–74.
8. Florence, A. L., Gefken, P. R. & Kirkpatrick, S. W., Dynamic buckling of copper cylindrical shells. *Int. J. Solids Structures* **27** (1991) 89–103.
9. Seaman, L., Simons, J. W. & Shockey, D. A., A microcracking fatigue model for finite element analysis. In *Advances in Local Fracture/Damage Models for the Analysis of Engineering Problems,* ed. J. H. Giovanola & A. J. Rosakis. ASME AMD, Vol. 137, 1992, pp. 243–55.
10. Ritchie, R. O., Mechanisms of fatigue crack propagation in metals, ceramics, and composites: role of crack tip shielding. *Mater. Sci. Engng,* **A103** (1988) 15–28.
11. Hilsdorf, H. K. & Kesler, C. E., Fatigue strength of concrete under varying flexural stresses. *Proc. Am. Concrete Inst.,* **63** (1966) 1059–75.
12. Monismith, C. L., Epps, J. A., Kasianchuk, D. A. & McLean, D. B., Asphalt mixture behavior in repeated flexure. Institute of Transportation and Traffic Engineering Report No. TE 70-5, University of California, Berkeley.
13. Yimprasert, P. & McCullough, B. F., Fatigue and stress analysis concepts for modifying the rigid pavement design system. Research Report 123-16, Center for Highway Research, University of Texas at Austin.
14. Treybig, H. J., McCullough, B. F., Smith, P. & von Quintus, H., Overlay design and reflection cracking analysis for rigid pavements, 1. Development of new design criteria. Research Report FHWA-RD-77-66, Federal Highway Administration, Washington, DC, 1977.
15. Paris, P. C., Gomez, M. P. & Anderson, W. E., A rational analytic theory of fatigue. *Trends Engng* **13** (1961) 9–14.
16. Paris, P. C., The growth of fatigue cracks due to variations in load. Ph.D. Thesis, Lehigh University, Bethlehem, PA, 1962.
17. Curran, D. R., Seaman, L. & Shockey, D. A., Dynamic failure of solids. *Phys. Rep.* **147** (1987) 253–388.
18. Giovanola, J. H. & Kirpatrick, S. W., Applying a simple ductile fracture model to fracture of welded T-joints. In *Advances in Local Fracture/Damage Models for the Analysis of Engineering Problems,* ed. J. H. Giovanola & A. J. Rosakis. ASME AMD, Vol. 137, 1992, pp. 285–303.

19. Rots, J. G., Smeared and discrete representations of local fracture. *Int. J. Fracture* **51** (1991) 45–59.
20. Murakami, S., Kawai, M. & Rong, H., Finite element analysis of creep crack growth by a local approach. *Int. J. Mech. Sci.* **30** (1988) 491–502.
21. Peeters, F. J. H. & Koers, R. W. J., Numerical simulation of dynamic crack propagation phenomena by means of the finite element method. In *Fracture Control of Engineering Structures*, ed. H. C. Van Elst & A. Bakker. Engineering Materials Advisory Services, Warley, UK, 1987.
22. Horstemeyer, M. F., Damage of HY100 steel plates from oblique constrained blas waves. In *Advances in Local Fracture/Damage Models for the Analysis of Engineering Problems*, ed. J. H. Giovanola & A. J. Rosakis. ASME AMD, Vol. 137, 1992, pp. 233–42.
23. Swenson, D. & Ingraffea, A., A finite element model of dynamic crack propagation with an application to intersecting cracks. In *Numerical Methods in Fracture Mechanics*. Pineridge Press, Swansea, UK, 1987, pp. 191–204.
24. Holmes, B. S. & Colton, J. D., Application of scale modeling techniques to crashworthiness research. In *Aircraft Crashworthiness*, ed. K. Saczalski, T. Singley III, W. D. Pilkey & R. L. Huston. University Press of Virginia, Charlottesville, 1975, pp. 561–81.
25. Giovanola, J. H., Scaling of fracture phenomena. Poulter Laboratory Technical Report, SRI International, 1982.
26. Chaboche, J. L., Dang Van, K., Devaux, J. C., Marandet, B., Masson, S. H., Pelissier-Tanon, A., Pineau, A. & Rousselier, G., eds, *Séminaire International sur l-Approche Locale de la Rupture*. Conference Proceedings, Centre de Recherches EDF, Les Renardiéres, Moret-sur-Loing, France, 1986.
27. Mudry, F., Methodology and applications of local criteria for prediction of ductile tearing. In *Elastic–Plastic Fracture Mechanics*, ed. L. H. Larson. ECSC, EEC, EAEC, Brussels and Luxembourg, pp. 263–83.
28. Tvergaard, V., Influence of voids on shear band instabilities under plane strain conditions. *Int. J. Fracture* **17** (1981) 389–407.
29. Rousselier, G., A methodology for ductile fracture analysis based on damage mechanics: an illustration of local approach to fracture. In *Nonlinear Fracture Mechanics*, Vol II: *Elastic–Plastic Fracture*. ASTM STP 995, American Society for Testing and Materials, PA, 1989, pp. 332–54.
30. Pineau, A. & Joly, P., Local versus global approaches to elastic plastic fracture mechanics: Application to ferritic steels and a duplex stainless steel. In *Defect Assessment in Components, Fundamentals, and Applications*, ed. J. G. Blauel *et al.* Mechanical Engineering Publications, London, 1991, pp. 381–414.

Chapter 3

CRITERIA FOR THE INELASTIC RUPTURE OF DUCTILE METAL BEAMS SUBJECTED TO LARGE DYNAMIC LOADS

NORMAN JONES & WEI QIN SHEN†

Impact Research Centre, Department of Mechanical Engineering, The University of Liverpool, PO Box 147, Liverpool L69 3BX, UK

ABSTRACT

The dynamic inelastic failure of beams is examined using two different failure criteria. An elementary rigid–plastic method of analysis gives useful predictions for the overall features of the failure behaviour of beams subjected to uniform impulsive and impact loadings. However, a recent procedure using an energy density failure criterion provides greater detail of the behaviour and is capable of further development for a wide range of dynamic problems in the fields of structural crashworthiness, hazard assessment and impact engineering. Nevertheless, further experimental and theoretical studies are required to assess whether or not this criterion may be used with confidence for design purposes, except possibly for predicting the dynamic inelastic failure of beams and frames.

NOTATION

k, k_i	defined by eqns (2)–(4)
l	defined in Fig. 1
l_m	mean value of hinge length across beam thickness
q	Cowper–Symonds exponent (eqn (11))
t	time
t_f	duration of response
t_1, t_2	durations of first and second phases of motion
B	beam breadth
D	Cowper–Symonds coefficient (eqn (11))

† Current address: School of Systems Engineering, University of Portsmouth, Anglesea Building, Anglesea Road, Portsmouth, Hampshire PO1 3DJ, UK.

D^{P}_{bL}, D^{P}_{mL}, D^{P}_{sL}	bending, membrane and shear work absorbed at the left-hand side of the impact point made dimensionless with respect to the total plastic work absorbed at time t at the left-hand side of the impact point
D^{0}_{b}, D^{0}_{m}, D^{0}_{s}	bending, membrane and shear work made dimensionless with respect to the total work absorbed plastically
D^{*}_{b}, D^{*}_{k}, D^{*}_{m}, D^{*}_{p}, D^{*}_{s}	bending, kinetic, membrane, total plastic and shear work made dimensionless with respect to the initial kinetic energy
G	mass of striker
H	beam thickness
I^*	dimensionless impulse $(\rho/\sigma_0)^{1/2}V_0$
$2L$	beam span
M_0, N_0, Q_0	fully plastic capacities of a beam cross-section for a bending moment, a membrane force and a transverse shear force acting alone
S	distance of impact point from left-hand support
S^*	S/L
V_0	initial uniform impulsive velocity or initial velocity of a striker in the impact loading case
W, W_m	maximum permanent transverse displacement
W_s	transverse shear displacement
α	dimensionless hinge length (eqn (13))
β	defined by eqn (14)
β_c	critical value of β for a Mode III failure
ε	strain
ε_c	critical strain
ε_f	nominal failure strain
ε_{f0}	engineering failure strain (zero gauge length)
ε_{max}	maximum strain
$\dot{\varepsilon}_m$	mean strain rate
ε_p	defined by eqn (8)
θ	plastic work per unit volume
θ_c	critical value of θ
θ_{cm}	defined by eqn (6)
λ	$2GV_0^2L/\sigma_0BH^3$
ρ	density of material
σ_d	dynamic flow stress
σ_{d0}	dynamic uniaxial yield stress

σ_0	static uniaxial yield stress
ϕ	defined by eqn (15)
Ω	plastic work absorbed in a plastic hinge
Ω_c	critical value of plastic work absorbed in a plastic hinge
Ω_s	plastic work absorbed in a plastic hinge through transverse shearing deformations
Ω^*	$\Omega/\sigma_{d0}BHl_m$
Ω_c^*	critical value of Ω^*
$(\dot{\ })$	$\partial(\ )/\partial t$

1 INTRODUCTION

A survey on the dynamic inelastic failure of beams was presented at the Second International Symposium on Structural Crashworthiness.[1] The beams were made from ductile materials, which could be modelled as rigid, perfectly plastic, and they were subjected to either uniformly distributed impulsive loads, as an idealisation of an explosion, or mass impacts to idealize dropped object loading. It was assumed that the energy imparted by the external dynamic load was significantly larger than the maximum amount of strain energy that could be absorbed by a beam in a wholly elastic manner. Thus large plastic strains were produced and material rupture occurred for sufficiently large dynamic loads.

It was observed that the different dynamic loadings in Ref. 1 may cause the development of different failure modes. The simplest failure modes were associated with a uniformly distributed impulsive loading. A rigid–plastic method of analysis[2] for this particular problem shows that membrane forces as well as bending moments must be retained in the basic equations for the response of axially restrained beams subjected to large dynamic loads that cause transverse displacements exceeding the beam thickness approximately. This is known as a Mode I response. If the external impulse is severe enough then the large strains that are developed at the supports of an axially restrained beam would cause rupture of the material, which is known as a Mode II failure. At still higher impulsive velocities, the influence of transverse shear forces dominates the response, and failure is more localised and occurs owing to excessive transverse shearing displacements (Mode III).

It is important to note that a Mode III transverse shear failure is more likely to occur in dynamically loaded beams than in similar statically loaded beams. The physical reason for this phenomenon is explained in Section 6.1 of Ref. 2. For example, infinite transverse shear forces are generated on the application of an ideal impulsive loading to a rigid–plastic beam with plastic yielding controlled by the bending moment alone, while transverse shear forces must remain finite in a similar static beam problem to equilibrate with the external static load. Thus, even a beam with a solid rectangular cross-section and a large length-to-thickness ratio can suffer a transverse shear failure under a dynamic loading (Mode III response), as observed by Menkes and Opat[3] and analysed in Ref. 4.

Paradoxically, it is observed in Ref. 1 that, despite the lower impact velocities of the mass impact case, the failure behaviour is much more complex than the impulsive loading case. The failure modes II and III discussed above for an impulsive loading also occur for a mass impact loading. However, it transpires that other more complex failure modes may develop. For example, a striker may cause an indentation on the struck surface of a beam, which, if sufficiently severe, may lead to failure. The impact velocity of a strike near to the support of a beam might not be sufficiently severe to cause a transverse shear failure (Mode III), but could distort severely a beam and cause rupture due to the combined effect of transverse shear force, membrane force and bending moment, as shown in Fig. 14 of Ref. 1.

It is evident[1] that the dynamic inelastic rupture of beams and other structures is an extremely complex phenomenon and that there is a pressing need for the development of a reliable criterion that can be used to predict the onset of material rupture for a wide range of engineering problems in the fields of structural crashworthiness and hazard assessment.

This chapter discusses some studies that have been published on the dynamic inelastic failure of structures since the Second International Symposium on Structural Crashworthiness.[1] In particular, the energy density failure criterion introduced in Ref. 5 for the dynamic rupture of beams is explored in some detail. To provide some background, Section 2 discusses several aspects of an elementary failure criterion.[4] The energy density failure criterion[5] is introduced in Section 3 and is used to examine the dynamic inelastic failure of fully clamped beams loaded impulsively and subjected to impact loads in Sections 4 and 5 respectively. Unlike the elementary failure criterion in Section 2, the energy density failure criterion retains the simultaneous influence of the bending moment, membrane force and the transverse shear force for all of the failure modes. The chapter finishes with a discussion and a conclusion in Sections 6 and 7 respectively.

2 OBSERVATIONS ON THE ELEMENTARY FAILURE CRITERION[1,4]

2.1 Tensile Tearing (Mode II)

A rigid, perfectly plastic method of analysis was introduced[4] to analyse the failure modes observed in the experimental tests of Menkes and Opat[3] on aluminium alloy 6061 T6 impulsively loaded beams. It was found that the failure mode II, which is due to rupture at the supports of an axially restrained beam, could be predicted from a rigid-plastic analysis that caters for the influence of membrane forces and bending moments, but neglects the effect of transverse shear forces in the yield condition. However, a rigid, perfectly plastic material idealisation has an unlimited ductility, and infinite strains are associated with idealised plastic hinges, whereas the strain in actual materials cannot exceed the rupture strain.

In order to overcome this limitation of rigid plastic analyses, it was noted that a plastic hinge in a beam has the shape shown in Fig. 1 when using plane strain methods of analysis. The plastic hinge in Fig. 1(a) develops for pure bending behaviour at zero transverse displacements of a beam, while the plastic hinge in Fig. 1(b) is associated with the onset of a membrane response when the maximum transverse displacement W equals the beam thickness H for a fully clamped beam.

It is evident from Fig. 1 that the length l of the plastic zone on the upper surface of a fully clamped beam changes from the beam thickness H when the maximum transverse displacement $W = 0$ to $2H$ when $W = H$. The establishment of a hinge length l in this way then allows the axial strains on the surface of a beam at a hinge to be evaluated when assuming that plane sections remain plane. Once the membrane state of a beam has been reached, the entire beam becomes plastic with an effective hinge length of $\frac{1}{2}L$, where $2L$ is the span of a fully clamped

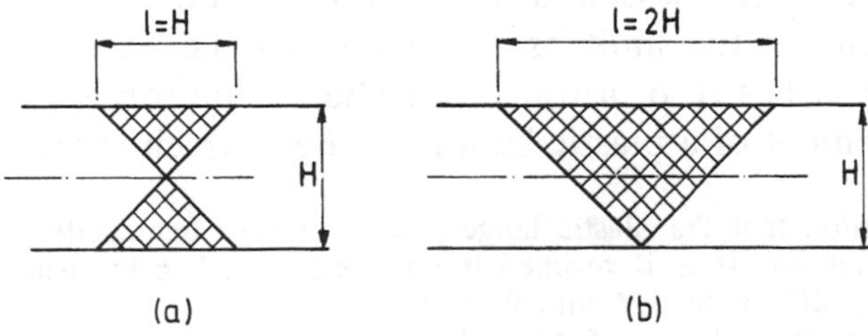

Fig. 1. Plastic hinge in a fully clamped beam: (a) when $W = 0$; (b) when $W = H$.

beam.† It is then a straightforward exercise[1,2,4] to evaluate the location and magnitude of the maximum strain for a given external loading so that the threshold conditions to cause a Mode II failure may be obtained when the total maximum strain equals the nominal rupture strain of the material, i.e.

$$\varepsilon_{max} = \varepsilon_f \tag{1}$$

This elementary procedure gave surprisingly good agreement with the corresponding experimental data of Menkes and Opat[3] when using $\varepsilon_f = 0{\cdot}17$ for the uniaxial nominal rupture strain of the aluminium alloy 6061 T6 beams.

The method outlined above was simplified by neglecting any material strain-hardening effects‡ and the influence of material strain-rate sensitivity both on the magnitude of the plastic flow stress and on the value of the rupture strain. In fact, the material strain-rate-sensitive effects should not be important for the aluminium alloy 6061 T6 beams studied by Menkes and Opat[3]. However, the uniaxial rupture strains of some aluminium alloys may change with an increase in strain rate, so that the threshold velocity for a Mode II failure would be different from that predicted using a static rupture strain.

Material strain-rate sensitivity could be important for the dynamic loading of the mild steel specimens, since the plastic flow stress may be enhanced significantly, while the rupture strain may decrease.[7] For the large plastic strains of interest in this chapter, it would be important to cater for the variation with strain of the coefficients in the Cowper–Symonds constitutive equations, since the influence of material strain rate sensitivity for mild steel reduces at large strains.[7]

2.2 Transverse Shear Failure (Mode III)

It was recognised by Symonds[8] that transverse shear slides or plastic shear hinges may develop in beams when large transverse shear deformations occur within a very short region of a rigid, perfectly plastic beam. In the limit as the length of the excessively deformed region approaches zero, a transverse shear slide forms by analogy with the development of a plastic bending hinge, which is an idealisation of

† The assumption that the plastic hinge length changes immediately from $2H$ to $\frac{1}{2}L$ when the membrane state is reached is removed in Ref. 6 by using a plastic hinge having a length $2W$ for $W > H$ until $W = \frac{1}{2}L$.

‡ An estimate of the influence of material strain hardening on the magnitude of the flow stress for the threshold conditions of a Mode II failure could be made by using a mean flow stress instead of the yield stress.

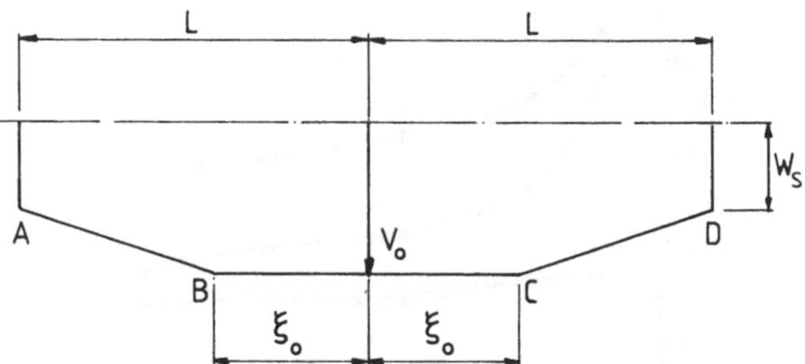

Fig. 2. Transverse velocity profile during the first phase of motion for a simply supported, impulsively loaded, rigid, perfectly plastic beam[2] with $Q_0L/2M_0 > 1{\cdot}5$.

rapid changes of slope across a short length of beam. Figure 2 shows the transverse velocity profile for a simply supported rigid, perfectly plastic beam a short time after it was subjected to an initial impulsive velocity[2] that was distributed uniformly across the entire span. Transverse shear slides have developed at both supports A and D, while the more familiar plastic bending hinges have formed at B and C. It was recognised in Ref. 4 that excessive values of the transverse displacements at the supports A and D in Fig. 2, for example, could lead to severance, which is known as a Mode III transverse shear failure. Thus a Mode III failure occurs at time t after impact when the total transverse shear displacement W_s at a particular location reaches a critical value, or

$$W_s = kH \tag{2}$$

where H is the beam thickness and $0 < k \leq 1$.

An inspection of the beam specimens tested by Menkes and Opat[3] reveals that the permanent transverse displacements of the beams, which had been severed through a Mode III failure, are relatively small. Thus it was suggested in Ref. 4 that a Mode III failure is governed mainly by the influence of transverse shear forces and bending moments, with membrane forces playing a minor role. It was assumed in Ref. 4 that $k = 1$ in eqn (2) for simplicity and in the absence of any data. This approximation led to theoretical predictions for the threshold impulse for a Mode III failure that gave reasonable agreement with the corresponding experimental values obtained by Menkes and Opat[3].

Experimental results have been reported in Ref. 9 on the double-shear failure of fully clamped aluminium alloy and mild steel beams with various thicknesses and struck across the entire span with impact velocities up to 11·6 m/s. These specimens had notches at the supports

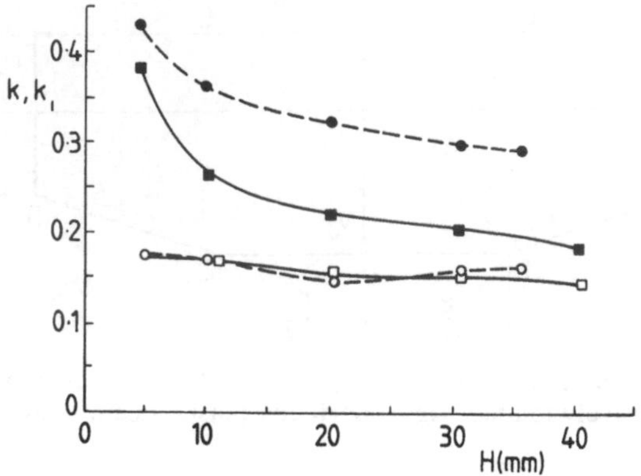

Fig. 3. Variation of k for complete severance and k_i for crack initiation with beam thickness H for dynamic double-shear tests:[9] ———, aluminium alloy; - - - -, mild steel; ■, ●, k for severance; □, ○, k_i for initiation.

to ensure that failure occurred across specified planes. Thus this experimental arrangement is different to that encountered usually in beams. Nevertheless it transpired that the corresponding value of k in eqn (2) for failure depends on the type of material and on the beam thickness, as shown in Fig. 3. The upper curve (k) for each material relates to complete severance, while the lower curve (k_i) corresponds to the initiation of a crack. The difference between the two curves for each material is large for thin beams, but decreases as the beam thickness increases. The values of k_i and k for these particular tests[9] are given by the empirical expressions

$$k_i = 0{\cdot}157 \tag{3a}$$

$$k = 0{\cdot}16 + 0{\cdot}03L/H \tag{3b}$$

for aluminium alloy and

$$k_i = 0{\cdot}165 \tag{4a}$$

$$k = 0{\cdot}27 + 0{\cdot}025L/H \tag{4b}$$

for mild steel, where $2L$ and H are the beam span and thickness respectively. Strictly speaking, eqns (3) and (4) are valid only for beams having geometries and material properties similar to the test specimens examined in Ref. 9, which all have the same span $2L$.

2.3 Comments

The Modes II and III failure criteria discussed in the previous two sections were first introduced[4] to explain the failure of impulsively loaded beams.[3] The theoretical procedure has also been used to examine the dynamic inelastic failure of beams subjected to mass impact loadings.[2,6] However, mass impact loadings may produce additional failure modes that would be more difficult to study using the elementary method of analysis outlined in Sections 2.1 and 2.2. It is quite likely that future studies will reveal still other failure modes in engineering structures subjected to various dynamic loads.

Evidently, it is important to seek a universal failure criterion that may be used to predict the rupture of a wide range of structures and that may be employed in theoretical analyses and incorporated into numerical schemes. One possible criterion is based on a critical density of plastic energy absorption. This criterion was introduced in Ref. 5 and is discussed in Section 3.

3 ENERGY DENSITY FAILURE CRITERION

In an attempt to obtain a universal failure criterion that could be used for a large class of dynamic structural problems, an energy density failure criterion was introduced in Ref. 5. It is assumed that rupture occurs in a rigid-plastic structure when the absorption of plastic work (per unit volume) θ reaches the critical value

$$\theta = \theta_c \tag{5}$$

where θ contains the plastic work contributions related to all of the stress components. The maximum possible value of θ_c is taken as

$$\theta_{cm} = \int_0^{\varepsilon_{f0}} \sigma_d(\varepsilon, \dot{\varepsilon}_m)\, d\varepsilon \tag{6}$$

where $\sigma_d(\varepsilon, \dot{\varepsilon}_m)$ is the dynamic engineering stress–strain curve, which is obtained from a dynamic uniaxial tensile test for a given $\dot{\varepsilon}_m$, and where ε_{f0} is the engineering rupture strain (zero gauge length).† In general, both ε_{f0} and σ_d could depend on the magnitude of the strain rate.[1,7] For a rigid, perfectly plastic material, eqn (6) simplifies to

$$\theta_{cm} = \sigma_{d0}\varepsilon_p \tag{7}$$

where

$$\varepsilon_p = \frac{1}{\sigma_{d0}} \int_0^{\varepsilon_{f0}} \sigma_d(\varepsilon, \dot{\varepsilon}_m)\, d\varepsilon \tag{8}$$

and σ_{d0} is the mean dynamic flow stress.

† The zero-gauge-length engineering rupture strain is calculated from $\varepsilon_{f0} = A_0/A_f - 1$, where A_0 and A_f are the original and the smallest final cross-sectional areas of a tensile test specimen respectively.

In the particular case of a rigid–plastic beam having a width B and a thickness H, eqn (5) gives the actual plastic work

$$\Omega = \Omega_c = \theta_c BHl_m = \sigma_{d0}\varepsilon_c BHl_m \tag{9}$$

absorbed at failure in a plastic hinge having an average length l_m across the beam thickness, where ε_c is a critical strain, the maximum value of which is ε_p given by eqn (8). Equation (9) may be recast into the dimensionless form

$$\Omega^* = \Omega_c^* = \frac{\Omega_c}{\sigma_{d0}BHl_m} = \varepsilon_c \tag{10}$$

The Cowper–Symonds constitutive equation[2] may be used to give the mean dynamic flow stress

$$\sigma_{d0} = \sigma_0\left[1 + \left(\frac{\dot{\varepsilon}_m}{D}\right)^{1/q}\right] \tag{11}$$

where D and q are material constants, σ_0 is the static uniaxial yield stress and $\dot{\varepsilon}_m$ is the mean strain rate throughout the response. Equation (11) is a reasonable approximation for those strain-rate-sensitive materials (e.g. mild steel) that have a stress–strain curve that approaches a rigid, perfectly plastic form with increasing strain rate. The mean strain rate $\dot{\varepsilon}_m$ in eqn (11) is estimated from the expression

$$\dot{\varepsilon}_m = \varepsilon_c/t_f \tag{12}$$

where t_f is the time when a structural failure occurs. It is not anticipated that eqn (12) would introduce any significant errors into the calculations, because the strain-rate-sensitive effects are highly non-linear for many materials. The approximations that are embodied in eqns (11) and (12) were introduced in Ref. 5 to simplify the analysis, but they could be removed in any future work if considered necessary.

The mean hinge length l_m across the beam depth in eqns (9) and (10) is taken as

$$l_m = \alpha H \tag{13}$$

Nonaka[10] used slip line theory to obtain the mean hinge length for the particular case of a fully clamped beam. His results may be used to show that $\alpha = 0{\cdot}5$ for the pure bending hinge in Fig. 1(a) and that $\alpha = 1$ when the bending behaviour is exhausted and the membrane state is reached, as shown in Fig. 1(b). However, it was observed when using the numerical analysis in Ref. 5 that the dimensionless plastic hinge length $\alpha = l_m/H$ in the experimental results of Menkes and Opat[3] on aluminium alloy 6061 T6 beams depends on the parameter

$$\beta = \Omega_s/\Omega \tag{14}$$

where Ω_s is the plastic work absorbed in a plastic hinge through shearing deformations and Ω is the total amount of plastic work absorbed at the same plastic hinge.

4 FAILURE OF AN IMPULSIVELY LOADED BEAM USING THE ENERGY DENSITY FAILURE CRITERION

4.1 Introduction

The fully clamped impulsively loaded beam, which was studied experimentally by Menkes and Opat[3] and theoretically with a rigid–plastic method of analysis in Ref. 4, was re-examined in Ref. 5 using the critical density failure criterion outlined in Section 3. It was remarked in Section 2.1 that a Mode II tensile tearing failure is obtained by considering the influence of the bending moment and axial membrane force in the yield condition, but neglecting the influence of the transverse shear force. However, the analysis in Ref. 5 retains the simultaneous influence of bending moment, membrane force and transverse shear force in the yield condition because it is possible that all three generalised stresses will contribute to the threshold conditions for both the Mode II and Mode III failures of actual structures.

4.2 Mode II Failure

The response of a fully clamped beam subjected to a uniformly distributed impulse consists of three phases of motion.[5] The first is extremely short, since it lasts for the duration of the dynamic pressure pulse, which is idealised with a rectangular-shaped pressure–time history. Thus the influence of finite deflections, or membrane forces, is neglected during the first phase of motion, and the transverse velocity profile is as shown in Fig. 4(a).

After the pressure pulse has been released, the plastic hinges at B and D start to travel towards the mid-span C during a second phase of motion, as shown in Fig. 4(b). If material failure has not occurred during the first or second phases of motion then a third phase of motion commences when the two travelling plastic hinges reach and coalesce at the mid-span. This final phase of motion has stationary plastic hinges at both supports A and E and at the mid-span C, as shown in Fig. 4(c).

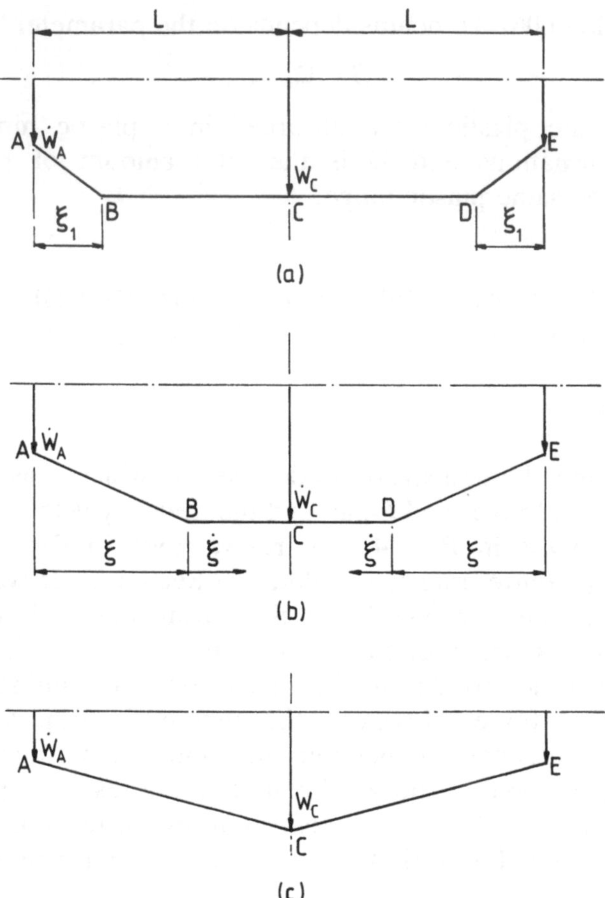

Fig. 4. Transverse velocity profiles for a fully clamped, impulsively loaded, rigid–plastic beam.[5] (a) First phase of motion with stationary transverse plastic hinges at A and E and stationary plastic hinges at B and D. (b) Second phase of motion with stationary plastic hinges at A and E and travelling hinges at B and D. (c) Third phase of motion with stationary plastic hinges at A, C and E.

This theoretical solution[5] was obtained with the aid of a Runge–Kutta numerical scheme and was found to be both statically and kinematically admissible. If at the cessation of motion ($t = t_f$), $\Omega < \Omega_c$ at every plastic hinge in the beam, where Ω_c is defined by eqn (9) and takes into account all the non-zero plastic work components, then the beam remains intact and, for a sufficiently large impulse, undergoes large plastic deformations without material rupture. This is known as a Mode I response, as noted in Section 1. In this case, eqn (12) cannot be used to obtain the mean strain rate $\dot{\varepsilon}_m$ during the response, because the final strain in a plastic hinge is $\varepsilon = \phi\varepsilon_c$, where $\phi \leqslant 1$. However, an estimate for ϕ is given by the ratio

$$\phi = \Omega/\Omega_c \tag{15}$$

so that eqn (12) is replaced by

$$\dot{\varepsilon}_m = \phi\varepsilon_c/t_f \tag{16}$$

where t_f is the response duration.

If $\Omega = \Omega_c$ at a plastic hinge at any time during the response then a Mode II response occurs, or, possibly, a Mode III response, as discussed in Section 4.3.

4.3 Mode III Failure

The numerical procedure outlined in Section 4.2 for a Mode II failure is the same as that employed in Ref. 5 for a Mode III transverse shear failure, because the calculations retain the bending moments, membrane forces and transverse shear forces in all of the basic equations, except during the short first phase of motion, when the influence of membrane forces was neglected. Only the influences of the bending moments and the transverse shear forces were retained throughout the elementary analysis[4] for a Mode III failure, which is outlined in Section 2.2.

It is proposed in Ref. 5 that a Mode III transverse shear failure occurs when $\Omega = \Omega_c$, as for a Mode II failure, but with $\beta = \beta_c$, where β is defined by eqn (14). The transition between a Mode II and a Mode III failure may also depend on other factors, including the shape of a beam cross-section. The actual value of β_c for a Mode III failure must be found from comparisons between the theoretical predictions and the corresponding experimental results, as discussed in Section 4.4.

4.4 Comments

The numerical calculations outlined in Sections 4.1–4.3 were compared[5] with some experimental results[3] on impulsively loaded fully clamped aluminium alloy 6061 T6 beams. The permanent transverse displacements at the mid-span relative to the final transverse displacement at the supports were measured from the photographs of the severed beams in the original reports from which Ref. 3 was drawn. The value of the dimensionless plastic hinge length α (eqn (13)) is then calculated from the numerical scheme with $\varepsilon_c = 0{\cdot}5$ so as to ensure that the predicted value of the permanent transverse displacement is equal to the corresponding experimental value. Also calculated at the same time is the associated value of β from eqn (14).

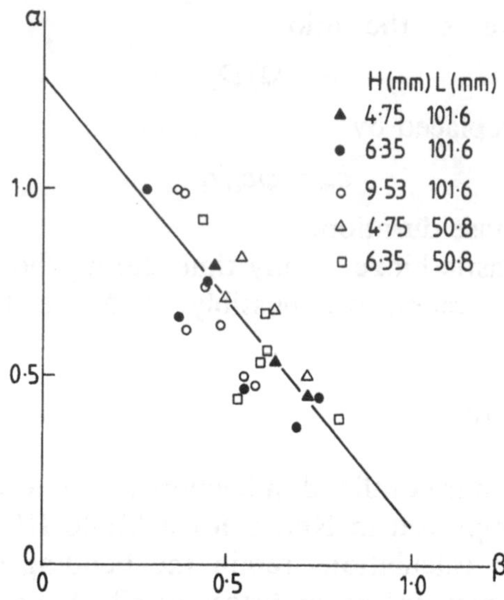

Fig. 5. Values of α and β from the experimental results of Menkes and Opat[3] for impulsively loaded beams according to the numerical analysis[5] (———, eqn (17)).

The above numerical results are plotted in Fig. 5 and may be approximated by the straight line

$$\alpha + 1{\cdot}2\beta = 1{\cdot}3 \tag{17}$$

Thus for $\beta = 0$ (i.e. no transverse shear deformations), $\alpha = 1{\cdot}3$, so that the plastic hinge length is 30% longer than the beam thickness. In fact, the experimental results of Menkes and Opat[3] do not extend to the ordinate in Fig. 5. The largest experimental value of the dimensionless hinge length is $\alpha \approx 1$, which has an associated value of $\beta = 0{\cdot}25$ according to eqn (17). It is evident from Fig. 5 and eqn (17) that α decreases as β increases, which is also anticipated from a physical viewpoint. The smallest value of the dimensionless hinge length is $\alpha \approx 0{\cdot}37$ for the test results of Menkes and Opat,[3] while the largest value of β is approximately 0·8. The semi-empirical equation (17) and the value of β calculated from eqn (14) were used to estimate the dimensionless hinge length α for all the calculations that retain the influence of the bending moment and the membrane and transverse shear forces, and which are reported in Ref. 5.

It should be noted that the permanent transverse displacements of the beams in Ref. 3, which were measured to construct Fig. 5 and to give eqn (17), might differ from the transverse displacements at the instant of rupture. In other words, if a beam fails before all of the initial kinetic energy has been absorbed in plastic deformations then the beam has a residual kinetic energy at severance. Some of this may be absorbed in further plastic flow as the beam continues to deform after severance until it reaches a permanent deformed state with an associated rigid body kinetic energy. A similar phenomenon was observed in Ref. 11 for the blast loading of a free–free, rigid, perfectly plastic beam. When the inertia loading is non-uniform, some of the external dynamic energy is absorbed in plastic hinges before the beam reaches a permanently deformed shape without any futher plastic deformation but with a residual rigid body kinetic energy.

A critical value of $\beta_c = 0{\cdot}45$, where β is defined by eqn (14), was suggested in Ref. 5 to mark the transition from a Mode II to a Mode III failure. Thus, summarising the theoretical results for an impulsively loaded fully clamped beam gives the following ranges of validity for the three failure modes:

$$\text{Mode I:} \quad \Omega < \Omega_c, \quad \beta < \beta_c \tag{18a}$$

$$\text{Mode II:} \quad \Omega = \Omega_c, \quad \beta < \beta_c \tag{18b}$$

$$\text{Mode III:} \quad \Omega = \Omega_c, \quad \beta \geqslant \beta_c \tag{18c}$$

where Ω contains all the contributions associated with the non-zero generalised stresses, and Ω_c is defined by eqn (9).

Table 1 of Ref. 5 shows a comparison between the numerical predictions[5] discussed in this section, the elementary theoretical predictions[4] outlined in Section 2 and the corresponding experimental results of Menkes and Opat.[3] The present numerical predictions give better agreement with the experimental results than the elementary theoretical predictions in Section 2 because the semi-empirical relation (17) has been employed in the calculations, and the simultaneous influence of the membrane forces, bending moments and transverse shear forces has been retained in the governing equations, as indicated in Figs 6 and 7.

The numerical results confirm various features of the Modes I, II and III responses that were predicted by the elementary analysis[4] discussed in Section 2. Naturally, the numerical results are more accurate, but both methods predict large residual kinetic energies in a beam after a Mode III failure and reveal that the response duration decreases progressively from a Mode I response through a Mode II rupture to a Mode III failure. In fact, the elementary theory predicts remarkable agreement with the response durations obtained from the numerical calculations. However, the numerical results in Figs 6 and 7 with $\varepsilon_c = \varepsilon_{fo} = 0{\cdot}5$ reveal that the transverse shear forces are also important for the transition between a Mode I and a Mode II failure. Moreover,

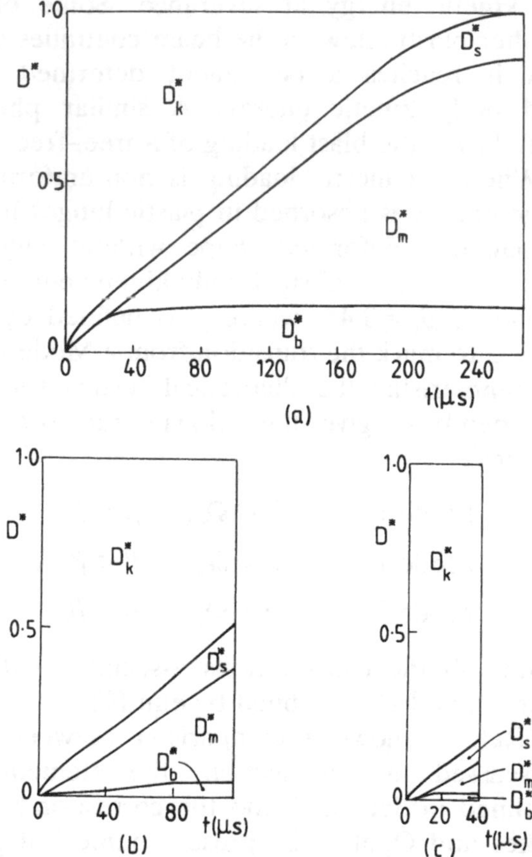

Fig. 6. Variation of the dimensionless energies with time in a fully clamped beam with $L = 101{\cdot}6$ mm, $B = 25{\cdot}4$ mm, $H = 6{\cdot}35$ mm, $\rho = 2686$ kg/m^3, $\sigma_0 = 279{\cdot}6$ N/mm^2, $D = 6500$ s^{-1}, $q = 4$ and $\varepsilon_{f0} = 0{\cdot}5$ that is subjected to uniformly distributed impulses of (a) $I^* = 0{\cdot}323$ (Mode I), (b) 0·638 (Mode II) and (c) 0·961 (Mode III). D_k^*, D_s^*, D_m^* and D_b^* are the kinetic energy and the plastic shear, membrane and bending work components made dimensionless with respect to the initial kinetic energy, and $I^* = V_0(\rho/\sigma_0)^{1/2}$ is the dimensionless impulse per unit area, where ρ is the density of beam material and σ_0 is the static uniaxial yield stress.

the elementary analysis exaggerates the importance of transverse shear forces at the threshold of a Mode III failure, as indicated by the influence of the membrane forces in Figs 6(c) and 7(d). It is interesting to note that the numerical analysis predicts that $k \approx 0{\cdot}6$ (see eqn (2)) both for a Mode II failure and at the onset of a Mode III failure. Nevertheless it should be remarked that the value of k depends on the critical value β_c as well as the magnitude of ε_c used in eqns (10) and (18).

The numerical method of analysis has been also used to obtain the

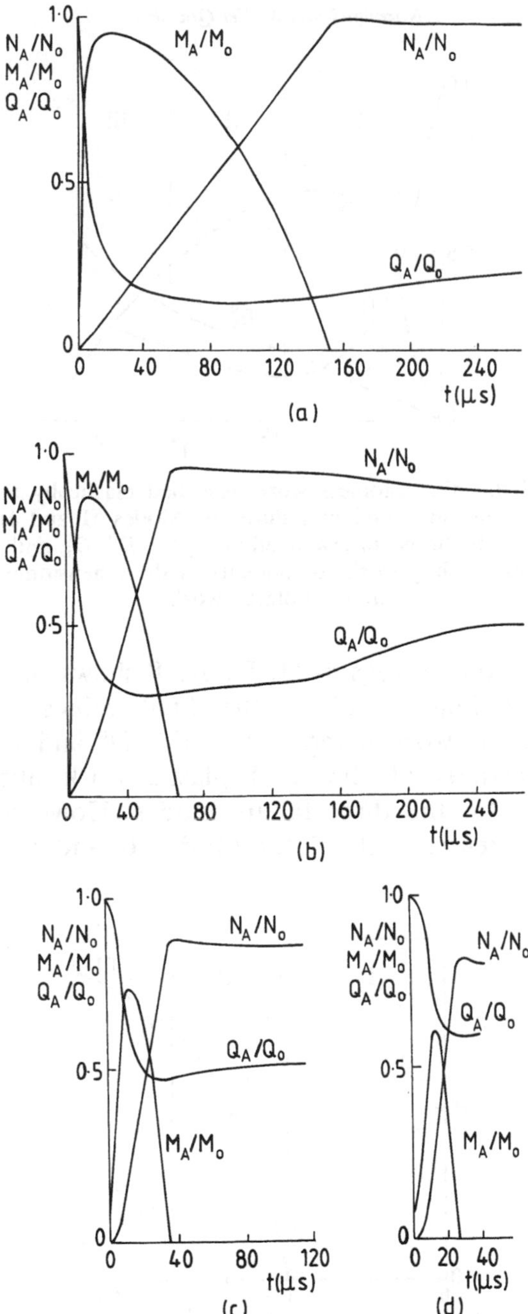

Fig. 7. Variation of the dimensionless bending moment M_A/M_0, transverse shear force Q_A/Q_0 and membrane force N_A/N_0 with time at the support A of the fully clamped beam in Fig. 4 when subjected to uniformly distributed dimensionless impulses of (a) $I^* = 0{\cdot}131$, (b) 0·323, (c) 0·638 and (d) 0·961. M_0, Q_0 and N_0 are the fully plastic capacities of the beam cross-section for a pure bending moment, transverse shear force and a membrane force acting alone. The dimensions and material properties of the beams are the same as those in Fig. 6.

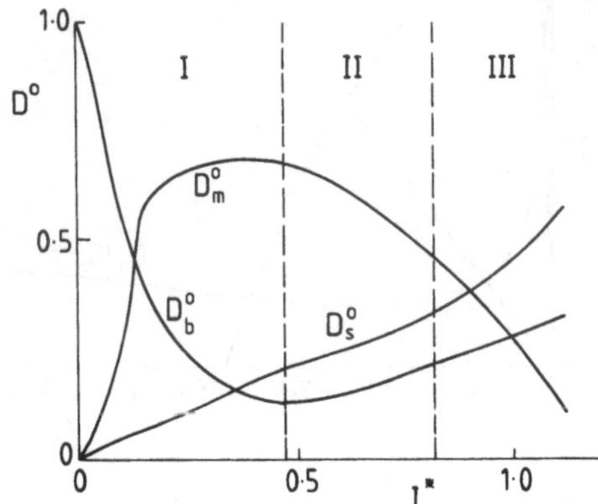

Fig. 8. Variation of the dimensionless work absorbed plastically at the cessation of motion for Mode I behaviour and at failure for Modes II and III responses with dimensionless impulse for the beam examined in Fig. 6. D^0_s, D^0_m and D^0_b are the plastic shear, membrane and bending work components made dimensionless with respect to the total plastic work.

theoretical predictions in Figs 8–11. Figure 8 shows the variation with the dimensionless impulse I^* of the final values of the various dimensionless plastic work components (D^0_b, D^0_s and D^0_m), which are expressed as fractions of the total plastic work absorbed, where $D^0_b + D^0_s + D^0_m = 1$ for the three failure modes. However, the residual kinetic energy increases with I^* for Modes II and III responses, as

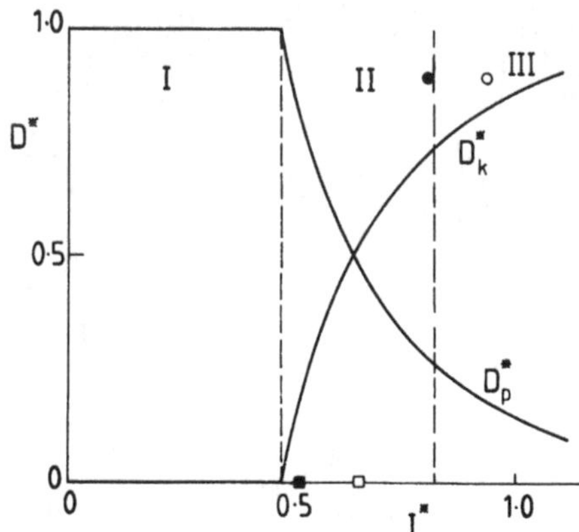

Fig. 9. Variation of the dimensionless residual kinetic energy D^*_k and the total energy absorbed plastically D^*_p at the cessation of motion for Mode I behaviour and at failure for Modes II and III responses for the beam examined in Fig. 6. Predictions of D^*_k from the elementary rigid, perfectly plastic analysis:[4] ■, □, transition from Mode I to Mode II for inscribing and circumscribing yield curves with $\varepsilon_f = 0{\cdot}17$ respectively. ●, ○, transition from Mode II to Mode III for inscribing and circumscribing yield curves, respectively.

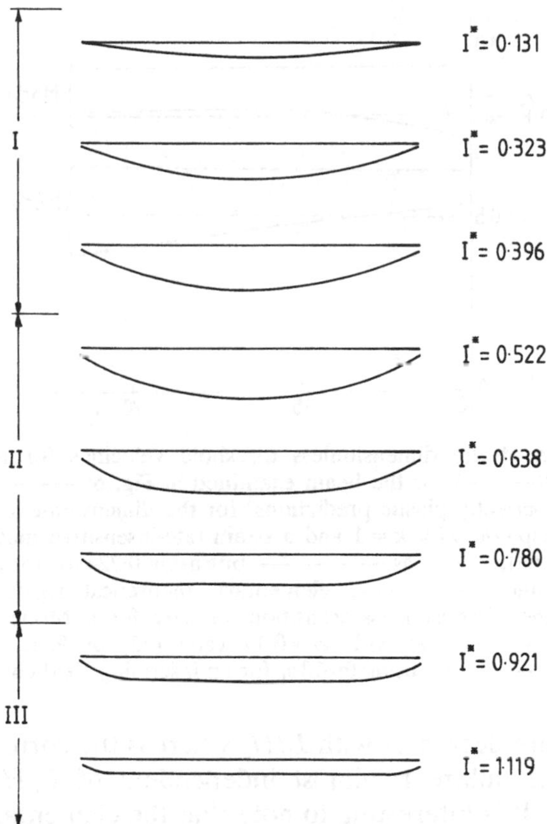

Fig. 10. Variation of the permanently deformed profile with the dimensionless impulse for the beam examined in Fig. 6. (The same scales have been used for the beam length and the permanent transverse displacement.)

shown in Figs 6(b, c) and 9. Thus the total plastic work used to make the work components dimensionless in Fig. 8 is a diminishing proportion of the initial kinetic energy as I^* increases. It is evident that $D_p^* = 1$ for a Mode I response, while $D_p^* \leqslant 1$ for Modes II and III behaviour.

Figure 10 illustrates the change of shape of the permanent beam profiles as the dimensionless impulse I^* increases. The features of these profiles are remarkably similar to those obtained experimentally[3] and to those found using the simple theoretical rigid, perfectly plastic method of analysis[4] discussed in Section 2. For example, the elementary rigid, perfectly plastic method of analysis[4] for a Mode III failure of a beam in Section 2.2 predicts that the central portion of length $2L - 3\sqrt{3}\,H$ remains straight, with the deformations concentrated within a distance $\frac{3}{2}\sqrt{3}\,H$ of each support.

It is evident from Fig. 11 that the dimensionless threshold velocity for

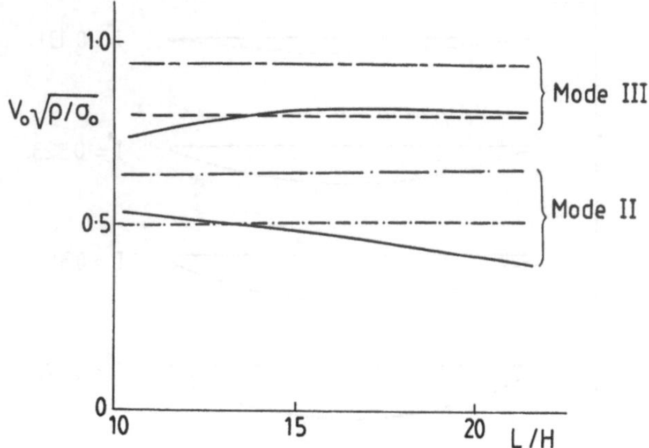

Fig. 11. Variation of the dimensionless threshold velocities for Modes II and III failures with L/H(———) for the beam examined in Fig. 6: — — — —, elementary theoretical rigid, perfectly plastic predictions[4] for the dimensionless transition velocity for a Mode III response with $k=1$ and a strain-rate-insensitive material (eqn (24) of Ref. 4, upper bound); - - - - -, as — — — — but using $0{\cdot}7245\sigma_0$ for an inscribing yield curve (lower bound); —·—·—, elementary theoretical rigid, perfectly plastic predictions[4] for the dimensionless transition velocity for a Mode II response of a strain-rate-insensitive material with $\varepsilon_f=0{\cdot}17$ (eqn (15) of Ref. 4, upper bound); · · · · · · · · ·, as —·—·— but using $0{\cdot}618\sigma_0$ for an inscribing yield curve (lower bound).

a Mode II failure decreases with L/H, whereas the corresponding value for a Mode III failure is almost independent of L/H for $L/H>15$ approximately. It is interesting to note that the elementary rigid–plastic analysis[4] also predicts a threshold velocity for a Mode III failure that is independent of L/H, as shown in Fig. 11. However, the elementary analysis[4] predicts a slight increase in the Mode II threshold velocity as the ratio L/H increases, which is in contradistinction with the results of the numerical method[5] in Fig. 11.

5 LOW-VELOCITY MASS IMPACT FAILURE OF A BEAM USING THE ENERGY DENSITY FAILURE CRITERION

5.1 Introduction

The energy density failure criterion introduced in Section 3 was used in Ref. 5 to study the failure modes of an impulsively loaded, fully clamped, rigid–plastic beam, as discussed in Section 4. This failure criterion predicted good agreement with the available experimental results on impulsively loaded beams and confirmed broadly the failure characteristics that were first observed using the elementary analysis[4] outlined in Section 2.

Tables 4 and 5 in Ref. 1 contain a comparison between the experimental results[12] for the failure of fully clamped beams, which were impacted by masses at various locations on the span, and the theoretical predictions of an extension[6] of the elementary method of analysis in Section 2 with $\varepsilon_c = 0{\cdot}19$, which was developed originally for the impulsive loading case. The tensile tearing Mode II failure predictions in Table 4 give reasonable agreement with the experimental results on the aluminium alloy 6061 T6 beams with enlarged ends in the clamps, whereas the theoretical predictions for the specimens with flat ends in Table 5 were less satisfactory. For convenience, the appendix to this chapter contains the predictions of the elementary theoretical analysis outlined in Section 2 for the Mode II failure of a fully clamped, rigid, perfectly plastic beam struck at the mid-span by a mass travelling with an initial velocity V_0. Various other failure modes were observed in Refs 1 and 12, but no other theoretical methods have been developed except for a Mode III failure in Ref. 6.

The critical energy density failure criterion in Section 3 was used in Ref. 13 to examine the failure of fully clamped beams subjected to mass impact loads. Again the numerical analysis retains the simultaneous influence of bending moments, membrane forces and transverse shear forces in all of the basic equations including the yield condition. This analysis is outlined in Section 5.2.

5.2 Outline of Analysis

Reference 13 examines the dynamic behaviour of a rigid–plastic beam that is fully clamped across a span of length $2L$ and struck at a distance S from the nearest support by a mass G travelling with an initial impact velocity V_0. The dynamic plastic response consists of three phases of motion. The first has two stationary plastic hinges at the impact point together with two travelling plastic hinges within the span, as shown in Fig. 12(a), where $S < L$. The first phase of motion continues until the travelling plastic hinge at B reaches the support A. A second phase of motion then follows, with a stationary plastic hinge at A and a plastic hinge still travelling at D in the section CE, as indicated in Fig. 12(b). An axial displacement u develops at the impact position owing to the loss of symmetry during this phase of motion, which is completed when the right-hand travelling plastic hinge D reaches the right-hand support E. The final phase of motion has four stationary plastic hinges, as shown in Fig. 12(c).

It is evident from the transverse velocity profiles in Fig. 12 that transverse shear sliding could continue underneath the striker through-

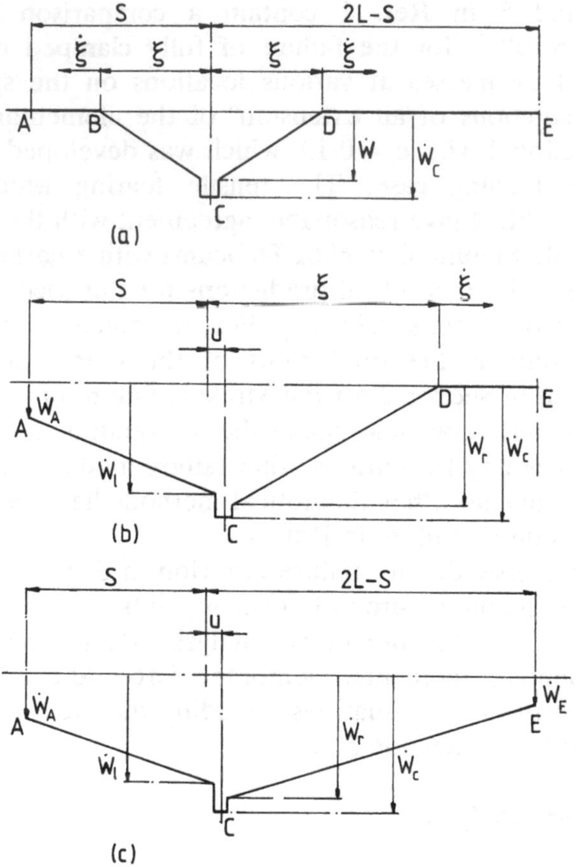

Fig. 12. Transverse velocity profiles for a fully clamped rigid–plastic beam subjected to a mass impact loading at a distance S from the left-hand support:[13] (a) first phase of motion; (b) second phase of motion; (c) third phase of motion.

out the three phases of motion and at the stationary plastic hinges that form at the supports during the second and third phases. Thus either motion ceases when all of the initial kinetic energy of the striking mass is absorbed in plastic deformations (i.e. Mode I), or else the ductility of the beam material is exhausted (i.e. Modes II or III) prior to the cessation of motion.

The above theoretical procedure[13] uses the semi-empirical equation (17), which was developed[5] from experiments[3] on impulsively loaded beams, and gives excellent agreement with the experimental results presented in Refs 12 and 14 for a Mode I response (i.e. large permanent ductile deformations) when the influence of strain rate effects are retained, as outlined in Section 3. Larger impact energies may cause failure, which is discussed in Section 5.3.

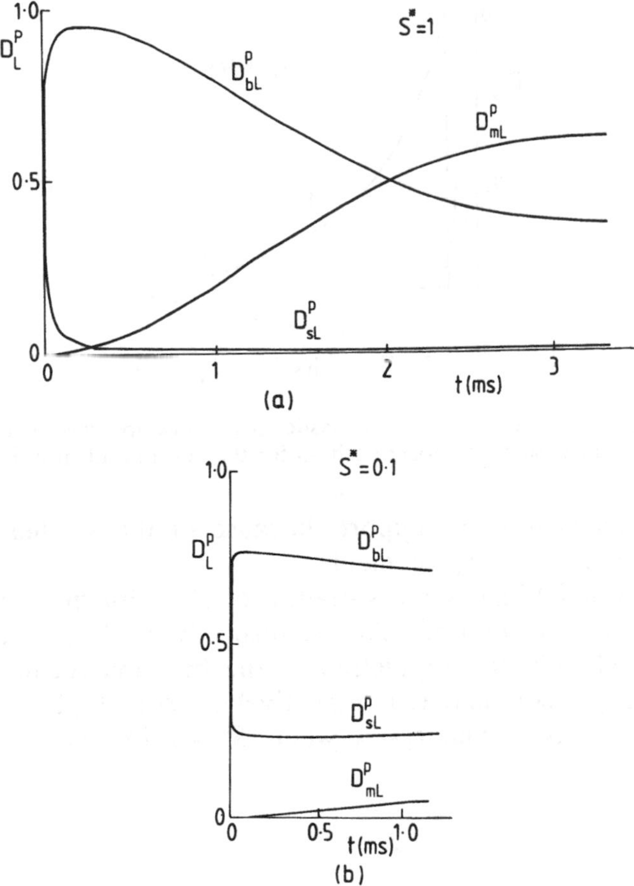

Fig. 13. Temporal variation of the dimensionless work in a fully clamped beam having the same dimensions and material properties as the beam in Fig. 6 and subjected to a mass $G = 5$ kg striking the beam with an impact velocity $V_0 = 4$ m/s at a position $S = LS^*$ from the left-hand support: (a) $S^* = 1$; (b) $S^* = 0{\cdot}1$. D^P_{bL}, D^P_{sL} and D^P_{mL} are the plastic bending, shear and membrane work components at the left-hand side of the impact point made dimensionless with respect to the total plastic work absorbed at time t at the left-hand side of the impact point.

Figure 13 shows the time variation of the dimensionless plastic work absorbed at the left-hand side of the impact position at C in Fig. 12 during a Mode I response for identical mass impacts at the mid-span and near to a support, where $S^* = S/L$. In this case, the plastic work components are made dimensionless with respect to the total plastic work absorbed at the left-hand side of the impact point at time t. Thus $D^P_{bL} + D^P_{sL} + D^P_{mL} = 1$ for all times throughout the response. As the deformation progresses, the membrane plastic work becomes the most important for a strike at the mid-span, but is always the least important

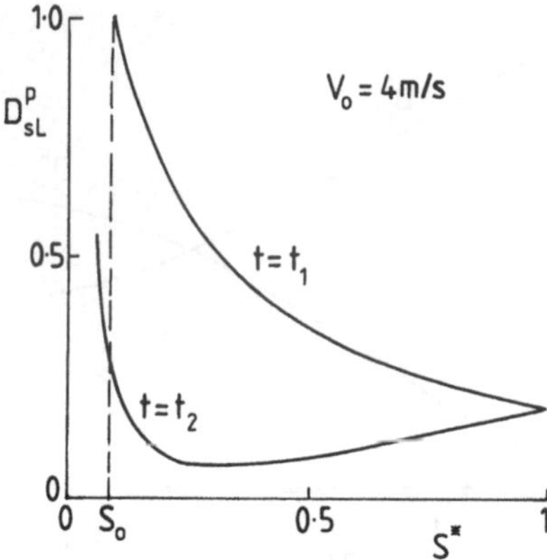

Fig. 14. Variation of the dimensionless plastic shear work absorbed at times t_1 and t_2 with the dimensionless impact position for the beam problem in Fig. 13.

for an impact near to a support, because of the smaller associated permanent transverse displacements.

Figures 14 and 15 show the variation of D^p_{sL} with the dimensionless impact position and impact velocity, respectively. The times t_1 and t_2 are associated with the completion of the first and second phases of motion in Figs 12(a) and (b) respectively. No first phase of motion occurs for $S^* \leq S_0$, while $t_1 = t_2$ when $S^* = 1$ for an impact at the mid-span.

5.3 Failure

Equation (10) shows that the dimensionless plastic work at failure equals a critical strain ε_c. It was found to be acceptable to take the

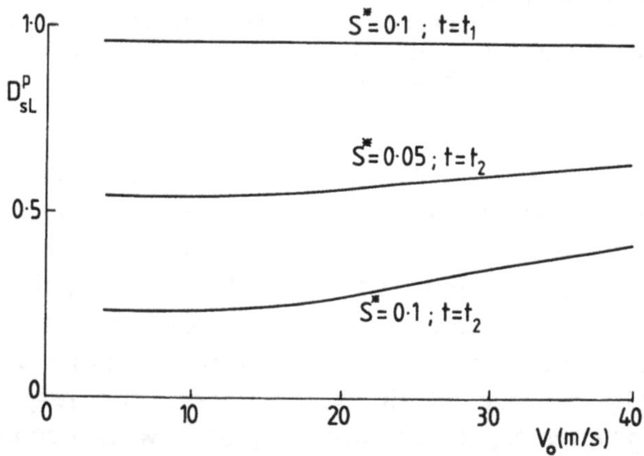

Fig. 15. Variation of the dimensionless plastic shear work at times t_1 and t_2 with the impact velocity for the beam problem in Fig. 13.

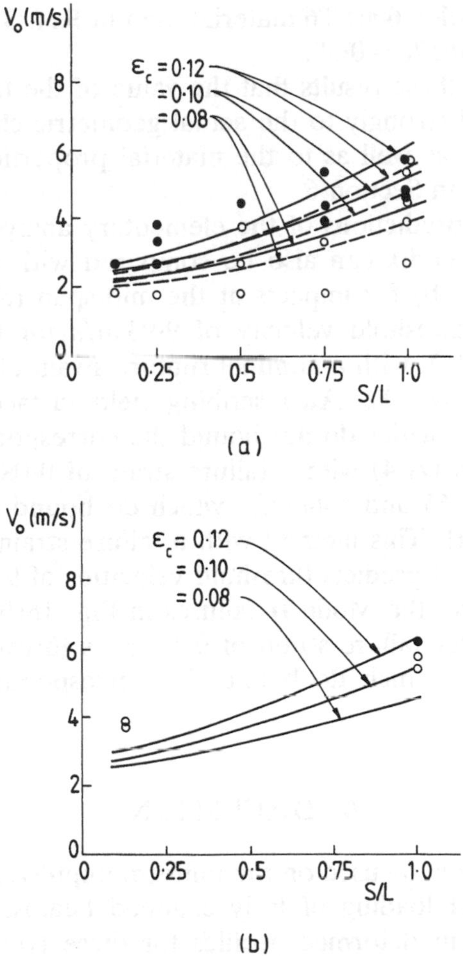

Fig. 16. Comparison between the theoretical predictions with several values of ε_c and experimental results for aluminium alloy beams subjected to mass impact loadings at various locations on a fully clamped span: ——, theoretical predictions[13] with $D = 6500\,s^{-1}$ and $q = 4$; - - - - -, theoretical predictions[13] for a strain-rate-insensitive material; ●, ruptured; ○, no failure. (a) Experimental results from Ref. 12 with $H = 3{\cdot}81$ mm. (b) Experimental results from Ref. 14 with $H = 6{\cdot}39$ mm.

dynamic uniaxial engineering rupture strain ε_{f0} as ε_c for the impulsive loading case.[5] However, the comparisons made in Fig. 16 indicate that the failure of the impact-loaded aluminium alloy beams in Refs 12 and 14 occurs for $\varepsilon_c < \varepsilon_{f0}$. In fact, $0{\cdot}08 < \varepsilon_c < 0{\cdot}12$, which is significantly smaller than the value $\varepsilon_c = \varepsilon_{f0} = 0{\cdot}5$ used to predict successfully the failure of aluminium alloy 6061 T6 beams subjected to an impulsive velocity loading.[5] ε_{f0} equals 0·66 (average of 0·529 and 0·785) at a strain rate of $140\,s^{-1}$ for the aluminium alloy specimens in Ref.14, with a nominal static rupture strain of 0·176. The nominal static rupture strain of the aluminium alloy 6061 T6 material used in Ref. 3 is 0·135,† so that $\varepsilon_{f0} \approx 0{\cdot}66 \times 0{\cdot}135/0{\cdot}176 = 0{\cdot}50$.

It appears from these results that the value of the failure strain ε_c in eqn (10) is related strongly to the actual geometric characteristics of a structural member as well as to the material properties. This observation is returned to in Section 6.

The theoretical predictions of the elementary analysis for a Mode II failure in the Appendix can also be compared with the experimental results in Figs 16(a,b) for impacts at the mid-span ($S^* = 1$). Equation (A.4) predicts a threshold velocity of 9·93 m/s for the experimental results[12] in Fig. 16(a), with a nominal rupture strain of $\varepsilon_f = 0{\cdot}19$ for the test specimens in Ref. 12. An inscribing yield surface gives 7·81 m/s. These threshold velocities do not bound the corresponding experimental results, but eqn (A.4) with a failure strain of 0·08 gives upper and lower bounds of 5·65 and 4·44 m/s, which do bound the experimental results in Fig. 16(a). This method with a failure strain of 0·176 for the specimens in Ref. 14 predicts threshold velocities of 9·22 and 7·25 m/s, which do not bound the Mode II failures in Fig. 16(b). However, eqn (A.4) with a smaller failure strain of 0·12 gives threshold velocities of 7·03 and 5·53 m/s, which do bound the corresponding experimental results.

6 DISCUSSION

Sections 4 and 5 have focused on the uniform impulsive loading and the low-velocity impact loading of fully clamped beams. Figure 17 compares the permanent deformed profiles for these two particular prob-

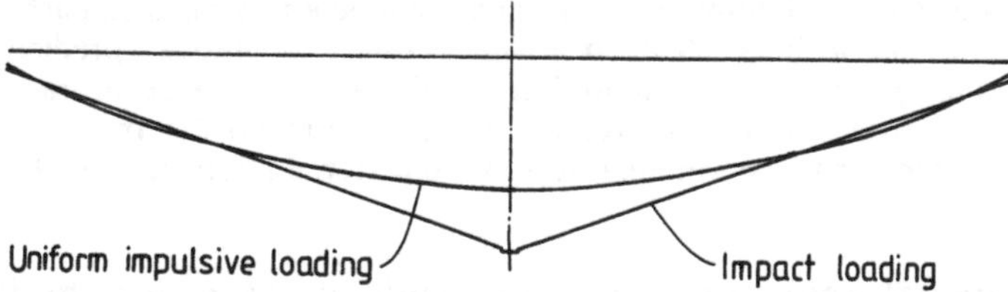

Fig. 17. Final profiles for identical fully clamped beams having the dimensions and material properties of the beam in Fig. 6 when subjected to a uniform impulsive loading over the entire span ($I^* = 0{\cdot}396$) or a mass impact loading at the mid-span ($G = 20$ kg and $V_0 = 8{\cdot}455$ m/s) having the same external energy and producing a Mode I response. (The same scales have been used for the beam length and the permanent transverse displacement.)

† A nominal rupture strain of 0·17 was used from Ref. 15 for the theoretical analysis in Ref. 4.

lems when both are subjected to the same external energy. The travelling plastic hinges cause a curved shape to be produced for an impulsive loading, but evidently the travelling plastic hinges exercise a negligible influence on the permanent profile for the mass impact loading case.

Some further interesting comparisons are made in Figs 18–20 between the uniform impulsive loading and impact loading cases having the same external energy. It is evident that the response duration is more than an order of magnitude longer for the impact loading, as found also from a comparison between the corresponding infinitesimal displacement cases.[2] Except for the timescale, the temporal variation of the dimensionless plastic bending, shear and membrane work absorbed throughout the entire beam are similar in Figs 18(a) and (b) for impact and impulsive loadings respectively, though the plastic membrane work exercises a more dominant influence for an impact loading.

A comparison is made between the dimensionless plastic work at the left-hand side of the mass impact point in Fig. 19(a) and at the support (A) of an impulsively loaded beam in Fig. 19(b). Again the dimensionless plastic membrane work is more important for an impact loading. The numerical results for these two cases are re-plotted in Fig. 20 to show the total dimensionless plastic work absorbed at both supports ($2D^*_A$) and on both sides of the striking mass ($2D^*_L$) in Fig. 20(a) for an impact loading and at both supports for the impulsive loading case in

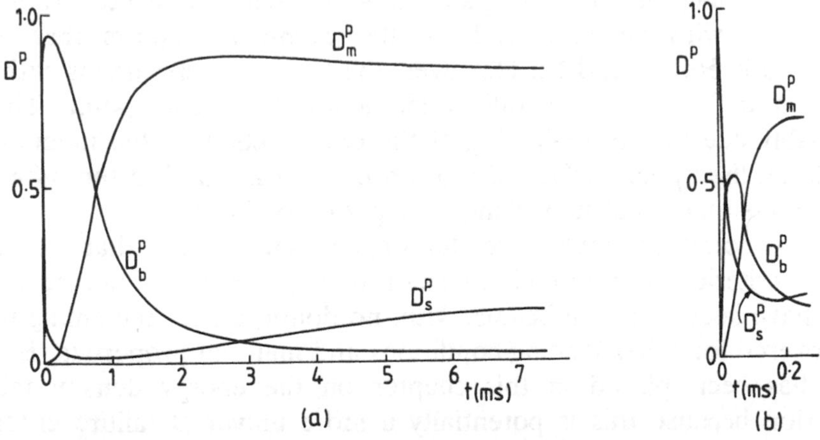

Fig. 18. Temporal variation of the dimensionless plastic work components absorbed in a fully clamped beam having the same dimensions and material properties as in Fig. 6 and subjected to (a) impact at the mid-span or (b) uniform impulsive loading having the same total external energies as that in Fig. 17. D^p_b, D^p_s and D^p_m are the plastic bending, shear and membrane work components absorbed up to time t in the beam and made dimensionless with respect to the total plastic work absorbed at t.

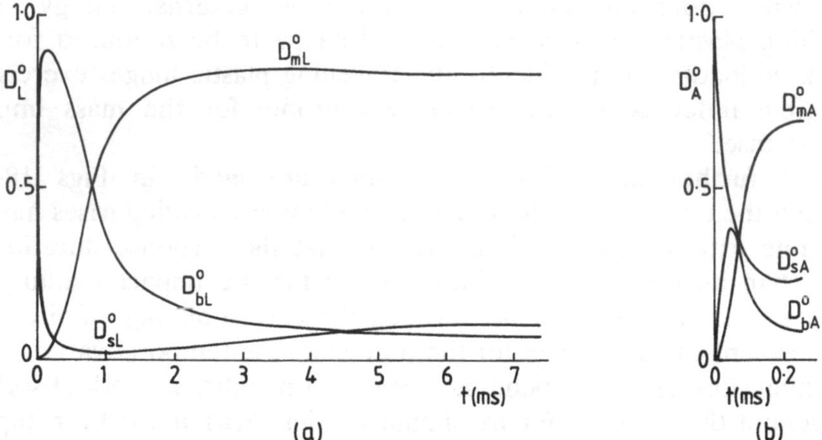

Fig. 19. Temporal variation of the dimensionless plastic work absorbed in a fully clamped beam having the same dimensions and material properties as in Fig. 6 and subjected to (a) impact at the mid-span or (b) uniform impulsive loading having the same total external energies as that in Fig. 17. D^0_{bL}, D^0_{sL} and D^0_{mL} are the plastic bending, shear and membrane work components at the left-hand side of the impact point for a mass impact loading made dimensionless with respect to the total plastic work at time t at the left-hand side of the impact point. Similarly, D^0_{bA}, D^0_{sA} and D^0_{mA} correspond to the left-hand support of an impulsively loaded beam.

Fig. 20(b). It is evident from Fig. 20(a) that considerably more plastic work is absorbed at the supports than at the striking position. This implies that a Mode II failure is more likely to occur at an ideal support condition than at the mid-span, as was indeed observed from the specimens with enlarged ends in the clamped supports that were reported in Refs 1 and 12. However, the beam test specimens with flat ends in Ref. 12 fail generally underneath the impact point. This is probably due to the weakening of the cross-section by the indentation made on the upper surface of a beam by a striker and to the influence of the less than ideal fully clamped support conditions.

It was noted, particularly for the impact loading case, that a durable failure criterion is required because of the large variety of failure modes that have been found in beams, with, no doubt, others remaining to be discovered for other loading conditions and material properties. Emphasis has been placed in this chapter on the energy density failure criterion because this is potentially a more universal failure criterion than the elementary ones outlined in Section 2, which nevertheless do give surprisingly good predictions in some cases. However, one difficulty with this energy criterion is the selection of the critical volume of material to be used in the calculations. In the present study, which is described further in Refs 5 and 13, the volume of material is selected as

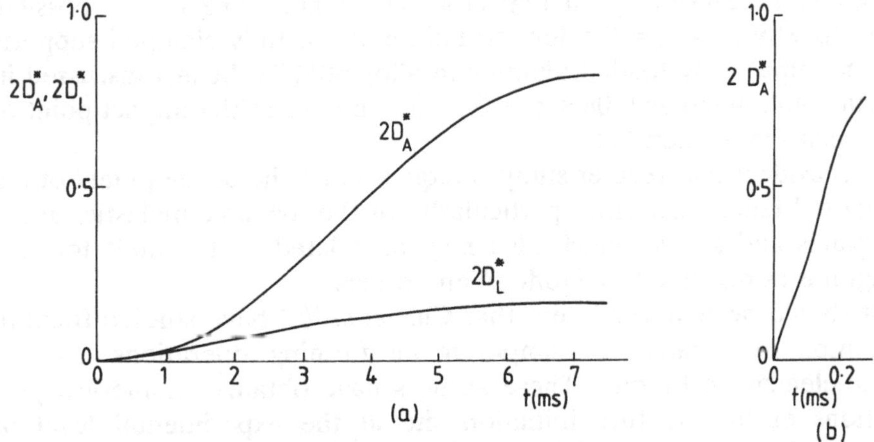

Fig. 20. Comparison between the total dimensionless plastic work absorbed (a) at both sides of the impact point ($2D_L^*$) and at the supports ($2D_A^*$) for the mass impact loading case; (b) at the supports for the uniform impulsive loading case ($2D_A^*$). The beams are subjected to the same external energies as that in Fig. 17 and have the same dimensions and material properties as the beam in Fig. 6.

αHBH, where α is the dimensionless hinge length from eqn (13). This choice of volume is consistent with the use of generalised stresses (i.e. bending moments, transverse shear forces and membrane forces) in the analyses of beams, frames, plates and shells. Nevertheless it is not clear whether or not this volume of material will be too large for some dynamic inelastic failure problems, since material rupture often occurs locally in the cross-section. In any event, the semi-empirical equation (17) requires some further study. It should be noted that the selection of α is required only for the current approximate theoretical analysis using generalised stresses. It would not be necessary to use eqn (17) and, therefore α, if a finite-element programme in terms of stresses was used to predict failure according to eqns (5) and (14) with the volume of an individual element used for the calculations.

The selection of the critical strain ε_c in eqn (10) is another difficult and important problem. It was noted in Section 5.3 that, when $\varepsilon_c = \varepsilon_{f0}$ in eqn (10), ε_{f0} is the engineering rupture strain for a zero gauge length, whereas ε_f in eqn (1) for the elementary theory in Section 2.1 is the nominal or regular engineering rupture strain. The choice of the appropriate value of ε_c in eqn (10) for failure requires some further study, since, apart from the material property, εV_f, it is likely to be affected by many other factors such as the details of the support conditions, stress concentration effects, material imperfections, geometric imperfections produced by the manufacturing process and indentations caused by an impactor. These are some of the possible reasons why $\varepsilon_c = \varepsilon_{f0} = 0{\cdot}5$ for the failure at the fully clamped supports of the impulsively loaded aluminium alloy 6061 T6 beams discussed in Section 4.4, whereas $0{\cdot}08 \leqslant \varepsilon_c \leqslant 0{\cdot}12$ for rupture at the impact point of the beams in Section 5.3.

It is evident that further study is required into the development of the proposed failure criterion, particularly for the dynamic inelastic failure of plates and shells, which also may be related to the limit forming diagrams as discussed by Duffey[16] and others.

It should be remarked here that Clift *et al.*[17,18] have studied fracture initiation in a range of simple metal forming operations using a finite-element technique. These authors have obtained numerical predictions of the fracture initiation site at the experimental level of deformation at fracture for nine different fracture criteria. This was achieved by locating the node in the mesh that had the largest accumulated value of the fracture criterion function. These predicted sites were then compared with the crack initiation sites observed experimentally. It transpires[17,18] that only the criterion based on the critical value of generalised plastic work per unit volume, as suggested by Freudenthal,[19] successfully predicted the fracture initiation sites found experimentally in all of the metal forming operations. Incidentally, Cockroft and Latham[20] have pointed out that the non-uniform stress distribution across the neck of a tensile test specimen[21] should be taken into account when estimating the plastic work at failure. However, it turns out the the failure criterion in Ref. 20 is not as successful as one based on Freudenthal's suggestion and that is also used in the present study.

The energy density failure criterion discussed in this chapter does retain the influence of material rate sensitivity, as indicated by eqn (11). However, all of the numerical predictions for failure that are presented in the various figures in this chapter have been obtained for an aluminium alloy 6061 T6, which is only a mildly strain-rate-sensitive material at the strain rates involved in the present class of problems. Further comparisons are required for strongly strain-rate-sensitive materials such as mild steel, for example, and some experimental data on beams have been published.[12] However, the variation of material strain-rate sensitivity with strain should be taken into account,[7] but is ignored in eqn (11). Furthermore, the rupture strain may change with strain rate,[1,7] which will influence the values of ε_f in eqn (1) and ε_c in eqn (10).

It has been shown in Ref. 22 that the energy absorbed per unit volume up to the tensile failure of carbon and stainless steels depends on the magnitude of the strain rate. Over a range of strain rates from 10^{-3} to $10^3\ s^{-1}$, the total energy absorbed increases with increase in the carbon content from 22% when $C = 0{\cdot}033\%$ to 54% when $C = 0{\cdot}21\%$. A similar phenomenon is observed for stainless steels. Over the same range of strain rates, the energy absorbed at failure does not change for a Ni content of 8·97% but increases 8% and 42% for Ni contents of 10·79% and 13·54% respectively.

Some recent experimental[23,24] and theoretical[25] studies have been reported on the dynamic inelastic failure of fully clamped circular plates subjected to uniformly distributed impulsive loads. The influence of the boundary conditions on the magnitude of the critical impulse is shown to be very important, which, in fact, has already been observed for beams.[12,26] This further complicates the predictions of an energy density failure criterion, but shows the importance of specifying carefully the exact details of the boundary conditions in both experimental tests and theoretical studies.

Holmes *et al.*[27] (Chapter 2 of this book) have used a local damage model in a DYNA 3D computer programme in order to predict the ductile fracture of a welded steel T-joint subjected to dynamic loading. The fracture initiation and the fracture propagation path through a structure are obtained using eqn (22) in Chapter 2. This criterion assumes that failure occurs when the integral of the equivalent plastic strain increments in a characteristic volume of the material equals a critical failure strain that is a function of the stress triaxiality (defined as the ratio of the mean stress to the equivalent stress). The critical failure strain function is determined from a series of notched tensile tests on specimens having various notch radii, and is shown in Fig. 11(b) of Chapter 2 as a function of stress triaxiality for several materials. It is evident that this method is a further development of the elementary Mode II tensile tearing failure criterion given by eqn (1) and discussed in Section 2.1. However, the characteristic volume of the material may be associated with microstructural dimensions, so that the criterion used by Holmes *et al.* in Chapter 2 is capable of examining non-geometrically similar scaling effects, as discussed by them and which is known to be significant for the dynamic plastic response of structures.[9,28,29]

Bammann *et al.*[30] have discussed in Chapter 1 of this book the background to the internal state variable type of criterion, which has been developed at Sandia National Laboratories, and which can be used to predict failure in ductile metal structures subjected to large dynamic loads. The constitutive model caters for the deviatoric deformations resulting from the presence of dislocations and the dilatational deformations and the ensuing failure that arise from the growth of internal voids. Strain-rate and temperature effects are retained in the model. The constitutive model incorporates various assumptions, and the values of the plasticity parameters can be obtained from tensile and compression tests on the material at various strain rates and temperatures, while further tests are required on notched specimens to obtain the damage parameters.

Bammann *et al.* (see Chapter 1) have incorporated this constitutive equation into a finite-element code and studied various dynamic problems, including the particular case of an aluminium 6061 T6 disc impacted by a hardened steel cylindrical rod. The theoretical results in Figs 20 and 21 of Chapter 1 for an impact velocity of 64 m/s show that the maximum effective plastic strain is about $\frac{5}{3}$ times larger than the failure elongation in a uniaxial tensile test. The finite-element analysis predicts initial failures for impact velocities within the range 84–89 m/s and plugging between 102–107 m/s, compared with experimental values of 79–84 m/s for initial failure and 92–106 m/s for complete plugging. The maximum effective plastic strain in the disc for an impact velocity of 89 m/s is about three times the corresponding uniaxial tensile failure elongation. No comparisons are made with the predictions of an energy density failure criterion, which is used in this chapter, or with the strain criterion utilised by Holmes *et al.*[27] in Chapter 2.

Ferron and Zeghloul[31] review in Chapter 4 of this book the conditions for localised necking of bars and various thin-walled structures. They also explore the stabilising effects of material strain hardening and strain-rate sensitivity and the destabilising effect of thermo-mechanical coupling on the predictions of sheet metal ductilities under in-plane loading. Chapter 4 is written largely from a metal-forming viewpoint, but it provides valuable insight for further studies in the field of structural failure.

Finally, it should be remarked that Zhu and Cescotto[32] have used an energy-equivalent damage evolution theory to examine the dynamic plastic failure of a circular plate struck by a mass travelling with an initial impact velocity of 15 m/s. This particular paper focuses on the modelling of contact-impact between two deformable bodies, so that no information is given on the experimental values of the various parameters in the damage theory and no comparisons are made with experimental test results.

7 CONCLUSIONS

This chapter has examined the dynamic inelastic failure of beams subjected to uniform impulsive and mass impact loadings. An elementary rigid–plastic method of analysis gives good predictions for various features of the failure behaviour. However, an energy density failure criterion is examined in some detail since a universal criterion is required in order to predict the dynamic inelastic failure of a broad class of structural impact problems. This failure prediction method is very promising, but more study is required before it becomes accepted generally and incorporated into finite-element programmes and other design methods. Finally, it should be noted that an acceptable criterion does not exist for the dynamic inelastic rupture of a broad class of structures, except possibly for beams and frames.

ACKNOWLEDGEMENTS

The authors are indebted to Mrs M. White for typing the manuscript and to Mr F. J. Cummins for preparing the tracings.

REFERENCES

1. Jones, N., On the dynamic inelastic failure of beams. In *Structural Failure*, ed. T. Wierzbicki & N. Jones. Wiley, New York, 1989, pp. 133–59.
2. Jones, N., *Structural Impact.* Cambridge University Press, 1989.
3. Menkes, S. B. & Opat, H. J., Broken beams. *Exp. Mech.*, **13** (1973) 480–6.
4. Jones, N., Plastic failure of ductile beams loaded dynamically. *Trans. ASME, J. Engng for Ind.*, **98** (B1) (1976) 131–6.
5. Shen, W. Q. & Jones, N., A failure criterion for beams under impulsive loading. *Int. J. Impact Engng*, **12** (1992) 101–21, 329.
6. Liu, J. H. & Jones, N., Plastic failure of a clamped beam struck transversely by a mass. In *Inelastic Solids and Structures*, ed. M. Kleiber & J. A. Konig. Pineridge Press, Swansea, UK, 1990, pp. 361–84.
7. Jones, N., Some comments on the modelling of material properties for dynamic structural plasticity. In *Mechanical Properties of Materials at High Rates of Strain*, ed. J. Harding. Institute of Physics Conference Series No. 102, Bristol, 1989, pp. 435–45.
8. Symonds, P. S., Plastic shear deformations in dynamic load problems. In *Engineering Plasticity*, ed. J. Heyman & F. A. Leckie. Cambridge University Press, 1968, pp. 647–64.
9. Jouri, W. S. & Jones, N., The impact behaviour of aluminium alloy and mild steel double-shear specimens. *Int. J. Mech. Sci.*, **30** (1988) 153–72.
10. Nonaka, T., Some interaction effects in a problem of plastic beam dynamics. Parts 1–3. *J. Appl. Mech.*, **34** (1967) 623–43.
11. Jones, N. & Wierzbicki, T., Dynamic plastic failure of a free–free beam. *Int. J. Impact Engng*, **6** (1987) 225–40.
12. Liu, J. H. & Jones, N., Experimental investigation of clamped beams struck transversely by a mass. *Int. J. Impact Engng*, **6** (1987) 303–35.

13. Shen, W. Q. & Jones, N., The dynamic plastic response and failure of a clamped beam struck transversely by a mass. *Int. J. Solids Structures* (in press).
14. Yu, J. & Jones, N., Further experimental investigations on the failure of clamped beams under impact loads. *Int. J. Solids Structures,* **27** (1991) 1113–37.
15. Jones, N., Dumas, J. W., Giannotti, J. G. & Grassit, K. E., The dynamic plastic behaviour of shells. In *Dynamic Response of Structures,* ed. G. Herrmann & N. Perrone. Pergamon Press, Oxford, 1972, pp. 1–29.
16. Duffey, T. A., Dynamic rupture of shells. In *Structural Failure,* ed. T. Wierzbicki & N. Jones. Wiley, New York, 1989, pp. 161–92.
17. Clift, S. E., Hartley, P., Sturgess, C. E. N. & Rowe, G. W., Further studies on fracture initiation in plane-strain forging. In *Applied Solid Mechanics—2,* ed. A. Tooth & J. Spence. Elsevier, Amsterdam, 1987, pp. 89–100.
18. Clift, S. E., Hartley, P., Sturgess, C. E. N. & Rowe, G. W., Further prediction in plastic deformation processes. *Int. J. Mech. Sci.,* **32** (1990) 1–17.
19. Freudenthal, A. M., *The Inelastic Behaviour of Solids.* Wiley, New York, 1950.
20. Cockcroft, M. G. & Latham, D. J., Ductility and the workability of metals. *J. Inst. Metals,* **96** (1968) 33–9.
21. Hill, R., *The Mathematical Theory of Plasticity.* Oxford University Press, 1950.
22. Kawata, K., Miyamoto, I. & Itabashi, M., On the effects of alloy components in the high velocity tensile properties. *Impact Loading and Dynamic Behaviour of Materials. Proceedings of Impact '87, Bremen,* Vol. 1, ed. C. Y. Chiem, H.-D. Kunze & L. W. Meyer. DGM Informationsgellschaft Verlag, 1988, pp. 349–56.
23. Teeling-Smith, R. G. & Nurick, G. N., The deformation and tearing of thin circular plates subjected to impulsive loads. *Int. J. Impact Engng,* **11** (1991) 77–91.
24. Nurick, G. N. & Teeling-Smith, R. G., Predicting the onset of necking and hence rupture of thin plates loaded impulsively—an experimental view. In *Structures Under Shock and Impact II,* ed. P. S. Bulson. Computational Mechanics Publications, Southampton and Boston/Thomas Telford, London, 1992, pp. 431–445.
25. Shen, W. Q. & Jones, N., Dynamic response and failure of fully clamped circular plates under impulsive loading. *Int. J. Impact Engng* **13** (1993) (in press).
26. Yu, J. & Jones, N., Numerical simulation of a clamped beam under impact loading. *Comp. Structures,* **32** (1989) 281–93.
27. Holmes, B. S., Kirkpatrick, S. W., Simons, J. W., Giovanola, J. H. & Seaman, L. Modeling the process of failure in structures. In *Structural Crashworthiness and Failure,* ed. N. Jones & T. Wierzbicki. Elsevier Applied Science Publishers, Barking, Essex, 1993, Chap. 2.
28. Atkins, A. G., Scaling in combined plastic flow and fracture. *Int. J. Mech. Sci.,* **30** (1988) 173–91.
29. Yu, T. X., Zhang, D. J., Zhang, Y. & Zhou, Q., A study of the quasi-static tearing of thin metal sheets. *Int. J. Mech. Sci.* **30** (1988) 193–202.

30. Bammann, D. J., Chiesa, M. L., Horstemeyer, M. F. & Weingarten, L. I., Failure in ductile materials using finite element methods. In *Structural Crashworthiness and Failure,* ed. N. Jones & T. Wierzbicki. Elsevier Applied Science Publishers, Barking, Essex, 1993, Chap. 1.
31. Ferron, G. & Zeghloul, A., Strain localization and fracture in metal sheets and thin-walled structures. In *Structural Crashworthiness and Failure,* ed. N. Jones & T. Wierzbicki. Elsevier Applied Science Publishers, Barking, Essex, 1993, Chap. 4.
32. Zhu, Y. Y. & Cescotto, S., A fully coupled elastoplastic damage modeling of contact-impact between two deformable bodies. In *Structures Under Shock and Impact II,* ed. P. S. Bulson, Computational Mechanics Publications, Southampton and Boston/Thomas Telford, London, 1992. 113–32.
33. Liu, J. H. & Jones, N., Dynamic response of a rigid-plastic clamped beam struck by a mass at any point on the span. *Int. J. Solids Structures,* **24** (1988) 251–70.

APPENDIX

Elementary Analysis for a Mode II Failure of a Fully Clamped Beam Struck at the Mid-Span

It is particularly straightforward to obtain the threshold velocity for the initiation of a Mode II tensile tearing failure for this case when the mass G of the striker is much larger than the beam mass.

The maximum permanent transverse displacement W_m at the mid-span of a beam that is fully clamped across a span $2L$ can be written in the form[1,33]

$$W_m/H = \tfrac{1}{2}[(\lambda + 1)^{1/2} - 1] \tag{A.1}$$

where $\lambda = 2GV_0^2L/\sigma_0BH^3$, H and B are the beam thickness and beam breadth respectively, σ_0 is the flow stress and V_0 is the initial impact velocity. Equation (A.1), which retains the influence of finite transverse displacements, simplifies to

$$W_m/H = \tfrac{1}{2}(\lambda^{1/2} - 1) \tag{A.2}$$

when $\lambda \gg 1$ or $W_m/H > 1$.

The transverse displacement profile associated with eqns (A.1) and (A.2) for large mass ratios is a simple mode form with stationary plastic hinges at both supports and underneath the striker. Thus an estimate for the maximum strain ε_{max} in a beam with $W_m/H \geqslant 1$ is given by eqn (12) in Ref. 4, or

$$\varepsilon_{max} = 2\left(\frac{H}{2L}\right)^2\left[\left(\frac{W_m}{H}\right)^2 + \frac{L}{H} - 1\right] \tag{A.3}$$

Substituting eqn (A.2) into eqn (A.3) and solving the resulting quadratic equation for $\lambda^{1/2}$ gives the dimensionless impact velocity

$$\lambda^{1/2} = 1 \pm 2\left[1 + 2\varepsilon_{max}\left(\frac{L}{H}\right)^2 - \frac{L}{H}\right]^{1/2} \tag{A.4}$$

in terms of the maximum strain ε_{max} in a beam. Equation (A.4) therefore provides the dimensionless threshold velocity for a Mode II tensile tearing failure when ε_{max} is replaced by the rupture strain ε_f for the material.

Equation (A.1) was derived using a square yield curve that circumscribes the exact or maximum normal stress yield curve. Another yield curve that is 0·618 times as large would completely inscribe the maximum normal stress yield curve. Equation (A.4) remains unchanged for this case, except that σ_0 should be replaced by $0{\cdot}618\sigma_0$. It appears reasonable to assume that the theoretical predictions according to the exact yield curve would lie between the predictions of the circumscribing and inscribing square yield curves.

Chapter 4

STRAIN LOCALIZATION AND FRACTURE IN METAL SHEETS AND THIN-WALLED STRUCTURES

G. FERRON & A. ZEGHLOUL
LPMM, ISGMP, URA CNRS 1215,
Ile du Saulcy, 57045 Metz Cedex 01, France

ABSTRACT

Localized necking in metal sheets is analysed in a variety of situations with emphasis on material properties and boundary conditions. The localization analyses under in-plane loading of the sheet are described and a quantitative assessment is made of the stabilizing effects of material strain-hardening and strain-rate sensitivity, and of the destabilizing effect of thermo-mechanical coupling. The dependence of the forming limits on the shape of the yield surface and the relationship between limit strains at necking and fracture strains are also discussed. Consideration of out-of-plane stretching conditions prevailing in actual forming processes brings into picture the interaction between material and geometry. The stabilizing effect of the increase in sheet curvature resulting from a lateral loading is outlined and a comparative cross-checking with pressurized tubes and shells is presented. The role of friction between the tools and the workpiece is also briefly discussed. It is concluded that the limit strains at necking likely depend on the forming process investigated.

1 INTRODUCTION

The onset and development of plastic instability in the form of a neck that precedes ductile fracture is an important factor limiting the ductility of thin-walled structures. From a basic standpoint, both the material properties and the boundary conditions of the process are controlling conditions that can lead to diffuse necking or to catastrophic thinning and failure.

Two main classes of boundary value problems will be tackled in this chapter. One deals with sheet-metal stamping processes, the other with pressurized thin-walled structures. While the global complications of structural problems have been recognized for a long time, most of the theoretical approaches to the forming limits in metal sheets assume idealized conditions of uniform in-plane loading. Thus efforts to obtain accurate predictions of sheet-metal ductility have stimulated numerous investigations in the field of metal plasticity. By contrast, much less attention has been given to the effects of geometry and boundary conditions in the analysis of the limits to ductility in actual sheet-metal forming operations.

The first part of the chapter reviews some classical conditions of instability based on bifurcation analyses for rigid–plastic materials. Thus the one-dimensional analysis of Considere[1] for diffuse necking under uniaxial tension and the two-dimensional analysis of Hill[2] for localized necking under biaxial tension are outlined, as well as the conditions derived by Mellor[3] for instability in thin-walled tubes under combined tension and internal pressure, and in spherical shells under internal pressure. These various situations demonstrate the profound influence of geometry on the instability strains.

The second part of the chapter is devoted to the prediction of sheet-metal ductility under assumed in-plane loading. One of the most widely used approaches is the localization analysis introduced by Marciniak and Kuczynski.[4] The in-plane localization approach allows a quantitative assessment of the stabilizing effects of strain-hardening and strain-rate hardening, and of the destabilizing effect of thermo-mechanical coupling. However, predictions of available limit strains frequently diverge from experimental results, especially in the biaxial stretching range, corresponding to positive values of the two principal strains in the plane of the sheet.

Finally, the third part of the chapter focuses on the global analysis of some forming processes. Particular emphasis is given to the influence of boundary conditions. The limit strains expected in the plastic bulging of a sheet loaded by lateral hydrostatic pressure and in the hemispherical punch test are analysed. The results display a contrasting behaviour in comparison with the predictions of instability models assuming in-plane stretching.

2 BIFURCATION ANALYSES FOR BARS AND THIN-WALLED STRUCTURES

A large class of instability problems can be approached by means of bifurcation analyses assuming rigid–plastic material behaviour. These analyses predict the onset of necking, but do not provide quantitative information relating to the post-uniform behaviour. The rate of the strain concentration process is actually governed by, among other variables, material strain-rate sensitivity,[5–7] damage effects[8–11] and thermo-mechanical coupling.[12–15]

Bifurcation analyses, however, do provide a reasonable first-order approximation of the limit strains. They also allow a clear understanding of the effect of geometry in some problems where localization analyses or perturbation analyses using more elaborate descriptions of material behaviour are difficult to handle.

From a mathematical point of view, the onset of necking is associated with the bifurcation of the problem in velocities. Stated differently, this means that either unloading or continued plastic deformation can occur beyond some point corresponding to a condition of neutral loading, where material strain-hardening capacity just balances geometrical softening.

2.1 Considere's Analysis for Diffuse Necking in Tensile Bars

The analysis of Considere[1] deals with the usual uniaxial tensile test of a long bar. In a one-dimensional analysis, the onset of a neck affecting the cross-sectional area of the specimen occurs at the point of maximum load. The condition of instability, $\mathrm{d}(\sigma A) = 0$, where σ is the axial Cauchy stress and A is the cross-sectional area, can be expressed as

$$\frac{\mathrm{d}\sigma}{\sigma} + \frac{\mathrm{d}A}{A} = 0 \tag{1}$$

or, with $\mathrm{d}A/A = -\mathrm{d}\varepsilon$ resulting from the condition of plastic incompressibility,

$$\gamma = 1 \tag{2}$$

where $\gamma = \mathrm{d}\ln\sigma/\mathrm{d}\varepsilon$ is the work-hardening coefficient and ε is the logarithmic axial strain. For the power-type stress–strain law $\sigma = K\varepsilon^N$ (where N is the work-hardening exponent and K is a constant) the condition (2) becomes

$$\varepsilon = N \tag{3}$$

This form of instability is called diffuse necking. In the absence of more refined modeling, it should be noted that the spread of the neck along the tensile axis is unspecified. A parameter of paramount importance for the post-uniform behaviour is the material strain-rate sensitivity, often characterized by the strain-rate sensitivity coefficient $m = \partial \ln\sigma/\partial \ln\dot{\varepsilon}|_{\varepsilon}$, where $\dot{\varepsilon}$ is the axial strain rate. Positive m-values bring about a stabilizing effect in the load equilibrium equation, leading to less non-uniform strain distributions and increased ductilities as m increases (see e.g. the localization analysis of Hutchinson and Neale[5]). The results of uniaxial tensile ductilities collected by Woodford[16] are frequently utilized to check the predictions of different descriptions of the diffuse necking process in viscoplastic metals.

2.2 Hill's Analysis for Localized Necking in Sheets

The description given by Hill[2] allows the prediction of local through-thickness instability in biaxially loaded sheets. This instability takes the form of a necking band, as shown in Fig. 1. In a two-dimensional analysis, the onset of this neck is possible along a characteristic line of zero extension in the plane of the sheet ($\dot{\varepsilon}_{tt} = 0$, Fig. 1). Assuming by convention that $\dot{\varepsilon}_2 < \dot{\varepsilon}_1$, this condition enables the determination of the angle $\psi = (\mathbf{1}, \boldsymbol{n})$ defining the orientation of the band:

$$\psi = \mathrm{Arctan}\,(\sqrt{-\rho}) \tag{4}$$

where $\rho = \dot{\varepsilon}_2/\dot{\varepsilon}_1$ is the nominal strain-rate ratio. Furthermore, under the assumption of generalized plane stress, the equilibrium equations normally to the band are $(\partial/\partial x_n)(h\sigma_{n\alpha}) = 0$, $\alpha = n, t$, where $\sigma_{n\alpha}$ are Cauchy stress components and h is the thickness of the sheet. The conditions of instability, associated with the simultaneous stationarity of

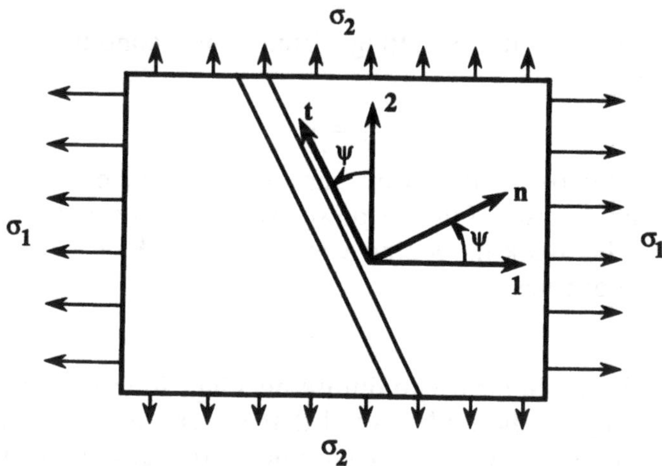

Fig. 1. Onset of a necking band along a line of zero extension, according to Hill's analysis for localized necking in sheets.

$\sigma_{nn}h$ and $\sigma_{nt}h$, can be written as

$$\frac{d\sigma_{nn}}{\sigma_{nn}}+\frac{dh}{h}=0 \tag{5}$$

and

$$\frac{d\sigma_{nt}}{\sigma_{nt}}+\frac{dh}{h}=0 \tag{6}$$

These conditions are verified simultaneously in the case of a constant stress ratio. Assuming further isotropic hardening and appropriate definitions of effective stress $\bar{\sigma}$ and effective strain $\bar{\varepsilon}$, the following identities are satisfied:

$$\frac{d\sigma_{nn}}{\sigma_{nn}}=\frac{d\sigma_{nt}}{\sigma_{nt}}=\frac{d\bar{\sigma}}{\bar{\sigma}} \tag{7}$$

and

$$\frac{d\varepsilon_3}{\varepsilon_3}=\frac{d\bar{\varepsilon}}{\bar{\varepsilon}} \tag{8}$$

where ε_3 is the thickness strain. Finally, observing that $dh/h = d\varepsilon_3$ and using the power law $\bar{\sigma} = K\bar{\varepsilon}^N$ (that is, $d\bar{\sigma}/\bar{\sigma} = N\,d\bar{\varepsilon}/\bar{\varepsilon}$) the conditions (5) and (6) become

$$\varepsilon_3 = -N \tag{9}$$

or, using the condition of plastic incompressibility,

$$\varepsilon_1 + \varepsilon_2 = N \tag{10}$$

It is evident from the condition for the orientation of the necking band that Hill's prediction of local necking is restricted to the deep-drawing range ($\varepsilon_2 < 0$). It is worth noting that eqn (10) is true irrespective of the shape of the yield locus. The effect of yield locus shape only comes from the different strain ratios $\rho = \varepsilon_2/\varepsilon_1$ corresponding to a given stress ratio $\alpha = \sigma_2/\sigma_1$. For instance, with Hill's quadratic yield criterion[17] with the assumption of isotropy in the plane of the sheet, the strain path is linked to the stress ratio by

$$\rho=\frac{(1+R)\alpha-R}{(1+R)-R\alpha} \tag{11}$$

where R is the anisotropy coefficient, identified as the width-to-thickness strain-rate ratio in a uniaxial tension test. According to Hill's analysis, the well-known beneficial effect of large R in deep-drawing operations should thus be ascribed to this dependence of the strain path on R.

3.3 Mellor's Analysis for Necking in Tubes and Spherical Shells

The conditions of instability for a thin-walled tube with closed ends under combined internal pressure and external tensile load P have been set up by Mellor[3] and re-examined by Hill.[18] Using the membrane approximation, the axial and hoop Cauchy stresses are respectively

$$\sigma_1 = (P + \pi r^2 p)/2\pi r h \tag{12}$$

and

$$\sigma_2 = pr/h \tag{13}$$

where r is the current mean radius and h is the current thickness. Under constant-stress-ratio conditions (i.e. with $P/\pi r^2 p = \text{constant}$) and assuming, as before, isotropic hardening and appropriate definitions of effective stress and strain, the stress and strain increments satisfy

$$\frac{d\sigma_1}{\sigma_1} = \frac{d\sigma_2}{\sigma_2} = \frac{d\bar{\sigma}}{\bar{\sigma}} \tag{14}$$

and

$$\frac{d\varepsilon_1}{\varepsilon_1} = \frac{d\varepsilon_2}{\varepsilon_2} = \frac{d\bar{\varepsilon}}{\bar{\varepsilon}} \tag{15}$$

where ε_1 and ε_2 are the logarithmic axial and hoop strains respectively. With $dr/r = d\varepsilon_2$ and $dh/h = -(d\varepsilon_1 + d\varepsilon_2)$, the conditions of instability, associated with the simultaneous stationarity of either p and $P/\pi r^2$, or P and $\pi r^2 p$, can be expressed in the form of the critical value of $\gamma = d\ln\bar{\sigma}/d\bar{\varepsilon}$ as a function of the stress ratio $\alpha = \sigma_2/\sigma_1$.[3] For the stress–strain law $\bar{\sigma} = K\bar{\varepsilon}^N$, the instability strains satisfy the simple expressions

$$\varepsilon_1 + 2\varepsilon_2 = N \tag{16}$$

and

$$\varepsilon_1 = N \tag{17}$$

Equation (16) relates to the condition of stationary of p and $P/\pi r^2$. This condition is reached first when the tube radius increases, i.e. when the hoop strain ε_2 is positive. Inversely, eqn (17) is associated to the stationarity of P and $\pi r^2 p$, which is attained first when ε_2 is negative.

Similar calculations can also be performed for a spherical thin shell under internal pressure p.[19] Then the membrane stresses are

$$\sigma_1 = \sigma_2 = pr/2h \tag{18}$$

where, as before, r is the current mean radius and h is the current thickness. With equations identical with (14) and (15), and with $dr/r = d\varepsilon_1 = d\varepsilon_2$ and $dh/h = -2\,d\varepsilon_1$, the surface strains at instability are

$$\varepsilon_1 = \varepsilon_2 = \tfrac{1}{3}N \tag{19}$$

2.4 Influence of Geometry on Instability Strains

The conditions of instability derived above are summarized in Fig. 2. They provide a correct first-order approximation of experimental results obtained under the different geometrical conditions assumed, for rate-insensitive or weakly rate-sensitive materials. This point is well documented in the literature for Considere's maximum load condition for uniaxial tensile ductilities of bars, and for Hill's local necking criterion in the case of sheet specimens of different width-to-gauge-length ratios pulled in tension. Data points for tubes under combined external load and internal pressure, taken from Ref. 20, are also plotted on Fig. 2. For these experiments performed under the conditions assumed in Mellor's analysis,[3] the agreement with theoretical predictions is good, except at plane-strain axial tension ($\varepsilon_2 = 0$), which is interpreted by the authors as the result of a lower work-hardening rate for this particular strain path. Nevertheless, the profound asymmetry with respect to the $\varepsilon_1 = \varepsilon_2$ axis observed in that case is an additional convincing proof of the geometrical effects.

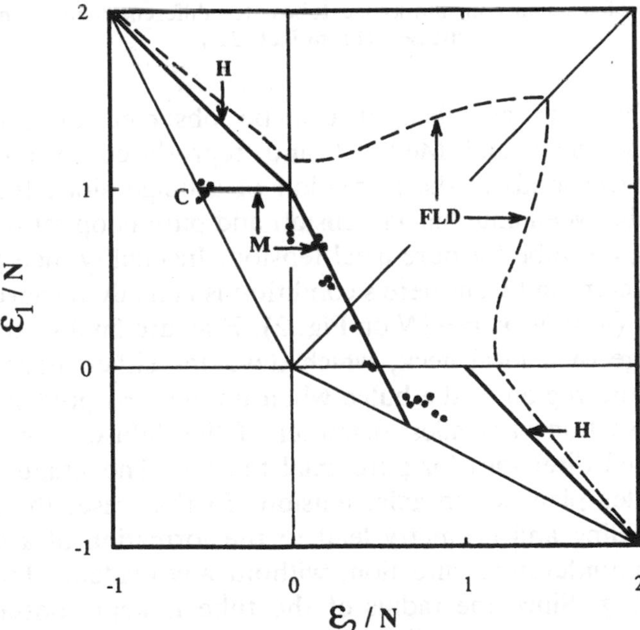

Fig. 2. Instability strains for different loadings and geometries: C, Considere's condition for diffuse necking in tensile bars; H, Hill's condition for localized necking in sheets; M, Mellor's conditions for necking in tubes. The points represent experimental data for tubes.[20] The dashed line represents a typical forming limit diagram for metal sheets.

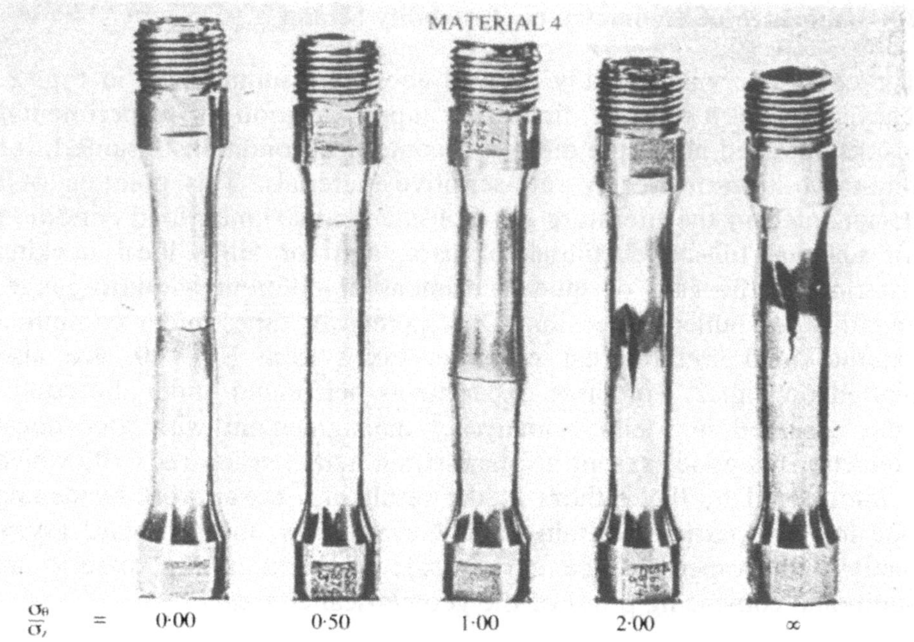

Fig. 3. Modes of instability in thin-walled tubes, for different ratios of hoop-to-axial stresses. (From Ref. 21.)

Still in the case of tubes, it can be observed on photographs presented by Jones and Mellor,[21] and reproduced in Fig. 3, that instability corresponds to the formation of a bulge along the tube for stress states between equi-biaxial tension and pure hoop tension, and of a neck along the tube for pure axial tension. Instability then presents a diffuse character, and Considere's condition is actually retrieved in pure axial tension ($\varepsilon_1 = N$, $\varepsilon_2 = -\frac{1}{2}N$ on Fig. 2). Fracture finally occurs as the terminal stage of a local neck, which develops either along the axial direction in the region of the bulge when hoop stress predominates, or in the cross-section of smallest diameter of the diffuse neck (along the circumferential direction) for pure axial tension. The situation is quite different under plane-strain axial tension. In that case, the combined effects of loading and geometry lead to the formation of a local neck along the circumferential direction, without any evidence for previous diffuse necking. Since the radius of the tube is kept constant during loading, the situation is actually much the same as that of a sheet under plane-strain tension, and Hill's condition for localized necking is retrieved ($\varepsilon_1 = N$, $\varepsilon_2 = 0$ on Fig. 2), with the neck aligned along the direction of zero extension.

A typical forming limit diagram (FLD) for metal sheets under biaxial tension is also shown for the purpose of comparison on Fig. 2. The FLD represents the plot of the limit values of permissible surface strains in experiments where almost-proportional conditions of straining are applied. These limits to ductility are set up by the development of local through-thickness thinning of the sheet. The analysis of the forming limits in metal sheets should thus be based on models for localized necking. While Hill's criterion for a local neck does provide a correct account of experimental FLDs for rate-insensitive or weakly rate-sensitive materials, none of the bifurcation models presented above enables the prediction of sheet-metal forming limits in the biaxial stretching range. The profound scatter observed between the limit strains in pressurized tubes and spherical shells on the one hand, and in biaxially stretched sheets on the other, should essentially be ascribed to the destabilizing effect of the decrease in membrane curvature experienced in the former case, as will be discussed in more detail in the last part of this chapter.

3 LOCALIZATION ANALYSES FOR METAL SHEETS

This section focuses on the prediction of the forming limits in metal sheets under assumed in-plane, proportional loading paths. The main difficulties are met in the stretching range, i.e. for positive values of the two principal surface strains in the plane of the sheet. With flow theory, and in the absence of a pointed vertex on the yield surface, forming limits can be predicted only by postulating the existence of initial defects in the sheet[4] or by making use of a perturbation analysis for studying the rate of growth of instabilities.[22]

Another approach, initiated by Stören and Rice,[23] is based on a bifurcation analysis utilizing deformation theory. As discussed by these authors, the equations of deformation theory can be used to simulate the destabilizing effect of a vertex on the yield surface. From a physical point of view, however, and assuming crystallographic slip as the fundamental mode of plastic deformation, the existence of vertices is possible only for rate-insensitive crystals, while it is ruled out for crystals having a positive strain-rate sensitivity (in which case the shear rates over the slip systems are determined without ambiguity[24–26]). A positive (even small) strain-rate sensitivity (and its counterpart, i.e. a negative temperature sensitivity) is quite a general feature in metals, where plastic flow is assisted by thermal activation. Possible existence of vertices is thus not considered in this review, and the reader is referred to Refs 23 and 27 for bifurcation analyses utilizing deformation theory or assuming vertices on the yield surface.

On the other hand, both the localization analyses and the perturbation analyses are somewhat troublesome in so far as the values of the initial size of the defect and of the critical rate of growth of the fastest unstable mode respectively are most often chosen in numerical simulations without obvious reference to physical considerations. A possible rationale for the defect introduced in the localization analyses can be founded on microstructural features relating to either inhomogeneous void distribution[8,9] or inhomogeneous particle distribution leading to inhomogeneous void nucleation.[10,11] Finally, wc present the broad lines of the localization analyses, which are much more developed in the literature.

3.1 General Features of the Localization Analyses

The localization analyses, first introduced by Marciniak and Kuczynski[4] and extended by Hutchinson and Neale[27,28] to the full range between uniaxial tension and equi-biaxial tension, are based on the consideration of a pre-existing defect in the form of a band of smaller thickness inclined at an angle $\psi = (\mathbf{1}, \mathbf{n}) = (\mathbf{2}, \mathbf{t})$ with respect to the principal directions of stress and strain rate outside the band (which coincide for an isotropic material, Fig. 4). These principal directions are further assumed to remain fixed, parallel to a material coordinate system. The sheet is thus assumed to consist of two regions, referred to as the neck and the bulk and denoted by the superscripts n and b respectively. The

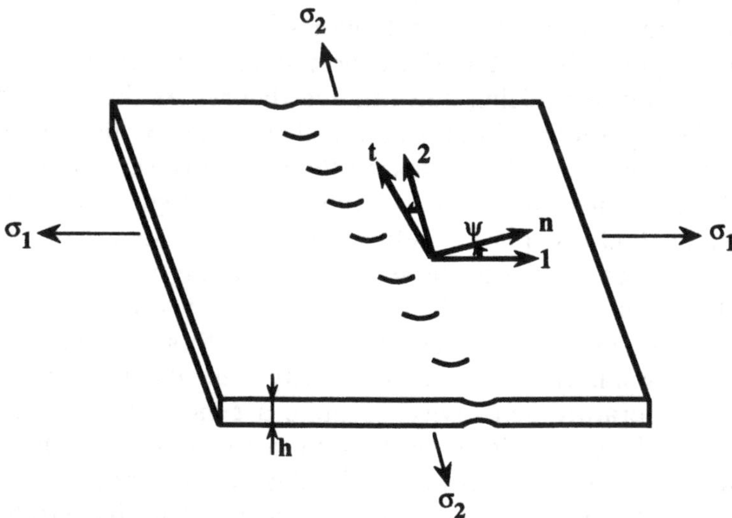

Fig. 4. Schematic representation of a linear thickness non-uniformity across a sheet.

size of the initial defect is defined by

$$f = \frac{h^{b}(0) - h^{n}(0)}{h^{b}(0)} \tag{20}$$

where $h(t)$ is the thickness at time t. In the general case $\psi(0) \neq 0$, the band rotates during straining and the current $\psi(t)$ value is related to the initial band inclination $\psi(0)$ by

$$\tan \psi = \exp(\varepsilon_1^{b} - \varepsilon_2^{b}) \tan \psi(0) \tag{21}$$

where ε_1 and ε_2 are the major and minor principal strains in the plane of the sheet respectively. The basic equations are given by the condition of compatibility

$$d\varepsilon_{tt}^{n} = d\varepsilon_{tt}^{b} \tag{22}$$

the equilibrium conditions within the assumption of generalized plane stress,

$$\sigma_{nn}^{n} h^{n} = \sigma_{nn}^{b} h^{b} \tag{23}$$

$$\sigma_{nt}^{n} h^{n} = \sigma_{nt}^{b} h^{b} \tag{24}$$

and the expressions of the stress and strain-increment components in the **n**, **t** axes. The choice of a particular theory of plasticity then enables the derivation of equations describing the localization process. With the J_2 flow theory, the localization equations under proportional straining ($\rho = d\varepsilon_2^{b}/d\varepsilon_1^{b} = \text{const}$) have been set up by Hutchinson and Neale[27] for rate-insensitive materials with the stress–strain law $\bar{\sigma} = K\bar{\varepsilon}^{N}$ and by the same authors[28] for rate-sensitive materials with the constitutive law $\bar{\sigma} = K\bar{\varepsilon}^{N}\dot{\bar{\varepsilon}}^{M}$ (where $M \equiv m = \partial \ln \bar{\sigma}/\partial \ln \dot{\bar{\varepsilon}}|_{\bar{\varepsilon}}$ is the strain-rate sensitivity coefficient). The forming limit is assessed to be reached when $\bar{\varepsilon}^{n}/\bar{\varepsilon}^{b} \rightarrow \infty$. The initial band inclination $\psi(0)$ is determined so as to minimize the forming limit. With the J_2 flow theory, $\psi(0)$ is found to be zero in the stretching range ($\varepsilon_2 \geqslant 0$); $\psi(0)$ is non-zero in the deep-drawing range ($\varepsilon_2 < 0$).

This analysis of localized necking in sheets is an extension of the defect approach to diffuse necking in tensile bars developed by Hutchinson and Neale.[5] In both cases, positive strain-rate sensitivities slow down the localization process, thus leading to enhanced ductilities. A fundamental difference between the two types of necking, however, comes from the fact that localized necking in sheets requires deviation of the strain path in the neck towards plane strain. Qualitative interpretation of this effect has been given by Sowerby and Duncan[29] in the case of positive strain-rate ratios ρ. While ρ is kept constant in the bulk, the compatibility condition ($d\varepsilon_2^{b} = d\varepsilon_2^{n}$ for $\psi = 0$) implies that the

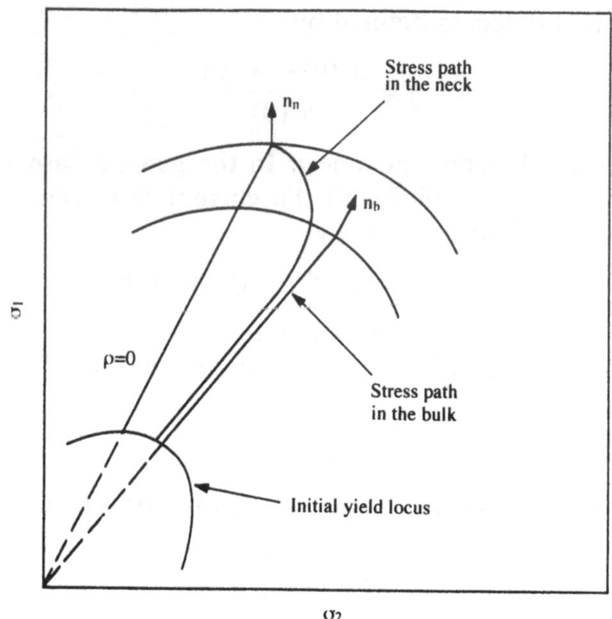

Fig. 5. Interpretation of resistance to necking associated with the deviation of the strain path in the neck towards plane strain. This deviation implies an increase in larger principal stress in the neck. (After Ref. 29.)

strain path in the neck steadily deviates towards plane strain as localization progresses ($\rho^n = d\varepsilon_2^b/d\varepsilon_1^n \to 0$). For a given value of effective strain in the neck, this deviation of the strain path would imply an increase in larger principal stress σ_1^n, as illustrated in Fig. 5. Stated differently, strain-path deviation brings about a stabilizing effect in the load equilibrium equation, $\sigma_1^n h^n = \sigma_1^b h^b$ for $\psi = 0$. This source of resistance to necking is the stronger the farther from zero is the nominal strain-rate ratio $\rho = \rho^b$. This behaviour both explains the general shape of the FLD (Fig. 2), where the forming limits, in terms of effective strain, are lower under plane-strain tension, and the coincidence of the calculated limit strains for diffuse necking under uniaxial tension and for localized necking under plane-strain tension, since no deviation of the strain path is experienced in either case.

The effect of strain-path deviation in the local neck is particularly explicit when comparing the localization equations obtained from linearized defect analyses performed in the two cases of diffuse necking in bars and localized necking in sheets. In the former case, the equation for the evolution of gradients in cross-sectional area is[6,7]

$$(1-\gamma-m)\,\delta\ln A + m\frac{\delta\dot{A}}{\dot{A}} + \gamma\,\delta\ln A(0) = 0 \tag{25}$$

and the equation for the evolution of thickness gradients in the case of localized necking in sheets is[15]

$$(1-\gamma F-C-mD)\,\delta \ln h+(C+mD)\frac{\delta \dot{h}}{\dot{h}}-\gamma G+\gamma F\,\delta \ln h(0)=0 \quad (26)$$

where δ denotes the Lagrangian difference or gradient between the neck and the bulk. The detailed calculations can be found in Ref. 15. The most important difference between the above equations comes from the substitution of the strain-rate sensitivity coefficient m in eqn (25) by the apparent value $C+mD$ in eqn (26). This apparent rate-sensitivity index includes the combined effects of strain-rate sensitivity and strain-path deviation. In the biaxial stretching range ($\psi=0$) and with the J_2 flow theory, it takes the value[15,30,31]

$$C+mD=\left[m+\frac{3\rho^2}{(2+\rho)^2}\right]\frac{(1+\rho)(2+\rho)}{2(1+\rho+\rho^2)} \quad (27)$$

This index thus increases from $C+mD=m$ under plane-strain tension up to a value as large as $m+\frac{1}{3}$ under equi-biaxial tension. The value of $\frac{1}{3}$, if it represented material strain-rate sensitivity, would correspond to a superplastic material. The effect of strain-path deviation is thus capable of explaining the high levels of effective strain achievable under equi-biaxial tension, even in work-hardened (unannealed) materials with low strain-hardening capacity and strain-rate sensitivity.

Otherwise, the FLD in the deep-drawing region was investigated in detail by Chan *et al.*[32] and Lian and Zhou.[33] Assuming in that case that the defect is aligned along Hill's direction of zero extension, a critical thickness strain criterion can be derived, where the limit thickness strain depends on N, M and f only. This criterion matches Hill's one (eqn (9)) for a strain-rate-insensitive material ($M=0$) without defect ($f=0$). In a general way, the localization analyses show that the FLD is practically insensitive to the shape of the yield surface in the deep-drawing range. This result contrasts with the predictions of the forming limits in biaxial stretching, which will be discussed next.

3.2 Influence of the Shape of the Yield Surface

Numerous investigations aiming at explaining the poor agreement that is often observed between experimental and calculated FLDs in the stretching range have been performed through the use of alternative plasticity models. Hill's quadratic yield criterion[17] along with the assumption of isotropic hardening has been utilized, either in the case of transverse isotropy[4] or in the general case of orthotropy.[34] The

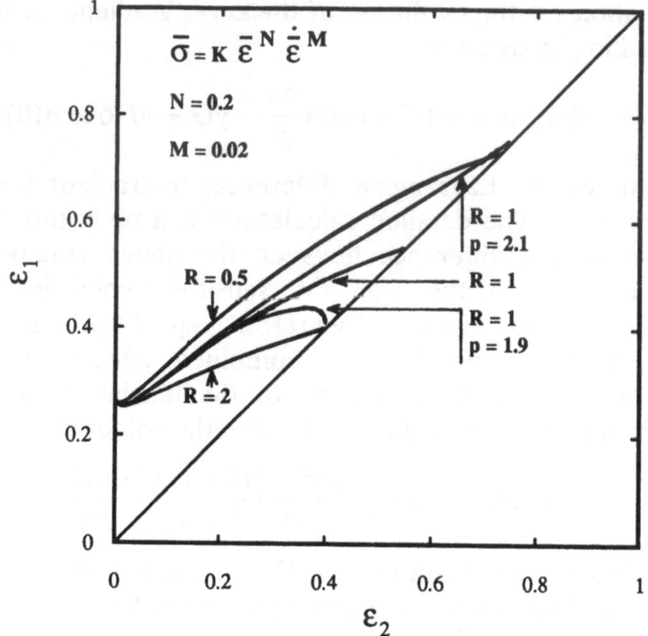

Fig. 6. FLDs computed according to von Mises (J_2) yield function ($R = 1$); Hill's quadratic yield function ($R = 0{\cdot}5$ and 2); and Hill's non-quadratic yield function ($R = 1$, $p = 1{\cdot}9$ and $R = 1$, $p = 2{\cdot}1$). Hill's non-quadratic yield function utilized is $(1 + 2R)(\sigma_1 - \sigma_2)^p + (\sigma_1 + \sigma_2)^p = 2(1 + R)\bar{\sigma}^p$.

predictions expected from non-quadratic yield functions proposed by Hill[35] and Bassani[36] have also been investigated,[37–40] and the effect of kinematic hardening has been analysed.[10,14,41]

Some of these predictions are shown in Fig. 6. Quite important variations in the forming limits are observed, which can again be understood as a result of the deviation of the strain path in the neck towards plane strain. In fact, this deviation is promoted by high values of the yield locus curvature between equi-biaxial tension and plane-strain tension, or, equivalently, by low values of the ratio of major plane-strain yield stress to equi-biaxial yield stress, σ_{ps}/σ_b. This is the case, for instance, for increasing values of the normal anisotropy coefficient R in Hill's quadratic yield function (with $\sigma_{ps}/\sigma_b = \sqrt{2(1+R)/(1+2R)}$). Still more drastic effects are observed for Hill's non-quadratic yield function[35] (Fig. 6), since in that case the local curvature at the point of equi-biaxial tension is either infinite or zero according to whether the exponent p in the yield function is respectively smaller or larger than 2.

An explicit account of the effect of yield locus curvature near equi-biaxial tension can be obtained by using the linearized localization approach of Ferron and Mliha-Touati[15] and Zeghloul *et al.*[31] The calculations, based on a parametric description for planar-isotropic yielding proposed by Budiansky,[42] show in particular that the apparent rate-sensitivity index $C + mD$ can be expressed under equi-biaxial tension in the form[31]

$$C + mD = m + \frac{1}{|C_b|} \tag{28}$$

where $|C_b| = -\sigma_b \, d^2\sigma_2/d\sigma_1^2|_{\sigma_1=\sigma_2=\sigma_b}$ represents a normalized measure of yield locus curvature at the point of equi-biaxial tension ($C_b = 3$ with the J_2 (von Mises) yield function, cf. eqn (27)). The destabilizing effect of sharp curvature is clearly apparent in eqn (28). Approximate expressions of the forming limits, calling upon knowledge of the local shape of the yield surface at the loading point, can also be found in Ref. 31.

3.3 Influence of Thermo-Mechanical Coupling

The thermally activated processes that control dislocation glide and hence the plastic flow properties of metals lead with few exceptions to a positive strain-rate sensitivity and a negative temperature sensitivity of the flow stress. Since most of the irreversible part of energy expended during straining is converted to heat, the more intense plastic work generated in the neck produces a higher temperature rise in that region, which in turn is compensated for by an auto-catalytic destabilizing effect corresponding to an increased strain rate for the load equilibrium equations to be verified.

In order to account for heat transfer, the coupled solution of equilibrium, compatibility and energy equations is required (see e.g. Refs 43 and 44 in the case of the sheet tensile test). Straightforward insight into the thermal effects, however, is obtained by observing that the examination of the isothermal and adiabatic situations permits upper and lower bounds respectively to be derived for the forming limits. Such analyses have been developed by Ferron[12] and Fressengeas and Molinari[13] in the case of diffuse necking in tensile bars, and by Dudzinski and Molinari[14] and Ferron and Mliha-Touati[15] for localized necking in sheets. Under adiabatic conditions, the energy equation reduces to

$$\dot{T} = \alpha_T \bar{\sigma} \dot{\bar{\varepsilon}} / c \tag{29}$$

where T is the temperature, c is the heat capacity per unit volume and α_T (the Taylor–Quinney coefficient) is the fraction of plastic work converted to heat. Using a Zener–Hollomon variable $Z = \dot{\bar{\varepsilon}} \exp(Q/RT)$ (where Q is an apparent activation energy and R is the gas constant) to account for the effect of temperature in the constitutive law, a modified form of eqn (26) is obtained under adiabatic conditions, where the most important change comes from the replacement of the work-hardening coefficient γ by the apparent value

$$\gamma' = \gamma - K \frac{\alpha_{\mathrm{T}} \bar{\sigma}}{c} \frac{mQ}{RT^2} \tag{30}$$

where the non-dimensional coefficient K depends on the strain path and on the yield function employed. This decrease in resultant work-hardening is the controlling factor of earlier flow localization under adiabatic conditions. Experiments[12] and calculations[45] concur in showing that the transition between quasi-isothermal and quasi-adiabatic conditions of straining for approximately 1 mm thick steel sheets occurs in air in the strain-rate range $\dot{\bar{\varepsilon}} \approx 10^{-4}$–$10^{-2}\,\mathrm{s}^{-1}$.

3.4 Comparisons between Experimental and Calculated Forming Limit Diagrams

The effects of yield surface shape and thermo-mechanical coupling are illustrated on Fig. 7, which compares experimental and calculated FLDs on aluminium. Attention has been given in this comparison to utilize results from in-plane stretching experiments[46] using flat-bottomed punches specially designed for approaching the geometrical conditions assumed in the calculations. This precaution is important, since the forming limits obtained in out-of-plane punch stretching experiments are significantly higher than under in-plane stretching.[47] Hecker and Stout[48] also reported experimental von Mises stress–strain curves under different stress states for aluminium, thus allowing us to estimate the shape of the yield locus under the assumption of isotropic hardening (Fig. 8). It is observed that the FLD calculated with Hill's quadratic yield function strongly overestimates the experimental forming limits in the stretching range (curve 1, Fig. 7). The agreement is much improved (curve 2, Fig. 7) by making use of a modified yield locus (curve 2, Fig. 8), which closely fits the data points estimated from Hecker and Stout.[48] In comparison with curve 1, this decrease in calculated forming limits in the stretching range in a direct consequence of the increase in yield locus curvature in the vicinity of equi-biaxial tension. In addition, the consideration of thermo-mechanical coupling (curve 3, Fig. 7) leads to a lowering of about 10% of the whole FLD when comparing adiabatic with isothermal conditions. Otherwise, particular significance should

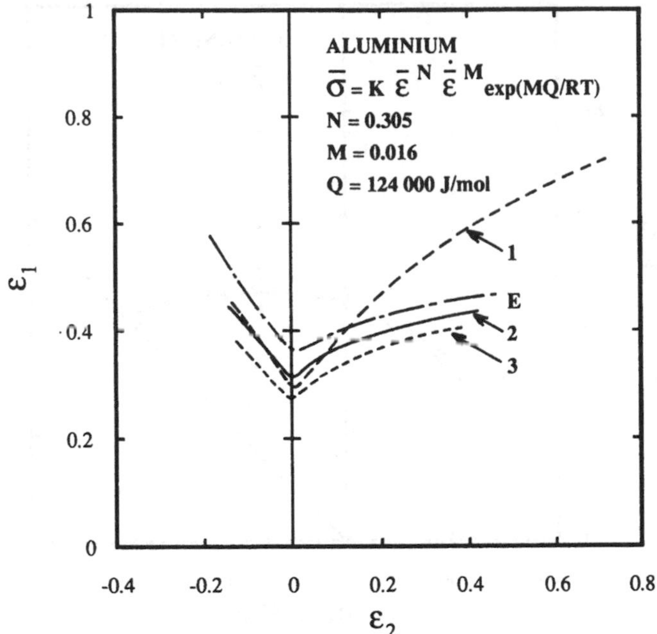

Fig. 7. Comparison of experimental and theoretical FLDs for aluminium. The FLDs are computed under the following assumptions: 1, Hill's quadratic yield function and isothermal conditions; 2, 'modified' yield function (curve 2, Fig. 7) and isothermal conditions; 3, 'modified' yield function and adiabatic conditions; E, experimental FLD (from Ref. 46).

not be attributed to the differences in the level of experimental and calculated FLDs, considering in particular the possible inaccuracy of N- and M-values adopted in the simulations.

Another comparison between experimental and calculated FLDs, taken from Chan,[49] is shown in Fig. 9. The Drucker yield function,[50] which contains the second and third invariants of the deviatoric stress tensor, is used in the simulations as a best fit of the crystallographic yield locus calculated with the Bishop–Hill[51] homogenization theory. Both for isotropic b.c.c. and f.c.c. polycrystals, this yield locus lies between those of Tresca and von Mises. The fairly sharp curvature exhibited by the Drucker yield locus near equi-biaxial tension is responsible for a substantial lowering of the forming limits computed in the stretching region, in comparison with the results of calculations using the von Mises yield criterion.

The forming limits are usually defined as the strain values outside the neck when stretching in the neck becomes unbounded, or, in practice, when the ratio of effective strains in the neck and in the bulk attains some prescribed value. Instead of this, Ghosh[52] consistently proposed

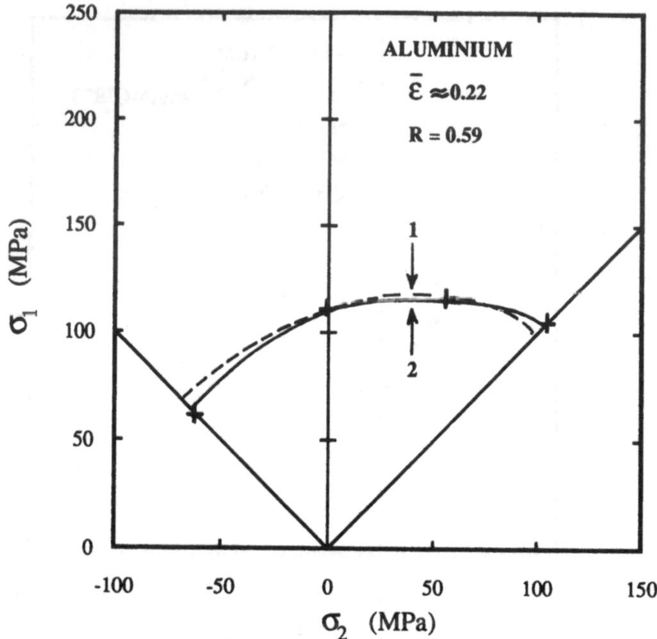

Fig. 8. Comparison of experimental and phenomenological yield surfaces: 1, Hill's quadratic yield function; 2, 'modified' yield function utilizing Budiansky's parametric representations;[42] +, data points estimated from Ref. 48.

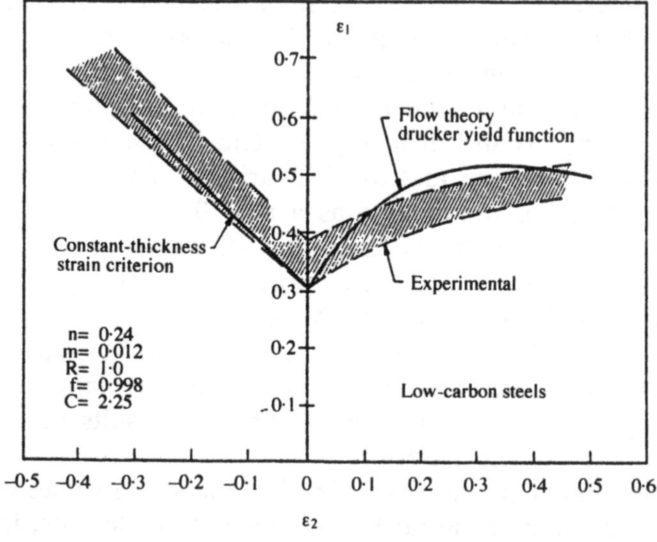

Fig. 9. Comparison of experimental and theoretical FLDs for steels: the theoretical FLD is computed with the Drucker yield function. (From Ref. 49.)

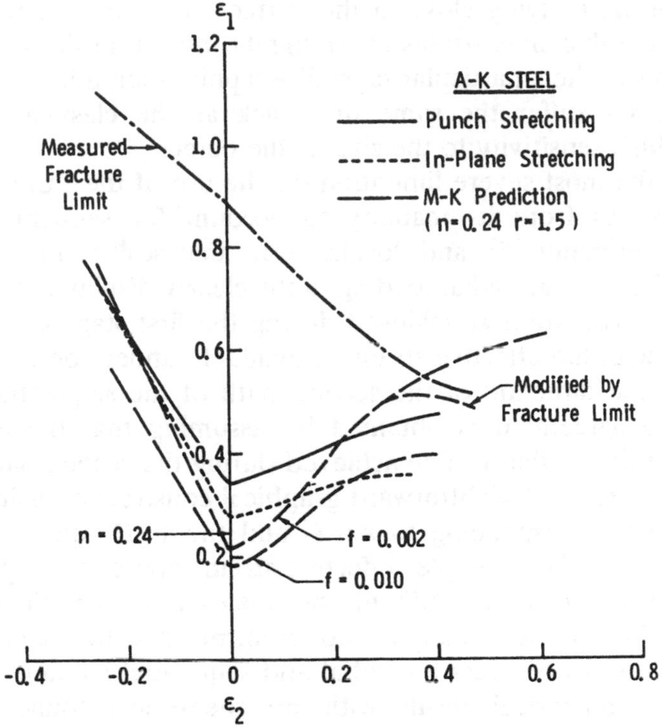

Fig. 10. Comparison of experimental and theoretical FLDs for steels: the theoretical FLDs are computed with Hill's quadratic yield function, with imperfection sizes of 0·002 and 0·010. Consideration of experimental fracture limit curve brings a better prediction of the FLD in punch stretching. (From Ref. 52.)

using measured fracture strains as a terminal condition for the stretching process. The predicted forming limits (Fig. 10) are not influenced by the condition of neck fracture near plane-strain tension, but are markedly lower in the vicinity of equi-biaxial tension, where the strain concentration process is slower and fracture strains are lower. This presumption that fracture terminates plastic flow near equi-biaxial tension assists a better prediction of the FLD obtained in punch stretching, and is consistent with the observation of a fracture-like failure with little visible necking in this region.[52]

3.5 Further Developments of the Localization Analyses

Some recent FEM simulations[53] tend to demonstrate the influence of a defect of finite length instead of an infinite groove. These FEM analyses may provide a more realistic account of material heterogeneity, associated for instance with a surface roughening effect resulting from colonies of grains lying close to the surface, and for which the strain paths and the thickness strains differ significantly from the values in the bulk, owing to their particular crystallographic orientation.[54] However, these analyses suffer the same drawback as the classical approach, namely a high sensitivity to the size of the defect.

Finally, the most severe limitation on the use of the FLD in a strain diagram comes from its inability to account for sequential loading paths. Experiments[55–57] and localization analyses[58,59] in various sequences of two-stage radial loading paths clearly display the prominent role of effective strain $\bar{\varepsilon}_{\mathrm{I}}$ achieved during the first stage of loading, as compared with the effective strain $\bar{\varepsilon}_{\mathrm{II}}$ available under continuous radial loading in the direction of the second path of the sequential loading. Reasonable agreement is obtained by assuming that the amount of effective strain $\Delta\bar{\varepsilon}$ that can be achieved during the second path is equal to $\Delta\bar{\varepsilon} = \bar{\varepsilon}_{\mathrm{II}} - \bar{\varepsilon}_{\mathrm{I}}$. A straightforward graphical construction using the von Mises ellipse corresponding to $\bar{\varepsilon} = \bar{\varepsilon}_{\mathrm{I}}$ and the FLD follows from this assumption.[55,56] This simple effective-strain approach supports the concept of an intrinsic FLD in the diagram of principal in-plane stresses.[60] Localization analyses do confirm that the same FLD is obtained in terms of stress for radial and sequential loadings.[58] In spite of being a numerical result with no theoretical foundation, this independence of the loading history is remarkable. A careful validation of this approach to the FLD in a stress diagram would require an accurate account of material behaviour under complex loading, instead of the simple isotropic hardening model assumed so far in the calculations.

Quite different approaches were developed by Triantafyllidis[61] and Chen and Gerdeen[62] to account for thickness effects associated with bending. On the basis of shell theory, these authors determined FLDs from the loss of ellipticity of the equilibrium equations. Instability criteria were also set up by Cordebois *et al.*[63] for different boundary conditions, including the influences of non-uniform stress fields and sheet curvature effects.

4 SHEET-METAL NECKING IN LABORATORY TESTS AND ACTUAL FORMING PROCESSES

4.1 Practical Aspects of Sheet-Metal Forming Operations

The main types of laboratory tests used to determine sheet-metal forming limits consist either in stretching blanks over a punch or by means of hydrostatic pressure for investigating the biaxial stretching range, or in pulling sheet specimens of different width-to-gauge-length ratios for studying the deep-drawing region. These tests reflect a quite general feature encountered in stamping operations. Deep drawing is expected in regions surrounding a punch, where the metal is forced 'inwards' by the punch. The limiting case of plane-strain stretching ($\varepsilon_2 = 0$) is observed in the side wall of a stamping. Under these circumstances, the curvature of the sheet either is small or does not evolve significantly during the sequence of interest. In contrast, biaxial stretching occurs in regions where ribs, domes or embossments are formed by the action of a punch or of a pressurized fluid. Deep drawing thus occurs without significant evolution in the curvature of the sheet, while biaxial stretching is observed in practice under conditions of out-of-plane loading. An important increase in curvature is thus experienced by the sheet in the latter case. In punch stretching operations, this geometrical effect is further complicated by the influence of friction, which restrains the displacement of the sheet on the punch.

A large number of defects may actually arise during a stamping operation. A detailed review on this subject can be found in Levaillant and Chenot.[64] Aside from structural effects such as wrinkling, defects may result from the amount of plastic deformation (cracks, cavities, orange peel surface), or from bad conditions of friction (scratches, coating damage). Limiting ourselves to necking, the bulge test and the hemispherical punch test are now examined as representative laboratory experiments.

4.2 Instability Strains in the Hydraulic Bulge Test

The bulge test consists of applying hydraulic pressure on one side of a sheet-metal workpiece, generally clamped along an elliptical contour. In the case of a circular contour, this test can be used to determine the stress–strain behaviour under equi-biaxial tension, by recording the pressure, and the extension and curvature at the pole by means of an extensometer and a spherometer.[65] In the general case of an elliptical contour, it is also utilized for determining sheet-metal forming limits under biaxial stretching. The bifurcation behaviour expected for rate-insensitive materials from the condition of maximum pressure is observed to correspond in practice to the development of a local neck spreading from the pole along the direction of the major axis.[66]

The two limiting cases of a long strip clamped at its two edges and of a circular sheet are first analysed in some detail. Results from FEM

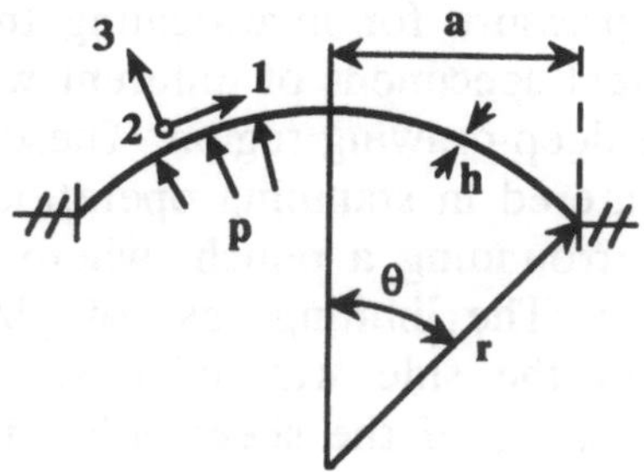

Fig. 11. Long strip under hydrostatic pressure.

simulations[67] will also be mentioned in the general case of an elliptical contour. In all cases, the membrane approximation is used.

An exact analytical solution can be obtained for the plane-strain plastic bulging of a strip of infinite length in one direction, of width $2a$ in the perpendicular direction, clamped at its two edges and loaded by lateral hydrostatic pressure p.[68] Up to instability, the bulge profile is circular and the stresses and strains are uniformly distributed along the bulge (Fig. 11). The hoop and longitudinal Cauchy stresses are respectively

$$\sigma_1 = pr/h \tag{31}$$

and

$$\sigma_2 = \alpha_{ps}\sigma_1 \tag{32}$$

where r is the current mean radius of the bulge, h is the current thickness and α_{ps} is the stress ratio under plane-strain conditions, which is a constant depending on the yield function utilized in the case of isotropic hardening. For the stress–strain law $\bar{\sigma} = K\bar{\varepsilon}^N$, and with $\mathrm{d}h/h = -\mathrm{d}\varepsilon_1$ resulting from the condition of plastic incompressibility, stationarity of p is found to be reached when

$$\frac{N}{\varepsilon_1} - 1 - \frac{1}{r}\frac{\mathrm{d}r}{\mathrm{d}\varepsilon_1} = 0 \tag{33}$$

where ε_1 is the logarithmic hoop strain. A simple analysis using an expansion of ε_1 in powers of a/r leads to the expression of $r^{-1}\,\mathrm{d}r/\mathrm{d}\varepsilon_1$ in the form[68]

$$\frac{1}{r}\frac{\mathrm{d}r}{\mathrm{d}\varepsilon_1} \approx -\frac{1}{2\varepsilon_1} + \frac{11}{10} \tag{34}$$

and the instability condition (33) finally yields

$$\varepsilon_1 \approx \tfrac{5}{21}(1+2N) \approx 0{\cdot}238(1+2N) \tag{35}$$

The effect of geometry is obvious when comparing eqn (33) with Hill's condition for localized necking under plane-strain (eqn (10) with $\varepsilon_2 = 0$) or with Mellor's conditions under plane-strain axial tension of a tube (eqns (16) and (17) with $\varepsilon_2 = 0$). In these cases where the effects of global changes in specimen geometry are absent, the limit strain $\varepsilon_1 = N$ is obtained. According to eqn (33), on the other hand, instability will be attained for ε_1-values larger or smaller than N respectively, according as the minimum value of the radius of the bulge ($r = a$ for the half-cylinder) is either not yet reached (for small N) or exceeded (for large N) at instability. This qualitative analysis is confirmed by eqn (35). For small N, the sharp decrease in r in the early stages of bulging has a strongly stabilizing effect, with an amount of stable stretching as large as $\varepsilon_1 \approx 0{\cdot}238$ for a perfectly plastic material. Hill's condition, $\varepsilon_1 = N$, is retrieved for $N = \ln \frac{1}{2}\pi \approx 0{\cdot}452$, in which case maximum pressure is obtained for the minimum value of the radius of the bulge, $r = a$. For larger N, instability occurs for $\varepsilon_1 < N$, owing to the destabilizing effect of the increase in r when the polar deflection exceeds a.

The plastic bulging for a circular contour was formerly treated by Hill[69] under the kinematical assumptions that the diaphragm deforms into a spherical membrane and that the velocity at each time and for any particle is normal to the instantaneous membrane profile. With these assumptions, the radius r of the membrane and the polar surface strain ε_1 are linked by the differential equation

$$\frac{1}{r}\frac{dr}{d\varepsilon_1} \approx \frac{3}{4} - \frac{1}{2\varepsilon_1} \tag{36}$$

As shown by Hill,[69] his kinematical assumptions provide the exact solution for a rigid–plastic Von Mises material obeying the stress–strain law $\bar{\sigma} = K \exp(\bar{\varepsilon})$. Using eqn (36) together with the condition of instability, $dp = 0$, for the stress–strain law $\bar{\sigma} = K\bar{\varepsilon}^N$, that is,

$$\frac{N}{\varepsilon_1} - 2 - \frac{1}{r}\frac{dr}{d\varepsilon_1} = 0 \tag{37}$$

the following polar surface strains at instability are derived:[69]

$$\varepsilon_1 = \varepsilon_2 \approx \tfrac{2}{11}(1 + 2N) \tag{38}$$

However, with the exception of the special exponential stress–strain law $\bar{\sigma} = K \exp(\bar{\varepsilon})$ and of some solutions obtained by means of a small-deflection analysis,[70,71] a numerical treatment is needed to obtain an accurate prediction of the evolution of the radius at the pole, or, directly, the polar strain at maximum pressure. A FDM analysis carried out by Zeghloul and Ferron[68] utilizing a variety of yield surfaces shows that, notwithstanding the sensitivity of the strain distribution and of its evolution to the shape of the yield surface, the polar strain at instability does not significantly depend on the yield function utilized. For all yield functions explored, the empirical relationship

$$\varepsilon_1 = \varepsilon_2 \approx 0{\cdot}26(1 + 1{\cdot}2N) \tag{39}$$

predicts the polar surface strain at instability with an accuracy of $\pm 0{\cdot}02$, in the range $0 \leqslant N \leqslant 0{\cdot}8$. These limit strains are substantially higher than those given by eqn (38), particularly for small N, which is due to the fact that Hill's assumption of a spherical bulge underestimates the curvature at the pole.

The stabilizing effect of the increase in membrane curvature during plastic bulging is evident when comparing eqn (39) with the result for a spherical shell (eqn (19)), which experiences the destabilizing effect of a decrease in curvature during pressurization.

FEM calculations for elliptical contours[67] further demonstrate the preponderant effect of geometry, since both the strain paths at the pole and the polar strains at maximum pressure are almost independent of

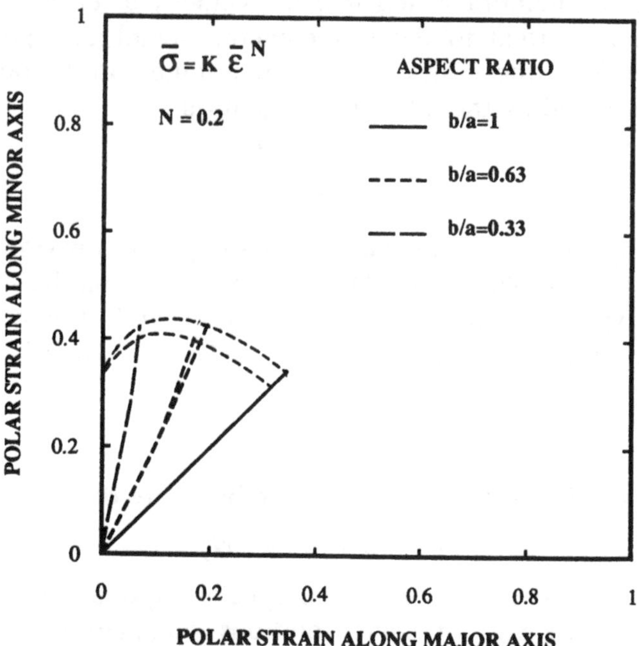

Fig. 12. Strain paths and polar strains at maximum pressure in the hydraulic bulge test, for different aspect ratios of the elliptical contour. The different yield functions utilized are not identified on the figure.

the normal anisotropy coefficient R in Hill's quadratic yield function. These results are summarized in Fig. 12.

4.3 Limit Strains in the Hemispherical Punch Test

Another test representative of biaxial stretching conditions, and frequently used to determine forming limits, is the hemispherical punch stretching of a blank clamped along a circular contour. In this experiment, the combination of geometrical and tribological effects leads to the development of a peak in radial strain distribution during the punch travel. Flow localization terminates in failure along some intermediate circumferential direction in the region of punch contact.[72–74] This experimental issue is reproduced in FEM simulations[74,75] without considering any initiator of instability, such as a defect or a perturbation.

The effects of tribological and rheological parameters are illustrated in Figs 13 and 14, which display the strain distribution obtained from FEM calculations[67] for different values of the Coulomb friction coefficient μ and the normal anisotropy coefficient R in Hill's quadratic yield function. The effect of friction (Fig. 13) obviously consists in

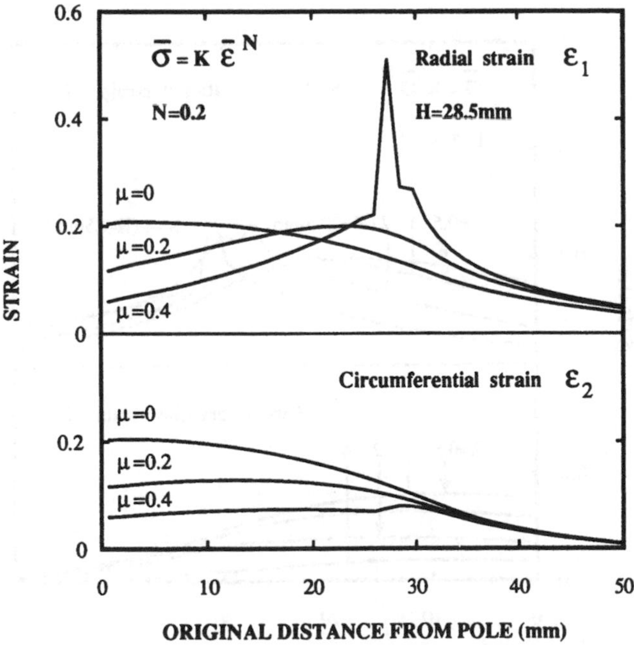

Fig. 13. Influence of the Coulomb friction coefficient μ on the distribution of radial strain ε_1 and circumferential strain ε_2 in hemispherical punch stretching.

restraining radial displacement along the punch. For a given punch travel, the polar strain thus decreases as μ increases, while radial stretching near the edge is almost unaffected by μ. Also, as μ increases, the radial strain peak that develops during stretching is located at larger values of the original distance from the pole, and its emergence is more rapid, leading to a more non-uniform distribution of radial strains ε_1. According to this behaviour, the strain path at the critical location of radial strain concentration r_0^c depends strongly on μ. It deviates either slowly towards plane strain for small μ, or very abruptly for large μ. This effect is further amplified by the decrease in circumferential strains ε_2 observed as μ increases. A wide range of strain paths between plane-strain tension and equi-biaxial tension is thus encompassed at r_0^c by decreasing μ, i.e. by improving lubrication at the punch contact.

The effect of R-value (Fig. 14) is explained by the fact that the ratio of major plane-strain yield stress to equi-biaxial yield stress decreases as R increases. Plane-strain stretching at the edge is thus comparatively enhanced, and resistance to equi-biaxial stretching at the pole is increased as R increases. Concerning the radial strain peak, the effect of increasing R is, on the whole, much the same as that of increasing μ. This can be understood as the result of the decrease in radial displacement as either μ or R increases.

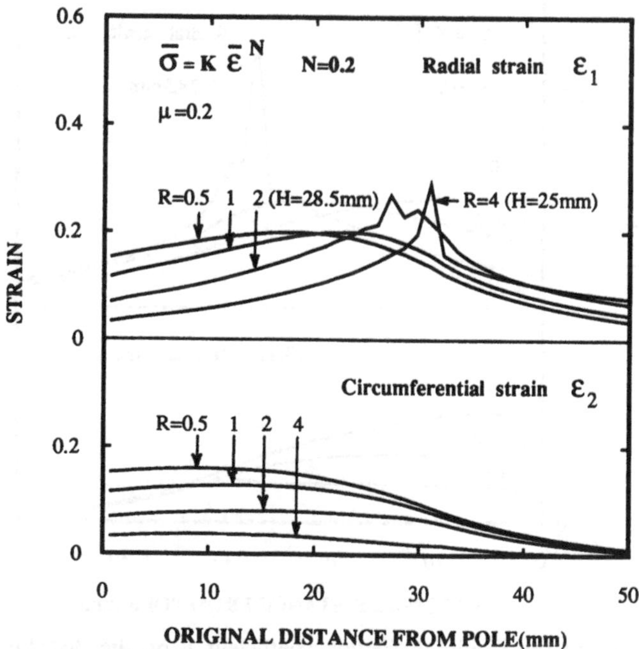

Fig. 14. Influence of the anisotropy coefficient R on the distribution of radial strain ε_1 and circumferential strain ε_2 in hemispherical punch stretching.

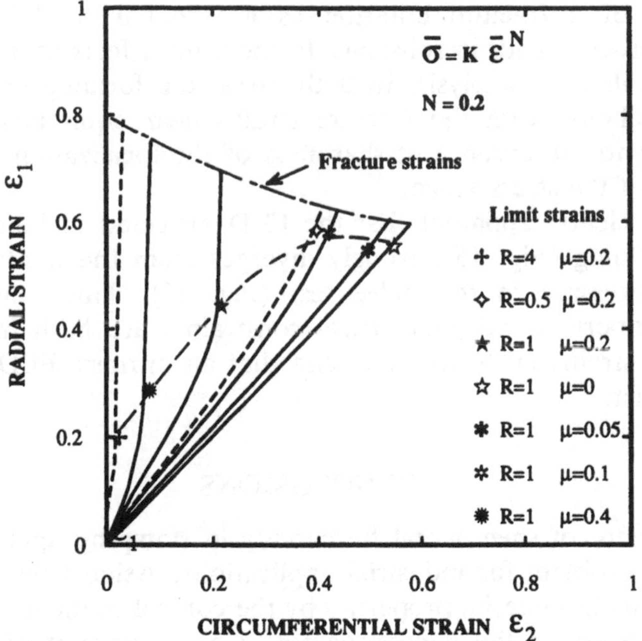

Fig. 15. Strain paths and fracture strains at the critical location of radial strain concentration for different μ and R, and corresponding limit strains in hemispherical punch stretching.

Figure 15 shows the strain paths and the fracture strains at the failure site, together with the limit strains at necking derived from the above simulations. Ductile fracture at the failure site has been assumed to occur according to the energetical criterion of Cockcroft and Latham,[76] and the limit strains at necking have been defined from the analysis of radial strain distribution for the nodes of the FEM mesh adjacent to the failure site, so as to approximate to experimental practice. The similar effects of friction and yield surface shape (or anisotropy) tend to bring the points of limit strains at necking over a single curve. This curve represents the FLD expected from punch tests performed under various conditions of lubrication on materials with different anisotropy coefficients.

The profound effect of yield surface shape on the FLD, obtained with in-plane analyses (Figs 6 and 7), thus appears to vanish in the out-of-plane experiment investigated. This drastic difference, however, does not preclude the detrimental influence of a large value of the anisotropy coefficient R in punch stretching, since the critical strain path for a given friction coefficient is nearer to plane strain for larger R (Fig. 15).

The use of a fracture criterion as a terminal condition of the stretching process leads in substance to the same effects as in Ghosh's[52] in-plane localization analysis. In both cases, the forming limits are in strong correlation with the fracture strains near equi-biaxial tension only, owing to the absence of sharpness of the localization process in this region of the stress space.

It is also clearly apparent that the FLD expected of hemispherical punch stretching (Fig. 15) strongly diverges from the limit strains at maximum pressure in the bulgc test (Fig. 12). This confirms that material properties and geometrical constraints are both involved in permissible strains up to necking, and that an intrinsic FLD probably does not exist.

5 CONCLUSIONS

The assessment of sheet-metal formability in stamping operations is a challenging problem for industrial applications. Aside from continued improvements in material properties by the control of inclusion content and optimization of the microstructure, a great amount of work has been devoted to the prediction of permissible strains before unacceptable necking occurs.

In attempts to define forming limits that would represent an intrinsic property of the material, many efforts have been aimed at analysing the idealized situation of a metal sheet loaded in its plane. While Hill's bifurcation analysis provides a simple and convincing explanation of localized necking in the range of negative minor strains in the plane of the sheet, a clear understanding of the onset of catastrophic thinning is still wanting for positive minor strains, in spite of significant advances in the knowledge of the plastic behaviour of polycrystalline metals. In particular, the defect introduced in the localization approaches is most frequently utilized as an adjustable parameter, and its size, generally of the order of one or a few percent, seems to be exceedingly large to be correlated with realistic material heterogeneities.

This difficulty, however, should be seen from the perspective that the assumption of in-plane loading may represent an oversimplification of boundary conditions in actual forming processes. In this regard, the limits to sheet-metal formability can be clarified by comparison with instability analyses in pressurized tubes or shells. The increase in curvature of metal sheets by the action of a punch or of a pressurized fluid, as compared with the decrease in curvature of pressurized tubes or shells, is a first-order structural effect capable of explaining the comparatively high values of the limit strains in stretched sheets.

Owing to geometrical effects, the existence of some criterion defining the success of a stamping with regard to the occurrence of necking, and which could be implemented in FEM codes, may thus be illusory. The intrinsic material limitation is more likely related to the terminal condition of fracture. The interaction between fracture strains and limit strains at necking may be either weak or strong according to whether the boundary conditions lead to the development of intense strain gradients well before fracture, or to still smooth gradients up to fracture. A transition from the first situation to the second seems to be expected when the stress state varies from plane-strain tension to equi-biaxial tension.

In spite of the criticisms that can be made of the forming limit diagram, this concept remains a useful guide for estimating material resistance to necking. However, analysis of the particular process of interest remains necessary to determine how the 'intrinsic' material ductility interferes with the geometrical constraints of the process. The forming limit diagram is also a valuable tool for assessing the performances of new sheet alloys designed for stamping operations.

Indeed, it would be an exciting challenge to put together the advances obtained in the fields of sheet-metal stamping and of impact loading of structural components, given the fairly wide gap existing at present between the two subjects. It comes readily to mind, for instance, that computations of the deformation state, deformation defects and residual stresses in, say, an autobody panel, can be performed to obtain detailed characterization of what will be the initial state of the structure before an impact. Formability studies have also generated significant progress in the knowledge of sheet-metal plasticity, which can be helpful in characterizing material resistance and residual ductility in a subsequent dynamic loading.

ACKNOWLEDGEMENT

The authors thank Dr C. Levaillant (CEMEF, Sophia Antipolis, France) for helpful suggestions and comments in preparing this paper.

REFERENCES

1. Considere, A., *Ann. des Ponts et Chaussées* **9** (1885) 574.
2. Hill, R. On discontinuous plastic states, with special reference to localized necking in thin sheets. *J. Mech. Phys. Solids*, **1** (1952) 19–30.

3. Mellor, P. B., Tensile instability in thin-walled tubes. *J. Mech. Engng Sci.*, **4** (1962) 251–6.
4. Marciniak, Z. & Kuczynski, K., Limit strains in the processes of stretch-forming sheet metal. *Int. J. Mech. Sci.*, **9** (1967) 609–20.
5. Hutchinson, J. W. & Neale, K. W., Influence of strain-rate sensitivity in necking under uniaxial tension. *Acta Metall.*, **25** (1977) 839–46.
6. Ghosh, A. K., A numerical analysis of the tensile test for sheet metals. *Metall. Trans.*, **8A** (1977) 1221–32.
7. Semiatin, S. L. & Jonas, J. J., *Formability and Workability of Metals.* American Society for Metals, Metals Park, OH, 1984, p. 159.
8. Jalinier, J. M. & Schmitt, J. H., Damage in sheet metal forming—II. Plastic instability. *Acta Metall.*, **30** (1982) 1799–809.
9. Ohno, N. & Hutchinson, J. W., Plastic flow localization due to non-uniform void distribution. *J. Mech. Phys. Solids,* **32** (1984) 63–85.
10. Needleman, A. & Tvergaard, V., Limits to formability in rate-sensitive metal sheets. In *Proceedings of ICM4, Stockholm,* ed. J. Carlsson & N. G. Ohlson. Pergamon Press, Oxford, 1983, pp. 51–65.
11. Tvergaard, V., Effect of yield surface curvature and void nucleation on plastic flow localization. *J. Mech. Phys. Solids,* **35** (1987) 43–60.
12. Ferron, G., Influence of heat generation and conduction on plastic stability under uniaxial tension. *Mater. Sci. Engng,* **49** (1981) 241–8.
13. Fressengeas, C. & Molinari, A. Inertia and thermal effects on the localization of plastic flow. *Acta Metall.*, **33** (1987) 387–96.
14. Dudzinski, D. & Molinari, A., Effect of anisotropic hardening and thermomechanical coupling on sheet formability. In *Proceedings of Considere Memorial on Plastic Instability, Paris,* ed. J. Salençon. Presses Ponts et Chaussées, 1985, pp. 25–34.
15. Ferron, G. & Mliha-Touati, M., Determination of the forming limits in planar-isotropic and temperature-sensitive sheet metals. *Int. J. Mech. Sci.*, **27** (1985) 121–33.
16. Woodford, D. A., *Trans. Am. Soc. Metals,* **62** (1969) 291.
17. Hill, R., A theory of the yielding and plastic flow of anisotropic metals. *Proc. R. Soc. Lond.* **A193** (1948) 281–97.
18. Hill, R., A theoretical perspective on in-plane forming of sheet metal. *J. Mech. Phys. Solids,* **39** (1991) 295–307.
19. Johnson, W. & Mellor, P. B., *Engineering Plasticity.* Ellis Horwood, Chichester, Sussex, 1985 p. 266.
20. Stout, M. G. & Hecker, S. S., Role of geometry in plastic instability and fracture of tubes and sheet. *Mech. Mater.*, **2** (1983) 23–31.
21. Jones, R. H. & Mellor, P. B., Plastic flow and instability behaviour of thin-walled cylinders subjected to constant-ratio tensile stress. *J. Strain Anal.*, **2** (1967) 62–72.
22. Dudzinski, D. & Molinari, A., Perturbation analysis of thermoviscoplastic instabilities in biaxial loading. *Int. J. Solids Structures,* **27** (1991) 601–28.
23. Stören, S. & Rice, J. R., Localized necking in thin sheets. *J. Mech. Phys. Solids,* **23** (1975) 421–41.
24. Hutchinson, J. W., Bounds and self-consistent estimates for creep of polycrystalline materials . *Proc. R. Soc. Lond.* **A348** (1976) 101–27.
25. Pan, J. & Rice, J. R., Rate sensitivity of plastic flow and implications for yield-surfaces vertices. *Int. J. Solids Structures,* **19** (1983) 973–87.

26. Peirce, D., Asaro, R. J. & Needleman, A., Material rate dependence and localized deformation in crystalline solids. *Acta Metall.*, **31** (1983) 1951–76.
27. Hutchinson, J. W. & Neale, K. W., Sheet necking—II. Time-independent behaviour. In *Mechanics of Sheet-Metal Forming*, ed. D. P. Koistinen & N. M. Wang. Plenum Press, New York, 1978, pp. 127–53.
28. Hutchinson, J. W. & Neale, K. W., Sheet necking—III. Strain-rate effects. In *Mechanics of Sheet-Metal Forming*, ed. D. P. Koistinen & N. M. Wang. Plenum Press, New York, 1978, pp. 269–85.
29. Sowerby, R. & Duncan, J. L., Failure in sheet metal in biaxial tension. *Int. J. Mech. Sci.*, **13** (1971) 217–29.
30. Ferron, G. & Molinari, A., Mechanical and physical aspects of sheet-metal ductility. In *Forming Limit Diagrams: Concepts, Methods and Applications*, ed. R. H. Wagoner, K. S. Chan & S. P. Keeler. TMS, Warrendale, PA 1989, pp. 111–51.
31. Zeghloul, A. & Ferron, G., Analytical expressions of the forming limits in biaxially stretched sheets. *Int. J. Mech. Sci.*, **32** (1990) 981–90.
32. Chan, K. S., Koss, D. A. & Ghosh, A. K., Localized necking of sheet at negative minor strains. *Mater. Trans.*, **15A** (1984) 323.
33. Lian, J. & Zhou, D., Diffuse necking and localized necking under plane stress. *Mater. Sci. Engng*, **111** (1989) 1–7.
34. Barata da Rocha, A., Barlat, F. & Jalinier, J. M., Prediction of the forming limit diagram of anisotropic sheets in linear and nonlinear loading. *Mater. Sci. Engng*, **68** (1985) 151.
35. Hill, R., Theoretical plasticity of textured aggregates. *Math. Proc. Camb. Phil. Soc.*, **75** (1979) 179–91.
36. Bassani, J. L., Yield characterization of metals with transversely isotropic plastic properties. *Int. J. Mech. Sci.*, **19** (1977) 651–60.
37. Parmar, A. & Mellor, P. B., Predictions of limit strains in sheet metal using a more general yield criterion. *Int. J. Mech. Sci.*, **20** (1978) 385–91.
38. Neale, K. W. & Chater, E., Limit strain predictions for strain-rate sensitive anisotropic sheets. *Int. J. Mech. Sci.*, **22** (1980) 563–74.
39. Ferron, G. & Mliha-Touati, M., Analysis of plastic instability in sheet-metal forming. In *Proceedings of ICM4, Stockholm*, ed. J. Carlsson & N. G. Ohlson. Pergamon Press, Oxford, 1983, pp. 649–55.
40. Lian, J., Zhou, D. & Baudelet, B., Application of Hill's new yield theory to sheet metal forming. *Int. J. Mech. Sci.*, **31** (1989) 237–47.
41. Tvergaard, V., Effect of kinematic hardening on localized necking in biaxially stretched sheets. *Int. J. Mech. Sci.*, **20** (1978) 651–8.
42. Budiansky, B., Anisotropic plasticity of plane-isotropic sheets. In *Mechanics of Material Behaviour*, ed. G. J. Dvorak & R. H. Shield. Elsevier Science Publishers, Amsterdam, 1984, pp. 15–29.
43. Semiatin, S. L., Ayres, R. A. & Jonas, J. J., An analysis of the non-isothermal tensile test. *Metall. Trans.*, **16A** (1985) 2299–308.
44. Kim, Y. H. & Wagoner, R. H., An analytical investigation of deformation-induced heating in tensile testing. *Int. J. Mech. Sci.*, **29** (1987) 179–94.
45. Raghavan, K. S. & Wagoner, R. H., Analysis of nonisothermal tensile tests using measured temperature distributions. *Int. J. Plasticity*, **3** (1987) 33–49.

46. Hecker, S. S., Factors affecting plastic instability and sheet formability. In *Proceedings of ICM4, Stockholm,* ed. J. Carlsson & N. G. Ohlson. Pergamon Press, Oxford, 1983, pp. 129–38.
47. Ghosh, A. K. & Hecker, S. S., Stretching limits in sheet metals: in-plane versus out-of-plane deformation. *Metall. Trans.*, **5A** (1974) 2161–4.
48. Hecker, S. S. & Stout, M. G., Strain hardening of heavily cold worked metals. In *Deformation, Processing and Structure,* ed. G. Krauss. ASM, St Louis, MO, 1983, pp. 1–46.
49. Chan, K. S., Effects of plastic anisotropy and yield surface on sheet metal stretchability. *Metall. Trans.*, **16A** (1985) 629–39.
50. Drucker, D. C., Relation of experiments to Mathematical theories of plasticity. *J. Appl. Mech.*, **16** (1949) 349–57.
51. Bishop, J. F. W. & Hill, R., A theory of the plastic distortion of a polycrystalline aggregate under combined stresses. *Phil. Mag.* **42** (1951) 414–27.
52. Ghosh, A. K., Plastic flow properties in relation to localized necking in sheets. In *Mechanics of Sheet Metal Forming,* ed. D. P. Koistinen & N. M. Wang. Plenum Press, New York, 1978 pp. 287–312.
53. Burford, D. A. & Wagoner, R. H., A more realistic method for predicting the forming limits of metal sheets. In *Forming Limit Diagrams: Concepts, Methods and Applications,* ed. R. H. Wagoner, K. S. Chan & S. P. Keeler. TMS; Warrendale, PA, 1989, pp. 167–82.
54. Chan, K. C. & Lee, W. B., Numerical computing of limit curves in the stretch forming of anisotropic sheet metals. In *Numiform 89,* ed. E. G. Thompson, R. D. Wood, O. C. Zienkiewicz & A. Samuelsson. A. A. Balkema, Rotterdam, 1989, pp. 413–18.
55. Grumbach, M. & Sanz, G., Influence des trajectoires de déformation sur les courbes limites d'emboutissage à striction et à rupture. *Mém. Sci. Rev. Métall.*, **11** (1974) 659–71.
56. Muschenborn, W. & Sonne, H., Einfluss des Formänderungsweges auf die Grenzformänderungen des Feinblechs. *Arch. Eisenhüttenw.*, **46** (1975) 597–602.
57. Kleemola, H. J. & Pelkkikangas, M. T., Effect of predeformation and strain path on the forming limits of steel, copper and brass. *Sheet Metal Ind.*, **54** (1977) 591–9.
58. Rasmussen, S. N., Assessing the influence of strain path on sheet metal forming limits. In *Proceedings of 12th Congress, IDDRG, Genova, 1982.* Associazione Italiana di Metallurgia, pp. 83–93.
59. Barata da Rocha, A. & Jalinier, J. M., Plastic instability of sheet metals under simple and complex strain paths. *Trans. Iron Steel Inst. Japan,* **24** (1984) 132.
60. Arrieux, R., Bedrin, C. & Boivin, M., Application d'un critère intrinsèque d'emboutissabilité à différents matériaux pour emboutissage. *Mém. Sci. Rev. Métall.* (1983) 685–92.
61. Triantafyllidis, N., The localization of deformation in finitely strained shells. In *Proceedings of Considere Memorial on Plastic Instability, Paris,* ed. J. Salençon. Presses Ponts et Chaussées, Paris, 1985, pp. 115–24.
62. Chen, P. & Gerdeen, J. C., Bending effects on forming limit diagrams for anisotropic sheets. In *Forming Limit Diagrams: Concepts, Methods and Applications,* ed. R. H. Wagoner, K. S. Chan & S. P. Keeler. TMS, Warrendale, PA, 1989, pp. 239–52.

63. Cordebois, J. P., Pierre, P. & Quagebeur, P., Formability of sheet material and geometrical aspects. In *Forming Limit Diagrams: Concepts, Methods and Applications,* ed. R. H. Wagoner, K. S. Chan & S. P. Keeler. TMS, Warrendale, PA, 1989, pp. 253–71.
64. Levaillant, C. & Chenot, J. L., Physical modelling and numerical prediction of defects in sheet metal forming. Presented at 2nd International Conference on Material Processing Defects, 1–3 July 1992, Siegburg.
65. Johnson, W. & Duncan, J. L., The use of the biaxial test extensometer. *Sheet Metal Ind.,* **42** (1965) 271–5.
66. Iseki, H., Murota, T. & Jimma, T., Finite element method in the analysis of the hydrostatic bulging of a sheet metal. *Bull. JSME,* **20** (1977) 285–91.
67. Zeghloul, A. & Ferron, G., ABAQUS analysis of the forming limits of metal sheets. In *Proceedings of ABAQUS Users' Conference,* Hibbitt, Karlsson & Sorensen, Inc., 1992, pp. 571–80.
68. Zeghloul, A., & Ferron, G., Limit strain predictions for out-of-plane stretching of sheet-metals. *Int. J. Plasticity* (in press).
69. Hill, R., A theory of the plastic bulging of a metal diaphragm by lateral pressure. *Phil. Mag.,* **41** (1950) 1133–42.
70. Hill, R. & Storakers, B., Plasticity and creep of pressurized membranes: a new look at the small-deflection theory. *J. Mech. Phys. Solids,* **28** (1980) 27–48.
71. Neale, K. W. & Chater, E., General solutions for the inelastic behaviour of pressurized membranes. *J. Appl. Mech.,* **54** (1987) 269–74.
72. Keeler, S. P. & Backofen, W. A., Plastic instability and fracture in sheets stretched over rigid punches. *Trans. ASM,* **56** (1963) 25–48.
73. Ghosh, A. K. & Hecker, S. S., Failure in thin sheets stretched over rigid punches. *Metall. Trans.* **6A** (1975) 1065–74.
74. Knibloe, J. R. & Wagoner, R. H., Experimental investigation and finite element modeling of hemispherically stretched steel sheet. *Metall. Trans.,* **20A** (1989) 1509–21.
75. Wang, N. M. & Budiansky, B., Analysis of sheet metal stamping by a finite-element method. *J. Appl. Mech.,* **45** (1978) 73–82.
76. Cockcroft, M. G. & Latham, D. J., Ductility and the workability of metals. *J. Inst. Metals,* **96** (1968) 33–9.

Chapter 5

IMPACT ON METAL TUBES: INDENTATION AND PERFORATION

W. J. STRONGE

Department of Engineering, University of Cambridge, Cambridge CB2 1PZ, UK

ABSTRACT

Impact by a compact missile can perforate a thin-walled tube as the result of a fracture process that depends on stretching, flexure and shear in the dented region around the impact point. As impact speed increases, inertia causes a decrease in the extent of the dented region, so flexure becomes more important than stretching in determining the minimum impact speed for perforation (ballistic limit) of any particular missile. The ballistic limit speed is also influenced by the nose shape of the missile, i.e. whether the nose is sharp and piercing, flat-faced and punching, or rounded and drawing out the surrounding indented material. These deformation features combine to finally give a biaxial strain criterion for initiation of fracture around the contact area. Fracture terminates the indentation process—a process that culminates in perforation of the tube wall by the missile. This article describes how dynamic deformation and subsequent perforation of thin-walled cylinders by blunt missiles depends on impact and structural parameters.

1 INTRODUCTION

Thin-walled metal tubes are commonly used for pipework that conveys fluids. Safety of pipework can depend on the ability of the tubes to withstand occasional accidental collisions. The impact threat is most serious in the case of free-flying missiles generated by explosive disintegration of nearby high-speed machinery. High-speed missiles generated in such an accident are often small in comparison with the diameter of the pipe; these missiles collide against pipework at speeds of as much as several hundred meters per second, weakening and possibly puncturing the thin-walled tubes. If the pipe is pressurized or it contains volatile fluids, an impact that punctures the pipe can initiate explosive disintegration, which propagates the sources of damaging missiles.

Small missiles damage thin tubes by local indentation of the region surrounding the impact point; if the tube is not punctured, internal pressurization may cause reverse bending that reduces the apparent indentation while further weakening the structure. The mode of perforation for a tube struck by a colliding missile ultimately depends on the size, shape and impact energy of the colliding missile in addition to the structural properties of the tube. For tubes with wall thickness that is appreciable in comparison with the diameter of the missile, rupture is the culmination of a perforation process involving local penetration of the missile into the wall and global denting of the surrounding region. Alternatively, if the tube is thin, a blunt missile causes significant denting and involves stretching of the deformed region; ultimately this results in rupture due to necking of stretched material around the periphery of the contact patch. In either case the deformation produced by high-speed impact is more localized than that which occurs owing to static loads.

2 STATIC INDENTATION BY BROAD WEDGES

In contrast to the localized indentation that results from high-speed impact of small missiles, a colliding body that has a diameter of the same order as the tube diameter produces indentation more like that from a wedge pressed across the tube. As a wedge presses into the tube, a lengthening contact line grows across the tube diameter; on both sides of the contact line, cross sections of the tube are ovalized within a distance of several tube diameters from the wedge. On either side of this line, a flattened region of triangular shape increases in size with increasing indentation. This simple mechanism of deformation was suggested originally by Pippard and Chitty.[1] This mechanism gives reasonable estimates of the force to produce any wedge indentation, as discussed in a review by Lukasiewicz.[2] The mechanism is initially attractive, because, as indentation increases, the increasing size of the flattened region develops as a consequence of almost solely flexural deformation. Reid and Goudie[3] pointed out, however, that there is a 'knuckle' at the periphery of the flattened region, and material entering this region is first sharply bent and later straightened by circumferential (hoop) tension in the flattened region; i.e. the mechanism is mostly but not entirely inextensional. Nevertheless, the effect of this tension on the fully plastic bending moment M_0 in the knuckle is assumed to be negligible. In relation to the effects of axial constraints (end conditions) or prestress, Reid and Goudie[3] and also Wierzbicki and Suh[4] consider the effect of an initial axial stress on the load carrying capacity of an indented cylindrical shell; i.e. they incorporate some effects of large deflections that cause stretching along the generators of the shell. Also, experiments on reduction of strength due to indentation by a wedge have been analyzed by Pacheco and Durkin[5] using a finite element structural model. Their paper contains some valuable information about the distribution of axial stress and circumferential (hoop) bending moment on tube cross sections near the indenting wedge.

The analyses outlined in this section have successfully employed relatively simple kinematic mechanisms to explain quasistatic transverse crushing of thin-walled tubes. These largely inextensional mechanisms of deformation result from contact forces generated by bodies that are broad enough to produce contact over a large part of the tube diameter. While these mechanisms are representative of tube deformation by large bodies, they do not represent the indented region that develops around a small colliding missile; the small missile produces localized indentation involving significant extensional deformation. This indentation and consequent perforation are the principal subjects of the present chapter.

3 STATIC INDENTATION BY SMALL PUNCHES

A small punch gives localized contact within only a short part of the tube circumference. Static deformation by a small punch results in only localized indentation and insignificant ovalization of sections outside the immediate contact region. The dented region is elliptical in plan view, with the major axis along a generator of the tube; the size of the dented region increases somewhat with indentation by the punch. Figure 1 compares the static indentation with the more localized deformation due to impact for two different thicknesses of thin-walled, welded steel tubing that had been annealed prior to indentation.

Measurements of denting in thin-walled aluminium alloy tubes with circular cylindrical cross section were obtained by Morris.[6] These tests were done on thin-walled tubing with a ratio of tube diameter D to thickness h of $D/h = 105$. In Fig. 2 the indentations for two different end constraints are compared; Fig. 2(a) is the indentation of a cylinder with free ends and length-to-diameter ratio $L/D = 3{\cdot}4$. For the same

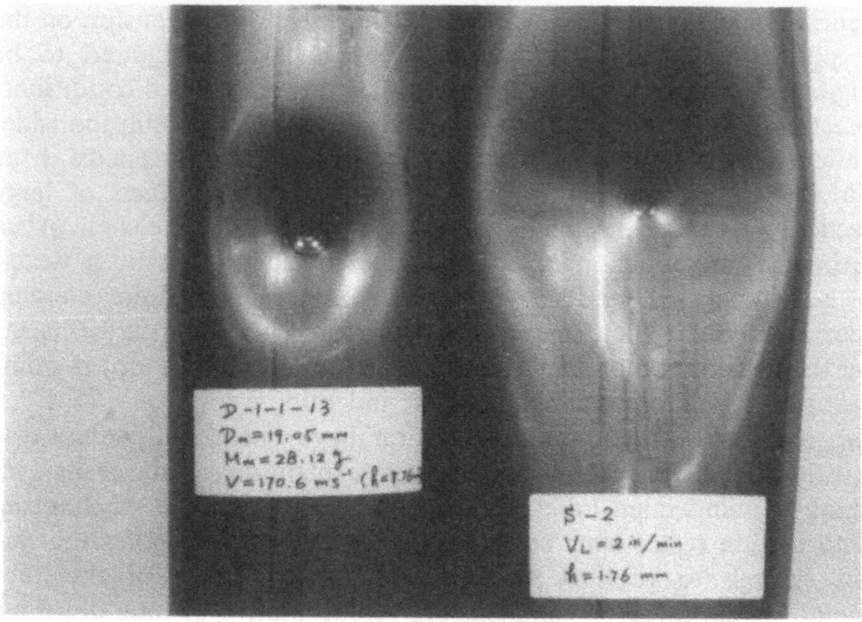

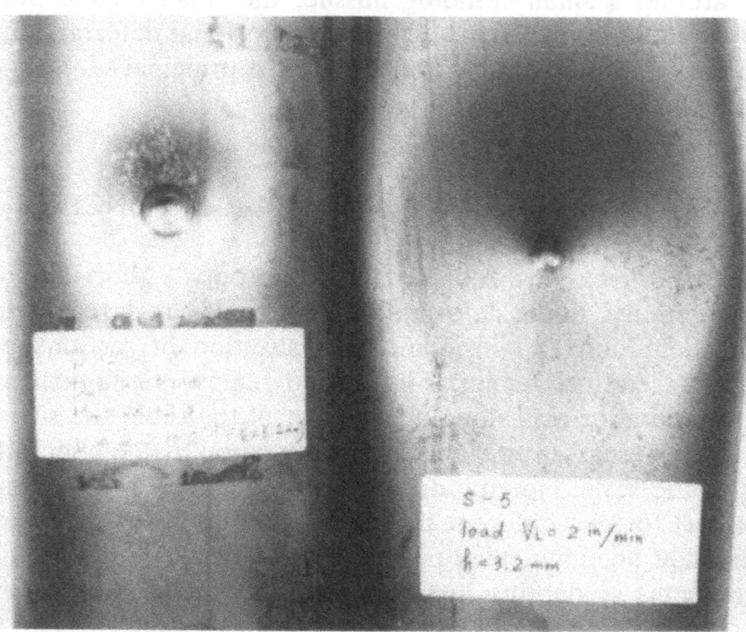

Fig. 1. Static (right) and dynamic (left) indentation of 102 mm diameter annealed mild steel tubes with thickness 1·7 mm (top) and 3·2 mm (bottom) pressed by a 19 mm spherical punch. The impact velocity for the left-hand specimens was roughly 80% of the speed required for perforation.

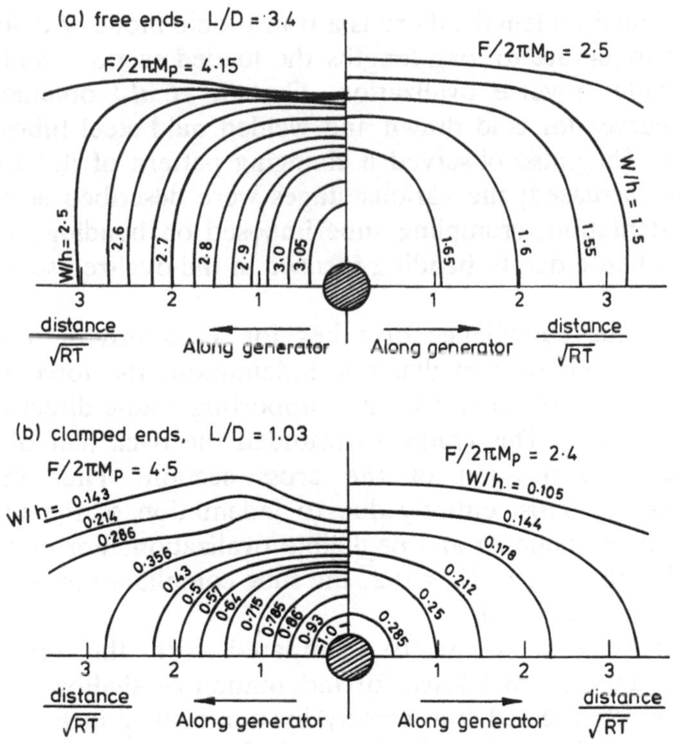

Fig. 2. Contours of constant deflection for statically indented tubes at two levels of punch force F/F_c. Tube (a) has free ends and $L/D = 3{\cdot}4$, while tube (b) has clamped ends and $L/D = 1{\cdot}03$.

punch, Fig. 2(b) shows the indentation of a cylinder with $L/D = 1{\cdot}03$; this cylinder has end restraints against both axial and normal displacement of the surface. At the same indentation force, the punch indentation with fixed ends is a small fraction ($\frac{1}{6}$–$\frac{1}{4}$) of the deflections of the free-ended cylinder. So the degree of end fixity (or length of free-ended specimens) has a major influence on the structural stiffness and extent of deformation. Indentation of rolled and welded mild steel tubes with free ends has shown that for $D/h = 31$ and a punch diameter ratio $d/D = 0{\cdot}125$ the tube compliance decreases with increasing length of tube up to $L/D = 6$.[7] The end constraints significantly affect local indentation if the punch is only a few tube diameters from the end.

Load–deflection curves for indentation of thin-walled cylindrical shells also show the effects due to overall bending along the length of the tube. Thomas, Reid and Johnson[8] performed experiments on simply supported aluminium and mild steel tubular beams with $L/D < 7$ and $40 \leqslant D/h < 190$. They observed three types of behaviour at moderately large deflections: (i) at short lengths the tube deforms as a ring; (ii) at medium lengths there is a transitional mode of deformation; and (iii) at moderate to long lengths the loaded section ovalizes while the ends suffer reverse ovalization. Corbett *et al.*[7] obtained force–deflection curves for cold drawn and welded mild steel tubes of three thicknesses. They also observed a changing pattern of deformation as indentation increased; the various stages were described as crumpling or local indentation, crumpling superimposed on bending, and finally structural collapse due to bending focused in the ovalized section under the punch.

To obtain the compliance of tubes for a pattern of deformation similar to that obtained in dynamic indentation, the force–deflection curves in Fig. 3 were obtained for a supporting cradle directly beneath the loaded section. This cradle surrounded the back half of the tube and restricted ovalization of the cross section. Thus the punch displacement is almost entirely due to indentation of the tube, since there is no beam bending and negligible ovalization. For punches with diameter d in the range $d/D < 0{\cdot}2$, the tube compliance is insensitive to the diameter of the punch.

These measurements can be compared with the upper bound estimate of static normal force for indentation of shallow rigid-plastic cylindrical shells pressed transversely by a circular punch. Morris and Calladine[9] extended the method they had used to analyze spherical shells; they represented the ellipsoidal indented region of a cylindrical shell by a small number of trapezoidal regions linked by generalized hinges. The analysis did not concern itself with stretching or flexure at these hinges, but instead used equilibrium of resultant forces with an appropriate distribution of fully plastic stresses. This method is useful if there is a reasonable estimate of stresses throughout the structure. The results of the analysis showed a strong dependence on the size of the punch, as shown in Fig. 4; these results agreed closely with experiments on short, thin-walled aluminium tubes ($L/D = 1$) with clamped ends. The punch forces obtained in the present experiments on long mild steel tubes are much more modest and do not exhibit this dependence on the size of the punch. The static collapse load F_c for the indentation of a cylinder is identical with that for a flat plate, $F_c = 2\pi M_0$.

The difference in initial load to initiate plastic deformation between these two sets of experiments is more due to differences in the shape of the punch face than to the end constraints. The present tests used flat-faced punches that initially contact the tube along a single generator, whereas the tests of Morris[6] used a concave punch face that conformed to the tube over the full cross-sectional area of the punch. For conforming punches pressing on short cylinders with negligible

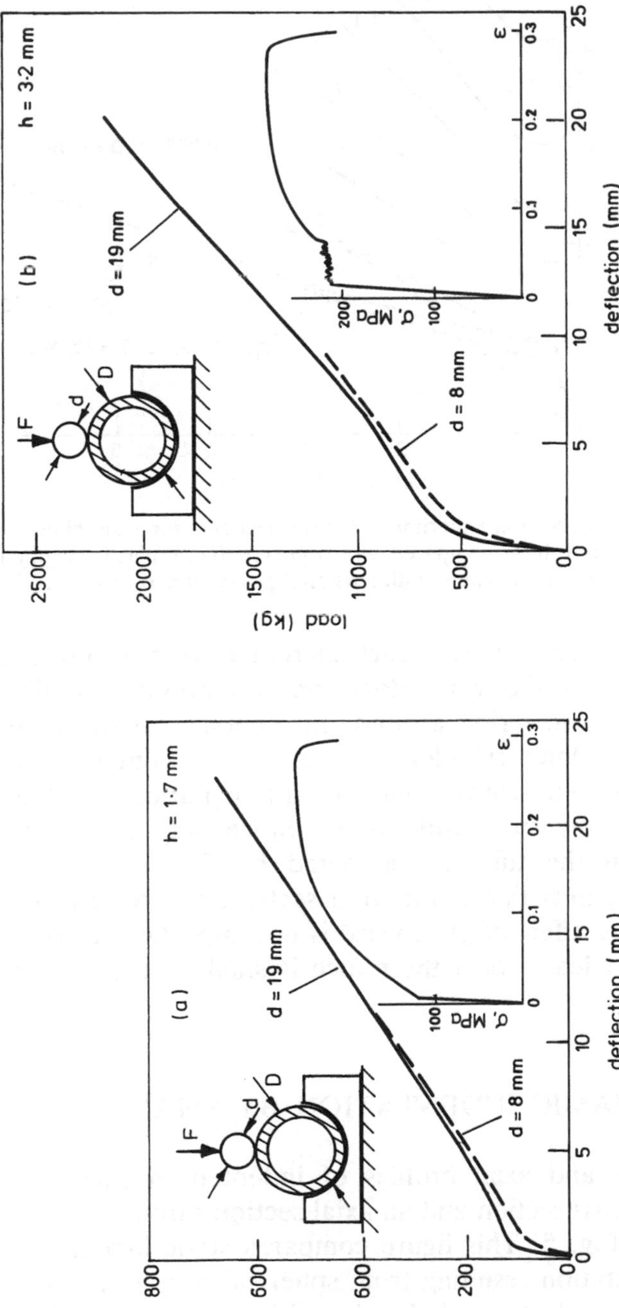

Fig. 3. Load–deflection and stress–strain curves for annealed mild steel tubes: (a) $h = 1{\cdot}7$ mm; (b) $h = 3{\cdot}2$ mm.

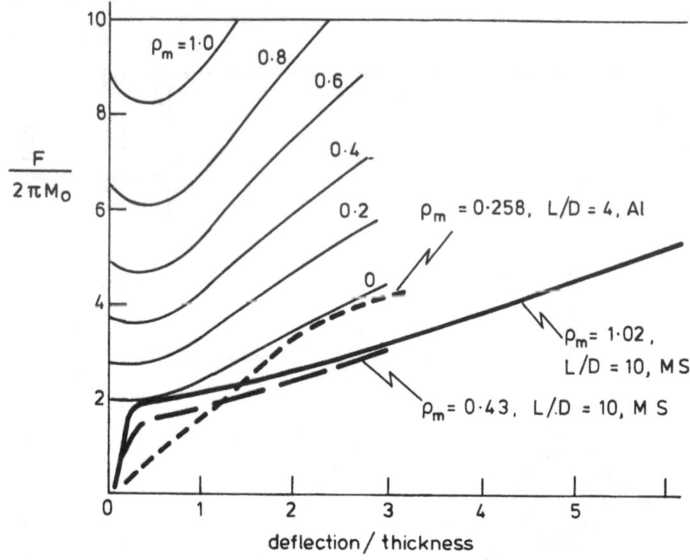

Fig. 4. Static load–deflection experimental measurements for aluminium and annealed mild steel tubes (bold lines) compared with rigid–perfectly plastic theory (light lines) for shallow shells. The nondimensional punch size $\rho_m = d/\sqrt{2DH}$.

deflection at the ends, during each increment of punch displacement a significant part of the work done on the cylinder is dissipated by bending in a narrow ring around the contact patch. In the present experiments on long cylinders, however, the contact patch is continually increasing in size with increasing indentation, and the deformed region is generally more widespread. Since only a small part of the work done on the tube is dissipated by deformation immediately adjacent to a punch that is far from sections where displacements are constrained, the effect of punch size is not important for compliance of these tubes—at least not if the punch is small in comparison with the tube diameter.

4 DYNAMIC INDENTATION BY SMALL MISSILES

Cross-sectional and axial profiles of indented thin-walled mild steel tubes on the cross section and an axial section through the impact point are shown in Fig. 5. This figure compares static indentation with the dynamic indentation resulting from spherical missile impact at a speed just below the *ballistic limit* V_*; i.e. this is the largest impact speed where this missile does not rupture the tube. For any particular missile colliding at a speed $V_0 < V_*$, the extent of indentation increases with V_0.

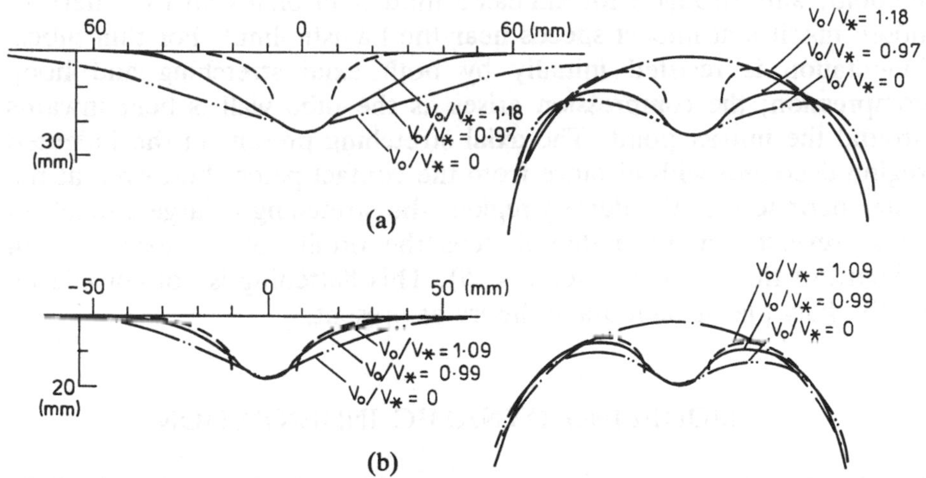

Fig. 5. Comparison of static and dynamic denting of 102 mm diameter annealed mild steel tubes with thickness 1·7 mm (a) and 3·2 mm (b). Dynamic denting resulted from impact of a sphere with diameter $d = 19$ mm and mass $G = 28{\cdot}1$ g striking at normal incidence with impact speed V_0.

Indented profiles for a somewhat larger impact speed are also shown. If the impact speed is larger than the ballistic limit $V_0 > V_*$, the extent of indentation decreases in both the hoop and axial directions as V_0 increases. Hence the size of the indented region is largest at the ballistic limit. The figure shows post-impact profiles for thickness ratios $h/D =$ 0·017 and 0·032; the thickness ratio influences both the extent of the deformed region and the change in radial curvature around the periphery of the contact patch. For the thicker tube, changes in shape of the final profile with relative impact speed are similar to those predicted by the fully plastic membrane theory of Wierzbicki and Hoo Fatt.[10]

There have been only a few transient measurements of the development of deformation in impacted tubes or pipes. Neilson, Howe and Garton[11] tested mild steel seamless pipe of diameter $D = 150$ mm and wall thickness $h = 7$ mm hit by a missile of diameter $d = 60$ mm. They placed strain gauges at a distance $r/d = 1{\cdot}2$ from the impact point on the generator passing through this point. At an impact speed $V_0 =$ 114 m s^{-1}, just below the ballistic limit, they recorded a maximum axial strain (tensile) $\varepsilon_z = 11{\cdot}4\%$ and a maximum hoop strain (compressive) $\varepsilon_\theta = -9{\cdot}4\%$. For a similar test at almost the same impact speed where the missile perforated the wall, the maximum axial strain at the same location was $\varepsilon_z = 4{\cdot}2\%$. In this case the hoop strain was first compressive and later tensile. Radial strains of 10–15% were also measured by

Palomby and Stronge[12] for annealed mild steel plates hit by spherical-nosed missiles at impact speeds near the ballistic limit. For thin tubes, indentation is resisted initially by both axial stretching and hoop compression; the compression arises as the tube wall is bent inwards around the impact point. The axial stretching present in the indented region decreases with distance from the contact patch, but, even at the outer periphery of the dented region, the stretching is large enough to cause reverse curvature that flattens the profile of the axial section inboard of this 'knuckle' (see Fig. 5). This flattening is not considered by the fully plastic membrane theory at present.

5 MODELING DYNAMIC INDENTATION

Localized indentation or denting of thin-walled cylindrical shells involves axial stretching and both axial and circumferential bending of the shell surface. Analytical solutions for these nonsymmetric plastic deformations are usually based on either an upper bound method using a kinematic approximation for the deformation field or a dynamic analysis that neglects the 'less significant' stress resultants. The accuracy of the former method depends on the approximation used for the velocity field; the estimate is accurate if the actual stresses in the assumed plastically deforming regions are close to yield for the entire deformation period. Palomby and Stronge[13] developed an algorithm for extending the kinematic approximation based on modes to include geometric effects of large deflections. The *evolutionary mode method* is an iterative procedure that finds the evolution of the mode shape as a function of deflection for any initial loading on a structure.

Alternatively, consideration of the most significant stress resultants leads to an analogy between radial deflections of the cylindrical shell and transverse deflections of a beam; for a perfectly plastic material, the radial deflections of the shell are analogous to transverse deflections of a rigid–plastic beam on a rigid–plastic foundation. For axisymmetric deformations of thin cylinders, the circumferential normal stress resultant is directly analogous to the foundation pressure. Calladine[14] extended this analogy to deal with localized nonsymmetric deformations in a long elastic cylinder. In this case the dominant forces are axial and hoop bending; in the present analog they take the roles of beam bending moment and foundation pressure respectively. This modeling presumes that the ends are sufficiently distant so as not to affect stresses in the dented region; moreover, the shell is thin ($h \ll D$), so circumferential (hoop) stretching is negligible.

The beam-on-foundation model is based on the following assumptions.

(i) A uniform rigid–perfectly plastic beam with fully plastic bending moment M_0 rests on a rigid-plastic foundation that provides a reaction q_0 per unit length opposing transverse motion.
(ii) Beam deformations are concentrated in two plastic hinges; one at the edge of the contact region with the colliding missile and the other travelling away from the contact region.
(iii) Only transverse displacements are considered; i.e. axial velocities and accelerations are assumed to be negligible while axial force is uniform.
(iv) The shell has increasing stiffness due to axial tension in the indented region; this large deflection effect is brought in by a membrane factor f_n that increases the fully plastic moment.

The beam has mass per unit length ρ and fully plastic bending moment M_0 that are representative of cross-sectional properties for the axial strip of shell passing through the contact patch; the foundation reaction q_0 is obtained from the transverse force per unit length due to hoop bending required to ovalize the tube in a uniform ring mode of deformation.

We consider a long rigid–plastic beam on plastic foundation that is struck by a rigid missile of mass G and width h; the model is illustrated in Fig. 6. The development of indentation in response to this impact can be represented by a transient phase of motion for the beam; during this phase, rotating rigid segments of the beam connect the central decelerating segment under the colliding missile with the undeforming outer segments of the long beam. The length of the rotating rigid segment is continuously increasing until motion stops. If the velocity of the missile and contact segment are $V(t)$, the beam has a transverse velocity field $\dot{W}(x, t)$, where

$$\dot{W} = \begin{cases} V & (-b \leqslant x \leqslant 0) \\ (1 - x/\lambda)V & (0 \leqslant x \leqslant \lambda) \\ 0 & (\lambda \leqslant x) \end{cases} \tag{1}$$

The corresponding accelerations can be obtained by differentiation:

$$\dot{W} = \begin{cases} \dot{V} & (-b \leqslant x \leqslant 0) \\ (1 - x/\lambda)\dot{V} + x\dot{\lambda}V\lambda^2 & (0 \leqslant x \leqslant \lambda) \\ 0 & (\lambda \leqslant x) \end{cases} \tag{2}$$

Beyond the travelling hinge $x > \lambda$ the shear force vanishes. Hence for

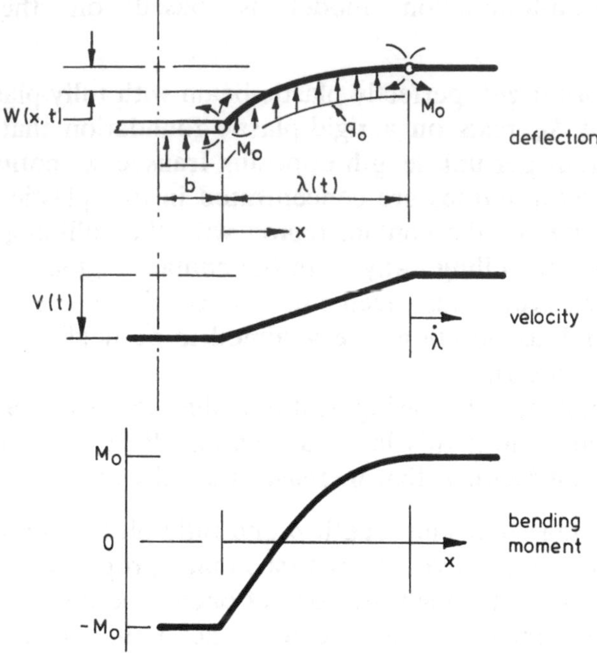

Fig. 6. Idealized model of rigid–perfectly plastic beam on plastic foundation: deflection, velocity field and bending moment distribution.

half the beam the rate of change of translational momentum and the moment of moment about the origin $x = 0$ give the equations of motion:

$$(\rho b + \tfrac{1}{2}G)\dot{V} + \int_0^{\lambda} \rho \ddot{W}\, \mathrm{d}x = -(b + \lambda)q_0 \tag{3a}$$

$$\int_0^{\lambda} \rho \ddot{W} x\, \mathrm{d}x = 2M_0 - \tfrac{1}{2}\lambda^2 q_0 \tag{3b}$$

With the acceleration field (2), these equations of motion can be integrated to obtain two first-order equations in V and λ:

$$\begin{aligned} (G + 2\rho b + \rho\lambda)\dot{V} + \rho\dot{\lambda}V &= -2(b + \lambda)q_0 \\ \rho\lambda^2\dot{V} + 2\rho\lambda\dot{\lambda}V &= 12M_0 - 3\lambda^2 q_0 \end{aligned} \tag{4}$$

This pair of first-order equations has been solved numerically using Runge–Kutta quadrature.[15]

5.1 Effects of Large Deflection

As deflections of a cylindrical shell become large in comparison with the shell thickness h, forces due to stretching in the indented region add substantially to the stiffness of the shell. The effect of this additional stiffness can be incorporated into the analysis by assuming that the increase in length due to increased inclination of the rigid segment between two plastic hinges on either side of the colliding missile develops solely by stretching at the hinges. This gives a rate of elongation $\dot{e}$ that is related to the central deflection $W_0 = W(0, t)$ and the rate of increase in deflection $\dot{W}_0$ by

$$\dot{e} = W_0 \dot{W}_0 / \lambda \tag{5}$$

Noting that the rate of rotation at these same hinges is $\dot{\theta} = \dot{W}_0/\lambda$, we obtain a rate of energy dissipation $\frac{1}{2}\dot{\Gamma}$ caused by bending and stretching in each half of the beam:

$$\tfrac{1}{2}\dot{\Gamma} = 2M\dot{\theta} + N\dot{e}$$

where the elongation force N and bending moment M at a fully plastic hinge are related by the yield condition

$$\left|\frac{M}{M_0}\right| + \left(\frac{N}{N_0}\right)^2 = 1 \tag{6}$$

Here the fully plastic bending moment M_0 and fully plastic axial force N_0 depend on the uniaxial yield stress Y. For a rectangular cross section of width b and depth h, they can be expressed as

$$M_0 = \tfrac{1}{4} Ybh^2, \qquad N_0 = Ybh = \frac{4M_0}{h} \tag{7}$$

For the yield condition (6) there is an associated flow rule for plastic deformation:

$$\frac{\mathrm{d}M}{\mathrm{d}N} = -\frac{2NM_0}{N_0^2} = -\frac{\dot{e}}{2\dot{\theta}} \tag{8}$$

from which it follows that $\dot{e} = hN\dot{\theta}/N_0 = hN\dot{W}_0/\lambda N_0$. Together with the elongation rate (5) this gives an axial force

$$N/N_0 = W_0/h$$

Consequently, the yield condition (6) gives a bending moment M at the deforming hinges:

$$\left|\frac{M}{M_0}\right| = 1 - \left(\frac{W_0}{h}\right)^2$$

Note that if $W_0 > h$, the conditions above give a corner for the yield surface where $M = 0$ and $N = N_0$; hence the dissipation rate for the entire beam can be obtained as

$$\dot{\Gamma} = \begin{cases} 4M_0\left[1+\left(\dfrac{W_0}{h}\right)^2\right]\dfrac{\dot{W}_0}{\lambda} & (W_0 \leq h) \\ \dfrac{8M_0W_0\dot{W}_0}{h\lambda} & (W_0 > h) \end{cases} \tag{9}$$

This can be compared with the dissipation rate for the centrally loaded beam with negligible stretching, $\dot{\Gamma} = 4M_0\dot{W}_0/\lambda$. Hence the effect of axial force at large deflections is represented by the product of the fully plastic bending moment M_0 and a membrane factor f_n, where

$$f_n = \begin{cases} 1 + W_0^2/h^2 & (W_0 \leq h) \\ 2W_0/h & (W_0 > h) \end{cases} \tag{10}$$

This factor is incorporated into the previous beam-on-foundation model by multiplying the fully plastic bending moment M_0 by f_n in eqn (3b). A similar large-deflection factor can be appended to the hoop flexural stiffness q_0.[16]

Indentation of thin-walled tubes predicted by this beam-on-foundation model has been compared with experimental results for two thicknesses of mild steel tube by Palomby[17] (Fig. 7). These measurements were made in drop hammer tests where a missile of mass $G = 12{\cdot}5$ kg and diameter $d = 12{\cdot}7$ mm hit the tube at an impact speed

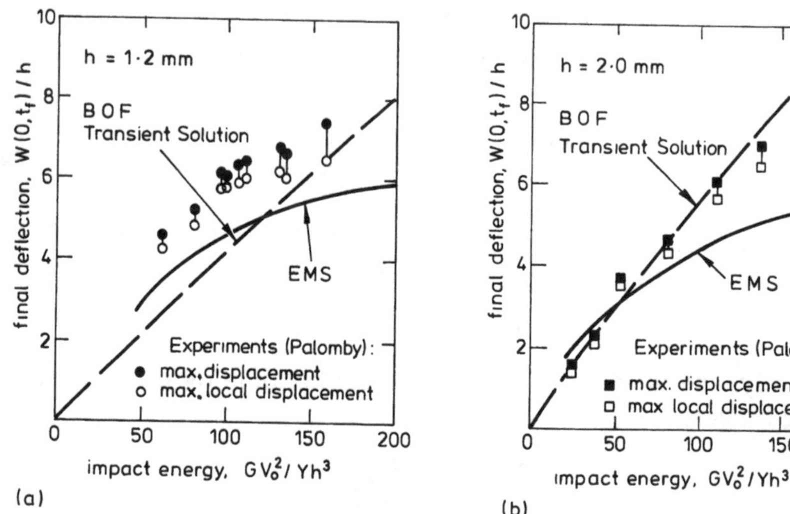

Fig. 7. Comparison of final centre deflection calculated with beam-on-foundation (BOF) and evolutionary mode (EMS) methods and the experimental measurements of denting in low-speed collisions ($V_0 < 6$ m s^{-1}) made by Palomby.[17]

in the range $2\,\mathrm{m\,s^{-1}} < V_0 < 6\,\mathrm{m\,s^{-1}}$. Tubes of two different wall thicknesses were tested; these thicknesses were 1·2 and 2·0 mm. The uniaxial flow stress of the cold-drawn, seamless tubes at approximately 2% strain was measured as 535 and 320 MPa respectively. All tubes had a nominal diameter of 50·7 mm. At each impact energy $\frac{1}{2}GV_0^2$ in the figure there are two experimental points: one for the global or overall deformation and a smaller value for the local indentation.[17] The measured final deflection under the colliding missile is compared with the results of calculations that used either the large-deflection beam-on-foundation model or the evolutionary mode method. These calculations assumed an effective beam width $b = \sqrt{\frac{1}{2}hD}$. From this comparison, it seems that the flow stress employed for the thin tubes (1·2 mm) was too large. Furthermore, when the deflection becomes as large as $W_0/h \geq 8$, the beam-on-foundation model underestimates the stiffening effect of axial tension.

The experimental results in Fig. 7 were for low impact speeds, where inertia of the shell wall was insignificant in comparison with that of the missile. For lighter missiles colliding at larger impact speeds, a fully plastic membrane model also has merit. The membrane or plastic string-on-foundation model neglects the initial phase of axial flexural stiffness and assumes that the strip of cell wall passing through the contact patch has an axial stress equal to the fully plastic limit at all times. This gives a generalized plastic 'hinge' where inclination of the 'string' is discontinuous; the hinge travels away from the impact site at a characteristic speed $c = \sqrt{\sigma_0/\rho}$. This contrasts with the beam-on-foundation model, where discontinuities in rate of rotation travel outward at a speed that varies. At large deflections these two models have the same stress field, but they adopt different velocity approximations. The propagating discontinuity in inclination of the string-on-foundation model is directly proportional at every instant to a discontinuity in transverse velocity at the hinge. Hoo Fatt and Wierzbicki[18] obtain a solution for this model that is valid in an intermediate range of impact speeds by assuming uniform velocity between the hinges. For the tests shown in Fig. 5, this approximation predicts too large and extensive an indentation. More representative deformation parameters are currently being developed for this model.

6 PERFORATION OF THIN TUBES

6.1 Deformation and Fracture Development

Indentation around blunt-nosed missiles causes axial tension and hoop compression in the indented region for central deflections, $W_0/h > 4$.

The hoop stress resultant is initially compressive as the indented cross section is ovalized, but at deflections larger than $W_0/h > 4$ this stress resultant can also become tensile after the width of the indented region stops increasing. It is difficult to obtain measurements of the deformation in the indented region, however, and only a few have been reported.

Schwer *et al.*[19] indented very thin-walled 6061-T6 aluminum tubes by means of explosive loading over a small spot. The unpressurized tubes ruptured initially owing to localized necking that ran circumferentially through the loaded region. An elastic–plastic finite element analysis of the dented region using DYNA3D and eight-node brick elements indicated that at failure the axial strains in the loaded region were 8·5–12·5%. They found that in a biaxial strain state the aluminum necked when the sum of the hoop and axial strains reached approximately 11–14%. This ultimate strain criterion may depend on the particular loading path followed in the experiments.[12]

Missile impact tests on seamless, cold-drawn mild steel pipes by Neilson *et al.*[11] gave measurements of a peak axial strain of 11·4% and peak hoop strain −9·7%; these measurements were made somewhat outside the contact patch but on the generator passing through the impact point. A DYNA3D analysis of a collision with impact speed $V_0 = 114\ \mathrm{m\,s^{-1}}$ that is very close to the ballistic limit gave peak plastic strains greater than 20% on the inside of the pipe in the hoop direction at the periphery of the blunt-nosed missile[24]. The location and orientation of these largest normal strains is directly correlated with first fractures that were observed at necks on the periphery of the contact region.

Photographs of axial sections of 3·2 and 1·7 mm thick annealed mild steel tubes hit by spherical steel balls are shown in Fig. 8. The impact speeds were 229 and $185\ \mathrm{m\,s^{-1}}$ respectively; in both cases this is very close to the ballistic limit. These impacts resulted in centre deflections of 20 and 24 mm for thick and thin tubes respectively. The photographs show cracks around the perimeter of the contact patch in a region undergoing a final stage of neck formation. In the necked region a fracture that initiated on the distal surface is propagating through the tube to the impact surface.

Middle surface strains in the dented region can be estimated from the change in thickness. Figures 9(a,b) show final strain normal to the surface in the same test specimens shown in Fig. 8. Both the hoop and axial strains show a concentrated peak in strain at the periphery of the contact region, where necking is well advanced. For these impacts of blunt missiles ($d/D = 0{\cdot}19$) on thin tubes ($h/D = 0{\cdot}017$ or $0{\cdot}032$), the

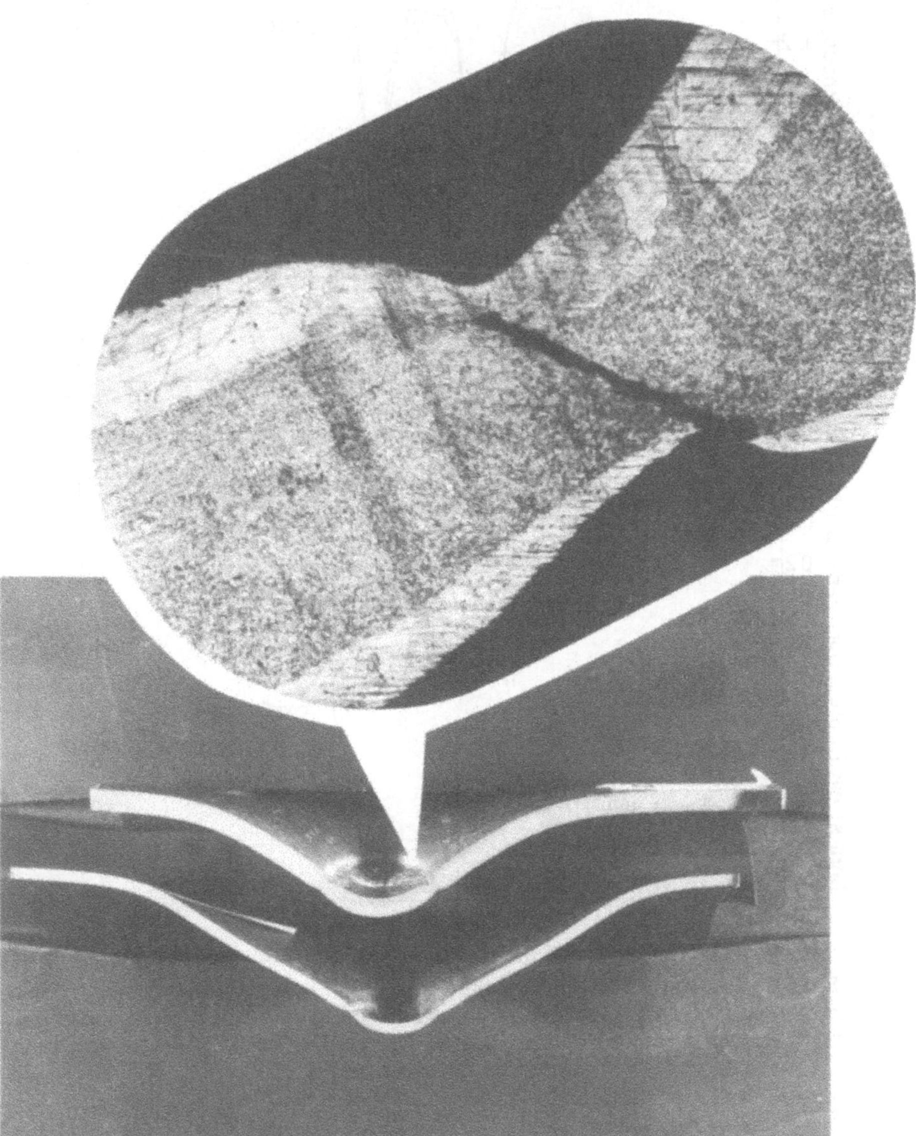

Fig. 8. Axial sections of annealed mild steel tubes with thickness $h = 3{\cdot}2$ and $1{\cdot}7$ mm hit by a spherical missile with diameter $D = 19$ mm and mass $G = 28{\cdot}1$ g. The missile hits normally at an impact speed just below the minimum speed for perforation. A magnified picture of the necked region where cracks are developing is shown in the auxiliary photograph.

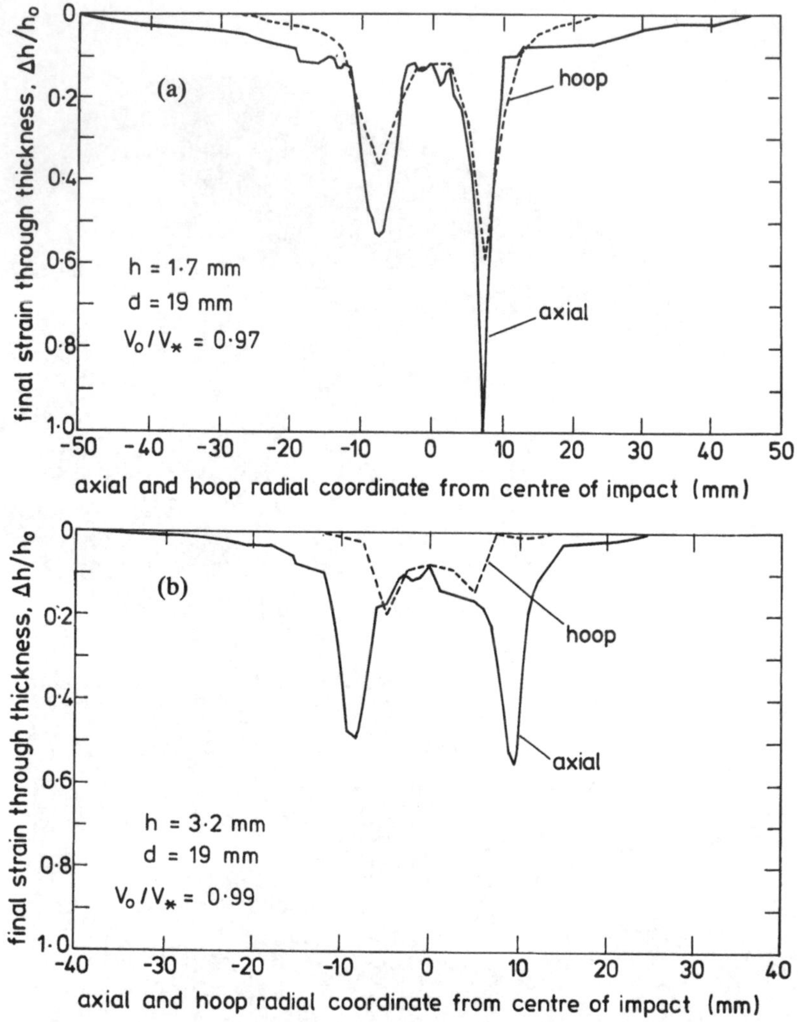

Fig. 9. Final engineering strain through thickness of 1·7 mm (a) and 3·2 mm (b) thick tubes hit by a spherical-nosed missile at an impact speed just below the ballistic limit V_*. Necking is well advanced at the edge of the contact area, while friction limits stretching at the centre.

largest strain is in the axial direction—this direction is normal to the initial fracture surfaces that form around the edge of the contact region. Necking develops in this region where the through thickness strain is 10–15%. Palomby and Stronge[12] also reported in-plane strains of 10–15% where necking initiated in an annealed mild steel plate struck by a blunt missile at the ballistic limit impact speed. Their results brought forth the biaxial nature of the necking failure criterion. They showed that a large intermediate principal strain can reduce the strain at necking by as much as one-third; in this case the intermediate principal strain is in the direction of the circumference of the contact patch.

6.2 Ballistic Limit for Round-Nosed Missiles

The ballistic limit speed V_* is the minimum impact speed where there is a 50% probability of perforation; for a blunt missile hitting any particular tube, this speed is practically identical with the speed where a crack penetrates through the full thickness of the wall. Since perforation follows from a fracture process, the ballistic limit depends ultimately on the development of deformation around the contact region.

For risk analysis or other vulnerability calculations, it is helpful to have formulae for the ballistic limit in terms of structural and impact parameters. For this purpose there are several empirical equations that have been suggested; Stronge[20] compared both historical and modern expressions for V_* and, better yet, the missile kinetic energy at the ballistic limit speed $K_* = \frac{1}{2}GV_*^2$ with data on perforation of tubes by spherical missiles. On this basis, he suggested the following empirical expression for the kinetic energy at the ballistic limit $K_*(J)$ of spherical-nosed missiles with diameters in the range $d = \{6{\cdot}35\text{ mm}, 12{\cdot}7\text{ mm}\}$ perforating cold-drawn mild steel tubes with wall thickness in the range $h = \{1{\cdot}2\text{ mm}, 3{\cdot}2\text{ mm}\}$:

$$K_* = Ch^{1{\cdot}63}d^{1{\cdot}48} \tag{11}$$

where C is a dimensional constant with units of J mm$^{-3{\cdot}11}$. Subsequently, additional data giving satisfactory agreement with this expression for tubes with thickness $1{\cdot}66\text{ mm} < h < 5{\cdot}0\text{ mm}$ have been obtained by Corbett *et al.*[7] They have pointed out, however, that this expression does not apply to rolled and welded tubing, which is much more ductile than stock that is cold-drawn.

It is tempting to obtain a dimensionless expression for the kinetic energy $K_*/\sigma_y hd^2$ as a function of the effective thickness h/d and the size of the missile d/D. Here the aim is to eliminate the effect of material properties represented by the uniaxial yield stress σ_y. This attempt has not been successful in eliminating effects of material properties in materials of different ductility, however, since the patterns of deformation at the ballistic limit are not geometrically similar. To be successful, any such relationship must explicitly depend on an additional material parameter, but at present there is insufficient data to evaluate an improved empirical expression. In part, this is due to lack of a consistent definition for flow or yield stress in a material like mild steel that has substantial strain-hardening. This makes it impossible to directly compare measurements by various research groups of non-dimensional variables that depend on material properties.

6.3 Effect of Nose Shape on Ballistic Limit

The perforation process described in the preceding sections is representative of the response of thin-walled tubes to round-nosed or blunt missiles. These missiles collide against the tube at a single point, and the contact patch spreads across the face of the missile as the depth of indentation increases. Thus, early in the indentation process, the rounded nose shape reduces the curvature at the periphery of the contact region. Later, when the deflection of the contact region is large and the axial force becomes significant, this force induces a rate of stretching at the edge of contact. The stretching at any material point is limited, however, because the short segment that is stretching is continually moving outward from the initial impact point.

In contrast, a nondeforming flat-faced missile has a constant contact area during the perforation process. At the periphery of the contact area, this results in contact pressure that is large enough to induce localized penetration of the missile into the wall. Near the edge of the penetration zone, there is a band of large through-thickness shear strain. Furthermore, the edge of the contact patch does not move over the surface, so both stretching and bending remain concentrated in this same narrow band of material throughout the indentation process. Consequently, fractures develop around the contact patch of a flat-faced colliding missile at a smaller indentation or deflection of the impact point than occurs for similarly placed fractures in the case of more rounded nose profiles. These fractures can initiate from either the impact or distal surface of the tube if the missile is flat-faced.

Despite these differences in localized deformation around the contact area, the ballistic limit of a tube struck normally by a flat-faced missile is only a little less than it is for a missile with hemispherical nose shape. Shadbolt *et al.*[21] showed that, in fact, there is an intermediate nose curvature where the energy required to initiate fracture due to combined bending, stretching and shear is a maximum. This maximum separates fracture processes termed plugging from those described as discing. Plugging is a fracture that is predominately due to shear, while discing results from a tensile mode of failure. Figure 10 shows that the variation of the ballistic limit with nose radius is largest for more ductile materials such as annealed mild steel, but quite insignificant for

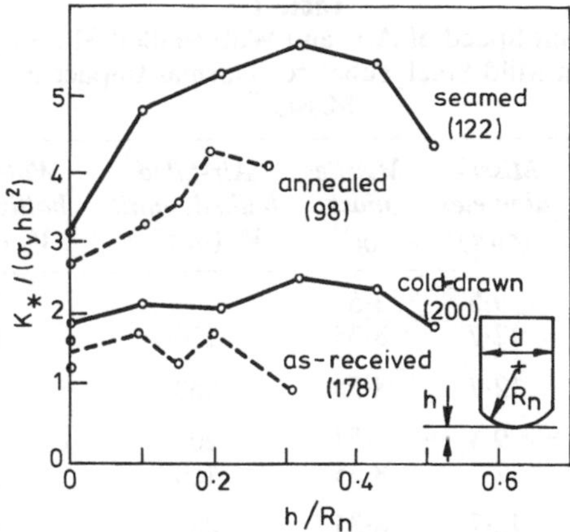

Fig. 10. Variation of missile kinetic energy at ballistic limit K_* with the missile nose radius R_n for both seamed and cold-drawn mild steel tubing. The solid lines are data of Corbett *et al.*,[7] while the dashed lines are data of Palomby and Stronge.[12] Numbers in parentheses indicate the Vickers hardness of the respective tubes.

cold-drawn mild steel tubes. Since the flow stress of the cold-drawn tubes is almost twice as large as that of rolled and welded tubing, the large difference in perforation energy as a result of heat treatment for these tubes is primarily due to the difference in ductility or ultimate strains.

6.4 Effect of Internal Fluid on Ballistic Limit

If the tube is filled with an incompressible liquid, this provides substantial 'added mass' that effectively stiffens the wall and resists denting. With an internal liquid, denting is less widespread and curvature around the contact patch is larger in magnitude when perforation occurs. Table 1 gives the ballistic limit speed for spherical missiles perforating cold-drawn mild steel tubes at a normal angle of obliquity.

Pressurization of liquid within a tube might also decrease the ballistic limit by prestressing the wall and thereby reducing the failure strain. Neilson *et al.*[11] obtained the ballistic limit of 150 mm diameter steel pipes with 7·2 mm wall thickness hit by 60 mm diameter flat-ended cylinders. For internal pressurization of a water-filled pipe that gave a circumferential wall prestress of as much as $\sigma_\theta = 0{\cdot}25\sigma_y$, they found no effect of pressurization. Nevertheless, the perforation energy was reduced from 30 kJ to approximately 22 kJ by additional inertia of the water. Schwer *et al.*[19] tested gas-filled thin-walled aluminium cylinders pressurized to roughly $\sigma_\theta = 0{\cdot}67\sigma_y$. For explosive loading over a small

Table 1
Ballistic Limit Speed of Air- and Water-Filled 51 mm Diameter Cold-Drawn Mild Steel Tubes for Normal Impact by Spherical Missile[22]

Tube thickness (mm)	*Missile diameter (mm)*	*Missile mass (g)*	*Air-filled ballistic limit* V_* *(m s⁻¹)*	*Water-filled ballistic limit* V_* *(m s⁻¹)*
1·2	9·5	3·53	—	230
1·2	12·7	8·35	150	190
1·8	12·7	8·35	160	240
2·1	9·5	3·53	200	285
2·1	12·7	8·35	160	250
3·3	12·7	8·35	285	—

spot on the surface, they obtained little effect of pressure on the impulse required to burst the shell, even though the pattern of deformation was changed. Overall, it seems that perforation or burst is much more strongly influenced by density of the filling fluid than any internal pressurization that may be present.

7 CONCLUSIONS

Perforation of thin-walled tubing due to impact by missiles at subordnance-range velocities is the culmination of a process of deformation. The pattern of deformation depends on the size of the colliding missile in comparison with the diameter and wall thickness of the tubing. The deformation pattern for large missiles ($d/D > 0{\cdot}5$) covers almost all of the tube diameter; in this case the rate of deformation with increasing indentation can be represented fairly accurately by a triangular flattened region on either side of the impact site. For small missiles ($d/D < 0{\cdot}25$), however, indentation is more localized and complex. The current beam and string-on-foundation analogies for impact by small missiles have too few generalized stress and strain variables to accurately represent plastic dissipation in the indented region. Further developments are needed to improve the representation of the localized deformation field and obtain equivalent mass and stiffness parameters that vary with longitudinal distance from the impact site.

Rupture due to impact of small missiles occurs during a final stage of deformation. Rupture is strongly influenced by details of deformation in the wall adjacent to the contact patch. These details depend on the ratio of wall thickness to missile diameter, h/d, the nose shape and material properties; they control whether rupture of thin-walled tubes develops due to plugging, discing or tensile tearing. At present a biaxial ultimate strain criterion has been successful in predicting rupture that develops underneath a colliding missile.[12,23] Use of this criterion is based on a forming limit diagram and an estimate for the radial and circumferential components of strain at the periphery of the contact patch; this is where fractures initiate that can grow into discing or plugging modes of perforation.

REFERENCES

1. Pippard, A. J. S. & Chitty, L., Experiments on the plastic failure of cylindrical shells. *Civ. Engng War,* **3** (1948) 2–29.
2. Lukasiewicz, S., Inelastic behaviour of shells under concentrated loads. In *Inelastic Behaviour of Plates and Shells,* ed. L. Bevilacqua, R. Feijoo & R. Valid. Springer-Verlag, Berlin, 1986, pp. 537–67.
3. Reid, S. R. & Goudie, K., Denting and bending of tubular beams under local loads. In *Structural Failure,* ed. T. Wierzbicki & N. Jones. Wiley, New York, 1989, pp. 362–3.
4. Wierzbicki, T. & Suh, M. S., Indentation of tubes under combined loading. *Int. J. Mech. Sci.,* **30** (1988) 229–48.
5. Pacheco, L. A. & Durkin, S., Denting and collapse of tubular members—a numerical and experimental study. *Int. J. Mech. Sci.,* **30** (1988) 317–331.
6. Morris, A. J., Experimental investigation into the effects of indenting a cylindrical shell by a load applied through a rigid boss. *J. Mech. Engng Sci.,* **13** (1971) 36–46.
7. Corbett, G. G., Reid, S. R. & Al-Hassani, S. T. S., Static and dynamic penetration of steel tubes by hemispherically nosed punches. *Int. J. Impact Engng,* **9** (1991) 165–90.
8. Thomas, S. G., Reid, S. R. & Johnson, W., Large deformations of thin walled circular tubes under transverse loading—I. *Int. J. Mech. Sci.,* **18** (1976) 325–33.
9. Morris, A. J. & Calladine, C. R., Simple upper-bound calculations for the indentation of cylindrical shells. *Int. J. Mech. Sci.,* **13** (1971) 331–43.
10. Wierzbicki, T. & Hoo Fatt, M. S., Deformation and perforation of a circular membrane due to rigid projectile impact. In *Dynamic Response of Structures to High-Energy Excitations,* ed. T. L. Geers & Y. S. Shin. ASME PVP Vol. 225, New York, 1991, pp. 73–83.
11. Neilson, A. J., Howe, W. D. & Garton, G. P., Impact resistance of mild steel pipes: an experimental study. UK Atomic Energy Authority Report AEEW-R 2125, Winfrith, 1987.
12. Palomby, C. & Stronge, W. J., Blunt missile perforation of thin plates and shells by discing. *Int. J. Impact Engng,* **7** (1988) 85–100.
13. Palomby, C. & Stronge, W. J., Evolutionary modes for large deflections of dynamically loaded rigid–plastic structures. *Mech. Structures & Mach.,* **16** (1988) 53–80.

14. Calladine, C. R., Thin-walled elastic shells analysed by a Rayleigh method. *Int. J. Mech. Sci.,* **13** (1977) 515–30.
15. Yu, T. X. & Stronge, W. J., Large deflections of a rigid-plastic beam-on-foundation from impact. *Int. J. Impact Engng,* **9** (1990) 115–26.
16. Yu, T. X., Large deflection response of thin tubes to impact. Cambridge University Engineering Department Report CUED/C-MECH/TR44, 1988.
17. Palomby, C., Dynamic deformation and perforation of ductile cylindrical shells. Ph.D. Dissertation. University of Cambridge, 1988.
18. Hoo Fatt, M. S. & Wierzbicki, T., Impact damage of long plastic cylinders. In *Proceedings of 1st International Offshore and Polar Engineering Conference, Edinburgh, 1991,* pp. 172–182.
19. Schwer, L. E., Holmes, B. S. & Kirkpatrick, S. W., Response and failure of metal tanks from impulsive spot loading: experiments and calculations. *Int. J. Solids Structures,* **24** (1988) 817–33.
20. Stronge, W. J., Impact and penetration of cylindrical shells by blunt missiles. In *Metal Forming and Impact Mechanics,* ed. S. R. Reid. Pergamon Press, Oxford, 1985, pp. 289–302.
21. Shadbolt, P. J., Corran, R. S. J. & Ruiz, C., A comparison of plate perforation models in the sub-ordnance impact velocity range. *Int. J. Impact Engng,* **1** (1983) 23–49.
22. Ma, X. & Stronge, W. J., Spherical missile impact and perforation of filled steel tubes. *Int. J. Impact Engng,* **3** (1985) 1–16.
23. Duffey, T. A., Dynamic rupture of shells. In *Structural Failure,* ed. T. Wierzbicki & N. Jones. Wiley, New York, 1989, pp. 161–91.
24. Neilson, A. J., A Dyna3D calculation for impact on a pipe target. UK Atomic Energy Authority Report AEEW-M 2058, Winfrith, 1983.

Chapter 6

COMPOSITE STRENGTH AND ENERGY ABSORPTION AS AN ASPECT OF STRUCTURAL CRASH RESISTANCE

C. M. KINDERVATER & H. GEORGI

Institute for Structures and Design, DLR, Pfaffenwaldring 38–40, Stuttgart 80, Germany

ABSTRACT

Composite materials offer considerable potential and design flexibility for high-performance structures. Also, controlled microfragmentation processes provide the capability to absorb kinetic energy during vehicle crashes. This chapter briefly overviews the state of the art of composite crash resistance. A basic understanding of the terminology, the energy absorbing mechanisms, and crushing performance of composite materials and generic structural elements is presented. Based on this fundamental knowledge, a design philosophy for composite airframe crash resistance is outlined, and more complex airframe components such as subfloor beams, intersections of beams and bulkheads (cruciforms), floor boxes, and frame structures are investigated with regard to strength and crash behaviour. It is generally assumed that composite structures have equal or even better energy absorption performance compared to metals. However, particular design features have to be taken into account during early design stages. As full-scale tests are expensive, component crash tests are a powerful tool to enable crash behaviour prediction of a composite structure by developing and verifying empirical and semiempirical procedures and to provide a verification and calibration database for more detailed crash prediction methods.

1 INTRODUCTION

1.1 Scope of Composite Materials

Composite materials offer high potential for *tailored designs* by a wide variety of matrices and fibres, various preforms, and laminate architecture, i.e. fibre orientation and stacking sequence of the single laminas. Typical reinforcements are glass, aramid, and carbon fibres. Property improvements and lower prices, especially with carbon fibres, could be observed over the years. Additionally, new fibre types such as high-performance polyethylene fibres (density 0·97 g/cm^3) are now available. Besides the classical thermosetting polymers such as polyester, vinyl ester, phenolic and epoxy resins, thermoplastic systems with continuous or short fibre reinforcements have came into use, owing to faster and cheaper fabrication and better recycling possibilities. Also, very interesting technical textile preforms such as 3D-woven fabrics, multi-axial knittings and braidings, and 3D-woven complex shaped profiles have been developed.

1.2 Benefits of Composites

The major advantages of composites are high specific strength and stiffness properties in combination with design flexibility. This results in lightweight, high-performance structures mainly used in aerospace applications. Even the extremely high temperatures up to 1500°C during reentry of space vehicles can be handled with fibre-reinforced structural ceramics. Also, the car and sporting-goods industries take advantage of the superior behaviour of composite materials.

1.3 Energy Absorption Capability

Composite materials also have considerable potential for absorbing kinetic energy during a crash. The composite energy absorption capability offers a unique combination of reduced structural weight and improved vehicle safety by higher or at least equivalent crash resistance compared with metal structures. *Crash resistance* covers the *energy absorbing capability* of crushing structural parts as well as the demand to provide a *protective shell* around vehicle occupants (structural integrity). This basic principle of occupant crash protection has been used in the automotive field since the early 1950s.[1] Crash safety has meanwhile become a well-established car design requirement. In aeronautics, the first structural design requirements for better crash protection were established for military helicopters and light fixed-wing aircraft in the form of the *Aircraft Crash Survival Design Guide*[2] and the *MIL-STD-1290A*.[3] For all other aircraft categories, at least the requirements for crash resistance of seats have recently been improved, and further progress can be expected in the future.

1.4 Status of Crash Investigation

Basic research work on composite structural crash resistance can be attributed mainly to special teams, whose work is summarized respectively in Refs 4–11, 12–16, 17–27, 28–38, 39, 40, 41, 42 and 43, and for more special or project-type investigations in Refs 44–50. Some dissertations have treated various problems of composite energy absorption behaviour.[51–56]

The two major focus areas are automotive and aeronautical applications. In acronautics, in particular composite helicopter airframe structural crashworthiness is considered,[13,28,32–34,39–43] owing to the increasing application of primary composite structures in helicopter airframes and already existing crash requirements. The material systems considered are higher-performance types such as epoxy resins reinforced with glass fibres and, increasingly, aramid and carbon fibres or hybrids thereof. In the automotive area reinforced polymers must satisfy a complex set of design requirements—among others the crash energy absorption management in the front-end and side structures of cars.[8,19,56,57] However, the decisive factors in automotive applications are the material costs compared with steel, and quick reproducible fabrication procedures such as sheet molding compounds (SMC), high-speed resin transfer molding (HSRTM), and thermoplastic stamping. Therefore the major reinforcement is made by glass fibres in combination with cheaper duroplastic resin systems such as polyester and vinyl ester (VE), and thermoplastic systems such as polyamides (PA) and polyethylene–terephthalate (PET). A certain cost penalty would probably be accepted if a significant performance improvement could be achieved, e.g. reduction of noise, vibration or ride harshness.

Most polymer composite crash behaviour investigations deal with tubular structures having various cross-sectional shapes, geometries, materials, and laminate architecture. It was recognized that microfragmentation mechanisms with brittle composite laminates under compression loading resulted in energy absorption capabilities and force–deflection characteristics equivalent to or even better than comparable metal structures. Metal structures collapse in buckling and folding modes. Crash energy is thereby absorbed by material yielding and plastification along the folding lines. The microfragmentation of composites is sensitive to many parameters. A key problem is a prescribed initiation behaviour of the crushing process, which can be performed by a so-called *trigger mechanism*.

This chapter briefly overviews the state of the art, basic understanding, and available database of the energy absorption properties of composite materials and structural elements, and then focuses on the design philosophy and investigations of more complex composite airframe components. As full-scale tests are expensive—especially with composite structures—analytical, semiempirical, or empirical crash property prediction methods and design concepts must be further developed and verified by a broad test database. Also, global and detailed crash prediction and simulation methods (KRASH, CRASH-CAD, PAM-CRASH, DYNA-3D, etc.) are based upon reliable knowledge of the crushing behaviour of the matcrials and individual components of the structure. In many cases the load–deflection characteristics of the single components can be superimposed to gain the overall crash response of a vehicle with sufficient accuracy. Thus component crash tests are a powerful tool to provide the crash prediction of a composite structure.

2 STATE OF THE ART

2.1 Valuation Criteria for Crushing Behaviour

A terminology has been developed to describe and compare the energy absorption performance of collapsing or crushing material specimens or structural parts. The valuation criteria are derived basically from the force–deflection curve and the absorbed energy, which is the area under this curve. The crushing characteristics typically have three phases: (1) initiation; (2) collapse by crushing or folding; and (3) 'bottoming out' or compacting. Figure 1 summarizes the criteria used in the present chapter.

Specific energy E_s. The specific energy relates the absorbed energy to the mass of the absorber or structure, and is therefore an important criterion for lightweight designs. Often only the crushed mass of the absorber is taken into account. However, in more complex structures and where clamping devices are used the crushed mass is difficult to determine, and therefore the complete mass of the structure and devices should be taken into account. For some reasons it might be of interest to relate the absorbed energy to the outer volume of the absorber structure and determine the *energy dissipation density*. High values are then required for compact absorbers.

Mean crushing stress. The crushing stress is calculated by the average crush force (F_{avg}) divided by the original cross sectional area of the

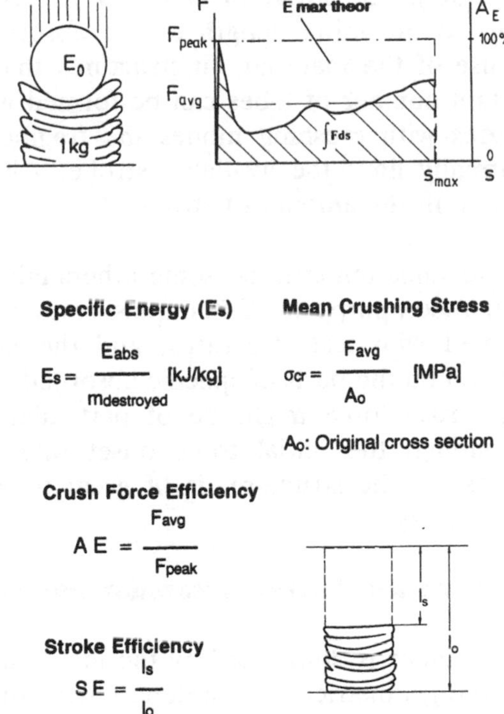

Fig. 1. Valuation criteria for crushing behaviour.

absorber. High values are required to cushion heavy loads. Also, the *specific crushing stress* is often used in the literature. This value is derived from the crushing stress by division with the material density. However, the specific crushing stress and specific energy are the same if the collapsibility (*stroke efficiency*) is not taken into account. The crushing stress is theoretically limited by the compression strength of the laminate.

Crush force efficiency (*AE*). This value relates the average crush force F_{avg} to the maximum force F_{peak} of the crush characteristic. The highest force occurs normally at the initiation phase. An absorber with a rectangular-shaped force–deflection curve (ideal–plastic) and theoretical maximum energy absorption has a crush force efficiency of 100%. Lower values indicate high initial force peaks or force serrations during crushing. In some publications the expression 1/AE is also used and is called the *load uniformity* (LU). This value has the disadvantage that it does not have a defined upper limit.

Stroke efficiency (*SE*). The stroke efficiency is the ratio of the stroke at 'bottoming out' to the initial length of the absorber. High ratios indicate efficient use of the material. In structures that fail in folding modes only a certain number of lobes can be formed within the initial length. In structures with crushing modes the fragmented and compacted debris normally limit the available stroke. Also, clamping or trigger devices can limit the amount of stroke.

Besides the above valuation criteria, some others might be helpful or necessary for comparison purposes. Often quasi-static and crash impact loading is performed with test structures, and therefore a static-to-dynamic ratio defined on the basis of specific absorbed energy, crushing stress, or average crush force might be of particular interest. Also, under dynamic loading, the initial force onset rate, i.e. the initial compression stiffness of the structure, is of great importance for the deceleration–time history.

2.2 Composite Failure and Energy Absorption Mechanisms

Structural crash resistance is dominated by the material behaviour, the structural design, and geometrical parameters. Owing to the brittleness, i.e. low strain to failure, of composites reinforced by glass or carbon fibres, material plastification is almost not apparent—only to a small amount in some matrix materials and some other synthetic reinforcing fibres such as aramid or high-performance polyethylene fibres. Basic fracture modes of unidirectional composites under various loading conditions are described in the literature.[27,56] Fractures occur in fibres, in the matrix, and at the fibre/matrix interface in tension, compression, and shear parallel and normal to the fibre direction. On the laminate level, delaminations caused by shear or buckling of single layers or groups thereof also occur. Fracture mechanics describes crack propagation as occurring as one of Modes I–III. The energy absorption mechanisms in composites, however, cannot be characterized by such simple models.

Crushing on the micromechanical level is very complex, and combinations of many mechanisms are apparent.[27,54] The key mechanism of high composite energy absorption under crash loading is a strength-controlled formation of a microcrack pattern which spreads out locally in the laminate. The formation of such a propagating crushfront must be activated by a stress concentration where cracks first occur.

The morphology of the crush front generated by a so-called *trigger* or crush initiator determines the level of crush force. The crush effect at a

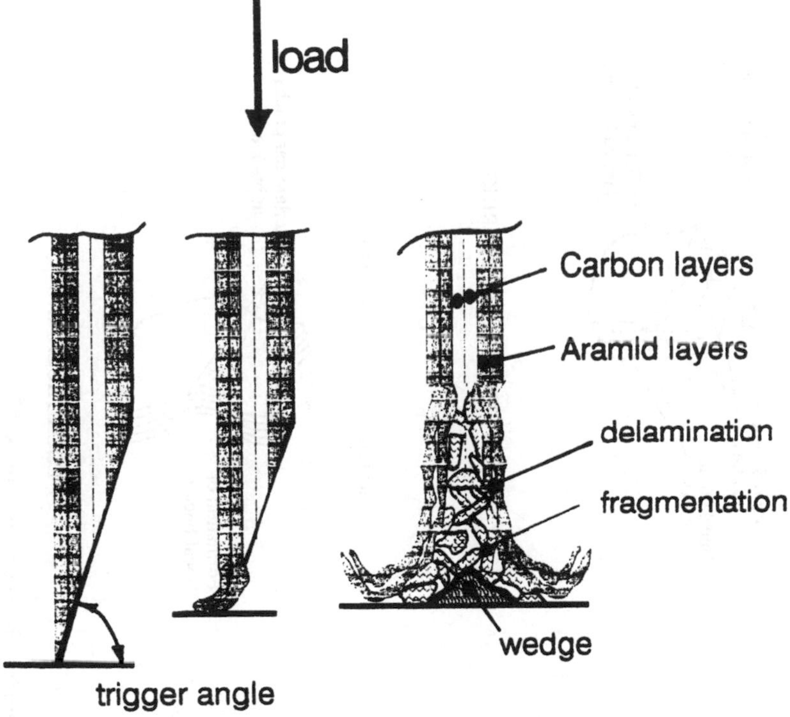

Fig. 2. Crush-front formation initiated by a chamfered laminate tip.

chamfered tip of a carbon/aramid-fibre/epoxy laminate and the stabilisation of the crush front is shown in Fig. 2. This complex sequence of local cracking and fragmentation is controlled by the material properties, the laminate's architecture, and the geometrical specimen or structural configuration. Instability-dominated failures caused by Euler and shell buckling generally lead to low energy absorption. Friction also plays an important role as an energy absorption process.[23] Frictional forces act between the broken and sliding parts within the crush front and also between the contact surface and the sliding fronds. On the macro- and microscopic levels energy absorbing mechanisms are described as 'splaying and fragmentation modes'.[27] Another set of basic crushing modes, especially with tubular absorbers, is given in Refs 16 and 54. These modes are transverse shearing, (2) lamina bending, and (3) local buckling (Fig. 3). In addition, fourth mode, 'brittle fracturing', is discussed as a combination of (1) and (2). Microfragmentation and friction in the crush front generate a 'quasi-plastification' energy dissipation process in composite laminates. A microscopic view into a crush front of a carbon/epoxy tube is shown in Fig. 4. Failure in 'local

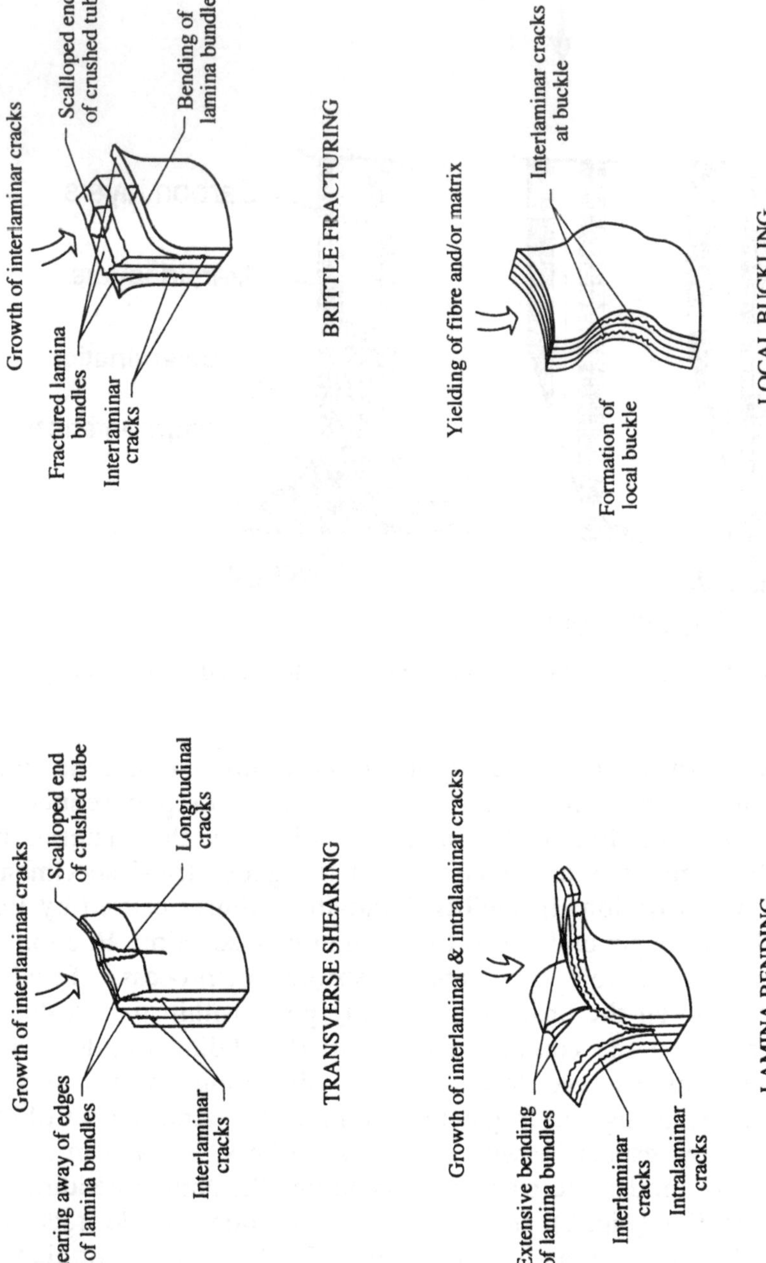

Fig. 3. Typical basic crushing modes observed with tubes.[16,54]

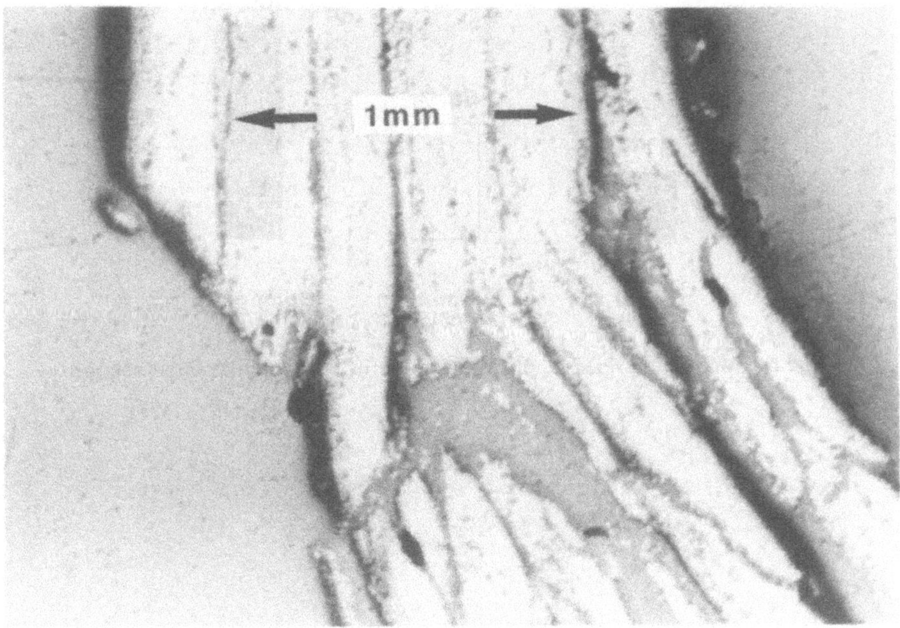

Fig. 4. Microscopic view of a crush front of a ±45° carbon/epoxy tube.

buckling' occurs in material compositions with high failure strain matrices and/or fibres (aramid, high-performance polyethylene).

2.3 Energy Absorption of Generic Composite Elements

To develop a database on the energy absorption behaviour of composite generic elements, many quasi-static and dynamic crush tests (drop tower) were performed to understand how composites absorb energy and how the numerous lamination parameters influence the crushing performance. The specific energy absorption of tube structures with circular cross section and comparable geometrical dimensions made out of steel, aluminium, and filament-wound ±45° carbon fibre/epoxy are compared in Fig. 5. The aluminium tubes show only 54% and the steel tubes 31% of the composite tubes' specific energy. The large potential of energy absorption performance of composites is more evident in Fig. 6, where composite and aluminium tubes are compared in terms of specific energy and crush force efficiency with other structural elements such as honeycombs, stringer and integrally-stiffened elements, sandwich designs, and cruciforms.

A tube with a circular cross section (composite and metal) is obviously the most favorable shape for energy absorption. Composite

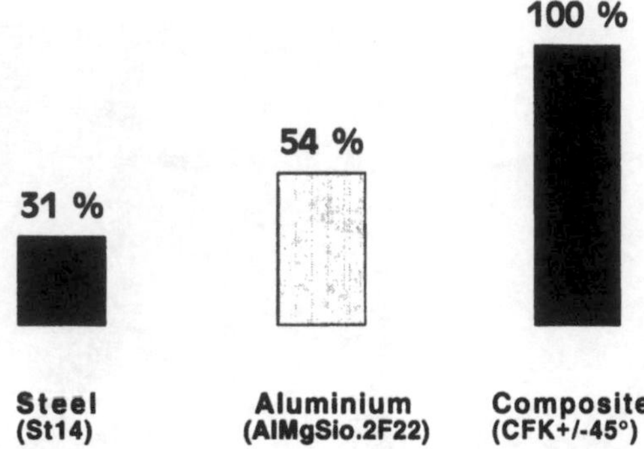

Fig. 5. Specific energies E_s of tube structures with circular cross section.

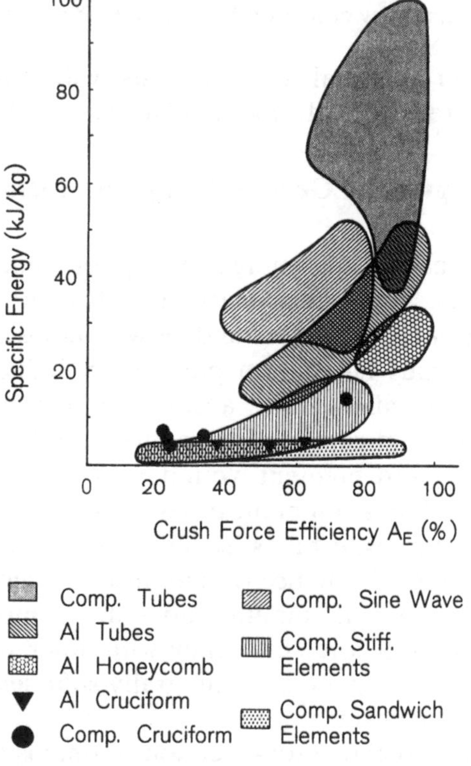

Fig. 6. Crushing performance of generic structural elements.

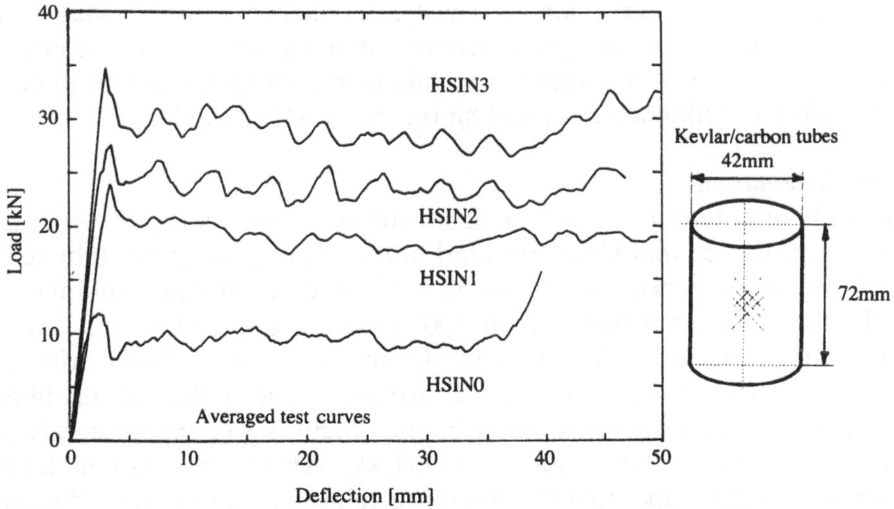

Fig. 7. Force-deflection characteristics of epoxy–carbon/aramid-fibre hybrid tubes.

tubes can achieve specific energies above 100 kJ/kg in combination with 95% crush force efficiency. Additionally, the force–deflection characteristics are almost 'ideal elastic–plastic'. Figure 7 shows a set of curves from crush tests with hybrid epoxy/carbon–aramid-fabric tubes having an increasing share of carbon fibre reinforcement. Also, the specific energies for composite sine-wave webs crushed in the web direction are high: the values range between 25 and 55 kJ/kg. With stringer-stiffened elements, the stringers predominately contribute to the energy absorption. Low specific energies are achieved with sandwich and cruciform elements, where instabilities and global folding can occur. Specific designs are then needed to improve crushing. It is evident that more complex structural assemblies have decreased crushing performance compared to tubes. Composite elements are generally superior compared with metal configurations.

Because of the particular importance of the tube as a structural element and energy absorber, some influencing parameters of the crushing process are briefly overviewed. Comprehensive contributions to tube crushing are reported in Refs 5, 6, 27, 29, 51–54, and 56.

Effect of Material Properties

Ranking for the reinforcing fibres can be given as carbon < glass < aramid fibres or other synthetic fibres. Some interesting hybrids have shown also high crushing performance compared with the pure carbon reinforcement.[35] For the matrix (resins) the specific energy tends to increase in the order phenolic < polyester < vinyl ester < epoxy.[6] A linear dependence of the specific energy upon the resin tensile strength and modulus is also reported.[6] The failure strains of fibres and matrix have different influences, depending on the crushing mode.[54]

Fibre Orientation

Tube ply orientation can have a significant effect on energy absorption. There is evidence that fibres oriented in the loading direction (tube axis 0°) increase the energy absorption capability, especially in combination with inner and outer hoop layers (90° orientation).[27] Also, only hoop layers at the outside (0°/90°), pure 90° and ±45° tubes,[29] and orientations ($\pm\theta/0°$) produce high crush forces.[54] The influence of fibre orientation is more evident when brittle fracturing is apparent (glass- and carbon-fibre reinforcements, hybrids). When tubes fail in local buckling (aramid- and polyethylene-fibre reinforcements), the influence of the fibre orientation is not as large. If the tube is designed primarily as a load-carrying structural element and energy absorption is only the secondary aspect, the considered loading direction dominates the selection of fibre orientation.

Tube Geometry

Energy absorption with regard to the cross-sectional shape increases in the order rectangular < square < circular for tubes with fabric lay-up.[6] Other important parameters are the ratio of circular-tube diameter D to wall thickness t or the ratio of square-tube width W to wall thickness t. As the tube D/t and W/t decrease, energy absorption capability increases.[54] Tube cross-sectional and length dimensions have to be selected so as to avoid Euler and global or local shell buckling.

Trigger Mechanism

The chamfered tube end is widely used and simple to fabricate. The trigger has two functions: (1) to initiate stable propagating crushing and (2) to reduce initial peak loads. With glass-cloth/epoxy tubes under compression a dependency upon the chamfer angle was reported.[26] However, in brittle fracturing modes chamfer angles between 30° and 45° can be recommended for stable crush initiation. Also, metal cone triggers[29] and metal internal mandrel triggers[24] initiate stable crushing. Comparison of identical glass-reinforced tubes with bevel and tulip triggers showed advantages for the tulip trigger.[11] With the geometry of the tulip trigger, the force–deflection curve can be 'tailored' to a predetermined shape. With tubes failing in local buckling, the trigger has no or much less influence compared to brittle fracturing.

Rate Effects
Rate effects result from structural inertia effects and strain-rate sensitivity of the mechanisms that control the crushing process. Therefore fibre or matrix strain-rate sensitivity can influence crushing. Owing to the different testing methods (hydraulically driven machines and drop tower), trends are difficult to interpret. In hydraulic machines the strain rates are held constant, whereas in a drop tower the more realistic crash process is simulated, i.e. the strain rates are high at initial impact velocity and zero at maximum stroke. However, higher or equal energy absorption is observed with epoxy resins and various reinforcing fibres.[6,29,15,54] The rate sensitivity for polyester glass-reinforced pultrusions in general is positive,[9] but very large decreases of 30% have also been reported.[47] An increase of 50% was observed for high-performance polyethylene fibre reinforcement and a thermoplastic polyamid matrix.[35,37]

Temperature
The resin properties control the energy absorption, and the effect can be directly related to the matrix temperature behaviour because most of the reinforcing fibres do not change mechanical properties within the temperature range considered (−40°C to 120°C).

Non-Axial Loading
The crushing process is very sensitive to non-axial loading. With carbon/epoxy tubes, inclinations of load of 5° already cause catastrophic failure instead of crushing. Depending on the effect of friction, off-axis and angled crushing have to be considered separately.[10] Cone structures are better suited for non-axial loading, and are very efficient absorbers.[20,25,50]

3 COMPOSITE AIRFRAME CRASH RESISTANCE

3.1 Design Philosophy and Requirements

In crash accidents with a high vertical component of the impact velocity the crash loads have to be absorbed by structural deformation. For the control of decelerative loads on occupants, the type of aircraft will affect the crashworthy design approach which is an optimization process for a certain aircraft category (Fig. 8). A systems approach should be applied whenever possible, involving the landing gear, the subfloor and the high mass retention structure and seat systems with tuned energy

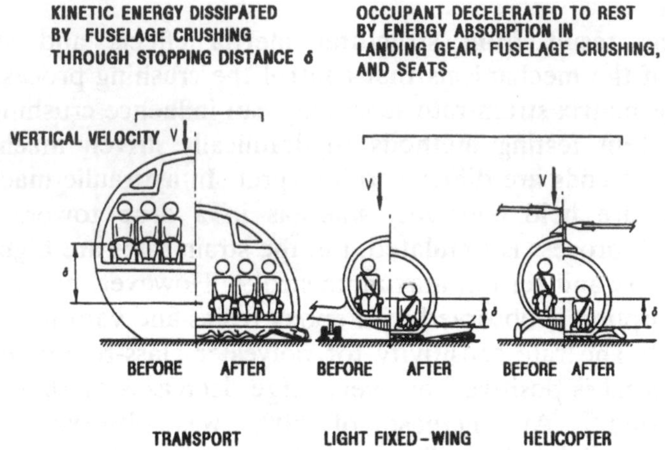

Fig. 8. Energy absorption concept for various aircraft.

absorbing characteristics. However, light fixed-wing general aviation aircraft, small passenger airplanes, and helicopters, especially with retracted landing gear, have little crushable airframe structure. Such designs typically consist of a framework of longitudinal beams and lateral bulkheads covered by the outer skin and cabin floor. The total structural height is often only about 200 mm.

The design of intersections of beams and bulkheads (cruciforms), and the beam webs contribute essentially to the overall crash response of a subfloor assemblage (Fig. 9). Under vertical crash loads cruciforms are 'hard-point' stiff columns, which create high decelerative peak loads at the cabin floor level and cause dangerous inputs to the seat/occupant system. Figure 10 shows a subfloor crush characteristic having a moderate initial stiffness and then a slightly increasing crush-force level. This was determined to be 'ideal' from parametric crash simulation studies using the computer code KRASH.[58] Equivalent crush curves can also be recommended for the front-ends of cars.

The basic requirements for crash-resistant subfloors can be summarized as follows:

- uncritical distribution of ground reaction and seat loads;
- limitation of the decelerative forces by structural deformation with a 'controlled load' concept;
- maintenance of cabin floor structural integrity;
- for minimization of cost and weight penalties, a dual function structural concept: load-carrying capability for normal operation and energy absorption for crash cases.

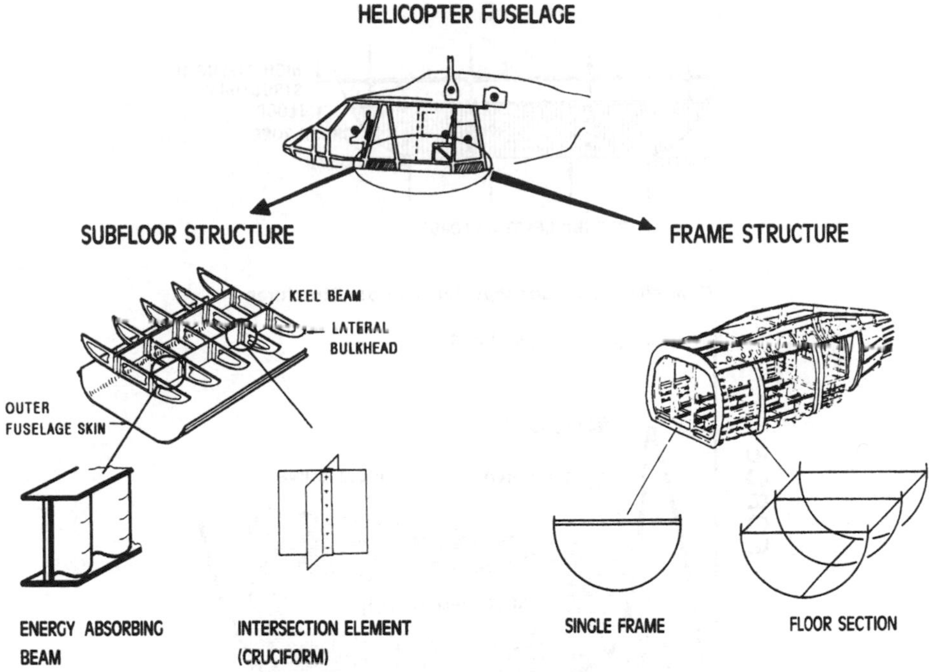

Fig. 9. Airframe substructures.

Frame and shell structures above the cabin floor are crucial elements for high mass retention (transmissions, engines, rotor hubs, etc.) and for providing a survivable volume in a crash sequence. Plastically deformable frames offer the possibility of load-limiting concepts for large overhead masses.

Composite aircraft structure must be designed carefully to assure crash-resistant features. Compared with metals, totally different design concepts have to be developed and verified. Figure 11 demonstrates an essential advantage of crushing composite structures. The force–deflection curves of three equal steel tubes that fail in very regular folding modes have a force that varies by up to 100 kN between peaks and troughs. A vehicle stopped with such an absorber would also have a highly undulating deceleration–time history. This can be dangerous to life if the deceleration peak levels and their duration are above human tolerance limits. Crash structures that fail by folding modes have only low average crush-force levels compared with the force maxima. The crushing characteristic of a glass-fibre-reinforced tube with comparable axial load-carrying capability has a 'smooth' crush curve with serrations in the range of 10–20 kN. Therefore the actual crush-force level is almost identical with the average crush force gained over the complete

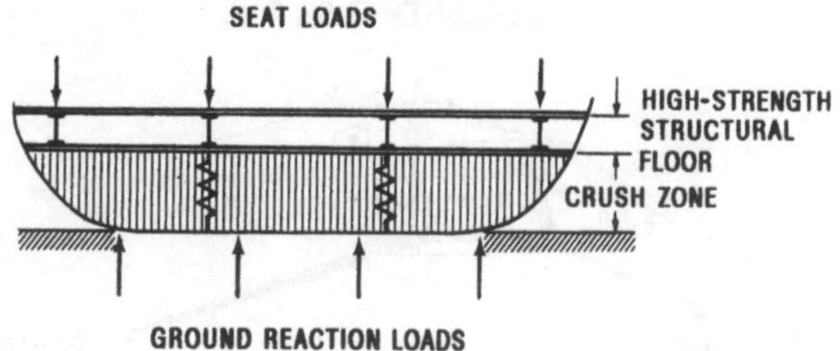

'Controlled Load Concept' for Subfloor Structures

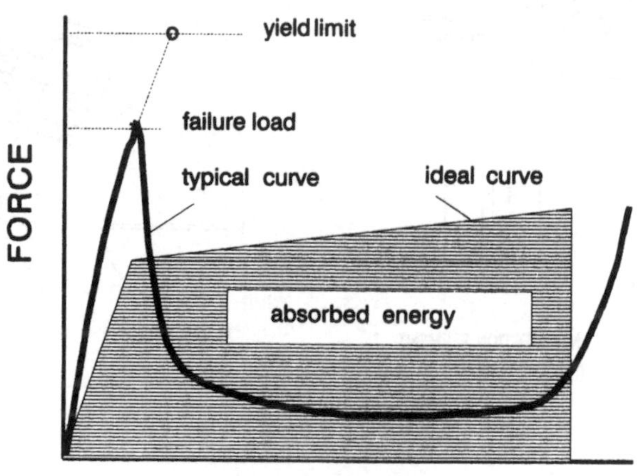

Fig. 10. Crash-resistant subfloor design.

stroke. This results in a very effective and smooth deceleration–time history.

High energy absorption with composites has been obtained for compressive loadings where brittle fracturing of the composite into sublaminates occurs. Under tensile or bending loads, structural integrity may be lost at initial fracture, and energy absorption can be low. To guarantee post-crush structural integrity, composite structures have to be *hybridized* with tougher fibres such as aramid (Kevlar) or high-performance polyethylene. Thus, with only limited crash response information available, the further need exists to examine generic composite structures under crash loadings, and to use superimposing principles to predict the crash response of complex full-scale airframe structures.

3.2 Structural Response of Sine Wave Beams

Sine wave beams are the most efficient subfloor design concepts yet evaluated: they combine high load-carrying capability, high energy

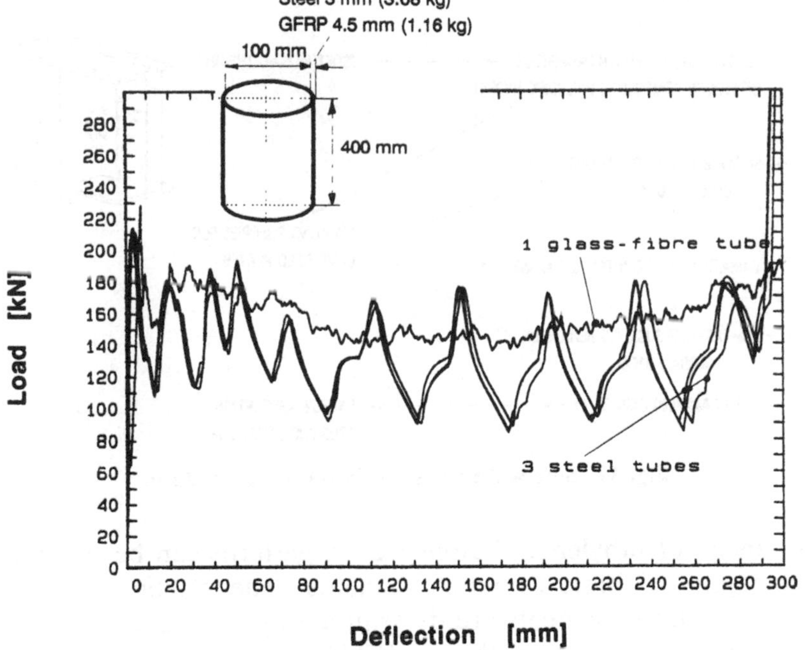

Fig. 11. Crushing of steel and GF/vinylester tubes.

absorption in the web direction, and good structural integrity by using hybrid lamination techniques. An experimental/analytical programme was performed to investigate systematically the structural response of sine wave beam sections.[38] The approach includes very elementary research concerning hybrids, the use of buckling theories for laminated plates (shear/compression buckling), and structural component crush testing for analysis validation and determination of energy absorption performance. Some elements of this work will be discussed in more detail in the following sections. Figure 12 shows the dimensions of the tested beams and the optimization methods.

Soft Mosaic Theory

Hybrid lamination techniques offer a large potential for crash resistance performance in terms of providing high energy absorption in combination with post-crush structural integrity. In performing load-carrying capability parametric studies, not all relevant stiffness and strength parameters of interesting hybrid material systems can always be determined by experiments. Therefore the *soft mosaic theory* was developed[59] to determine on an analytical basis the elastic and strength properties of woven composites including intraply woven hybrid fabrics containing carbon and synthetic fibres such as aramids or high-

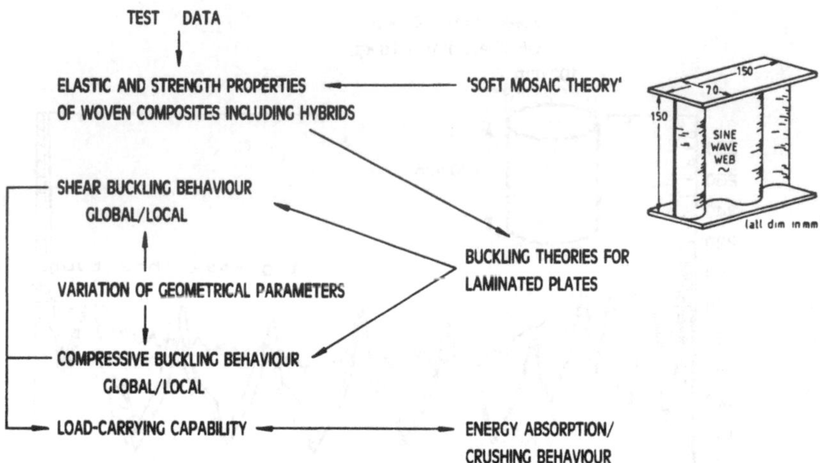

Fig. 12. Sine wave webs: optimization procedure.

performance polyethylene. Various weave patterns can be modeled by an assemblage of basic (mosaic) elements. The elastic and strength properties of those elements are determined by consequent application of the classical lamination theory (CLT) and a modified *Tsai–Wu* failure criterion. The undulation of warp and fill threads, one-dimensional reinforcement and resin-rich areas within the basic element (corners) can be taken into account. The prediction of the elastic properties of woven composites fits test data within a 5% range (Fig. 13).

Shear Buckling Behaviour

A finite element analysis (FEA) was performed on the shear buckling behaviour of various sine wave web configurations.[60] The effects of geometrical parameters such as free buckling length (spar height), laminate thickness, opening angle of the wave elements, and hybridization were investigated. The basic web laminate consisted of a 50/50 intraply woven carbon/Kevlar fabric (Interglas 2/2 Köper 98355) and an epoxy resin. The radius of the wave element was held constant at $R = 20$ mm. Buckling critical combinations of geometrical parameters are small laminate thickness, large spar height, and an opening angle of the tangent circular ring section (wave element) between 90° and 120°. A certain optimum is an opening angle of 150° in combination with all analysed laminate thicknesses and spar heights. An increase of the angle up to 180° does not result in most cases in a higher buckling load but it does lead to higher weight due to increasing cross section (Fig. 14).

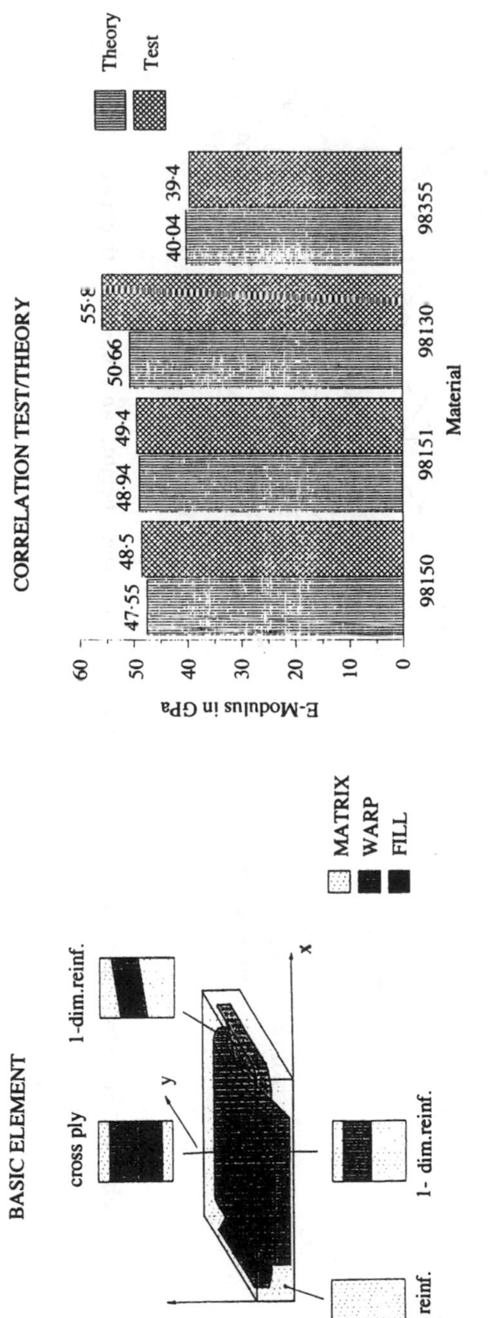

Fig. 13. Soft mosaic theory: optimizing sine wave webs.

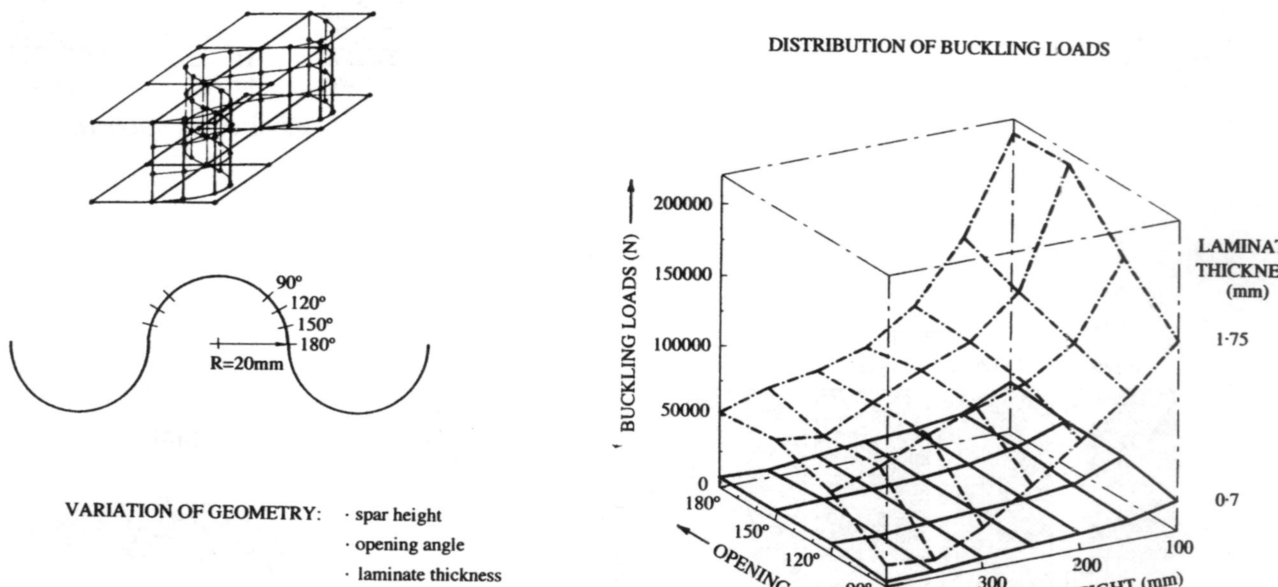

Fig. 14. Buckling loads depending on opening angle, spar height, and laminate thickness.

Influence of hybridization. Pure carbon-fibre-reinforced laminates under compression loading can have extremely high energy absorption capability, but disintegrate completely into small laminate fragments. Hybridization with tougher fibres such as Kevlar or high-performance polyethylene (Dyneema SK 60/DSM) provides post-crush structural integrity, but results in lower laminate stiffness due to the lower compression moduli of the synthetic fibres. Also, the aspect of weight reduction by hybridization is always interesting, especially with Dyneema SK60, which has a specific density close to 1 g/cm^3. Hybridization can be realized by alternating laminate stacking sequence or by using intraply woven fabrics (Fig. 15).

For the hybridization study the opening angle of 150° and the laminate thickness $t = 1{\cdot}4$ mm are held constant (seven laminate layers).

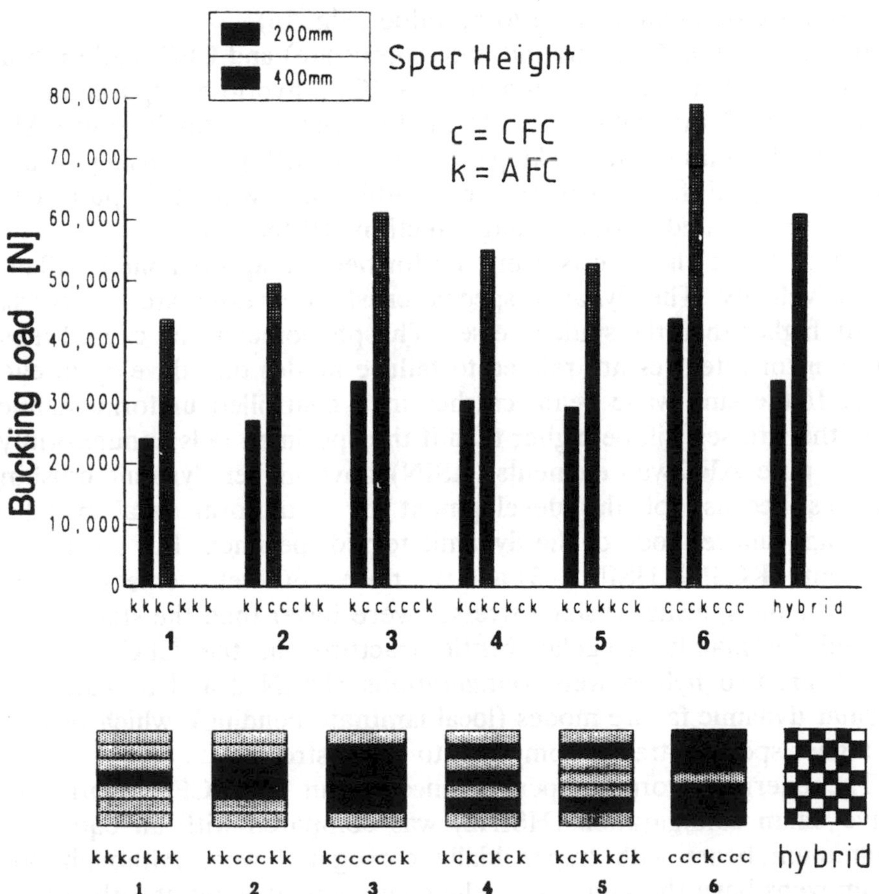

Fig. 15. Influence of laminate hybridization on shear buckling behaviour.

Two spar heights ($h = 200$ and 400 mm) are considered. The sine wave web with the lowest Kevlar share of the laminate (version 6) shows the highest shear buckling loads. However, the post-crush integrity of this laminate is not satisfactory. Acceptable shear buckling stiffness can be achieved with a laminate's Kevlar share between 30 and 50% (version 3 and intraply woven hybrid). Those webs provide under crash loading a good post-crush structural integrity. Higher percentages of Kevlar still improve the integrity, but lead to severe reduction in shear stiffness.

Crushing Characteristics

An I beam section with a sinusoidal-shaped web typically shows under compression (crushing) a high initial peak failure load followed by a sharp load drop. During the fragmentation and folding of the web laminate, however, the crush-force level is almost constant with some serrations around an average force value (Fig. 16).

In Fig. 17 pure AFC (aramid-fibre composite) and CFC (carbon-fibre composite) beam webs and various AFC/CFC hybrid configurations are considered. Hybridization is performed by either alternating pure AFC and CFC laminas in the stacking sequence (KCSIN) or by using intraply woven AF/CF fabric laminas in combination with 1–3 pure CFC midlayers oriented in the loading direction (HSIN 1–3).

All dynamic drop tests were performed at approximately 10 m/s initial velocity. The dynamic specific crushing stresses are not consistently higher than the static stresses. The phenomena that caused these stress inconsistencies are related to failure modes that develop in each case. If the sine wave beam crushes in a controlled uniform manner then the stresses will be higher than if the specimen fails nonuniformly.

The pure AFC-web elements (KSIN) have higher dynamic crushing stresses because of the development of a uniform local buckling (folding) failure mode of the dynamic tested specimen. For the Hybrid elements (KCSIN, HSIN 1–3) and the pure CFC-web element (CSIN), the dynamic specific crushing stresses were lower than the static values caused by mainly irregular brittle fractures in the CFC portions. However, two hybrid web configurations (HSIN 2 and 3) had more regular dynamic failure modes (local laminate bending), which resulted in higher specific stresses compared to static stresses.

The energy absorption performance of an AFC/CFC hybrid sine wave beam configuration (HSIN2) was compared with an equivalent aluminium beam with trapezoidally corrugated web. Although both beam webs have the same mass, the composite element absorbed twice the energy (Fig. 18).

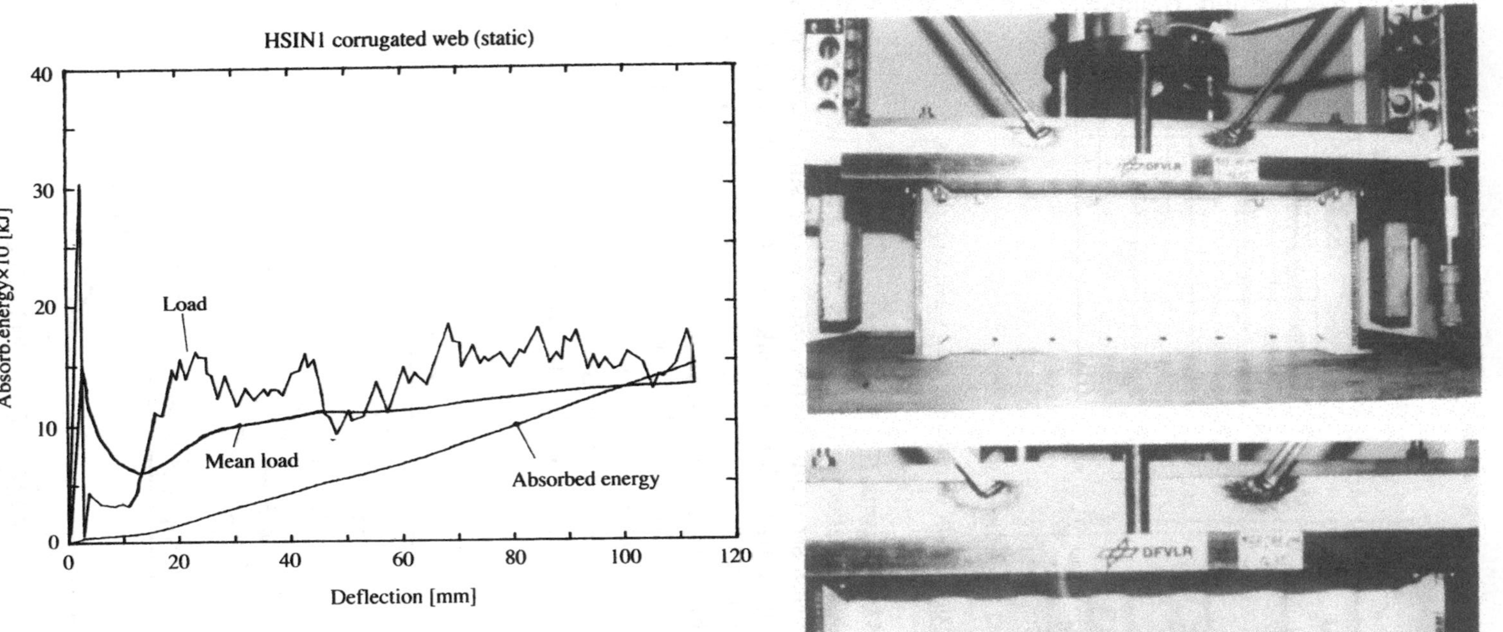

Fig. 16. Hybrid sine wave web: typical load–deflection curve.

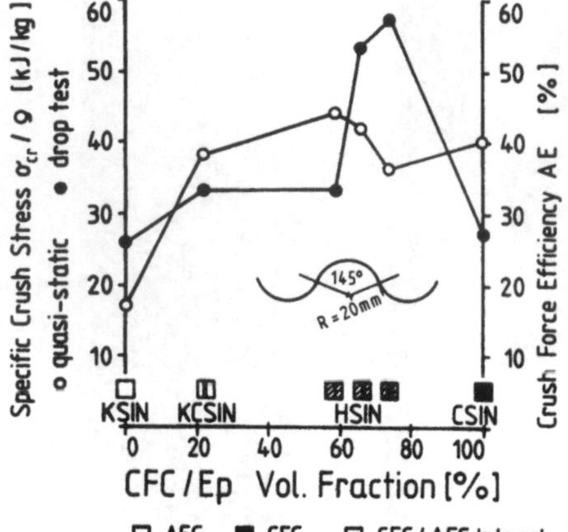

TEST DATA							
SYMBOL	SPEC. NO.	STATIC			IMPACT		
		F_{avg} (N)	L U	E_{sp} (kJ/kg)	F_{avg} (N)	L U	E_{sp} (kJ/kg)
	KSIN	11406	3 0	4 3	15600	2 4	24.4
	KCSIN	15908	2.6	32.5	13658	3 8	270
±45	CSIN±45	18644	3 4	12	--	--	--
	CSIN	18140	3 9	34 5	14348	6 1	23 5
1	HSIN1	15520	21	32.9	12327	3 4	27 8
2	HSIN2	19485	2 4	30 6	22800	2 3	48 0
3	HSIN3	21338	2.5	33.4	28931	21	45 8

Fig. 17. Comparison of different sine wave web designs.

Influence of Trigger Mechanisms

The shape of the load–deflection curve is important to the crash response of the subfloor structure. The *untriggered* sine wave hybrid beam (HSIN 2) shows a static initial peak of 48 kN, followed by an almost constant average force level of about 20 kN. These values result in a crushing force efficiency of only 40% (Fig. 19).

When trigger slots are used in the bottom of the web, the initial peak load is reduced to 22 kN, which is below the average crushing force of 23 kN for that particular test. The initial compression stiffness is not affected by the trigger slots. With the triggered configuration, a

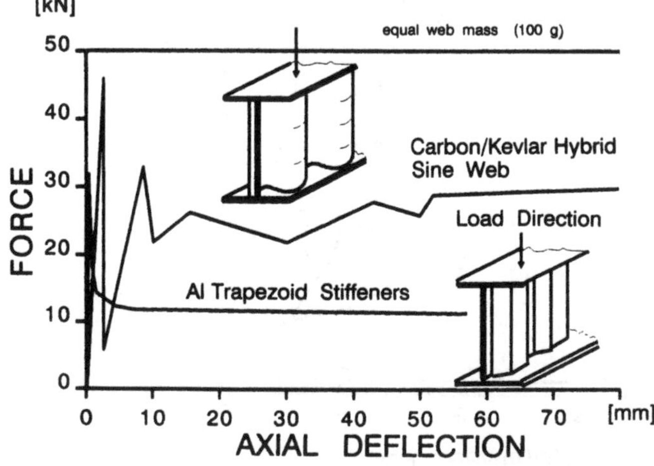

Fig. 18. Comparison of composite and aluminium beams.

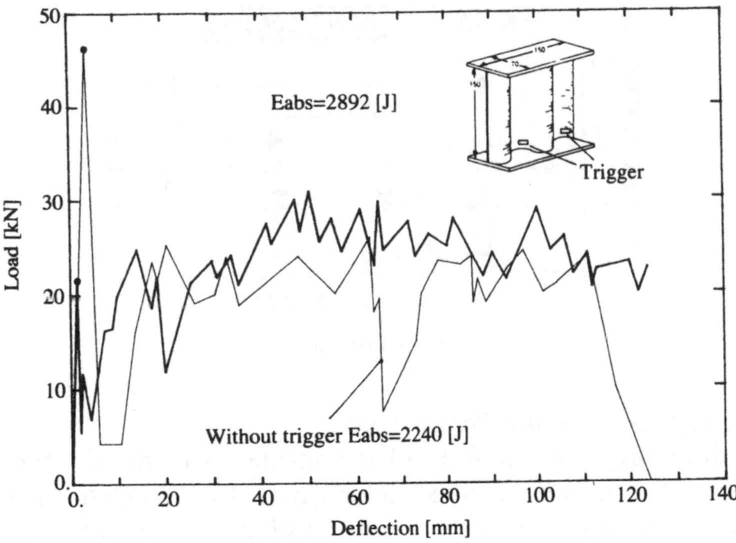

Fig. 19. Influence of slot trigger.

crushing force efficiency of about 100% is achieved (almost ideal elastic–plastic response). However, notch-type triggers must be considered very carefully, because they also severely reduce the beam's shear-load-carrying capability. Crushing initiators should preferably be embedded smoothly and uniformly in the laminate architecture, i.e. there should be less compression stiffness to the laminate in the trigger zone. A J-shaped connection to the lower beam cap could also be a promising solution for initial peak load reduction and stable crush initiation.

Inclined Impacts

Under vertical crash conditions, pitch and roll attitudes of the aircraft also have to be considered. Sine wave beam sections were tested under simulated 15° pitch and 10° roll conditions, in accordance with some requirements for helicopter crashworthiness (Fig. 20).[2,3]

For the study a CFC/AFC web hybridized by the stacking sequence was used (KCSIN). The drop-off in specific energy was found to be only 28% for the worst case—the 10° roll condition. The reduction of the specific energy for the simulated 15° pitch was only 15%. All tested parts were found to retain their structural integrity. In a test of a structural assemblage, e.g. the subfloor, the influence of inclined impacts on sine wave web crushing can be expected to be even lower compared with the single-element testing.

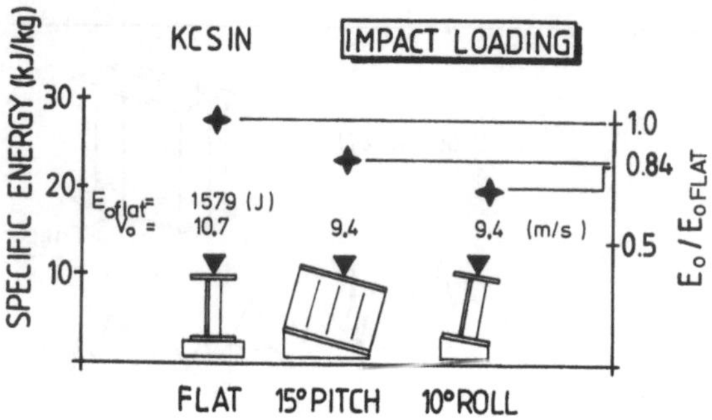

Fig. 20. Inclined impacts.

Prediction of the Crushing Performance

A procedure suggested in Ref. 13 is considered to predict the energy absorption of sine wave beams under quasi-static loading in terms of the specific crushing stress of a structural element, i.e. the web, by the summation of area-weighted specific crush stresses of characteristic elements (Fig. 21). This procedure then allows determination of the crushing response of webs by means of much simpler tube or tube-segment testing. The comparison of the specific crush stress of a carbon/epoxy sine wave web (opening angle 180°) and a tube with the same laminate layup (Gr/E) and radius shows very good correlation (Fig. 22).

The same procedure was applied for I-beams with CFC/AFC-hybrid sine wave webs (HSIN 1–3). The characteristic elements of those webs are tangent circular ring elements (wave segments) with an opening angle of 145°. Segments as well as tubes with the same HSIN laminates and radius were investigated. The comparison of the web's specific crush stresses with the tubes and segments shows an overestimation of the static crush response between 22 and 44%, whereas the differences between tube and segment are only about 4%. The discrepancy can be explained by different boundary conditions and failure modes of the two test series.

The 180° Gr/E web does not have a cap on top; both web and tube

Weighted-area rule[13]

$$\left(\frac{\sigma}{\rho}\right)_{\text{se}} = \sum_{i=1}^{n}\left[\frac{A_{i\,\text{ce}}}{A_{\text{se}}}\left(\frac{\sigma}{\rho}\right)_{i\,\text{ce}}\right]$$

Reduced formula for sine wave webs

$$(\sigma/\rho)_{\text{se}} = (\sigma/\rho)_{\text{ce}}$$

Fig. 21. Prediction of crushing behaviour.[13]

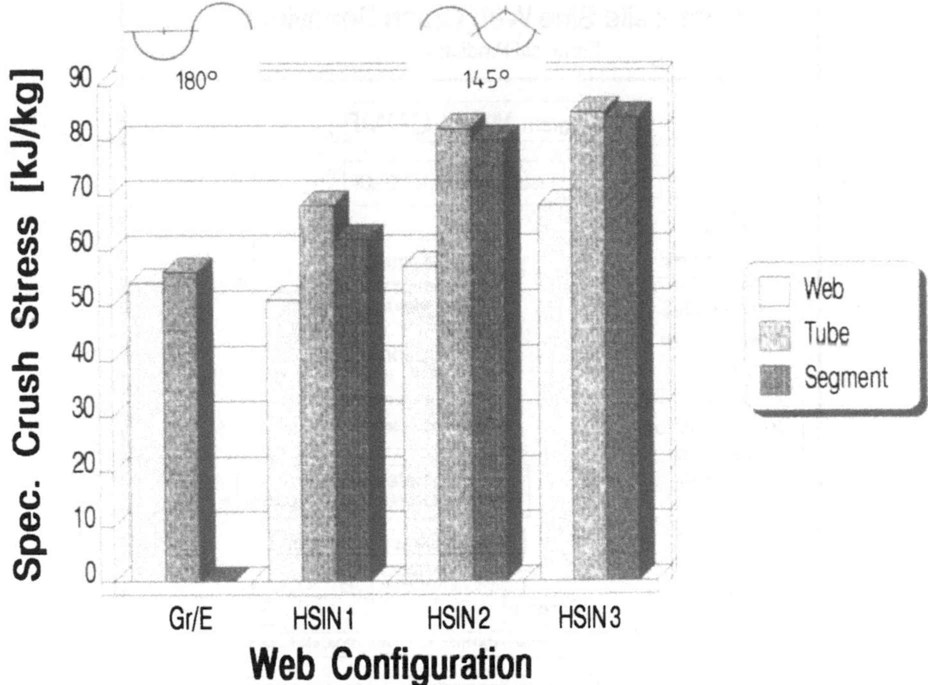

Fig. 22. Prediction for tubes, sine wave webs, and segments.

are triggered by a chamfered end. The structural (web) and the characteristic (tube) element fail over the the entire cross section by brittle fracturing and transverse shearing of the laminate. Web structures assembled of elements with lower opening angles fail only partly in this mode. At the intersections of the tangular ring elements with changing curvature a less efficient laminate bending mode is apparent. Therefore, for proper approximations of specific crush stresses of sine wave beams with other than 180° opening angle, an experimental correction factor is suggested.

Crushing Prediction Based on Tests

Based on a sine wave beam section test series (static and dynamic crushing), a linear extrapolation method is applied to approximate the crushing response of webs with similar laminate layup but different geometrical parameters such as web length, spar height, opening angle, laminate thickness, and trigger angle.[61]

The method combines the classical lamination theory (CLT) with experimental results. It is found that the peak failure loads and average crush forces are basically linear functions of the laminate's compression

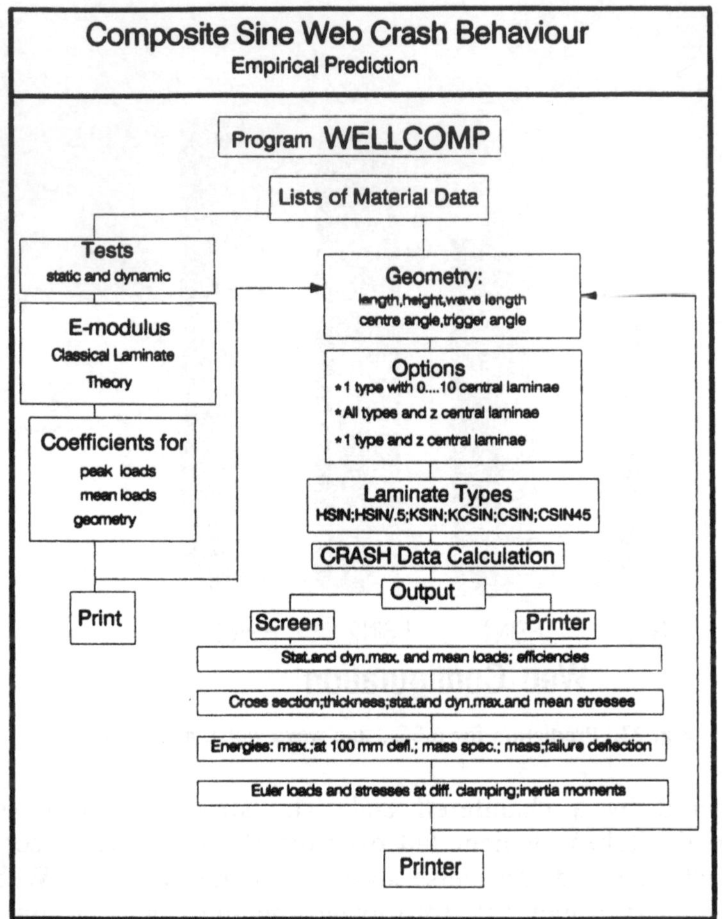

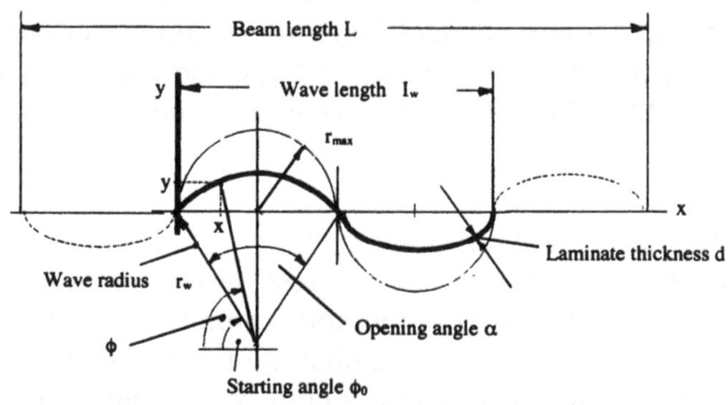

Sine wave web geometry

Fig. 23. DLR PC program WELLCOMP.

modulus, which can be calculated by the CLT. Using the empirical load coefficients (peak/average) for the various web laminates, other web configurations up to a compression laminate stiffness of 50 GPa can be approximated. The whole procedure is managed on a PC under DOS by a program named *WELLCOMP*. Figure 23 describes this program.

The output data contain the static and dynamic peak/mean crush force, the crush-force efficiencies, absorbed and specific energies at various deflections, web masses, moments of inertia, and Euler loads and stresses at various clamping conditions.

Figure 24 shows the linearized test results for the failure load and the mean load. An example of the approximate prediction of a load–deflection curve, based on this diagram, is illustrated in Fig. 25.

For strengthening and validation of the procedure, further experimental crushing data are needed. The empirical load coefficients can then be easily corrected within the program.

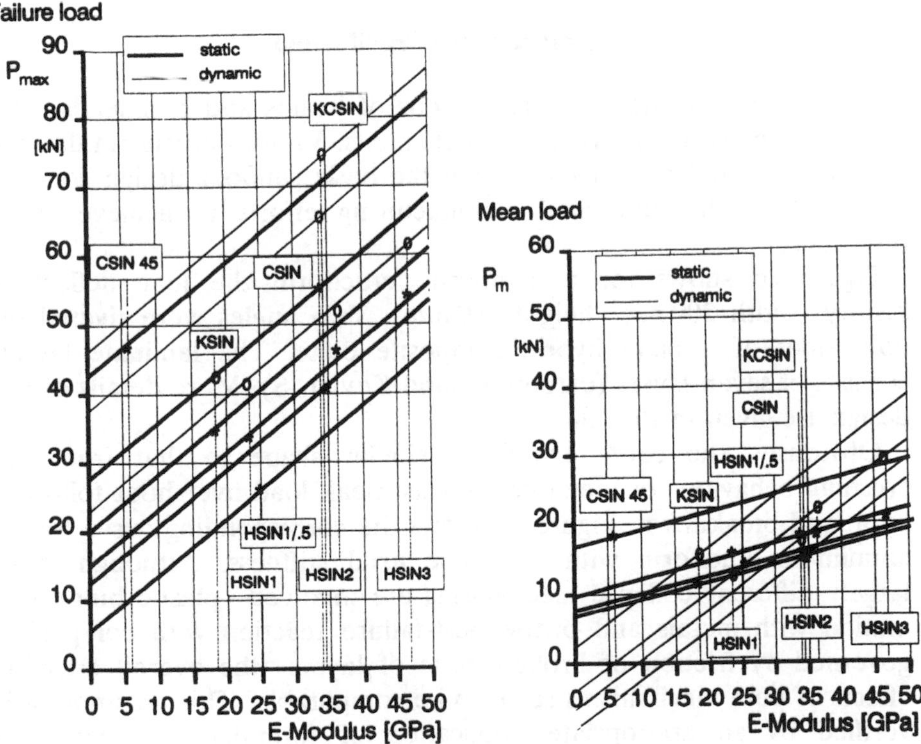

Fig. 24. Determination of load coefficients.

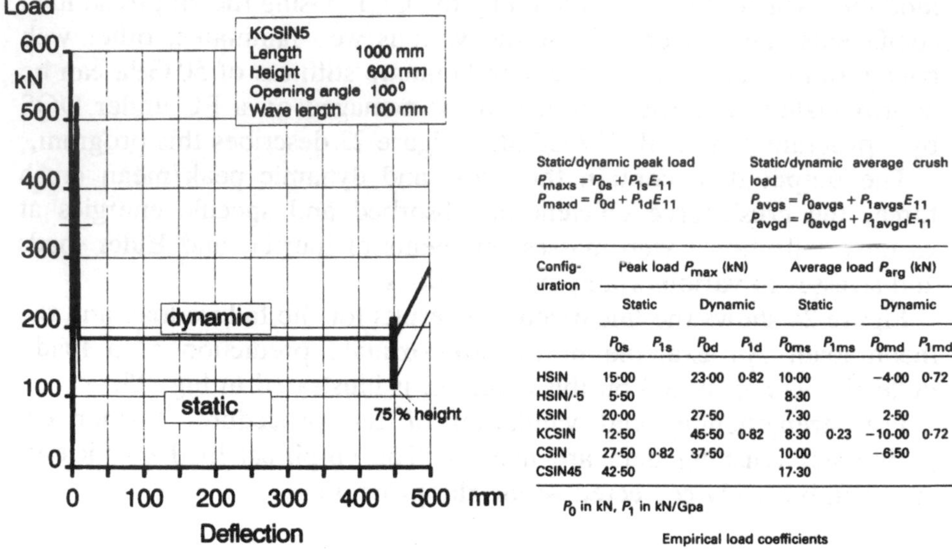

Static/dynamic peak load

$P_{maxs} = P_{0s} + P_{1s}E_{11}$

$P_{maxd} = P_{0d} + P_{1d}E_{11}$

Static/dynamic average crush load

$P_{avgs} = P_{0avgs} + P_{1avgs}E_{11}$

$P_{avgd} = P_{0avgd} + P_{1avgd}E_{11}$

Configuration	Peak load P_{max} (kN)				Average load P_{arg} (kN)			
	Static		Dynamic		Static		Dynamic	
	P_{0s}	P_{1s}	P_{0d}	P_{1d}	P_{0ms}	P_{1ms}	P_{0md}	P_{1md}
HSIN	15·00		23·00	0·82	10·00		−4·00	0·72
HSIN/·5	5·50				8·30			
KSIN	20·00		27·50		7·30		2·50	
KCSIN	12·50		45·50	0·82	8·30	0·23	−10·00	0·72
CSIN	27·50	0·82	37·50		10·00		−6·50	
CSIN45	42·50				17·30			

P_0 in kN, P_1 in kN/Gpa

Empirical load coefficients

Fig. 25. Approximate load–deflection curves.

3.3 Aircraft Subfloor Intersections (Cruciforms)

Cruciforms are formed by intersections of beams and bulkheads, and represent typical floor structure subelements. Various cruciform designs were investigated in order to retain the basic subfloor design concept and to limit the additional manufacturing efforts to achieve crash resistance.

Figure 26 shows for a composite cruciform the load–deflection history (width 200 mm, height 190 mm). Two angles were riveted on both sides of a plain hybrid composite plate. The laminate lay-up consists of seven layers (carbon, C, and Kevlar, S). More details of the design are given in Ref. 34.

Obviously, the curve exhibits a similar shape to that from the crushing behaviour of metal cruciforms: peak load overshoot followed by a load breakdown. Figure 27 shows the corresponding curve for an aluminium cruciform with the same axial stiffness. Although there largely different collapse mechanisms are involved—plastic hinge formation with metals and brittle post-failure reaction with composites governed by friction of broken parts of layers—the overall crushing effects of both mechanisms seem to be comparable. This assumption is verified by an appropriate application of formulae for metals to composites. For the *failure load*,[62]

$$P_v = \xi\sigma_v(F_w + \lambda F_H)$$

where ξ is the rivet reduction factor, σ_v the failure stress, F_w the

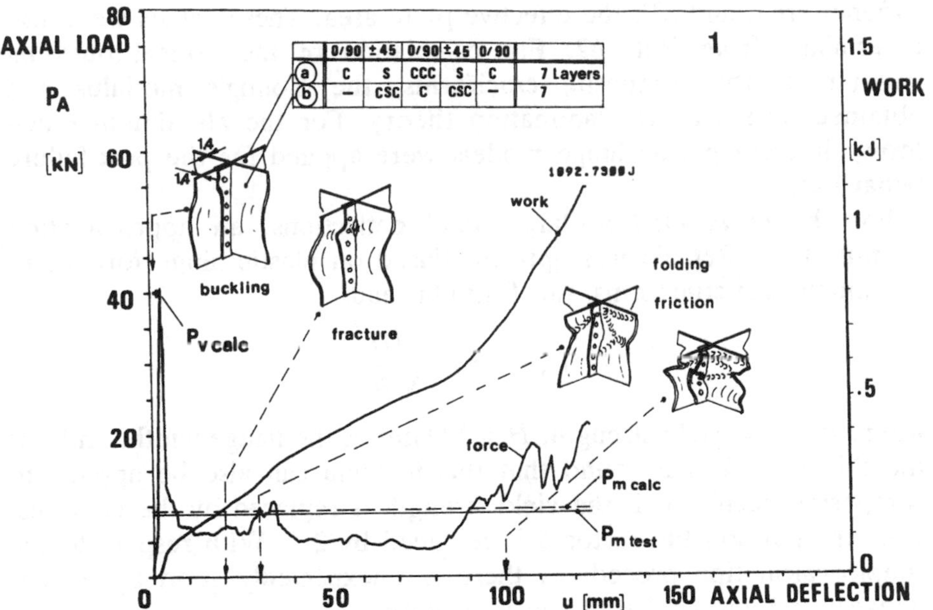

Fig. 26. Hybrid composite cruciform: test and calculation.

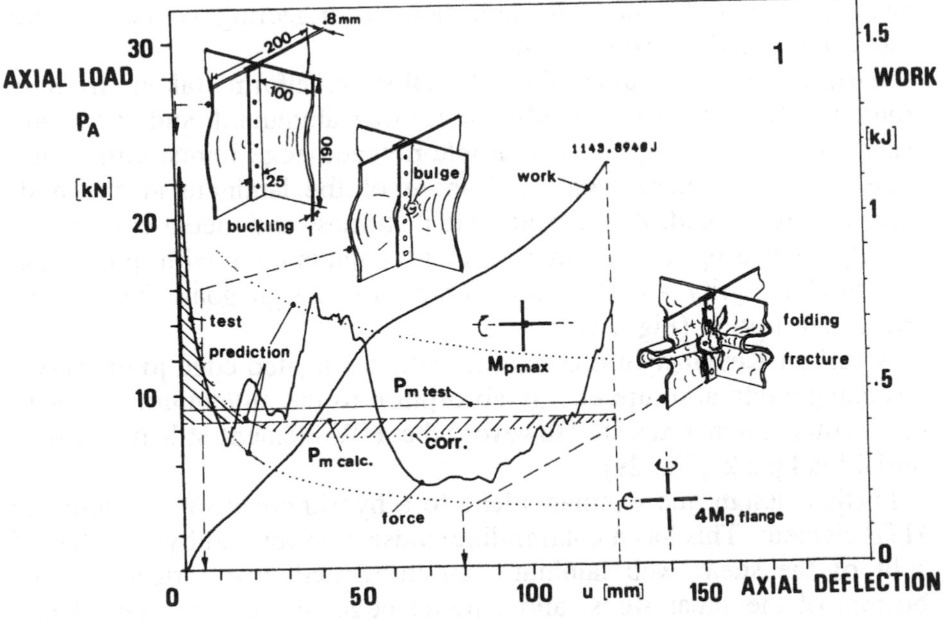

Fig. 27. Corresponding aluminium cruciform.

stiffener area, and λF_H the effective plate area. These values were also determined from Ref. 62. For calculation of the critical buckling strength of the composite cruciforms, the Young's modulus was obtained from classical lamination theory. For the aluminium cruciform, different plastic hinge models were applied for the post-failure behaviour.

For the *mean load* at large axial deflections, an approximation formula from Ref. 63 was applied, based on plastic hinge formation. For aluminium cruciforms, this formula reads

$$P_m \approx 5\sigma_F h^2 \sqrt{\frac{H}{h}}$$

where σ_F is the yield strength, $H = 100$ mm is the flange width, and h is the thickness. It was found that this formula can also be applied to composite cruciforms if the yield strength is replaced by the compressive strength and the factor 5 is replaced by 2·5. With respect to the simple calculation procedure, there is unexpectedly good agreement between test and analytical approximation.

Peak Load Triggering

Maximum load is determined by the stiffness of the cruciforms, and may be intolerable for crew and passengers in the case of a crash. Stiffness reduction is possible if thereby the capability of governing the operational loads is not affected. Several triggering concepts were tested, and are discussed below.

Starting from an aluminium baseline cruciform taken from a commuter-type aircraft (flat subfloor!), other aluminium and composite cruciforms having single- and multiple-notched edge joints, corrugated edge joints, and increasing AFC share of the laminate at the midsection were studied. A moderate initial stiffness and then a constant or slightly increasing crush force level in combination with post-crush structural integrity are the major cruciform design goals for vertical energy absorption (Fig. 28).

A carbon/Kevlar-hybrid cruciform with corrugated edge joints (HW) has comparable absolute energy absorption to the aluminium part with one center notch (AlN1). However, the HW element still has a high initial load peak (Fig. 29).

Further design improvements lead to a hybrid cruciform variant: the HTP element. This has a column-like midsection formed by a Y-shaped split of the shear web laminate, an integrated bevel trigger at the bottom of the shear webs, and tapered edge joints at the keel beam attachment. The keel beam and shear web laminates have a J-shaped

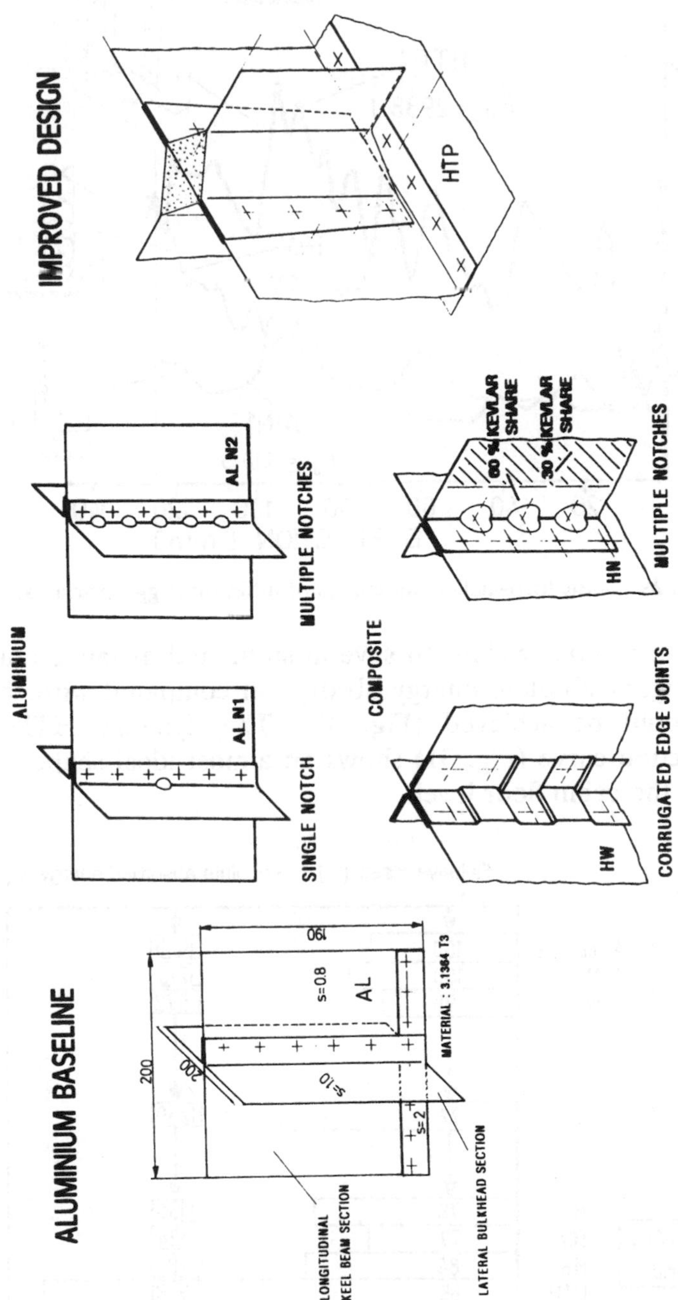

Fig. 28. Trigger variants of aluminium and composite cruciforms.

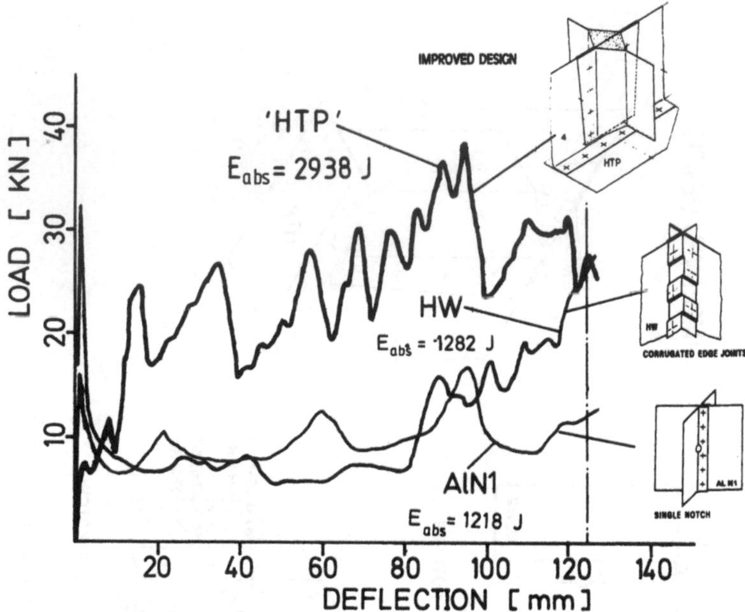

Fig. 29. Static load–deflection curves of different trigger variants.

connection to the outer skin. In several static and dynamic crush tests 3–4-times higher absolute energy absorption compared with the other elements could be achieved (Fig. 30). The dynamic HTP-element force–deflection curve (Fig. 31) shows an almost ideal shape for crush response at the cabin floor level.

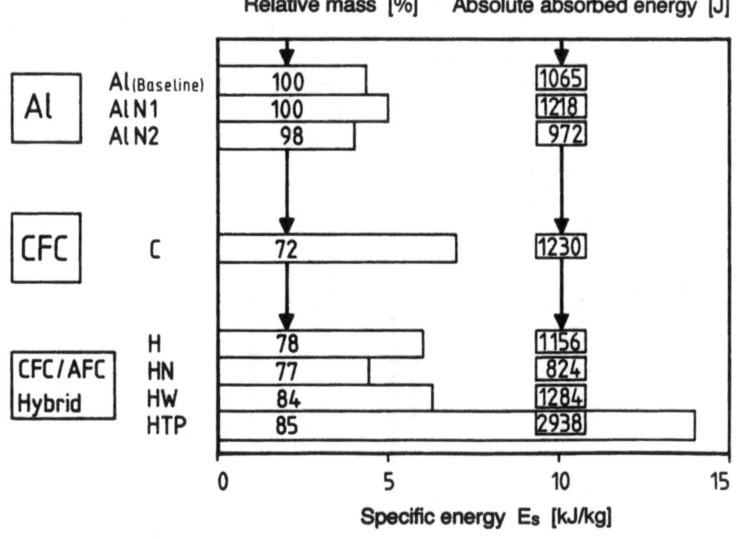

Fig. 30. Energy absorption of cruciforms.

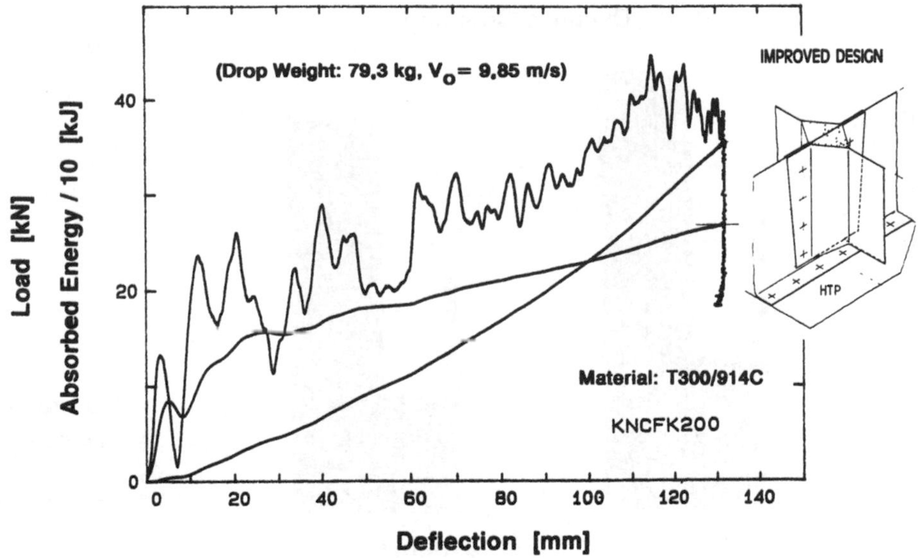

Fig. 31. HTP cruciform: dynamic load–deflection curve.

Compared with static loading, a slight drop in the dynamic crush performance (absolute energy, specific energy, crush force efficiency) is caused by a less regular folding and fracturing of the midsection, which provides the major part of the energy absorption with the HTP element (Fig. 32). The attached shear web laminates and the keel beams outside the midcolumn simply bend away in the single-element testing without making any considerable contribution to the energy absorption process.

From the cruciform test series, the following conclusions for cruciform crash resistant designs can be drawn.

- Multiple notching for peak failure load reduction results in low energy absorption.
- Post-crash structural integrity can be achieved by hybridization, for example, using a mixture of carbon and Kevlar laminates.
- Pure carbon cruciforms have high energy absorption and high weight savings (30%) compared with aluminium, but disintegrate completely during crushing. Hybrid elements have weight savings between 15–20%.
- Composite cruciforms show the same or even much higher (HTP element) absolute energy absorption compared with metal elements.

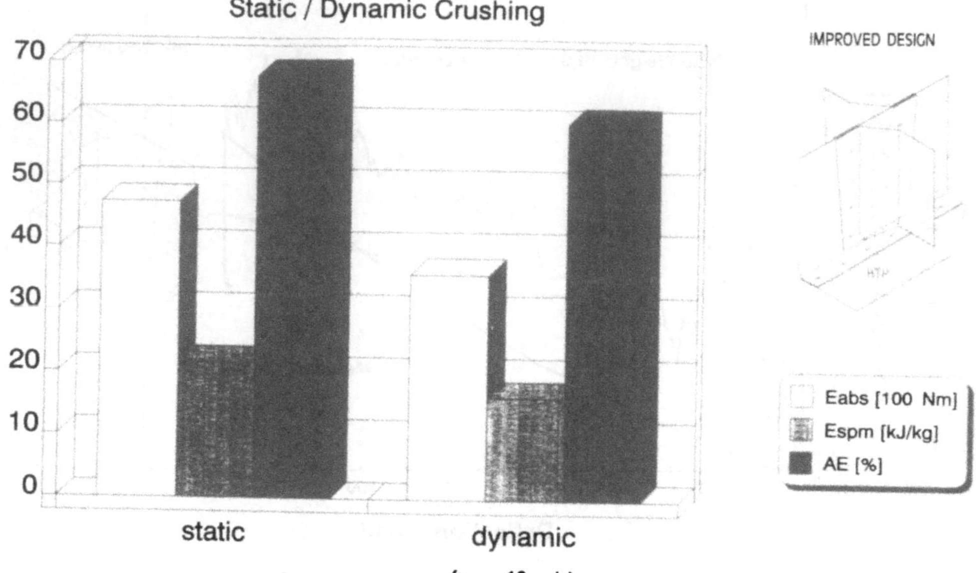

Fig. 32. HTP cruciform crushing performance.

3.4 Crash-Resistant Composite Subfloor Box

Based on the HTP cruciform test results, a crash-resistant CFC/AFC-hybrid floor box was designed in accordance with an aluminium baseline structure of a commuter airframe (Fig. 33). The box represents a part of the midsection of the subfloor structure, and the two keel beams carry the inner seat rails of the left and right seat rows. Three test articles were fabricated in autoclave technique and prepared for one static and two vertical drop tests ($v_0 = 10$ m/s). The two keel beams together with the inner and outer shear webs form four HTP cruciform-type elements. The keel beams and the shear webs are connected to the outer skin with J-shaped laminates at the bottom. All parts are bonded and riveted together. For testing, the cabin floor structure is simulated by an aluminium clamping plate.

The static test results for the floor box compared with the fourfold test values of a single HTP-element are shown in Fig. 34. The box fails first at 80 kN, a second peak occurs at 140 kN, and then a fairly constant average crushing force of about 100 kN develops up to a stroke of 135 mm. The superposition of the crush characteristics of four single HTP cruciforms up to 80 mm stroke shows that 80% percent of the energy is absorbed by folding and fracturing of the cruciforms, and about 20% by the beams and bulkheads between. Those failed by

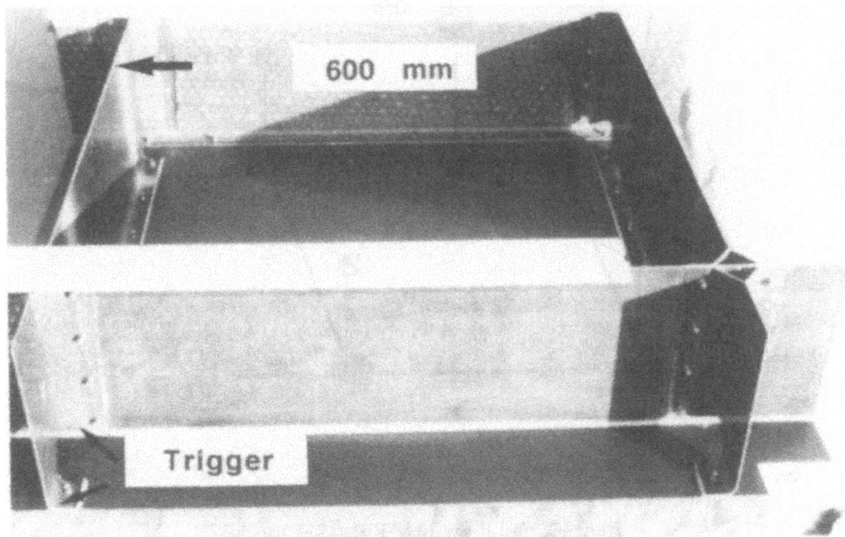

Fig. 33. Subfloor box design.

buckling and complete fracturing along the middle of their height. Therefore, for further crushing, the superimposed cruciform curve is even higher, and thc overall structural integrity of the box is poor. Considerable potential for crushing improvements of the keel beams and bulkhead sections is recognized, i.e. integral stiffening between the cruciform zones (beads, sine wave design) and toughening the webs'

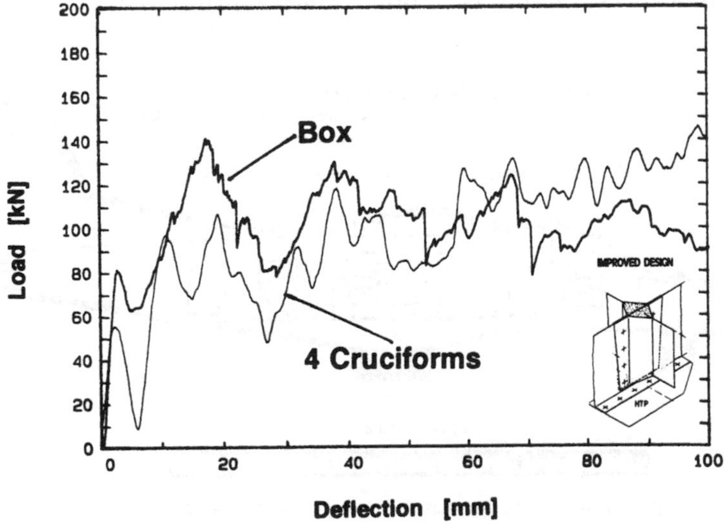

Fig. 34. Crush characteristics of subfloor box and superimposed cruciforms.

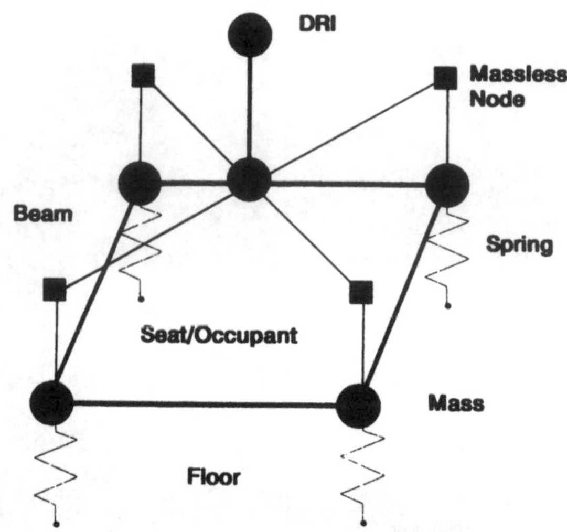

Fig. 35. Subfloor box KRASH model.

midsections by placing the Kevlar share at the bottom and top of the laminates.

For further evaluation of the design concept, a subfloor box KRASH model (Fig. 35) was generated using the composite HTP and aluminium cruciform (ALKN and ALN1) crush characteristics as nonlinear external spring inputs. The keel beams and shear webs are modelled as beam elements, and the mass distribution is represented by lumped masses. The model also contains a seat structure and an occupant as well as a DRI element. The latter is a single mass–spring–damper model and

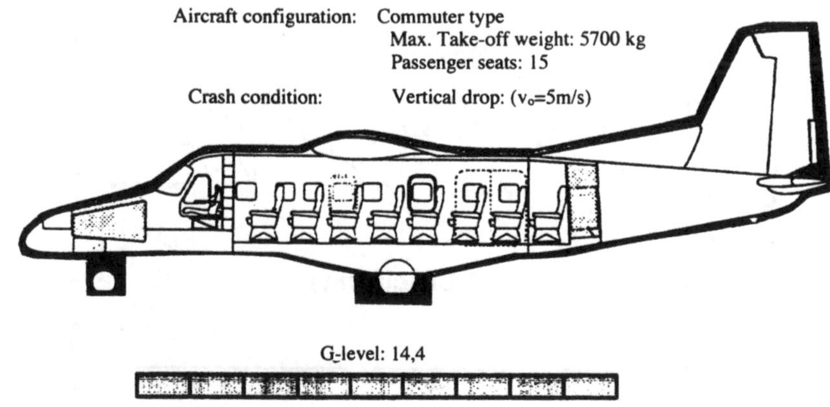

Fig. 36. Commuter aircraft crash case.

gives a rough idea of the probability of spinal injury by correlation with accident and sled test data.

The study is related to the crash case of a commuter-type aircraft (total gross mass 5700 kg) where the basic aluminium subfloor structure is replaced by 9 floor boxes containing 20 cruciform elements (Fig. 36). A pure vertical drop with 5 m/s initial impact speed is considered. The simulation results (Fig. 37) show that the aluminium subfloor bottoms

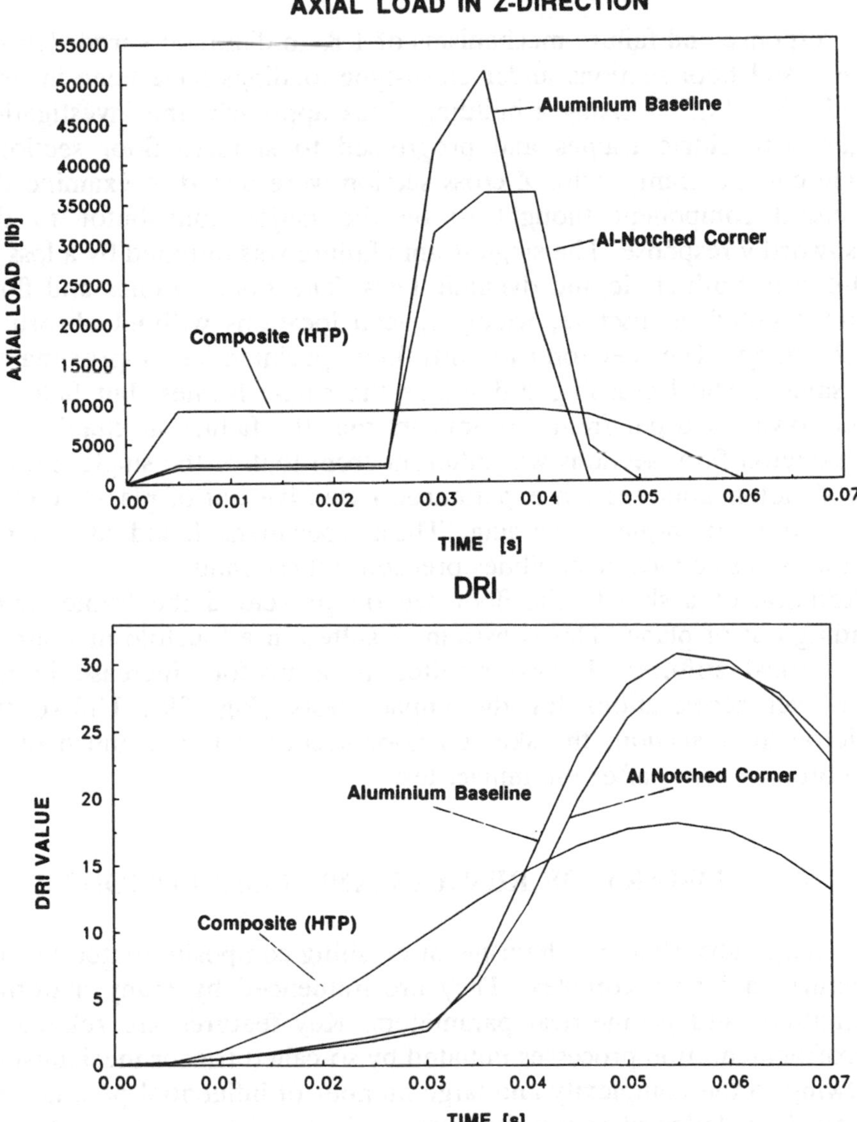

Fig. 37. Commuter aircraft crash case: selected KRASH simulation results.

out at 25 ms whereas the composite subfloor can absorb the crash energy at a floor acceleration level of about 14–15g. The DRI-level is about 18, which indicates less than 5% spinal injury potential. The DRI with the metal subfloor is about 30, caused by the bottoming out, and indicates above 50% potential of spinal injury. The next step will be complete aircraft modelling for crash simulations.

3.5 Fuselage Frame and Floor Sections

The response and failure mechanisms of 1·83 m diameter carbon/epoxy frames and floor sections under crash-type loadings have been investigated (Fig. 38).[36,42] Using a building-block approach, the investigation began with single frames and progressed to skinned floor sections. Single circular frames with Z cross section were tested to examine the structural component thought to be the major contributor to the crashworthy response. The single-frame failure was initiated by a loss of stability in both static and dynamic tests. The single frames and floor sections failed at discrete, widely spaced locations without absorbing much energy. The skeleton (no skin) floor specimen had approximately the same vertical crushing stiffness as the single frames, but failed at much lower load per frame. It appears that the failure mechanism for the skeleton floor sections was different from that of the single frames. The skeleton floor section experienced extensive out-of-plane bending in addition to in-plane bending. These specimens failed at notches without signs of local instabilities preceding these failures.

Addition of a skin to the floor section prevented the frames from bending out of plane. This constraint resulted in a fourfold increase in the vertical stiffness. It also resulted in a twofold increase in the floor-level accelerations for the impact tests (Fig. 39). Unlike the skeleton floor section, the skinned floor section retained much of its structural integrity after the impact test.

4 SUMMARY OF RESULTS AND CONCLUSIONS

The energy absorbing mechanisms of crushing composite materials and structures are very complex. They are influenced by many material, lamination, and geometrical parameters. Key features are related to microfragmentation processes initiated by so-called trigger mechanisms.

Owing to the complexity and large number of influential parameters, current knowledge of composite crushing is based on experiments. Up to now, detailed analytical modelling has been rather limited, and

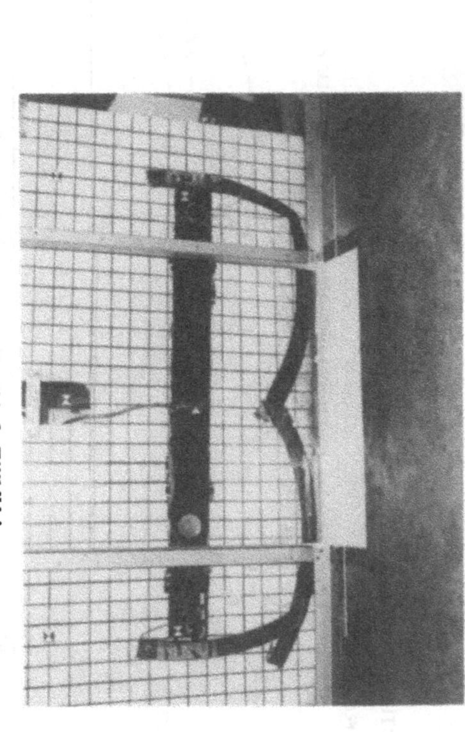

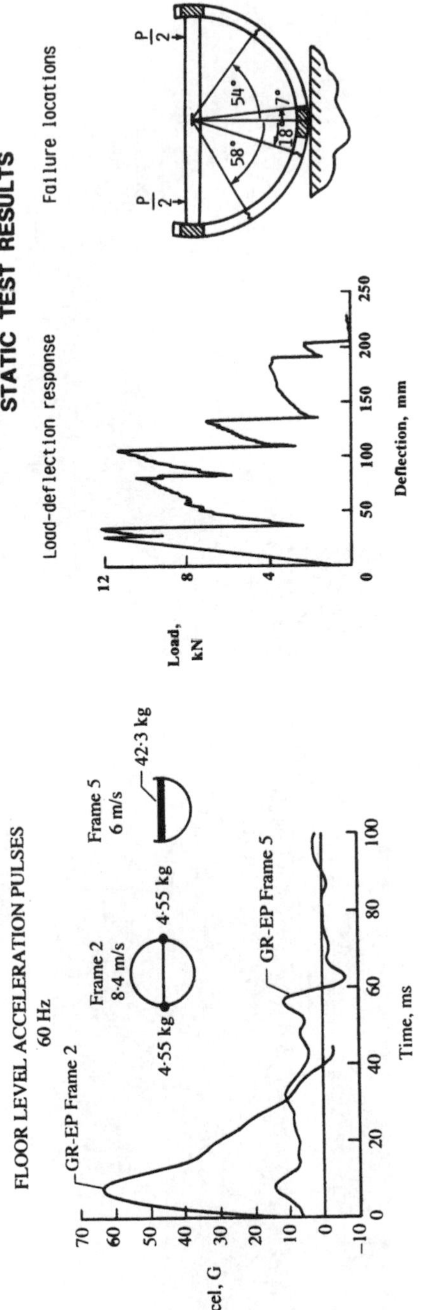

Fig. 38. Fuselage frame crash testing.[36,42]

SKELETON FLOOR SECTION AFTER 20 FPS IMPACT

SKINNED FLOOR SECTION AFTER 20 FPS IMPACT

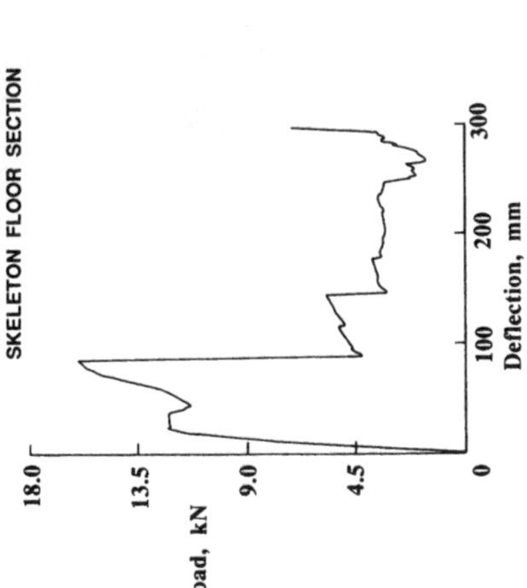

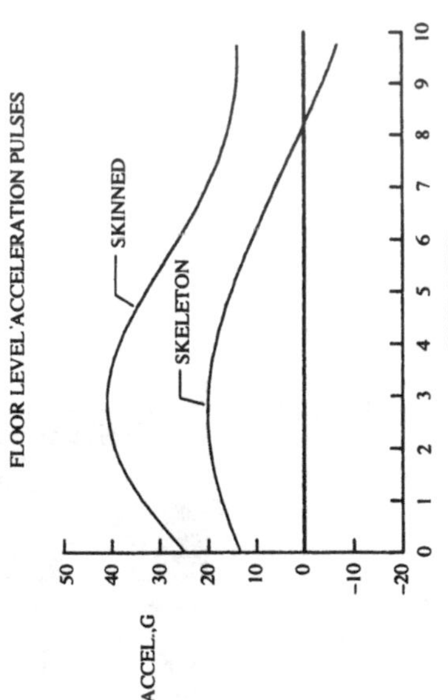

Fig. 39. Fuselage floor section crash testing.[36,42]

improvements can be expected in the future. Sophisticated models should take into account frictional effects, visco-elastic and visco-plastic behaviour of the constituents, and the microcracking response of the material.

Compared with metals, composite energy absorbing structures require different design concepts, which have to be further developed and verified for specific applications. A dual-function concept, i.e. load-carrying capability in combination with excellent energy dissipation by controlled post-failure behaviour, should be applied whenever possible. This should be included within the first design phases. Minor modification but effective concepts could limit cost and weight penalties.

Microfragmentation of composites generates smooth deceleration–time histories in crashing vehicles. Folding mechanisms, which typically occur with metals and tough composite material systems, produce highly undulated crush curves. This results in deceleration peaks and low crush-force efficiencies. Structural post-crash integrity is achieved by hybridization techniques, i.e. a mixture of brittle and tougher material systems.

Typical airframe substructures are subfloor beams, bulkheads, cruciforms, and frames. Such structures are important candidates for first design optimizations and the generation of a crash behaviour database.

Subfloor beams with corrugated webs (sine wave beams) are very efficient structures with respect to crushing and load bearing capability. Sine wave beam trigger designs need further emphasis for improvements, and should always be analysed or tested for structural performance. Also, the application of thermoplastic material systems is of great interest for improvements in manufacturing efforts.

Up to now, the predictions of sine wave beam and cruciform crushing have been based on tests, and have allowed initial estimates of peak and average crush force levels. Further developments of the procedures need more data input.

HTP elements are very efficient cruciform design concepts. Considerable weight savings could be achieved, and the absolute energy absorption performance is the same as or even much higher than for the counterpart aluminium structures.

Cruciform areas in HTP design absorb about 80% of the energy when placed into a subfloor box structure. Crushing and structural integrity between the cruciform zones are poor and have to be improved. Under simulated vertical crash loads, a simple KRASH model demonstrates the superior energy absorption management of composite floor boxes compared with the metal baseline design.

Experience in fuselage frame crushing is rather limited, and offers a broad field for further investigation. With CFC structures, major problems are catastrophic bending failures and loss of structural integrity. The latter can be improved by connections to the outer skin and hybridization with tougher materials.

For future vehicles—cars and aircraft—crash requirements will be of increasing importance with respect to operation licenses. Ultralight structural concepts will be required to account for low pollution, and effective engine concepts with respect to environmental aspects. Therefore composite crushing behaviour will be a matter of greater interest for the manufacturers. A large number of different composite element configurations have been crash tested in the past, and further tests must be carried out. Development of computer codes and approximation formulae must be continued. However, unrealistic emphasis on exaggerated accuracy should be avoided owing to the typical scatter of response of crashing structures.

REFERENCES

1. Barényi, B., Kraftfahrzeug, insbesondere zur Beförderung von Personen. Patentschrift Nr 854 147, Deutsches Patentamt, 1952.
2. Desjardins, S. P. *et al.*, Aircraft crash survival design guide. USAAVSCOM TR 89-D-22A-E, Vols I–IV, December 1989.
3. Military standard light fixed wing and rotary-wing aircraft crash resistance. MIL-STD-1290 A (AV), 24 September 1988.
4. Thornton, P. H., Energy absorption in composite structures. *J. Composite Mater.*, **13** (1979) 247–62.
5. Thornton, P. H. & Edwards, P. J., Energy absorption in composite tubes. *J. Composite Mater.*, **16** (1982) 521–45.
6. Thornton, P. H., Harwood, J. J. & Beardmore, P., Fiber-reinforced plastic composites for energy absorption purposes. *Compos. Sci. Technol.*, **24** (1985) 275–98.
7. Thornton, P. H., The crush behaviour of glass fiber reinforced plastic sections. *Compos. Sci. Technol.*, **27** (1986) 199–223.
8. Thornton, P. H. & Jeryan, R. A., Composite structures for atuomotive energy management. Presented at Autocom'87, 1–4 June 1987, Dearborn, MI.
9. Thornton, P. H., The crush behaviour of pultruded tubes at high strain rates. *J. Compos. Mater.*, **24** (1990) 594–615.
10. Czaplicki, M. J., Robertson, R. E. & Thornton, P. H., Non-axial crushing of E-glass/polyester pultruded tubes. *J. Compos. Mater.*, **24** (1990) 1077–100.
11. Czaplicki, M. J., Robertson, R. E. & Thornton, P. H., Comparison of bevel and tulip triggered pultruded tubes for energy absorption. *J. Compos. Sci. Technol.*, **40** (1991) 31–46.

12. Farley, G. L., Energy absorption of composite materials. *J. Compos. Mater.*, **17** (1983) 267–79.
13. Farley, G. L., A method of predicting the energy-absorption capability of composite sub-floor beams. NASA TM 89088, March 1987.
14. Farley, G. L., Crash energy absorbing sub-floor beam structure. *J. Am. Helicopter Soc.*, **32** (1987) 28–38.
15. Farley, G. L., the effects of crushing speed on the energy-absorption capability of composite tubes, *J. Compos. Mater.*, **25** (1991) 1314–29.
16. Farley, G. L. & Jones, R. M., Crushing characteristics of continuous fiber-reinforced composite tubes. *J. Compos. Mater.*, **26** (1992) 37–50.
17. Hull, D., Energy absorption of composite materials under crash conditions. In *Progress in Science and Engineering of Composites*, Vol. 1, ed. T. Hayashi, K. Kawata & S. Umekawa. ICCM-IV, Tokyo, 1982, pp. 861–70.
18. Hull, D., Axial crushing of fibre-reinforced composite tubes. In *Structural Crashworthiness*, ed. N. Jones & T. Wierzbicki. Butterworths, Guildford, 1983, pp. 118–35.
19. Vogt, H., Beardmore, P. & Hull, D., Crash Energie Absorption mit faserverstärkten Kunststoffen im Karosseriebau. *Kunststoffe als Problemlöser im Automobilbau.* VDI-Gesellschaft Kunststofftechnik, VDI-Verlag, Düsseldorf, 1987, pp. 239–68.
20. Price, J. N. & Hull, D., Axial crushing of glass fibre–polyester composite cones. *Compos. Sci. Technol.*, **28** (1987) 211–30.
21. Price, J. N. & Hull, D., The crush performance of composite structures. In *Composite Structures*, ed. I. H. Marshall. Elsevier, Amsterdam, 1987, pp. 2.32–2.44.
22. Berry, J. & Hull, D., Effect of speed on progressive crushing of epoxy–glass cloth tubes. In *Mechanical Properties at High Rates of Strain* (*Institute of Physics Conference Series* No. 70), ed. J. Harding. Institute of Physics, Bristol, 1984, pp. 463–70.
23. Fairfull, A. H. & Hull, D., Energy absorption of polymer matrix composite structures: frictional effects. In *Structural Failure*, ed. T. Wierzbicki & N. Jones. Wiley, New York, 1989, pp. 255–77.
24. Hull, D. & Coppola, J. C., Effect of trigger geometry on crushing of composite tubes. In *Materials and Processing—Move into the 90's*, ed. S. Benson, T. Cook, E. Trewin & R. M. Tuner, Elsevier, Amsterdam, 1989, pp. 29–38.
25. Pafitis, D. G. & Hull, D., Design of fiber composite conical components for energy absorbing structures. *Sampe J.* **27** (1991) 29–34.
26. Sigalas, I., Kumosa, M. & Hull, D., Trigger mechanisms in energy-absorbing of glass cloth/epoxy tubes. *Compos. Sci. Technol.*, **40** (1991) 265–87.
27. Hull, D., A unified approach to progressive crushing of fibre-reinforced composite tubes. *Compos. Sci. Technol.*, **40** (1991) 377–421.
28. Kindervater, C. M., Quasi-static and dynamic crushing of energy absorbing materials and structural elements with the aim of improving helicopter crashworthiness. In *Proceedings of 7th European Rotorcraft and Powered Lift Aircraft Forum, Garmisch-Partenkirchen, 8–11 September 1981.* Deutsche Gesellschaft für Luft- und Raumfahrt (DGLR), Paper No. 66.
29. Kindervater, C. M., Energy absorbing qualities of fiber-reinforced plastic tubes. In *Proceedings of AHS National Specialist Meeting on Composite*

Structures, Philadelphia, 23–25 March 1983. American Helicopter Society, Paper JV-5.

30. Bannerman, D. C. & Kindervater, C. M., Crashworthiness investigations of composite aircraft subfloor beam sections. In *Structural Impact and Crashworthiness*, Vol. 2, ed. J. Morton. Elsevier, Amsterdam, 1984, pp. 710–22.
31. Kindervater, C. M., Compression and crush energy absorption behavior of composite laminates. In *Proceedings of E/MRS Conference, Strasbourg, 1985*, pp. 289–97.
32. Bannerman, D. C. & Kindervater, C. M., Crash impact behaviour of simulated composite and aluminium helicopter fuselage elements. *Vertica*, **10** (1986) 201–11.
33. Kindervater, C. M., Gietl, A. & Müler, R., Crash investigations with sub-components of a helicopter lower airframe section. In *Energy Absorption of Aircraft Structures as an Aspect of Crashworthiness, Luxembourg.* AGARD Conference Proceedings No. 443, 1988, pp. 11.1–11.18.
34. Kindervater, C. M., Georgi, H. & Körber, U., Crashworthy design of aircraft subfloor structural components. *Energy Absorption of Aircraft Structures as an Aspect of Crashworthiness, Luxembourg.* AGARD Conference Proceedings No. 443, 1988, pp. 12.1–12.24.
35. Kindervater, C. M. & Scholle, K. F. M. G. J., Energy absorption behaviour of filament-wound Dyneema SK60/epoxy tubes. In New Generation Materials and Processes. Proceedings of 9th International SAMPE Conference, Milano, 1988, ed. F. Saporiti, W. Metari & L. Peroni, pp. 277–92.
36. Boitnott, R. L. & Kindervater, C. M., Crashworthy design of helicopter composite airframe structure. In *Proceedings of 15th European Rotorcraft Forum, Amsterdam, 12–15 September 1989.* European Helicopter Society, Paper No. 93.
37. Kindervater, C. M., Energy absorption of composites as an aspect of aircraft structural crash resistance. In *Developments in the Science and Technology of Composite Materials, ECCM4, Stuttgart, 25–28 September 1990,* ed. J. Füller, G. Grüninger, K. Schulte, A. R. Bunsell & A. Massiah, pp. 643–651.
38. Kindervater, C. M., Composite structural crash resistance. In *Proceedings of KRASH Users' Seminar, DLR Stuttgart, 3–5 June 1991.*
39. Sen, J. K. & Dremann, C. C., Design development tests for composite crashworthy helicopter fuselage. *SAMPE Q.* (Oct. 1985) 29–39.
40. Cronkhite, J. D., Haas, T. J., Berry, V. L. & Winter, R., Investigation of the crash impact characteristics of advanced airframe structures. USARTL-TR-79-11, Sept. 1979.
41. Cronkhite, J. D. & Berry, V. L., Investigation of the crash impact characteristics of helicopter composite structure. USA AVRADCOM-TR-82D-14, Feb. 1983.
42. Boitnott, R. L., Fasanella, E. L., Calton, L. E. & Carden, H. D., Impact response of composite fuselage frames. SAE 871009, 1987.
43. Boitnott, R. L. & Fasanella, E. L., Impact evaluation of composite floor sections. SAE 8910118, 1989.
44. Foye, R. L., Swindlehurst, C. W. & Hodges, W. T., A crashworthiness test for composite fuselage structure. In *Fibrous Composites in Structural*

Design, ed. E. M. Lenoe, D. W. Oplinger & J. J. Burke. Plenum Press, New York, 1980, pp. 241–57.

45. Provensal, J. & Cosatti, B., Dissipation d'énergie dans les structures en materiaux composite. VDI-Bericht Nr 369, 1980.
46. Kirsch, P. A. & Jahnle, H. A., Energy absorption of glass–polyester structures. SAE Paper 810233, 1981.
47. Schmuesser, D. W. & Wickliffe, L. E., Impact energy of continuous fiber composite tubes. *J. Engng Mater. Technol.*, **109** (1987) 72–7.
48. Mamalis, A. G., Manolakos, D. E. & Viegelahn, G. L., Crashworthy behaviour of thin-walled tubes of fiberglass composite materials subjected to axial loading. *J. Compos. Mater.*, **24** (1990) 72–91.
49. Gupta, V. B., Mittal, R. K. & Goel, M., Energy absorbing mechanisms in short-glass-fibre-reinforced polypropylene. *J. Compos. Sci. Technol.*, **37** (1990) 353–69.
50. Fleming, D. C. & Vizzini, A. J., The effects of side loads on the energy absorption of composite structures. *J. Compos. Mater.*, **26** (1992) 486–99.
51. Keal, R., Post failure energy absorbing mechanisms of filament wound composite tubes. Ph.D. Thesis, University of Liverpool, 1983.
52. Berry, J. P., Energy absorption and failure mechanisms of axially crushed GRP tubes. Ph.D. Thesis, University of Liverpool, 1984.
53. Fairfull, A. H., Scaling effects in the energy absorption of axially crushed composite tubes. Ph.D. Thesis, University of Liverpool, 1986.
54. Farley, G. L., Energy absorption capability of composite tubes and beams. Ph.D. Thesis, Virginia Polytechnic Institute and State University, 1989.
55. Zhou, W., Crash impact behavior of graphite/epoxy composite sine-wave webs. Ph.D. Thesis, Georgia Institute of Technology, 1989.
56. Maier, M., Experimentelle Untersuchung und Numerische Simulation des Crashverhaltens von Faserverbundwerkstoffen. Dissertation, University of Kaiserslautern, 1990.
57. Beardmore, P. & Johnson, C. F., The potential for composites in structural automotive applications. *J. Compos. Sci. Technol.*, **26** (1986) 251–81.
58. Hienstorfer, W., Crashsimulationsrechnungen und Bauteilidealisierung für einen Luftfahrzeugunterboden. *Z. Flugwiss. Weltraumf.* **11** (1987) 221–9.
59. Mai, U. & Hienstorfer, W. G., Die 'Soft Mosaik Theorie': Eine Methode zur Bestimmung der elastischen Kenngrößen und des Festigkeitsverhaltens von Gewebelaminaten. DGLR-Jahrestagung 1988, Darmstadt 20.–23.9.1988, Jahrbuch 1988 I.
60. Läpple, M., Schubtragverhalten von FVW-Wellstegen—Analytische Untersuchung mit Hilfe der FE-Methode. DGLR-Jahrestagung, Darmstadt, 21.–23.09.1988.
61. Georgi, H., Composite Wellstege—Prognose des Crashverhaltens. DLR Int. Bericht, IB 435 9/91, 1991.
62. Becker, H. & Gerhard, G., Handbooks of structural stability. NACA TN 3781–3785, 1957.
63. Hayduck, R. J. & Wierzbiecki, T., Extensional collapse modes of structural members. NASA Conf. Proc. 2245, 1982.

Chapter 7

CRASH RESPONSE OF COMPOSITE STRUCTURES

E. HAUG & A. DE ROUVRAY

Engineering Systems International SA, 94578 Rungis-Cedex, France

ABSTRACT

This chapter deals with the numerical simulation and prediction of the crash response of components and structures made of fiber-reinforced composite materials. While extensive crashworthiness simulation and prediction of metallic structures is by now well introduced into the everyday practice of virtually every car manufacturer, algorithms for the numerical crashworthiness prediction of components, and structures made of plastics and fiber-reinforced composites, are just emerging. For this reason, the literature on composite structural damage investigations is reviewed. Descriptions of the geometric and physical modelling of composite walls are given. Methods for the calibration of material laws using coupon tests are described. Based on the calibrated laws, validations of the numerical method are achieved by comparing simulations and experiments of simple, but representative, components made of composites. On passing these tests, the numerical methodology is ready to be applied to the crashworthiness prediction of full-scale composite structures. Calibration, validation and predictive simulations are described for the example of a full composite car crash simulation, using a finite element model with 22 000 composite multilayered thin-shell and sandwich-core brick elements. The success of the methodology is demonstrated and the validity of predictive industrial crashworthiness simulation of complex multimaterial composite structures is underlined.

1 BACKGROUND

This chapter may be considered a continuation of our contributions to the proceedings of the two previous conferences in this series.[1,2].

In Ref. 1 we discussed the just-emerging industrial application of the implicit and, in particular, explicit finite element time discretization methods to numerical simulations of static and dynamic crashworthiness test on structures and components made of thin-walled ductile sheet metal. At that time it became possible to numerically analyze simple structural components and structural subassemblies of arbitrary shape for crash events by using triangular and later underintegrated quadrilateral thin-shell finite elements, together with the diagonal mass explicit time integration technique, of the discretized nonlinear dynamic equations. It was hardly feasible, however, to analyze structural models that exceeded 2000–3000 thin-shell finite elements, because of the excessive CPU computer times spent and storage required with the departmental Vax computers or CDC mainframes then available.

This situation changed dramatically when the first Cray-1 'super'-computers became accessible by industry, such as the Cray-1 of the state-operated IAGB Institute (Munich), which was hired to host a frontal crash simulation on a VW Polo car by Volkswagen in 1985/86. The successful prediction of the acceleration–time history at this car's passenger compartment level, with the emerging industrial crash simulation code PAM-CRASH, using a thin-shell and beam finite element model of about 5600 elements, constituted one of the major breakthroughs towards the now massive industrial use of numerical crash simulation for the design of transport vehicles. Of particular importance was the fact that the job could be run overnight, enabling the engineers to analyze the results during a working day and to resubmit a variation of the analysis for another night run. In time, many major car companies in Europe, and worldwide, justified the acquisition of their own supercomputers by the need to perform competitive crash simulations, and nowadays industrial crash computations of transport vehicles with metallic structure have become standard and are fully integrated into the modern transport vehicle design cycle.

In Ref. 2 we discussed the application of numerical methods to predict fracture and failure of brittle and composite materials. The discussed methods were then applied to the prediction of subcritical material damage, such as distributed matrix cracking before fiber failure in fiber-reinforced composites. The discussions also included crack propagation up to composite material failure in tension, compression and bending, including fiber rupture, of composite material test pieces. The geometries considered comprised smooth specimens, center- or edge-notched unidirectional, multilayered or woven fabric coupons, compression coupons and three- and four-point bending specimens. The macrocracking behavour of brittle ceramics was studied by simulation of compact tension test pieces.

An important part of this work reflected a long-standing collaboration between ESI Paris and DLR Braunschweig under contract with ESA/ESTEC of the European Space Agency. In order to better understand the subcritical damage and ultimate failure behaviour of fiber-reinforced composite materials, such as carbon–epoxy (CE) composites, ESTEC requested DLR Braunschweig to create experimental tensile and compressive fracture maps of CE test pieces. Simultaneously, ESI was asked to perform numerical simulations on representative test pieces to investigate the initiation and progression of subcritical and critical matrix and fiber damage up to and including total specimen failure with fiber breaks. This dual numerical and experimental approach fostered the general progression of models for the numerical simulation of fracture and damage of bi-phase composite materials on the one hand, and a much better understanding of the physical test results on the other. For, if accompanied by numerical simulations using detailed calibrated models, insight into the damage, fracture and failure mechanisms of the composite stackups is largely enhanced. Such stackups should be considered structures rather than materials, and the simulation can easily extract behaviours and results that may remain hidden when relying exclusively on physical testing. Once reliable composite damage and fracture models are available, another benefit of simulating coupon tests resides in the fact that parametric studies and optimizations of stackups, notch geometries, etc., can be made in the computer. This can reduce the amount of physical experiments to the necessary minimum. It is therefore not surprising that physical testing campaigns now tend to be accompanied by numerical simulations, leading to a genuine partnership between the numerical analyst and the physical experimenter.

In this chapter we report on an industrially successful combination of the work and the numerical methodologies we have outlined and elaborated in our two previous contributions in this series.[1,2] The discussion centers around one of the first full-scale industrial crash simulations, and comparison with full-scale experiments, of a prototype passenger sports-car cabin structure, entirely made of composite materials.

The sports car was fabricated by the Japanese petroleum-refining Tonen Corporation as a demonstration prototype car to promote the use of composite materials in the automotive industry. They built two cabins, one equipped with front and rear suspension, steering assembly, rear engine, seats, outer skin and wheels, and the other cabin unequipped, for utilization in static tests for structural stiffness, dynamic vibration tests and, ultimately, for frontal- and side-impact destructive crash tests. The principal aim of the crash tests was to demonstrate the basic capacity of the structure, made of new materials, to effectively absorb kinetic energy. The basic feasibility of numerical crashworthiness analysis and prediction of complex, realistic, structures made of composites also had to be demonstrated. For, if car manufacturers are to employ these new materials, they wish to be able to simulate the response of structures made of such materials in the same way as they can rely today on static and dynamic FE simulation codes for the analysis of conventional steel automotive structures, including complete full car crash simulations.

For this prototype car, the material was arranged into a basic sandwich wall structure, with multilayered composite sandwich facings and an aramid hexagonal honeycomb core. The facings were made of Tonen's pitch-based FT500 high-strength/high-modulus unidirectional carbon–epoxy plies, plus Kevlar K49–epoxy woven fabric plies and Toray pan-based T300 carbon–epoxy woven fabric plies. This basic sandwich wall structure was used throughout the cabin structure of the prototype two-seater sports car, with different sandwich-core thicknesses and with various reinforcing layers added locally to the basic ply stackup.

Before going into a detailed description of the above mentioned exemplary full-scale composite car crash simulation, an overview of crash energy absorption capacity and mechanisms of generic composite components, such as monowalled multilayered circular and rectangular composite tubes, beams and box columns, is given. The few examples of full car or partial subassembly crash tests found in the literature are cited, and the usefulness of the so-far absent numerical simulation methods is underlined.

The modelling of composite wall material stiffness, damage, fracture and failure behaviour, with special emphasis on continuous fiber-reinforced unidirectional plies and plies made of woven fabric, using a bi-phase composite material description, is then reviewed, including sandwich wall structures. Implicit and explicit finite element codes for analysis are mentioned. In the application sections the material models are calibrated on a specific series of coupon tests for the purpose of modelling the composite car cabin of Tonen's prototype sports car. Once calibrated, the models are first validated on the crash behaviour of typical subassemblies of the car structure, which were isolated, built, crash-tested and simulated. The simulations with the calibrated models were compared with the detailed experimental results, carefully recorded for each tested component, which allowed the validity of the elaborated simulation models to be ensured by adjustment and fine tuning for optimal mesh densities and through thickness modelling of the sandwich walls and of material parameters, etc.

Once calibrated and validated, the numerical models are applied to simulate the full frontal crash and a concentrated pole-side impact test on the full car model. The finite element model of the analyzed symmetric half of the full cabin had about 22 000 shell and brick elements, which is within the range of the now commonly used models for cars made of steel or aluminum.

2 CRASHWORTHINESS OF COMPOSITE COMPONENTS

This section contains remarks on the basic capacity and mechanisms involved for simple composite structural components to absorb kinetic energy, based on a study of the available literature. The interested reader is referred to Refs 3–18 for more detailed information, and only a summary overview can be given here.

2.1 Axial Compression Failure of Circular Cylindrical Composite Tubes

Numerous experimental studies have treated the subject of the specific energy absorption of centrally compression loaded composite tubes.[3–9] The tubes have been shown to perform at their best when global Euler column buckling and local wall buckling with corresponding bending collapse modes has been precluded, i.e. when the geometric, material and loading conditions were such that axial failure of the compressed tubes was characterized by the progression of a destructive zone of constant size near the loaded end, also called the crash 'process zone', or 'crash frond'.

The aim of the designer is to arrange the column material such that the destructive zone can progress in a stable fashion, the energy absorption is as high as possible, and a sustained high-level crushing force will develop with little fluctuation in amplitude as the process zone travels with the crush loading along the component axis. Further, the wall destruction should be initiated smoothly, by avoiding large initial peak resisting forces, which might cause global wall or column buckling rather than the beneficial local wall destruction mode. If these conditions are satisfied, the component can absorb a large amount of kinetic energy having a small mass, i.e. the component will have a high specific energy absorption capacity, which is energy absorbed divided by the component mass.

Reference 3, for example, gives an overview conducted by NASA Langley on the respective roles of fibers and of the matrix of composite materials in crash energy absorption, based on static crushing tests of tube specimens made of graphite/epoxy, Kevlar/epoxy and hybrid combinations of these materials. The influences of fiber and matrix constitutive properties and of the laminate structure were investigated experimentally. The fiber and matrix ultimate strain levels were found to significantly affect energy absorption, as well as the stacking sequence. Admixture of Kevlar fibers in hybrid materials was found to preserve structural integrity after crushing.

Reference 4 presents results obtained by General Motors on impact energy absorption of continuous fiber composite tubes, fabricated from graphite/epoxy, Kevlar/epoxy and glass/epoxy composites. Graphite/epoxy tubes had higher specific energy absorption for axial crush loads than tubes made from the other materials for the same ply stackup. Angle ply tubes made of graphite and glass/epoxy exhibited brittle failure modes with fiber splitting and ply delamination, while the more ductile Kevlar/epoxy tubes failed in accordion buckling modes similar to soft steel or aluminum tubes.

Reference 5 gives a general introduction to the subject of composite tube crushing, and also compares the failure mechanisms of axially crushed metallic tubes with the failure mechanisms of composite tubes. Figure 1(a) shows catastrophic and progressive failure modes of brittle material and composite tubes and the corresponding low- and high-amplitude crushing force–displacement responses. Figure 1(b) shows the behaviour of a ductile metal tube. All figures are schematic, and illustrate the basic types of tubular crushing load responses. While catastrophic failure must be avoided by structural measures because of its insufficient energy absorption capacity, progressive failure of composite tubes can absorb considerable amounts of energy with low overall mass of material involved. The post-peak force response of properly crushed composite tubes is seen to be uniformly high when a crushing process zone develops near the loaded end of the tube. This zone steadily precedes the load along the tube axis towards the fixed end of the tube. On the other hand, the force response of a collapsing ductile metal tube is less uniform, which is due to the repeated formation of diamond shaped, accordion or 'dog-bone' shaped elasto-plastic buckles that will initiate, progress and compact themselves repeatedly.

Figure 2 gives an overview of the desired progressive composite tube crushing failure mode. In Fig. 2(a) the necessity to trigger the beneficial progressive tubular crushing mode is underlined. Figure 2(b) schematically depicts the sources of energy absorption, while Fig. 2(c) is the corresponding schematic force–displacement curve. Figure 2(d), finally, gives a close-up view of the crushing region of a multilayered composite tube wall under axial compression.

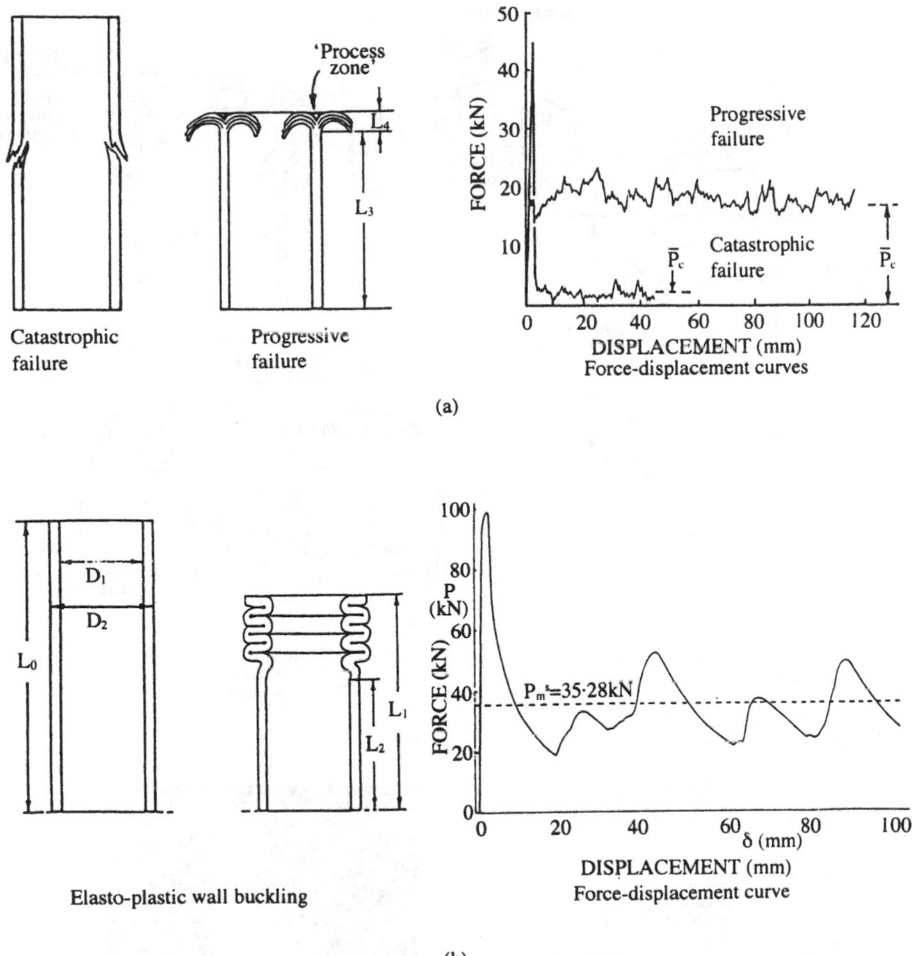

Fig. 1. Typical collapse modes and force–displacement curves for composite material tubes (a) and ductile metal tubes (b) under axial crushing loads. (After Ref. 5.)

References 6–10 discuss effects of speed of loading on progressive crushing of epoxy-glass cloth tubes,[6] the effects of specimen dimensions on the specific energy absorption of composite tubes,[7] the effect of friction between the tube specimen and the compression plate and in the crush zone,[8] and the effect of the trigger geometry on crushing of composite tubes.[9,10] All effects are described, based on numerous experiments carried out at the Department of Materials Science and Metallurgy, University of Cambridge, under Professor D. Hull. Reference 8 also gives a brief review of material, structural and testing parameters that affect the response of tubes and cones to axial compression. The interested reader is referred to the cited literature for more details.

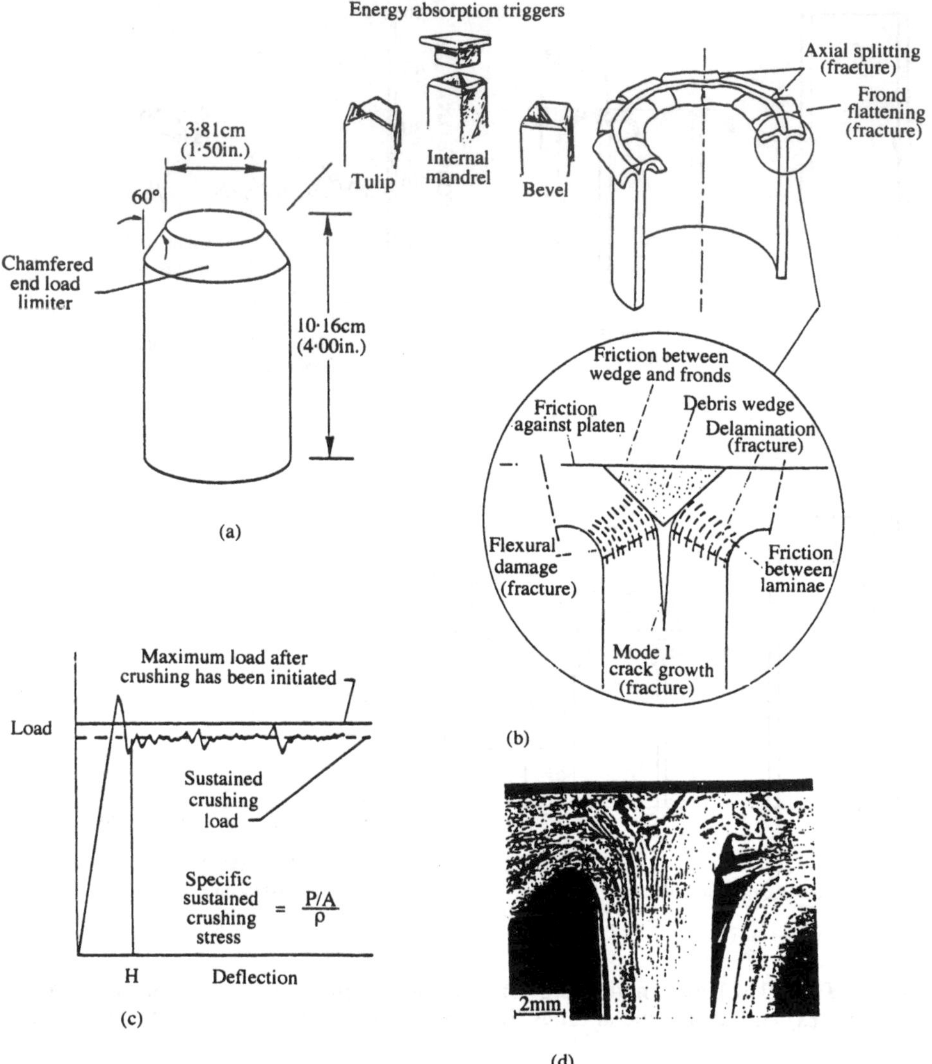

Fig. 2. Crash front development of axially crushed composite tubes (schematic). (a) Progressive crushing mode triggers (after Refs 3 and 11). (b) Representation of the crash zone, illustrating sources of energy absorption (after Ref. 8). (c) Load–deflection curve of composite tube specimen (after Ref. 3). (d) Crush region of composite tube wall under axial compression (after Ref. 11).

Figure 3 shows some experimentally obtained peak and post-peak force–deflection curves for various fiber-reinforced plastic tubes under axial compression, most with remarkably stable sustained crushing loads. The influence of the material (graphite, glass, Kevlar) is shown, including a far less uniform post-peak force level of a Kevlar tube that

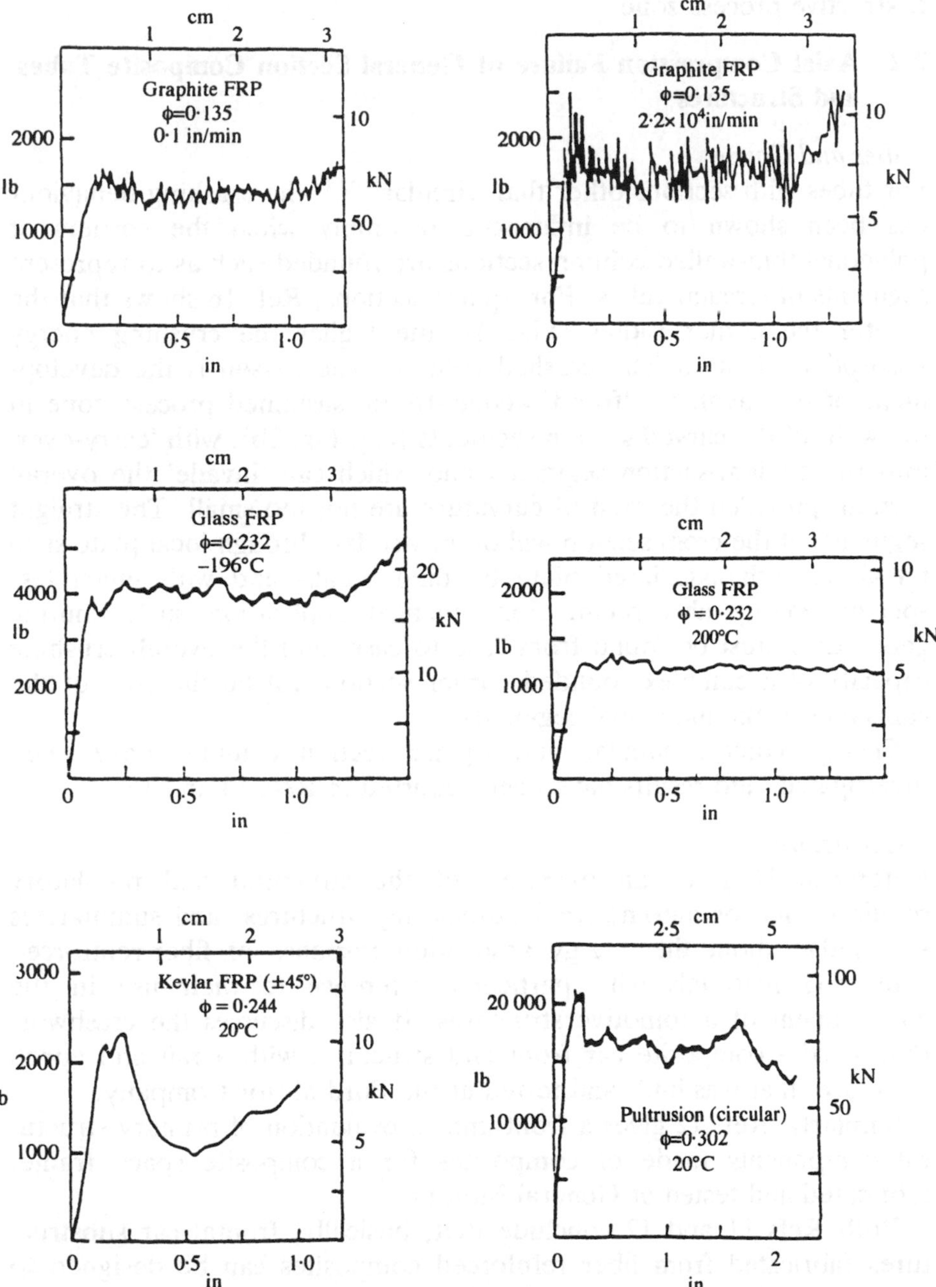

Fig. 3. Force–deflection curves for axially crushed composite tubes made of different materials (ϕ is the relative density). (After Ref. 11.)

exhibits elasto-plastic buckles without the formation of a localized destructive process zone.

2.2 Axial Compression Failure of General Section Composite Tubes and Structures

Tubes and Cones

For tubes with sections other than circular[11,12,16] the crushing behaviour has been shown to be influenced favorably when the corners of polygonal thin-walled column sections are rounded such as to represent segments of circular tubes. For square sections, Ref. 16 shows that the greater the corner radius (Fig. 4), the higher the crushing energy absorption of the axially crushed column. The reason is the development of the favorable 'frond–wedge–frond' sustained process zone in the walls of the curved section segments (e.g. Fig. 2b), with 'carry-over' into the straight-section segments and which can 'invade' the overall section, provided the radii of curvature are not too small. The straight segments of the cross section will otherwise fail through local plate strip buckling, with associated plate bending breaks and with much less specific energy absorption. For practical application such limiting geometries must be found from case to case, and the overall crushing capacity of a complex rounded corner section will be the sum of the capacities of the individual segments.

Finally, conical circular and square section columns have been investigated, and results have been reported in Refs 14 and 15.

Substructures

Reference 11 gives an overview of the structural and regulatory requirements for automotive load-bearing structures, and summarizes knowledge about the energy absorption response of fiber-reinforced composite materials with particular reference to their use in the environment of automotive structures. It also discusses the crashworthiness of a composite car front end structure, with a rail and apron structure, that was built and tested at the Ford Motor Company.

Similarly, Ref. 12 gives a front-impact evaluation of primary structural components made of composites for a composite space frame, fabricated and tested at General Motors.

Both Refs 11 and 12 conclude that, basically, frontal car substructures fabricated from fiber reinforced composites can be designed to manage crash energy absorption in a manner comparable to steel components. Finally, Refs 17 and 18 give general overviews of the impact response of structural composites and energy-absorbing composite structures in automotive and aerospace structures.

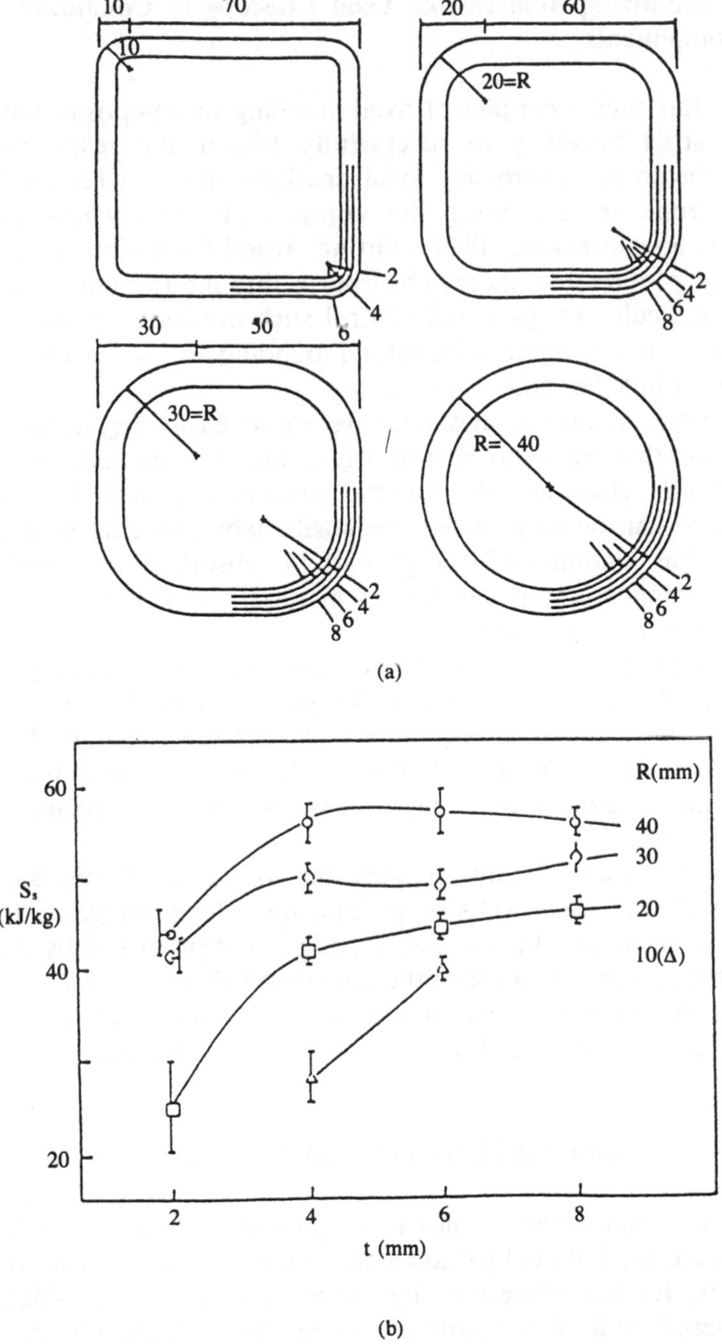

Fig. 4. Axial crush resistance of rounded-corner square-section composite tubes. (a) Cross-sectional geometry of tubes (all dimensions in mm). (b) Variation of specific energy absorption S_s with wall thickness t (all corner radii). (After Ref. 16.)

2.3 Summarizing Remarks on Axial Crushing of Composite Components

Most of the cited examples of axial crushing of composite tubes hinge on the strict necessity to successfully trigger the respective highly energy-absorbing progressive axial crushing mode, where a localized process zone travels along the component axis, which essentially provokes fiber breaking, fiber splitting, frond formation, matrix cracking, delamination and energy absorption through friction. In practice it may be difficult to trigger and control such modes when the loading is one sided, or combined with lateral bending, or when the triggering mechanism fails for other reasons.

Alternative failure modes that are more easily triggered and controlled, or that need no special triggering devices such as inversion tubes, bevels, chamfers, etc. may therefore be of practical interest. One such mode can be to provoke composite fabric tearing modes, where considerable amounts of energy can be absorbed by membrane or lateral tearing of composite fabrics in modes similar to the modes of tearing of a sheet of paper.

Membrane tearing modes can be obtained, for example, through 'petalling' deformation of composite plates or of the walls of components upon normal impact of a concentrated mass, or by straight tearing through lateral impact (Fig. 5). In each case membrane tensile stresses are created in the composite walls, leading to highly energetic tearing modes.

Recently, lateral tearing modes of composite fabric sheets, for example in sine wave webs of girders, have been triggered to absorb energy upon impact. In that case tearing is not provoked by membrane tensile stresses in the plane of the composite sheet, but by out-of-plane opposite shear loads, which cause a highly energetic tear to travel along with the applied loading. Figure 5 illustrates this approach.

3 MODELLING OF COMPOSITE WALLS

Figure 6 schematically shows a possible finite element three-dimensional brick model (or axisymmetric solid model for axisymmetric responses) for the numerical simulation of axial wall crushing for the multilayered wall of a composite component. Each ply in the wall stackup could be modelled by a layer of solid elements (Fig. 7a), and each element can have bi-phase material properties (Fig. 7d) or be described by other material models.

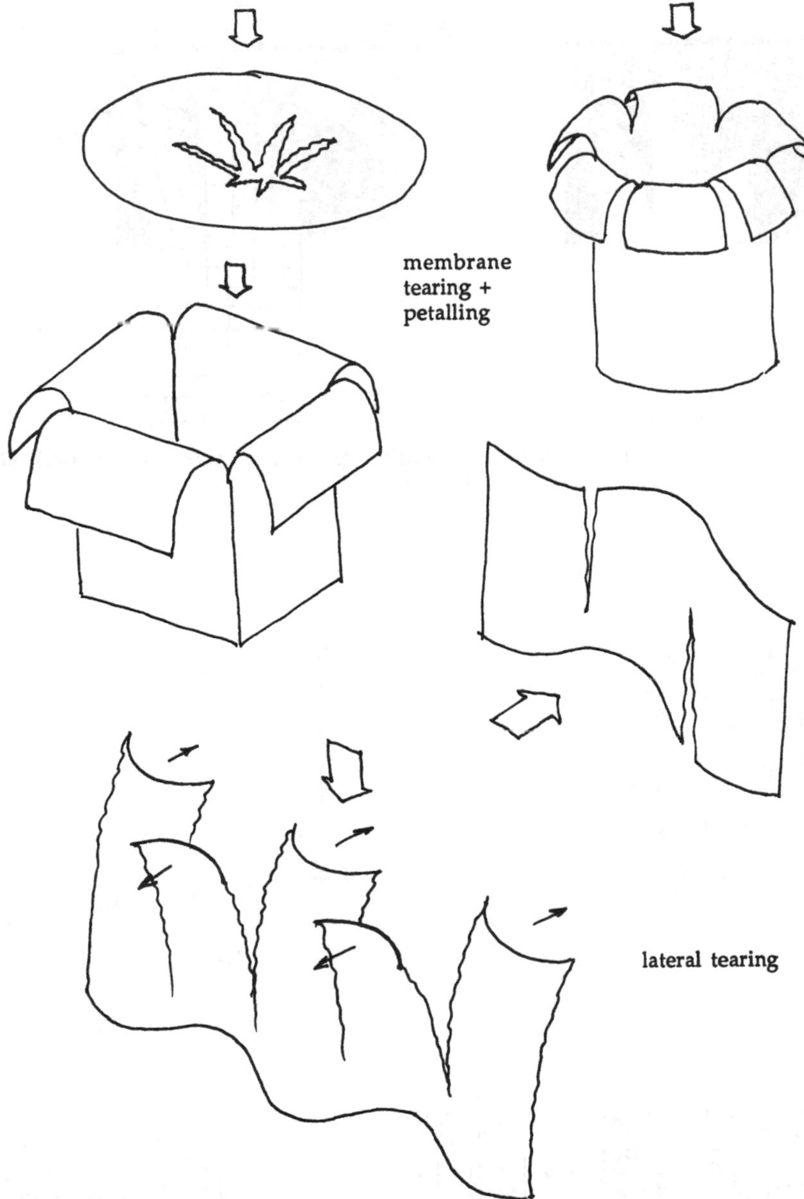

Fig. 5. Membrane + lateral tearing of composite fabric sheets.

Because of high cost, such detailed models cannot be used in a global crash simulation of a car structure. Such models can nevertheless be used to obtain the sustained mean crushing force levels, for example see Figs 2 and 3, for various composite wall stackups during the design phase, and therefore help designing the composite wall structure. If

Fig. 6. Detailed FE mesh for analyzing repetitive destructive cycle (slide model) (after Ref. 11).

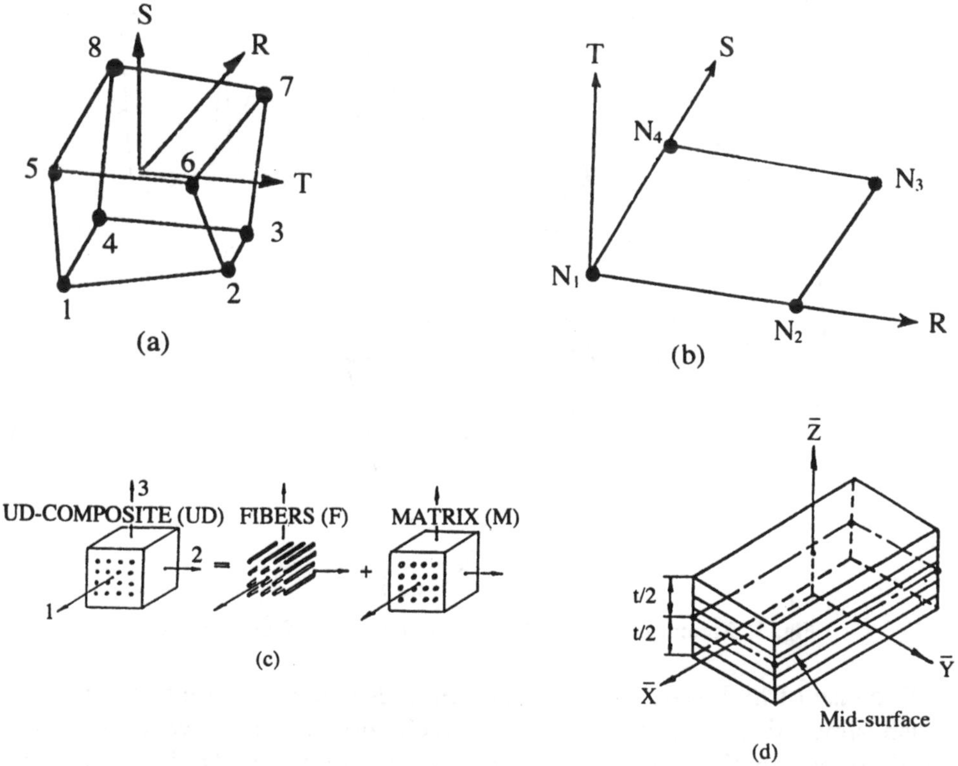

Fig. 7. Finite elements for composites: (a) solid element; (b) shell element (c) bi-phase material; (d) layered shell.

such models are not made, tests must be performed in order to obtain the mean wall crushing force experimentally.

3.1 Equivalent Shell Models

Figure 8 contains equivalent single-wall shell or sandwich-wall shell–brick–shell models of composite components, where the entire component wall stackup is represented by 'macro' shell elements, either in single-wall or sandwich-wall components.

These types of models can be used efficiently in a global composite component or in a car crash analysis. However, the equivalent elastic and inelastic (failure) properties of the shells must be calibrated beforehand, either on detailed axial crush simulations to evaluate the average component crushing force levels on models on the wall micro scale or by using corresponding experimental results.

Figure 7(b) shows a four-node shell finite element that can be used for this type of simulation, either as a monolayered or multilayered shell.

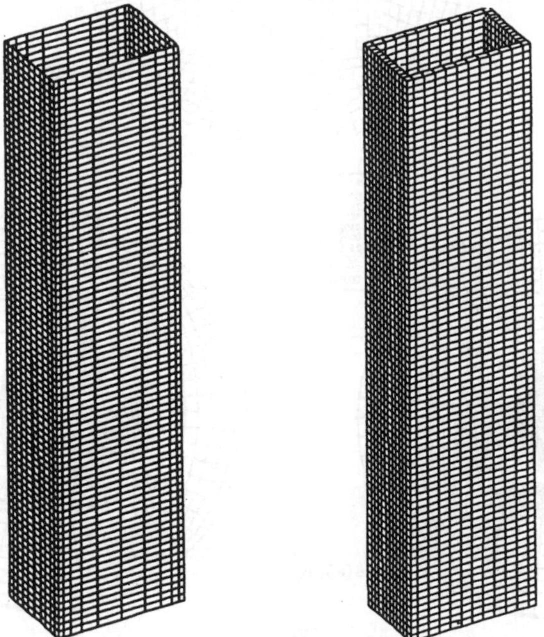

Fig. 8. Equivalent thin-shell and brick FE models of brittle-fiber composite tube (schematic).

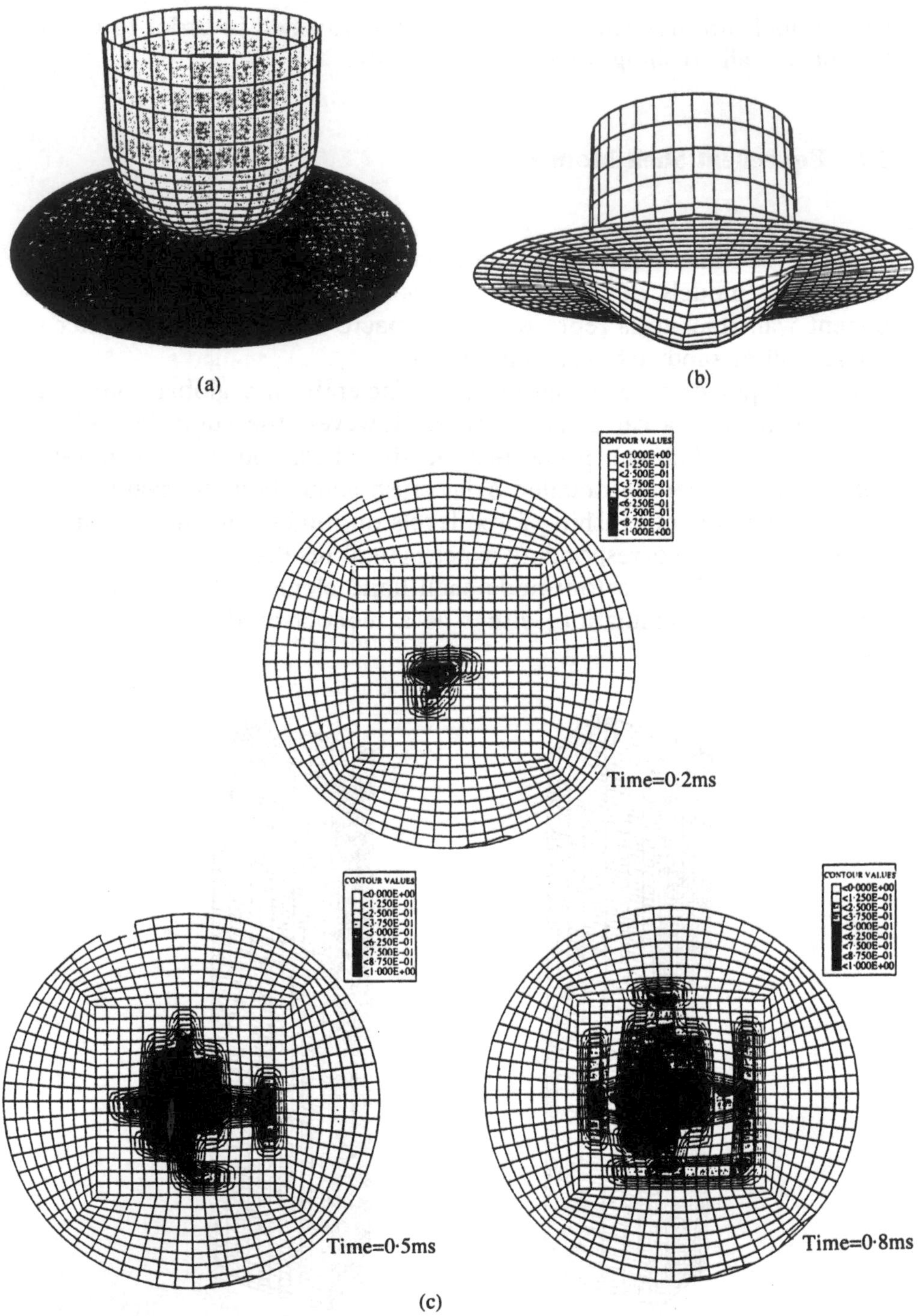

Fig. 9. SMC disc impacted by a rigid impactor (program PAM-CRASH): (a) undeformed mesh and geometry (top view); (b) deformed structure at 0·8 ms; (c) damage development in the SMC disc. (After Ref. 19.)

Monolayered Shell
The equivalent properties of a monolayered shell must be found through 'homogenization' of the real multilayered shell properties. Either the equivalent properties can be evaluated directly from coupon test results or detailed numerical models (e.g. using bi-phase brick elements) of the coupons can be made and the equivalent homogeneous shell element properties can be deduced from the detailed model simulations.

Figure 9 gives an example of an equivalent (random fiber-reinforced) monolayered shell model of a circular plate, penetrated by a rigid bullet (program PAM-CRASH).

Multilayered Shell
Figure 7(d) shows the thickness of a multilayered shell, which can consist of *N* layers of different bi-phase materials, with fibers oriented at angles with respect to a reference direction of the shell element midsurface. In this model each bi-phase layer (fibers plus orthotropic matrix) can fail individually according to its measured properties, which makes this element better suited for multimaterial multilayered stack-ups. Note that the model permits representation of one physical layer of a stackup by more than one shell layer (integration point), and, vice versa, one shell layer (integration point) may represent one or several physical layers.

The multilayered shell is also better suited to simultaneously represent *both* the correct elastic properties and failure behaviour for *in-plane loading* and for *plate bending,* which is difficult to achieve, for example, with the homogeneous equivalent shell model.

Sandwich Shell
Plates and shells made of a sandwich wall structure can be modelled effectively by a shell–brick–shell wall model. Each sandwich facing is represented by a thin shell, and the sandwich core by a brick element. The thin shells may be multilayered, and the core brick material properties can be orthotropic, so as to model honeycomb cores, or isotropic for foam cores.

3.2 Composite Material Models

The following material laws and failure descriptions appear best adapted to the description of composite failure, and they are available in the finite element simulation programs PAM-CRASH and PAM-FISS used for the worked examples.

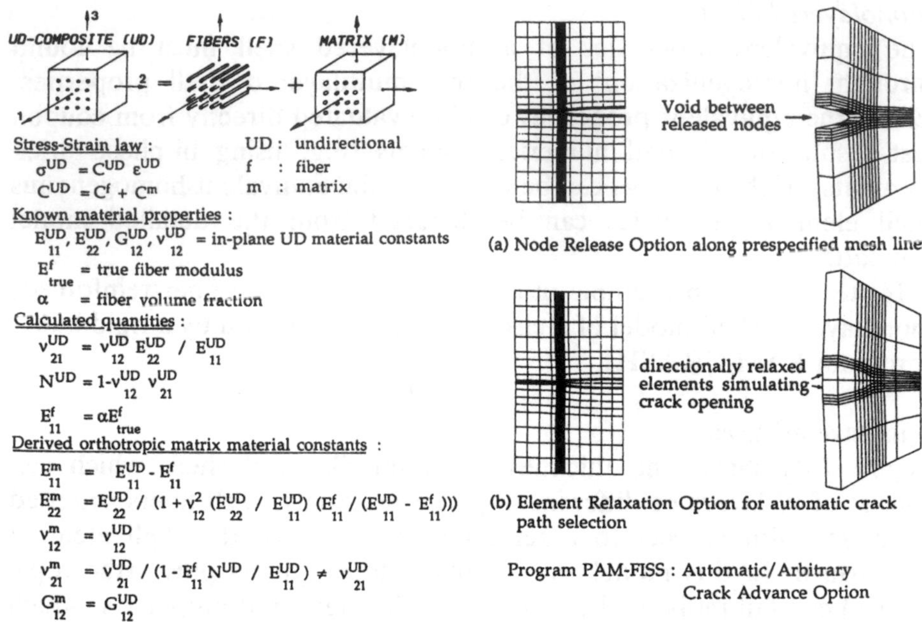

Fig. 10. The PAM-FISS/bi-phase rheological composite stiffness model.

Bi-Phase Model

Figure 10 gives an overview of the basic bi-phase material model incorporated into the finite element programs. In this model the orthotropic elastic properties of the matrix material minus the fibers (m) is deduced as indicated from the measured elastic properties of a unidirectional ply (UD), from the known or measured elastic properties of the fibers (f), and from the given fiber volume fraction (α). The figure also indicates ways of macro-crack propagation through solid finite element meshes, used in the fracture mechanics approaches of the finite element programs.

Fracturing Law

Figure 11 shows the characteristics of the modulus damage fracturing material law, useful to simulate subcritical, quasi-brittle, matrix micro-cracking damage in composites. The indicated 1D scalar damage is taken as the sum of damages due to the scalar hydrostatic volumetric strain (first invariant of the strain tensor, I_1), and a scalar deviatoric shear strain quantity (second invariant of the deviatoric strain tensor, J_2) in a general 3D situation. The threshold strains and damage measures for beginning (ε_i), intermediate (ε_1), and ultimate (ε_u), fracturing damage and the maximum damage are the required material parameters, which must be calibrated from tests.

(i) fracturing damage function

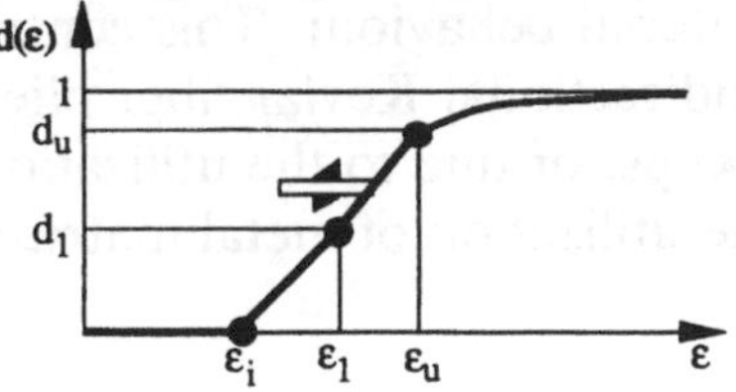

damage function $d(\varepsilon)$	range
$\dfrac{\varepsilon-\varepsilon_i}{\varepsilon_1-\varepsilon_i}\, d_1$	$\varepsilon_i < \varepsilon < \varepsilon_1$
$\dfrac{\varepsilon-\varepsilon_1}{\varepsilon_u-\varepsilon_1}(d_u-d_1) + d_1$	$\varepsilon_1 < \varepsilon < \varepsilon_u$
$1-(1-d_u)\,\dfrac{\varepsilon_u}{\varepsilon}$	$\varepsilon_u < \varepsilon < \infty$

ε = equivalent strain
ε_i = initial threshold strain
ε_1 = intermediate
d_1 = intermediate damage
E_o = initial elastic modulus
σ_u = residual constant stress level
ε_u = ultimate strain
d_u = ultimate damage

(ii) modulus damage

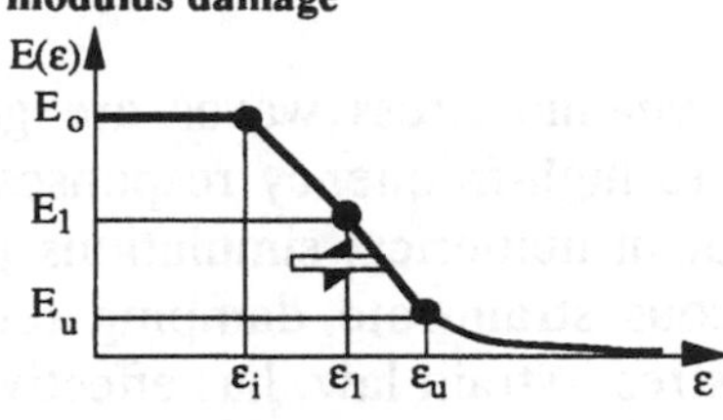

modulus damage
$E(\varepsilon) = (1 - d(\varepsilon))\, E_o$

(iii) stress-strain diagram

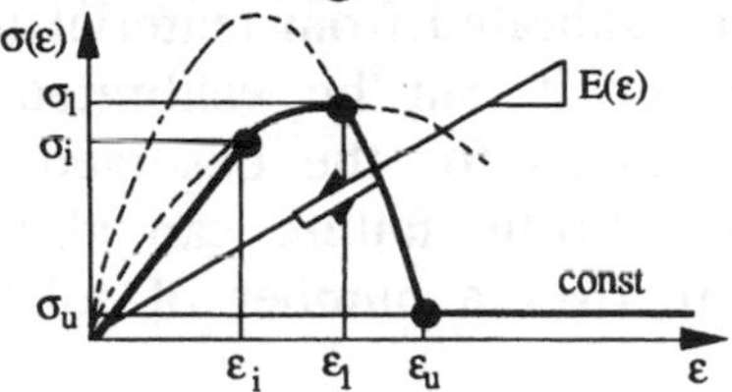

stress-strain relation $\sigma(\varepsilon)$	range
$E_o\,\varepsilon\left(1-\dfrac{d_1}{\varepsilon_1-\varepsilon_i}(\varepsilon-\varepsilon_i)\right)$	$\varepsilon_i < \varepsilon < \varepsilon_1$
$E_o\,\varepsilon\left(1-d_1-\dfrac{d_u-d_1}{\varepsilon_u-\varepsilon_1}(\varepsilon-\varepsilon_1)\right)$	$\varepsilon_1 < \varepsilon < \varepsilon_u$
$\sigma_u = E_o\,\varepsilon_u\,(1-d_u) = \text{const}$	$\varepsilon_u < \varepsilon < \infty$

(iv) equivalent scalar strain measures ($\underset{\sim}{\varepsilon}$ = strain tensor ; $\underset{\sim}{\delta}$ = unit tensor)

volume damage	$\varepsilon = \varepsilon_v = I_1 = \mathrm{tr}\,(\underset{\sim}{\varepsilon})$ (tr = trace)
shear damage	$\varepsilon = \varepsilon_s = \sqrt{J_2} = [1/2\ \mathrm{tr}\,(\underset{\sim}{\varepsilon} - 1/3\,\varepsilon_v\,\underset{\sim}{\delta})^2]^{1/2}$

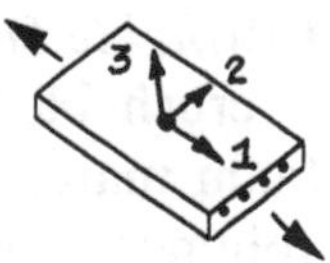

(v) calibration of damage parameters (ε,d) for orthotropic material on uniaxial coupon tests loaded in 1-direction (axial strain ε_{11})

phase	coupon test used for calibration	volume damage (v)*			shear damage (s)**		
		$(\varepsilon_{vi},0)$	$(\varepsilon_{v1},d_{v1})$	$(\varepsilon_{vu},d_{vu})$	$(\varepsilon_{si},0)$	$(\varepsilon_{s1},d_{s1})$	$(\varepsilon_{su},d_{su})$
matrix (m)	tension (t)	x	x	x	x	x	x
	compression (c)	x	x	x	x	x	x
fibers (f)	tension (t)	x	x	x			
	compression (c)	x	x	x			

* $\varepsilon_v^m = (1-\nu_{21}-\nu_{31})\,\varepsilon_{11}$; (1, 2, 3) = directions of orthotropicity ; $\nu_{21} = \nu_{12}\,E_{22}/E_{11}$

$\varepsilon_v^f = \varepsilon_{11}$ $\qquad \nu_{31} = \nu_{13}\,E_{33}/E_{11}$

** $\varepsilon_s^m = (\varepsilon_{11}/\sqrt{3})\,(1+\nu_{21}+\nu_{31}-\nu_{21}\nu_{31}+\nu_{21}^2+\nu_{31}^2)^{1/2}$

(vi) total damage

$d = d_v + d_s$ (the total damage is equal to the sum of the shear and volumetric damage)

Fig. 11. Bi-phase modulus damage fracturing law overview.

Plasticity
Standard von Mises strain-hardening plasticity models, available in the analysis programs, can be evoked to add a plastic contribution to the basically brittle fracturing composite material behaviour. This contribution may be due to an admixture of unidirectional Kevlar fiber plies or Kevlar cloth to brittle carbon-fiber stackups, or due to the utilization of hybrid weaves. It may also be due to the utilization of metal matrices in some metal–fiber composites.

Viscous Damping
When brittle materials fail suddenly, dynamic stress waves are generated, the propagation of which leads to high-frequency responses in the material, which must be damped out in numerical simulations just as in physical reality. An internal viscous strain-rate damping term, added to the elastic/fracturing bi-phase stress–strain law, has effectively damped out such spurious high-frequency responses in past simulations. Its internal damping coefficient must be calibrated from material tests (measures of sub-damage hystereses), or it can be calibrated by numerically reproducing dynamic test results for the observed frequency content. Shocks due to sudden brittle failure can also be attenuated by distribution of the failure over a number of solution cycles.

Crushable Foam
A crushable foam material model serves to simulate cellular collapse of foam materials in sandwich stackups. The collapse is established via a volumetric plasticity behaviour, which must be evaluated from material tests. The model also contains a conventional shear plasticity term.

Orthotropic Sandwich Core Model
The composite bi-phase model has been extended to a material model consisting of an orthotropic matrix with modulus damage and 'fibers' with non-symmetric non-linear tension–compression behaviour. This model is well suited to represent the elastic and crushing behaviour of hexagonal cell sandwich cores. Figure 12 shows this model applied to honeycomb sandwich cores, where the 'transverse' sandwich shear properties are taken by the orthotropic 'matrix' of the bi-phase material (G_{12}, G_{13}), whereas the 'thickness' properties are taken by the non-linear 'fibers' (E_{11}).

(a) ORTHOTROPIC ELASTICITY :

☐ : strong directions of honeycomb material ; E_{11} is due to fibers + matrix

$$\begin{bmatrix} \varepsilon_{11} \\ \varepsilon_{22} \\ \varepsilon_{33} \\ \varepsilon_{12} \\ \varepsilon_{23} \\ \varepsilon_{13} \end{bmatrix} = \begin{bmatrix} \boxed{1/E_{11}} & -\nu_{12}/E_{22} & -\nu_{13}/E_{33} & & & \\ -\nu_{21}/E_{11} & 1/E_{22} & -\nu_{23}/E_{33} & & 0 & \\ -\nu_{31}/E_{11} & -\nu_{32}/E_{22} & 1/E_{33} & & & \\ & & & \boxed{1/G_{12}} & & \\ & 0 & & & 1/G_{23} & \\ & & & & & \boxed{1/G_{13}} \end{bmatrix} \begin{bmatrix} \sigma_{11} \\ \sigma_{22} \\ \sigma_{33} \\ \sigma_{12} \\ \sigma_{23} \\ \sigma_{13} \end{bmatrix}$$

(b) HONEYCOMB CORE :

1 2 3

(c) CELL FAILURE MODES

(i) 'Fiber' phase :

σ_{11}

FIBER STRESS S
elastic, main curve only
elastic with unload curve active
elastic with reload curve active ($E < E_c$)
RELOAD CURVE
MAIN CURVE
UNLOAD CURVE
FIBER STRAIN E
S_1 S_2 S_3 S_4 S_5 S_6 S_7 S_8
E_1 E_2 E_3 E_4 E_5 E_6 E_7 E_8
E_c E_T

- axial crushing is represented by 'fibers' in the bi-phase model ; 'fibers' have nonlinear material behaviour ; curves are extrapolated along the slopes of the last curve segments.

(ii) 'Matrix' phase :

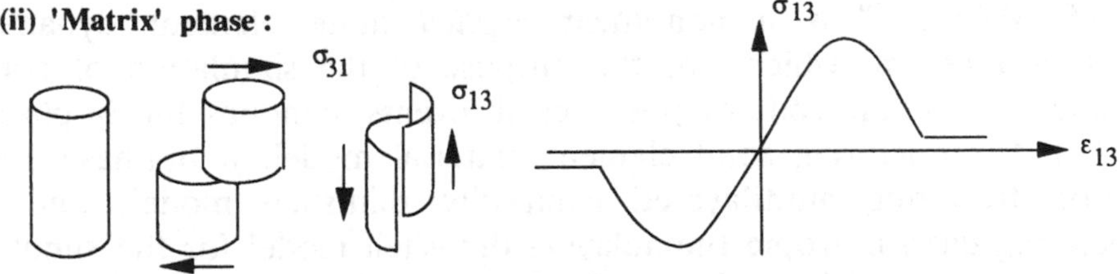

- transverse shear failure is represented by a matrix fracturing damage model ; for honeycomb matrix material the core in-plane properties (22, 33, 23) are usually very small as compared to the strong properties (11, 12, 13).

Fig. 12. Bi-phase sandwich brick model.

3.3 Analysis Tools

PAM-FISS

PAM-FISS is a non-linear *implicit* finite element (FE) quasi-static analysis program, which contains the unidirectional solid element bi-phase material model. This program also contains fracture mechanics analysis techniques, such as an automatic crack advance scheme, and several strain-energy release-rate calculation schemes (G-values). The fracture mechanics techniques have been applied successfully to matrix (resin) intralaminar (splitting and transverse) and interlaminar (edge delamination and outer ply blistering) crack advance.[19–25] A more involved damage mechanics technique, the original (D_c, r_c) material fracture criterion for critical damage over a characteristic distance, has been applied successfully to the simulation of the advance of fiber cracks in the critical plies of multilayered (ML) tensile test pieces.[2,23]

A matrix (resin) fracturing modulus damage and strain-softening material model, as well as a plasticity model for matrix damage, have been used to evaluate subcritical matrix damage in tensile test pieces. These laws have also been used for matrix compression damage with the associated fiber deconfinement and fiber buckling under compressive stress in composite fabric coupon compression and composite fabric coupon bending test simulations.

The PAM-FISS tool in conjunction with the bi-phase (fibers, matrix) material description is particularly well suited for quasi-static tensile, bending and compression, destructive and non-destructive coupon test simulations on the composite wall microscale, from which equivalent models on the wall macroscale can be derived.

PAM-CRASH

PAM-CRASH[26–29] is a non-linear *explicit* finite element dynamic analysis program, which, for the purpose of the simulation of continuous fiber-reinforced composite crash events, contains the bi-phase and matrix fracturing solid element material model, a bi-phase and matrix fracturing multilayered composite thin-shell model, and a fracturing quasi-isotropic (monolayer) thin-shell model for the simulation of dynamic failure (or crash) behaviour of short-random-fiber-reinforced composites (e.g. sheet moulding compounds, SMC). The bi-phase and fracturing material models are complemented by an internal viscous damping law and by a plastic material behaviour component, which can serve to dissipate energy from high-frequency oscillations, and which introduces quasi-plastic components encountered in many practical situations.

The crushable foam brick material model and the special honeycomb sandwich core brick model can serve to simulate foam and honeycomb (cellular) sandwich cores in sandwich FE models with multilayered or monolayered straight fiber, tissue fabric, or random-fiber-reinforced thin-shell or membrane facings.

The PAM-CRASH tool with its composite solid, thin-shell and membrane material models is particularly well suited for dynamic destructive tensile, compression and bending single-wall and sandwich coupon test simulations. The possibility to construct equivalent 'macro'-wall finite elements, calibrated on component crushing tests and/or on detailed PAM-FISS multilayered wall finite element analysis results, which absorb equivalent amounts of crushing energy, renders the PAM-CRASH code suitable for the simulation of component, subassembly and full structure composite crash events.

Orthotropic Material Representation

Using the bi-phase material model, the orthotropic character of the cloth and of the unidirectional composite layers of a stackup can be modelled in two principal ways: either using two material phases, namely fibers plus matrix ('classical' model), or using one material phase, namely an orthotropic matrix only ('modified' model).

In the first approach ('classical' bi-phase model) the material orthotropicity is represented primarily by the fiber phase. In the second case ('modified' bi-phase model) the orthotropic character of the fiber-reinforced material is represented by the suitably specified orthotropic constants of only the 'matrix' material, and no fiber properties need be specified. The first approach is usually more suited to represent unidirectional composite plies, while the second approach may be more convenient to represent, for example, cloth layers or a pair of cross-plies within one shell layer.

The second approach (orthotropic 'matrix' only), is well suited for modelling fracture of composite plies of a multilayer/multimaterial stackup via multilayered shell macro models, using the modulus damage fracturing law summarized in Fig. 11. In general, the resistance of a composite multilayered coupon is evaluated in uniaxial tensile and compressive tests, and the resistance to tensile loading is usually different from the resistance to compressive loading. For this reason, different initial, intermediate and ultimate volumetric and shear strain thresholds and associated damages may be specified in the most general form of the modulus damage fracturing law, as indicated in Fig. 11(v). Figure 11 and the calibration of the model parameters are discussed further in Section 5.2.

4 CRASH SIMULATION OF COMPOSITE COMPONENTS

4.1 Composite and Metal Crash Simulation

Composite structure crashworthiness simulation techniques can be evaluated against the background of metal structure crash simulation and the present state of the art of composite structure failure analysis and material models.

While the numerical crashworthiness prediction techniques of (visco-) elasto-plastic thin-shell metal structures have reached industrial maturity,[26–29] the crashworthiness simulation techniques of structures, the load-carrying and kinetic-energy-absorbing members of which are made of composite material, are less well established.[30,31] The reasons for this may be found mainly in the relative lesser utilization of composite materials on the large production scale of vehicles, the relative inexperience in engineering prediction of the damage behaviour of the relatively new composite materials, and in the added complexity of composite materials as compared with the basically isotropic/homogeneous material of metal structures.

As opposed to car body steel, composites are heterogeneous materials made of at least two different fiber and matrix phases. Composites tend to disintegrate when absorbing crash energy. The energy is absorbed mainly through fiber and matrix fracture, which may completely destroy the structural integrity of a composite component wall and of the component, whereas crushed metal structures rarely fracture but keep their structural integrity, even after large plastic buckling and folding deformation with associated plastic energy absorption takes place. Composites can be stacked up into different equivalent component wall materials, even when made from the same basic fiber and matrix material phases, and can differ in fiber volume fraction, random or straight fiber orientation, single-layer or stacked-up wall construction, woven fabric construction (single- or multilayered), sandwich construction (foam or honeycomb core) 3D weaves, etc. Finally, composites are often hybridized, that is component wall structures can be mixtures of carbon–epoxy and Kevlar–epoxy and glass–epoxy layers, or others, with straight fibers, randomly oriented fibers and fabrics.

The microscale of the composite wall construction (e.g. single-ply thickness 0·1–0·2 mm) makes it impossible at present to establish detailed numerical models for global crash simulation analyses to take into account the physical details of composite wall failure. The scale of thin-shell finite elements used in global structural crash simulations of metal structures is typically 5×5 mm and larger, and complex crash models of 30 000 elements and more have been analyzed successfully and economically.

Composite global crash models, for economy, must be of comparable size on the wall macroscale level. The mechanical and failure properties of equivalent thin shell elements must therefore be evaluated on the wall macroscale level when performing global crash simulations. These properties can be evaluated by laboratory tests, where composite coupon test pieces yield primary material parameters and where simple cylindrical section composite columns and composite plates are tested for axial and lateral crash loading.

4.2 State of the Art of Composite Structural Failure Analysis

Because of their inherent structural complexity, their mostly destructive crash energy absorption modes, their much scarcer application in everyday structures, composite safe life, fail safe and crashworthiness design and analysis are subjects that engineers are only now beginning to master.[3–18] Modern finite element techniques, proven solution techniques, modern fracture and damage mechanics material laws and testing techniques, together with economic high-speed and workstation computing power, have helped in recent years to open the subject to analytical treatment and to the prediction of complex composite damage and failure behaviour.[2,19–25]

Matrix Failure
Composite matrix failure in continuous fiber-reinforced composite stackups can occur in the forms of large, unconfined, matrix macrocracks, such as intralaminar 'splitting' cracks, 'transverse' cracks and interlaminar 'delamination' cracks. This can be treated according to the methods of classical brittle linear elastic fracture mechanics (LEFM) on the laminar scale. Matrix failure also exists in the form of microcracks, which occur, for example, near sharp stress risers such as notches and cutouts in tensile test pieces in multilayered stackups. Such cracks are subcritical to failure because they appear well before the first fiber cracks and can be viewed as a confined damage in the form of a fine network of matrix macrocracks that are arrested by the adjacent off-axis plies.

Because this damage is confined by the yet-unbroken fibers, subcritical confined matrix crack networks can be treated effectively using fracturing material models with strain softening and brittle fracturing unloading behaviour.

Fiber Failure
Fiber failure in long-fiber-reinforced composite stackups can occur as tensile failure, where the fibers of the load-carrying ply will break in tension, beyond their rupture limit, in undisturbed regions, or in regions of high stress concentrations, such as near notch roots. Fibers can also fail in compression, which in most cases is triggered by some form of fiber buckling due to the deconfining action of subcritical matrix damage, and which can also be attributed to the tensile failure of the regions under tension across bent fibers or fiber bundles.

While first fiber rupture in undisturbed regions of perfectly aligned fibers of equal strength may lead to unstable overall specimen failure, 'nucleated' fiber failure near regions of very high stress concentrations needs to 'grow' and spread over a certain 'characteristic' volume, before local fracture begins through 'coalescense' of individual fiber breaks. This can be studied using the relatively recent techniques of 'damage mechanics', which in principle can describe failure behaviour of materials where classical brittle linear elastic fracture mechanics does not apply.

Interface Failure
Failure of the interface between the fiber and the matrix phases, also called debonding with associated fiber pullout, is a consequence of fiber rupture. It can either be considered as a form of matrix failure, i.e. interface and matrix failure can be treated together in one and the same subcritical matrix damage material law. Owing to friction pullout, interface failure may add a quasi-plastic dispersive component to the quasi-brittle fracturing component in the material law used to describe matrix damage. Or it can be viewed as being part of the critical damage to be identified, which measures the nucleation, growth and coalescence of microvoids that leads to fiber fracture. In that case interface failure need not be treated explicitly when using the damage mechanics fracture criterion described below.

Note that other dispersive failure modes exist, for example, in random-fiber-reinforced monolayered composites. Such materials fail owing to matrix fracturing, deconfinement of the fibers, alignment of the randomly oriented fibers, still partially confined by the surrounding damaged matrix (with friction and pullout effects), and, finally, by tensile rupture of the aligned fibers.

4.3 Ductile, Brittle and Composite Component Crush Behaviour

In order to demonstrate the basic influence on crushing behaviour of ductile, brittle and mixed material properties, box columns made of such materials were crushed with the PAM-CRASH code.

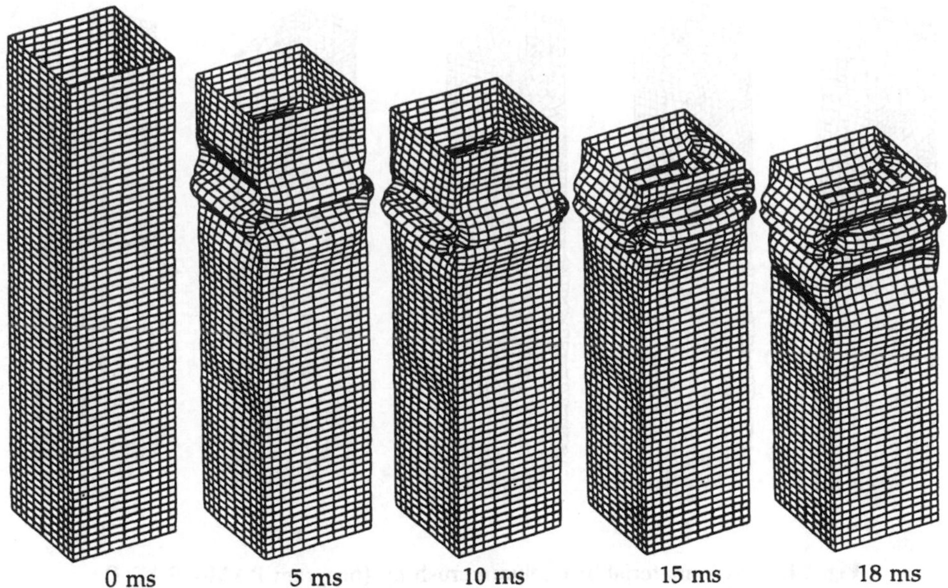

Fig. 13. Ductile material box column crushing (program PAM-CRASH).

Ductile Steel Columns

As a reference, Fig. 13 shows the initial geometry and deformed geometries of a perfectly ductile steel column, which is crushed at its upper end by applying a constant axial velocity. Typical 'dog-bone' antisymmetric elasto-plastic buckles are seen to develop. The small dent in the walls at about one-third of the initial distance from the bottom end is the beginning of a buckling lobe that did not fully develop, because of the drop in column force when the top end of the column started collapsing.

Column Made of Brittle Material

For comparison, Fig. 14 shows deformed pictures of a brittle box column at various times (1 ms corresponding to 1 cm crushing distance). The smaller side panels of the column are made of a linearly elastic material that never fails and the wider panels are made of the same material but are allowed to fail in a brittle fashion. Upon impact, a very regular pattern of linear elastic dynamic buckles is seen to develop first. As seen in the figure, the regular quasi-elastic pattern of dynamic buckles becomes irregular in the larger panels near 4 ms, which is due to a progressive brittle destruction of the wider panel material. At 5 ms, many elements have failed suddenly, which results in a very irregular deformed mesh plot and a sudden drop of the resisting column force

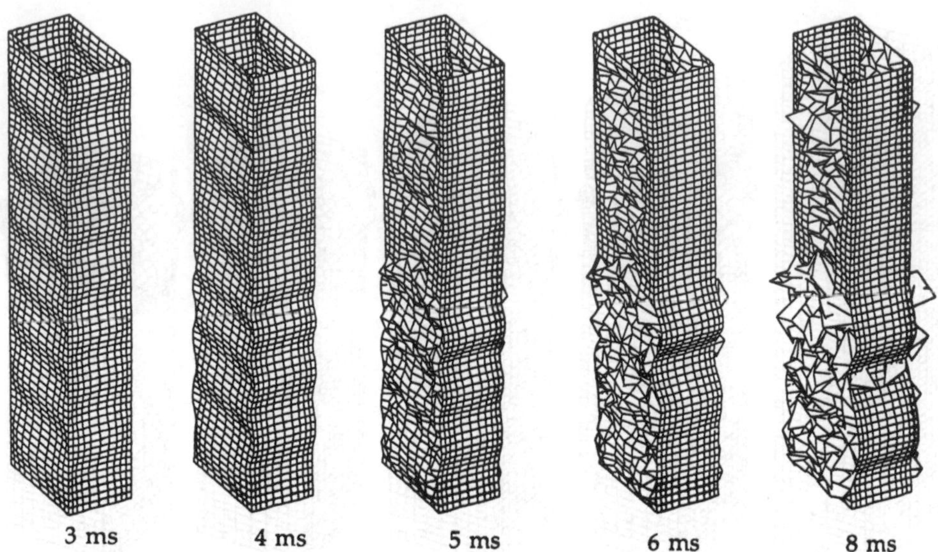

Fig. 14. Brittle material box column crushing (program PAM-CRASH).

from an average plateau value to a small value, as can be deduced from the associated axial-acceleration–time diagram. The force level remains very low from there on, which indicates severe column failure. With progressive failure, the buckling wavelength of the fully responding smaller side panels is seen to increase, which is due to the increasing lack of constraint coming from the failing larger side panels.

Monocoque Composite Components

The newly implemented multilayered/multimaterial monocoque thin-shell model of the PAM-CRASH code has been used to simulate the crushing of a thin-walled column that corresponds to subassembly 1 of Tonen's prototype composite passenger car, without the sandwich core. The monocoque material used is a symmetric stackup made of brittle Tonen FT500 pitch-based carbon-fiber cross-ply, more ductile DuPont aramid-fiber-based Kevlar 49 cloth, and of brittle Toray PAN-based carbon-fiber T300 cloth. In this first analysis the sandwich aramid honeycomb core was omitted.

A crushing analysis has been made with an applied constant velocity of the loaded column end (10 m/s). Figure 15(a) gives the material data stress–strain response. Figure 15(b) shows the deformation response at different times. The tendency of progressive plate bending failure with sharp hinge line formation of the component walls can be seen clearly in the deformed shape plots. Figure 15(c) shows the fracturing damage

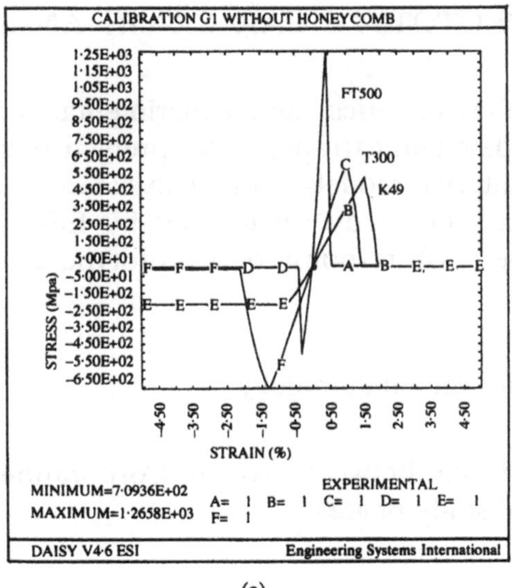

(a)

(b) (c)

Fig. 15. Multilayered composite column crush (program PAM-CRASH): (a) material properties; (b) crushing response; (c) damage in outer, middle and inner layers.

maps of the outer FT500 0°-layer, and the adjacent Kevlar 49 cloth and T300 cloth plies respectively, at 2 ms, i.e. 2 cm axial crushing distance. As expected, the very brittle FT500 outer layer shows the most damage, and damage decreases towards the midsurface of the symmetrized component wall.

5 COMPOSITE PROTOTYPE CAR CRASH ANALYSIS

This section describes the numerical and experimental methodology applied to calibrate, validate and extrapolate for predictive analyses the employed composite material models and analysis techniques. The reported work reflects a joint study carried out by ESI with Tonen Corporation in which the crash response of a prototype composite car was to be assessed.

5.1 Experimental/Numerical Methodology

The joint study carried out between Tonen Corporation and ESI comprised three essential study phases.

Calibration Phase 1
During this phase, ESI specified material coupon tests (tensile, compression, bending) that permit access to subcritical material data and data up to specimen failure. The coupon tests were performed by Tonen and the results exploited by ESI for the calibration of the material models used. For this purpose detailed FE models of the coupon tests were prepared and the material model parameters calibrated by comparison between the numerical and laboratory coupon experiments.

Validation Phase 2
Next, the calibrated material models were used to simulate identified benchmark crush tests of relevant composite sandwich subassemblies. These static and dynamic drop tests on sandwich plates and sandwich box beams were performed by Tonen and JARI, and the results of the experiments were compared with the numerical results. Coincidence was found to be quite good, and only minor adjustment of the least accessible material model parameters of the sandwich core material, calibrated in phase 1, had to be made.

Predictive Phase 3
Finally, the calibrated and validated numerical model was applied to predictive front- and side-impact analyses of the complete composite sandwich cabin. It was found that the indicated three-phase methodology of calibration, validation and prediction guarantees the best possible results.

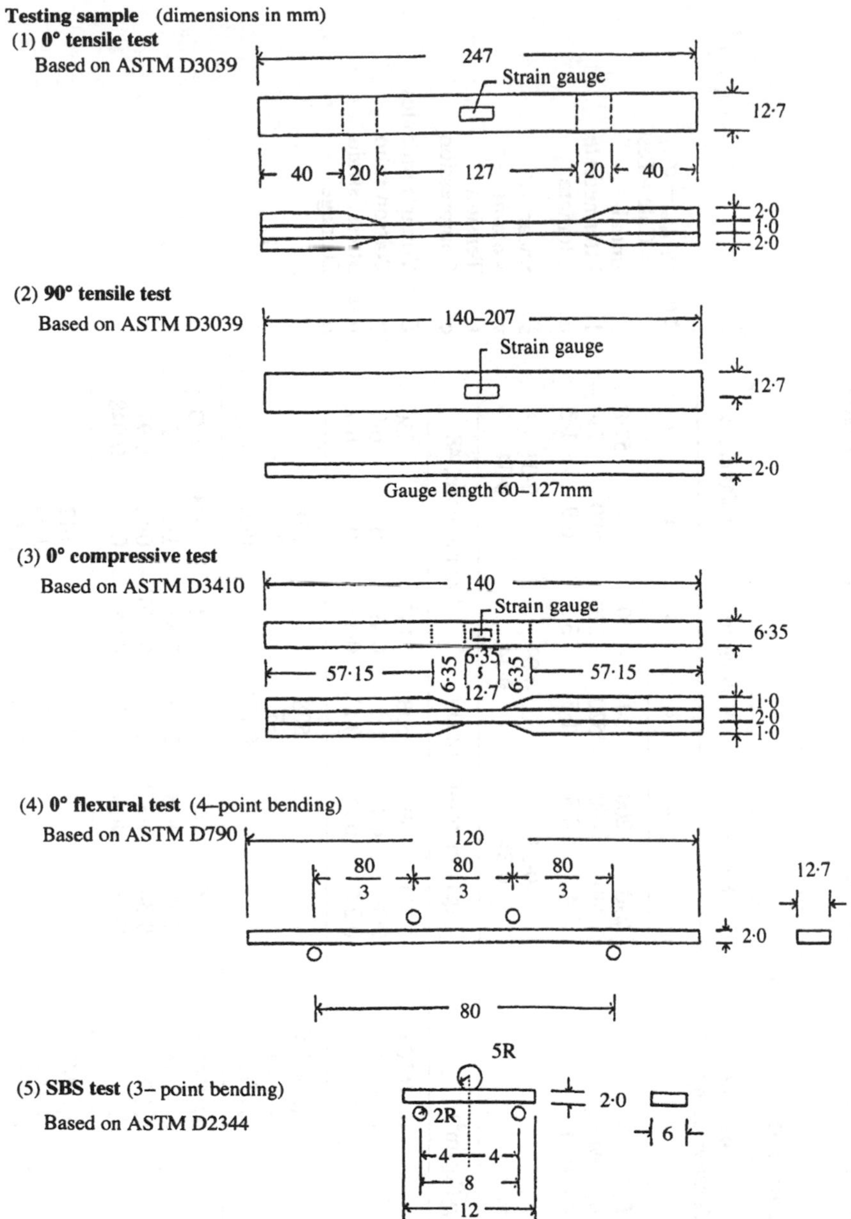

Fig. 16. Coupon Test Samples (after Tonen).

Table 1
Calibrated Composite and Honeycomb Material Properties used in Simulation Runs

(a) Facing Material Properties

		Material / *Tension/compression*	FT500		K49 cloth		T300 cloth	
			t	c	t	c	t	c
		E_0 (GPA)	290	200[a]	34	30	64	57·8[a]
		σ_{11u} (MPa)	1 240	400	520	220	590	710
		ε_{11u} (%)	0·4	0·3	1·5	3·5	0·9	1·3
		ν_{12}	0·34		0·09		0·04	
		ν_{13}	0·34		0·34		0·34	
		ρ (kg/m^3)	(g1) material, 1200			Aramid core, 48		
i	s	σ_{11} (MPa)	1 044	390	508	145	578	530
i	s	ε_{11} (%)	0·36	0·17	1·5	0·49	0·9	0·9
i	s	ε_s (%)	0·27[a]	0·2[a]	0·2[a]	0·375[a]	0·7	0·7
i	v	σ_{11} (MPa)			508		578	
i	v	ε_{11} (%)			1·5		0·9	
i	v	ε_v (%)			0·73[a]		0·37	
1	s	σ_{11} (MPa)	1 276	4	110	175	673	673
1	s	ε_{11} (%)	0·45	0·38	2·09	0·62	1·2	1·2
1	s	ε_s (%)	0·34[a]	0·37[a]	2·36[a]	0·476[a]	0·91	0·91
1	s	d_s	0·01	0·99	0·83[a]	0·05	0·05	0·048
1	v	σ_{11} (MPa)			110		612	
1	v	ε_{11} (%)			2·09		0·96	
1	v	ε_v (%)			0·86[a]		0·61[a]	
1	v	d_v			0·01[a]		0·05[a]	

$\Delta\sigma$ — $\sigma_{11\,max} - \sigma_{11u}$ = Calc.–Test
i — Initial
1 — Intermediate
u — Ultimate
s — Shear
v — Volume
t — Tension
c — Compression
E — Young's modulus
ν — Poisson ratio
σ, ε — Stress, strain
d — Damage

u	s	σ_{11} (MPa)	4	4	0	195	12	12
		ε_{11} (%)	0·5	0·45	2·9	1·03	1·94	1·94
		ε_s (%)	0·39[a]	0·42[a]	2·42[a]	0·8	1·5	1·5
		d_s	0·997	0·997	0·985[a]	0·375	0·99	0·99
	v	σ_{11} (MPa)			0		0	
		ε_{11} (%)			2·9		1·28	
		ε_v (%)			1·08[a]		0·87[a]	
		d_v			0·98[a]		0·99	
$\sigma_{11\,max}$ (MPa)			1 276	433	508	201	590[a]	710[a]
$\varepsilon_{11\,max}$ ($\sigma_{11\,max}$) (%)			0·4	0·25	1·5	3·5	0·96	1·1
$\Delta\sigma/\sigma_{11u}$ (%)			2·9	8·2	−2·3	−8·6	0·0	0·0

[a] Recalibrated values.

(b) Sandwich-Core Material Properties

Bulk (linear elastic)	
E_{11}	10 MPa
E_{22}	10 MPa
E_{33}	10 MPa
G_{12}	96 MPa
G_{23}	1 MPa
G_{31}	48 MPa
ν_{12}	0·01
ν_{23}	0·01
ν_{31}	0·01

Multiplier[a]	*Strain* *Stress*	*Cells*							
		Point (see Fig. 12)							
		1	2	3	4	5	6	7	8
1	ε (%)	−100	−80	−7·85	−4·7	0	0·5	1	1 000
	σ (MPa)	−3 000	−2·2	−2·2	−6·6	0	0·7	0·7	0·7
3	ε (%)	−100	−80	−7·85	−4·7	0	0·5	1	1 000
	σ (MPa)	−3 000	−6·6	−6·6	−19·8	0	2·1	2·1	2·1
5	ε (%)	−100	−80	−7·85	−4·7	0	0·5	1	1 000
	σ (MPa)	−3 000	−11·0	−11·0	−33·0	0	3·5	3·5	3·5
10	ε (%)	−100	−80	−7·85	−4·7	0	0·5	1	1 000
	σ (MPa)	−3 000	−22·0	−22·0	−66·0	0	7·0	7·0	7·0

[a] Used in parameteric studies.

5.2 Calibration of the Material Models

Figure 16 shows specimens that were fabricated from each of the involved materials and on which tensile, compression, bending and interlaminar shear tests have been performed. Similar test pieces were fabricated and tested for the multilayered stackups. The experimental responses of these test pieces were used as targets for calibration analyses with the PAM-FISS and PAM-CRASH codes, where detailed finite element models of the specimens were subjected to the respective loadings. Table 1 summarizes the basic material properties used in the computer runs for all materials involved. For the meaning of the given values see also Figs 10–12.

Sandwich Material

The basic materials used in Tonen's prototype passenger car cabin are multilayered composite sandwich facings made of high-stiffness and high-strength unidirectional TONEN FT500 pitch-based carbon-fiber/epoxy crossply (0, 90) layers, a layer of DuPont aramid-fiber-based Kevlar 49 cloth (0 * 90), and of Toray PAN-based T300 carbon/epoxy cloth (45 * −45) (Fig. 17a). The sandwich core material consists of aramid honeycomb material. The basic sandwich wall structure is provided with additional layers of reinforcements at strategic locations through the car cabin.

Material Model for Sandwich Facings

The models for each layer of the sandwich facing are bi-phase plane-stress material models, with which the rheological behaviour of the constituent monodirectional fiber and orthotropic plane stress matrix phases can be treated separately. Both material phases obey a brittle fracturing modulus damage description (Fig. 11), where the initial modulus E_0 is damaged by a damage function $d(\varepsilon)$ such that

$$E(\varepsilon) = E_0[1 - d(\varepsilon)]$$

where ε is a measure of total deformation. In Fig. 11(i) the damage function d is assumed to grow from values of zero up to an initial threshold strain ε_i, to values of one (or asymptotically reaching one, for non-vanishing ultimate strength) after an ultimate threshold strain ε_u, either linearly or nonlinearly. The degree of complexity of this function may be assumed, and the coefficients calibrated from basic coupon tests in tension, compression and bending. The non-linearity of the damage function is given at an intermediate strain ε_1 in the figure.

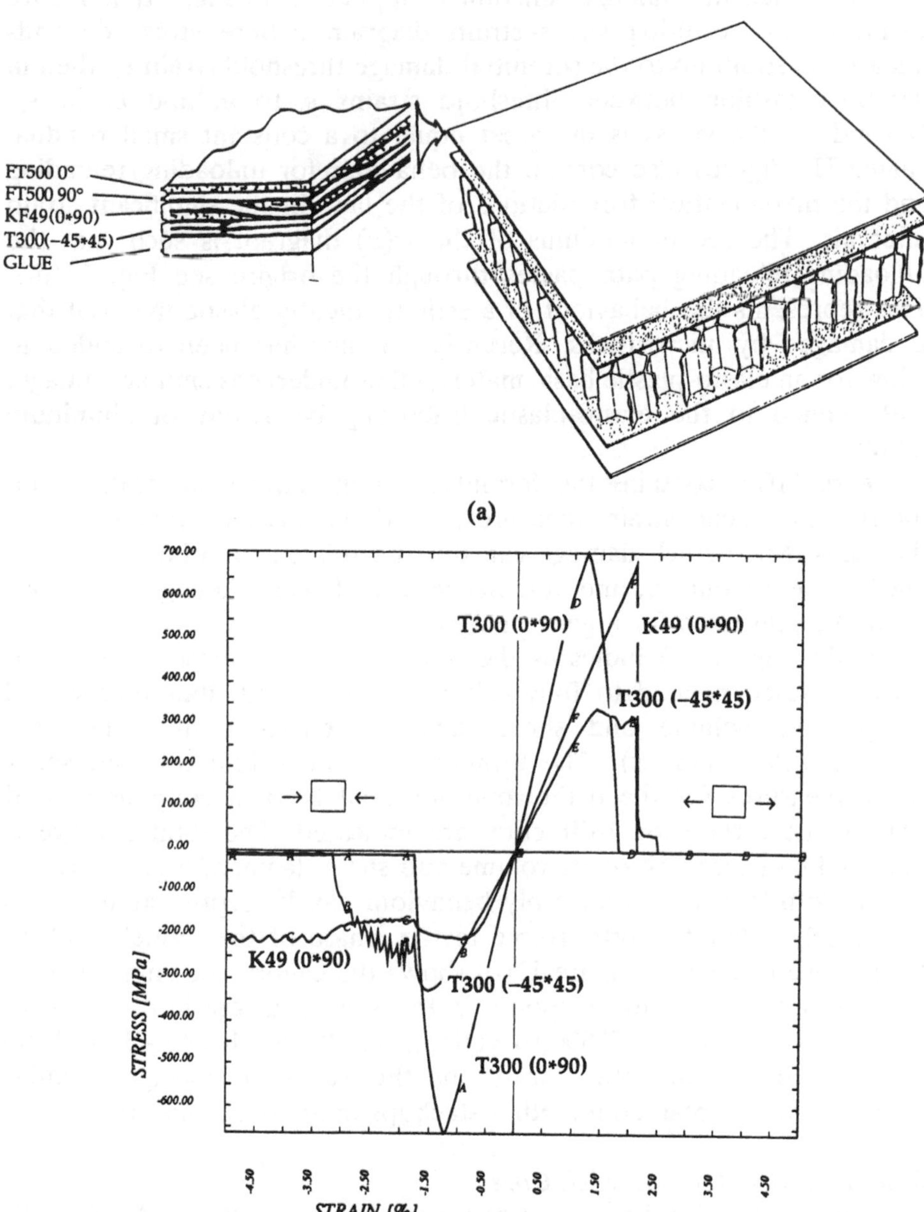

Fig. 17. Composite sandwich material used for Tonen's prototype sports-car cabin: (a) Basic composite sandwich wall structure. (b) Individual-ply tensile/compressive material response in basic sandwich facing stackup (program PAM-FISS).

Figure 11(ii) schematically shows the damage inflicted on the modulus when the damage function is applied as in Fig. 11(i). Figure 11(iii) is the resulting stress–strain diagram, where stress depends linearly on strain up to the set initial damage threshold strain ε_i, then in parabolic fashion between threshold strains ε_i to ε_1 and ε_1 to ε_u. Beyond ε_u, the stress is often set equal to a constant small residual value. The figures also contain the behaviour for unloading/reloading and the mathematical formulations of the laws in the significant strain intervals. The secant modulus of the $\sigma(\varepsilon)$ diagram is such that the unloading/reloading path passes through the origin; see Fig. 11(iii). This represents the behaviour of a strictly linearly elastic material that is damaged by microcracks. Recently this law has been extended to allow for an elasto-plastic basic material that undergoes similar damage, and applied to the elasto-plastic fracturing behaviour of aluminum alloys.

Figure 11(iv) contains the formulae for the equivalent scalar volumetric and shear strain measures, used as abscissa values in the diagrams. Microcrack damage can thus be related to volume strain ε_v and/or shear strain ε_s, and volumetric d_v and shear damage behaviour d_s can be calibrated for a given material.

Finally, Fig. 11(v) indicates the uniaxial coupon tension and compression tests, needed to fully calibrate matrix and fiber tensile and compressive volume and shear damage behaviour of orthotropic material (plies, fabrics). The formulae for the volumetric and shear strain measures for the orthotropic material for test coupons loaded only in the orthotropy 1-direction are indicated. The total damage is obtained by superposition of volume and shear damage; Fig. 11(vi).

For cloth layers the entire ply behaviour can be represented by the adequately calibrated orthotropic matrix phase of the model, and no fiber phase is needed. Figure 17(b) shows the calibrated, unsymmetric, tension and compression response in the specimen tensile direction of each ply of the basic FT500(0) carbon/K49(0 * 90) Kevlar-cloth/T300 (+45 * −45) carbon-cloth stackup of the sandwich facings. Similar responses were obtained for other stackups in areas of reinforcements.

Material Model for Sandwich Core

The model used for the aramid honeycomb core of the sandwich walls is a specialized 3D bi-phase solid model, where the 'fiber' phase is aligned with the axes of the hexagonal honeycomb cells. In this model the monodirectional 'fiber' properties can be given highly non-linear tension/compression unloading/reloading properties to represent the complex axial crushing response of the cells; see Fig. 12.

The transverse shear properties of the core material of a plate made of honeycomb sandwich material is represented by elastic-fracturing terms, with calibrated transverse shear modulus damage, in the 'matrix' phase of the adapted 3D bi-phase model. The in-plane stiffness and resistance terms of the honeycomb core material of the sandwich plates are assumed to be very small, and are represented by small numbers in the calculations.

Finite Elements
The described composite sandwich material models are built into multilayered Mindlin-type underintegrated four-node and C_0 compatible triangular shell and plate finite elements for the sandwich facings, and into eight-node underintegrated adapted bi-phase solid brick finite elements for the sandwich core; Fig. 7. To make up one 'sandwich' element, two multilayered thin-shell elements are connected to opposite faces of a brick element, the 'fibers' (cell directions) of which are assumed to be perpendicular to the facings.

5.3 Validation of the Material Models

The feasibility of the proposed numerical simulation technology is demonstrated on exploratory crushing analyses of composite subassemblies.

Geometry and Loading of Components
The analyzed components (Fig. 18) have the geometry of subassemblies 1 and 2 in Fig. 19(a). Each component axis is oriented vertically, with its lower end perfectly built in (displacements and rotations of all lower

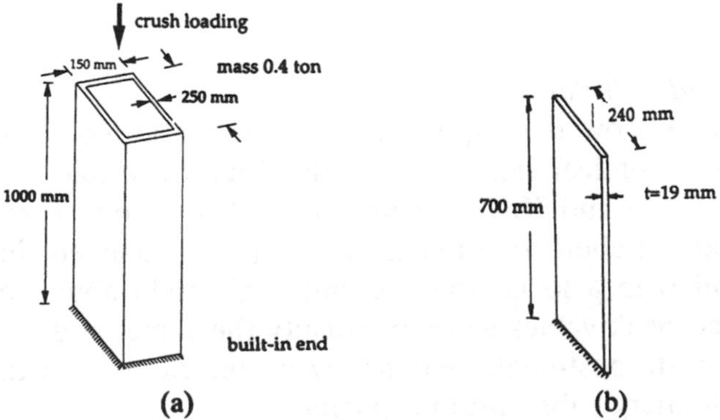

Fig. 18. Geometry of sandwich box column (a) and panel (b).

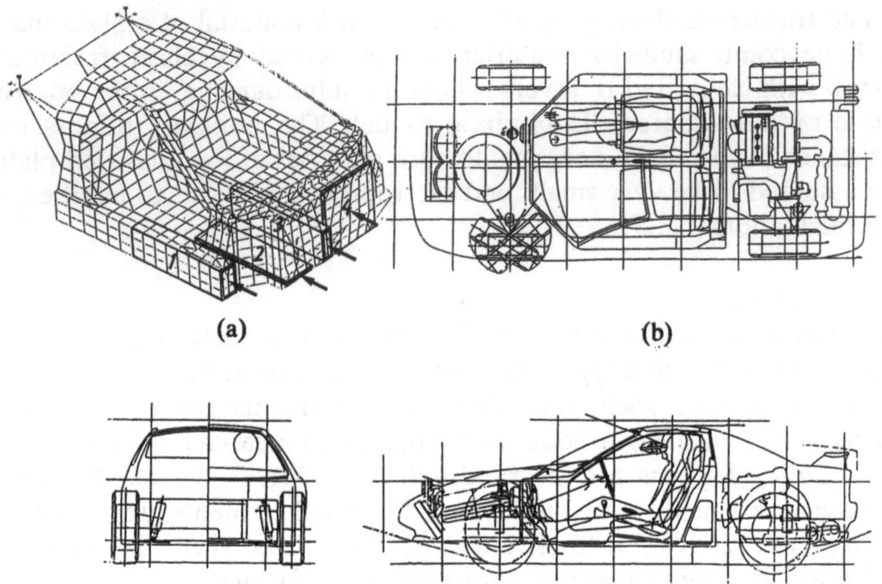

Fig. 19. Tonen's prototype car: (a) NASTRAN finite element mesh with subassemblies 1–4; (b) top, rear and side views.

end nodal points are constrained to zero in the FE mesh). In Figure 19(a) the schematic FE mesh of a composite prototype car (Fig. 19b) is shown with a certain number of subassemblies to be tested. The upper ends of the components in Fig. 18 are loaded by constraining all nodal points of the uppermost section to move with the same displacements in the axial direction, assigning a large mass (400 kg = 0·4 tons) to the constrained section, and by applying a vertical initial impact velocity to this mass or by applying a moving rigid wall to the upper column sections of infinite mass and constant velocity as in a quasi-static test simulation.

Impact on Sandwich Panel

Figures 20(a,b) show the experimental quasi-static load–displacement curves of two identical sandwich panels (Fig. 18b) made from a first stackup of carbon and Kevlar materials. While in one case a 'global' buckling mode at about mid-height of the plate occurred, in the other the panel failed in a local 'shear crumpling' mode above mid-height. The measured peak values were practically the same, and, as expected, the crushed plate post-peak response was inefficient from the point of view of crash energy absorption capacity.

Figure 20(c) shows deformations of several parametric simulations

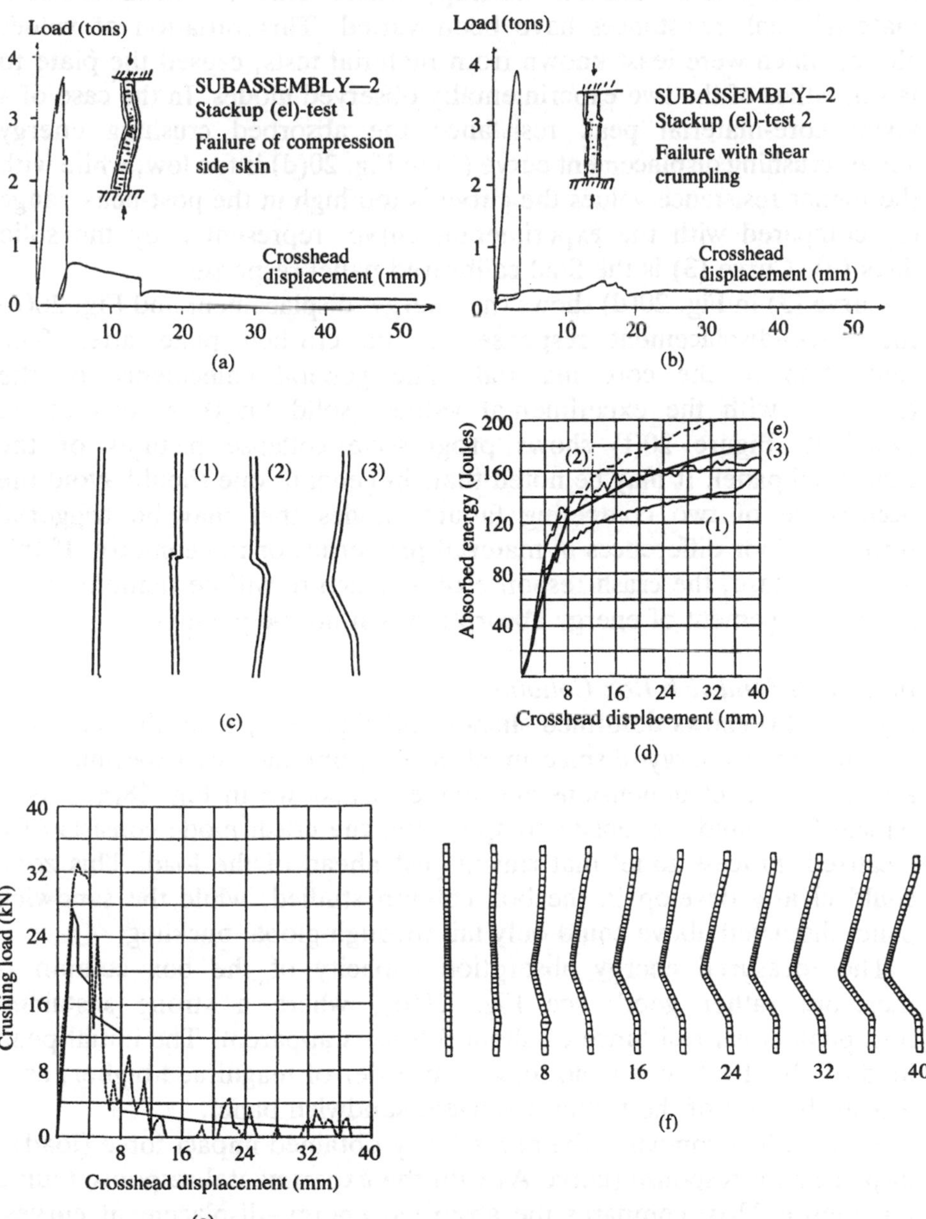

Fig. 20. Sandwich panel crush tests and simulations. (a) First crush test of panel (e1). (b) Second crush test of panel (e1). (c) Various simulated panel crushing modes (3 = final calibration). (d) Energy absorption curves for simulated panels (e = experiment; 3 = final calibration). (e) Final calibration panel load–displacement curves. (f) Final calibration panel crushing mode. (After Tonen; program PAM-CRASH.)

for a slightly less resistant stackup, where only the sandwich-core material peak resistances have been varied. This variation of values alone, which were least known from material tests, caused the plate to fail in either of the two experimentally observed modes. In the case of a lower core-material peak resistance the absorbed crushing energy versus crushing displacement curve (1) in Fig. 20(d) is too low, while with the higher resistance values the curve is too high in the post-peak range (2) compared with the experimental curve, represented by the solid lines (e). Curve (3) is the final calibrated panel response.

Curve (3) in Fig. 20(d) shows the energy–displacement and Fig. 20(e) the force–displacement response of the crushed plate after final calibration of the core material. The general coincidence of the responses with the experimental values (solid lines) is seen to be excellent. Figure 20(f) shows progressive collapse pictures of the calibrated panel. It may be noted that, in general, one should avoid the occurrence of two competing failure modes that may be triggered through minor differences in material properties or in geometry. If this is not the case, the crash response of a structure will be random and a clear management of energy absorption will not be possible.

Impact on Sandwich Box Column

Figure 21(a) shows deformed shapes and Fig. 21(b) load–displacement and absorbed energy–displacement curves, obtained by experiment on a box column of dimensions and loaded as shown in Fig. 18(a). As is typical for sound composite components, the crush mode consists of a localized 'process zone' that travels just ahead of the load. This zone could clearly develop in the box column studied, while the sandwich panel discussed above could only fail through global buckling.

The measured energy absorption capacity of the box column is therefore rather good; see Fig. 21(b), where a strong sustained post-peak crush resistance of about 3 tons is apparent. The initial peak load of about 6·5 tons is no longer an order of magnitude higher, as it was in the case of the rather inefficient sandwich panel.

Figure 22(a) compares the numerically obtained impact force (load)–displacement response (curve A) with the experimental response (curve B). Figure 22(b) compares the absorbed energy–displacement curves. The coincidence of the force response is seen to be rather good, which is also true for the energy response. Note, however, that the early numerical load response is too idealized, and reflects perfect column impact and fabrication conditions, while in reality the impact is not perfectly central and the built column contains material and geometrical imperfections.

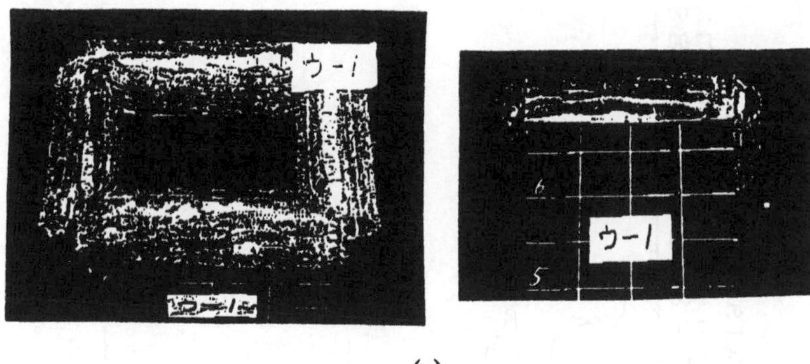

(a)

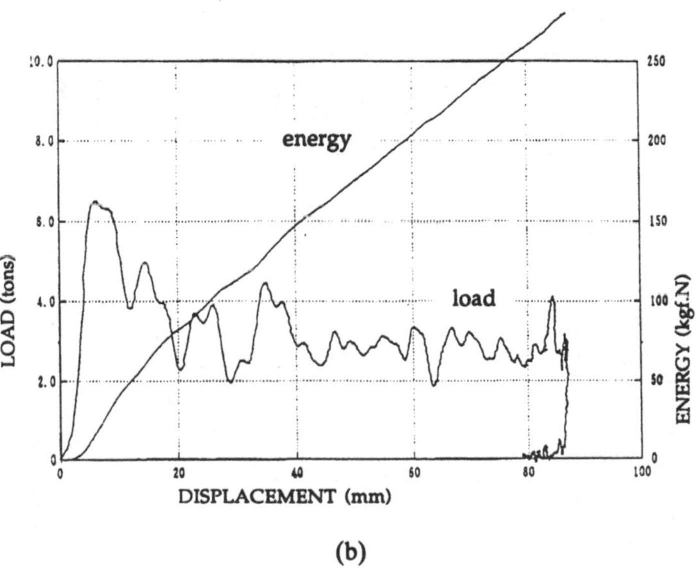

(b)

Fig. 21. Sandwich box column crush experiments. (a) Crushed process zone near impacted end. (b) Experimental load– and energy–displacement curves. (After Tonen.)

The results were obtained without any further recalibration of the basic material properties. Their good quality constituted a validation of the overall numerical model parameters, and the methodology can now be applied to full car models.

Figure 23(a), finally, shows the simulated crushing deformations and Fig. 23(b) the simulated average damage contours for the multilayered inner sandwich facing (back row) and outer sandwich facing (front row). For clarity, the orthotropic hexagonal cell core brick elements situated

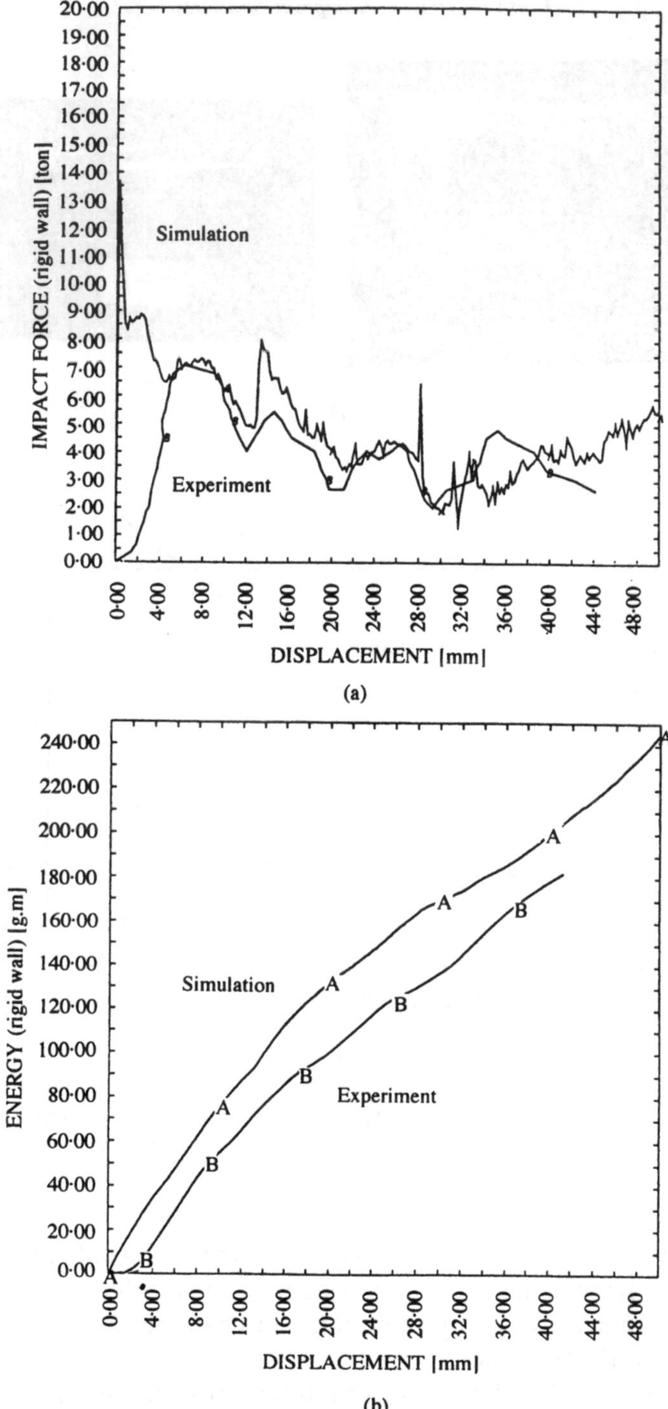

Fig. 22. Sandwich box column crush test and simulations (program PAM-CRASH): experimental and simulated impact force–displacement (a) and absorbed energy–displacement (b) responses.

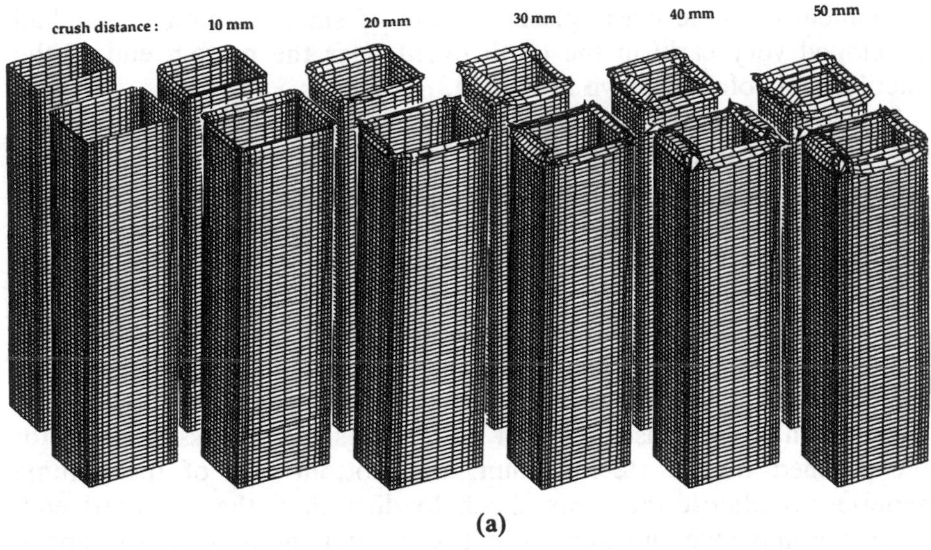

(a)

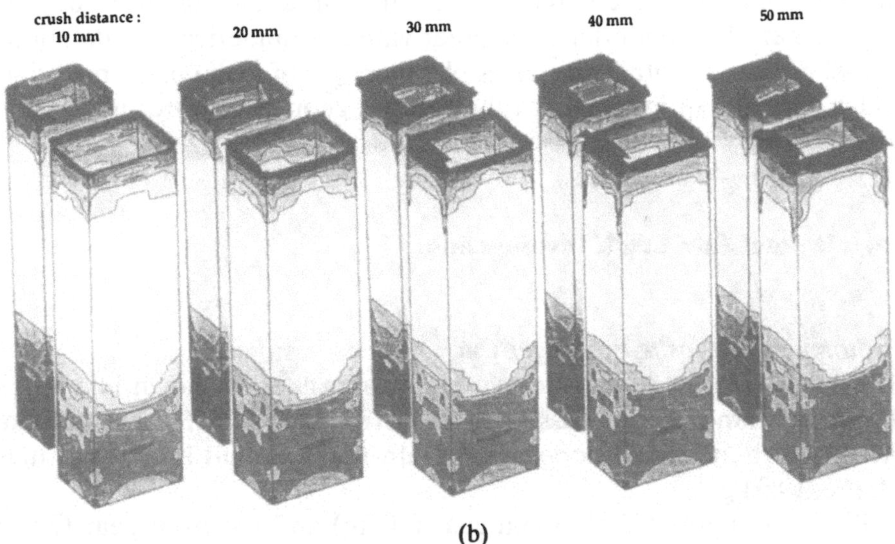

(b)

Fig. 23. Sandwich box column crush simulation (program PAM-CRASH): (a) deformed shapes of inner and outer facings; (b) average damage contours over inner and outer facings.

between the inner and the outer facings are not shown. The strong localization of the crushing deformations near the top-end 'process zone' is clearly evident (Fig. 23a). A short distance from the bottom end, however, a barely visible temporary dynamic buckle of small amplitude appears at 10 mm crushing displacement. This buckle may be considered equivalent to and has the same basic origin as the permanently visible elasto-plastic buckle of small amplitude that had developed very early in the crush event near the bottom end of the ductile steel column shown in Fig. 13.

Figure 23(b) shows average damage contours ($0 \leq d \leq 1{\cdot}0$), plotted over the mid-surfaces of the facings. The average damage is calculated, according to Fig. 11, from the damage of each constituent ply material (carbon plies, Kevlar plies, etc.), and gives a global overview of the residual strength status of each facing. The black areas indicate completely damaged material ($d = 1{\cdot}0$), as seen near the crushed top end of the column. Significant damage, however, also developed near the bottom end of the column, although in the later deformed pictures (Fig. 23a) no deformation can be seen. This damage developed early, and is seen to be constant for all later stages. As discussed above for the crushed ductile steel column, the bottom end of the column experiences almost the same shock loading than the impacted end. Since the impacted end starts to fail soon after the moment of impact, the column crush force is relieved and the initial damage will not grow any further. This underlines the great value of numerical simulation for the depiction of otherwise invisible damage of composite materials, which tend to 'spring back' to their initial geometries after undergoing non-destructive fracturing damage.

5.4 Frontal Car Crash Investigation

Prototype Composite Passenger Car

The layout of the composite two-seater sportscar is shown in Fig. 19. The mainframe of the passenger cabin is made 100% from carbon–Kevlar–carbon aramid honeycomb plate material and has a total mass of about 80 kg.

The rear engine (150 kg) and gear (30 kg) and the front gear (20 kg) are attached to the cabin rear and front walls at steel insets via tubular metallic subframes. The hood, engine cover and roof are also made of composite material. The car has been built and run on test tracks.

The material of the cabin component walls consisted in the mixed stackup sandwich facings of the type described earlier (see e.g. Fig. 17), and in facings made of $[(\mathrm{T300}(0*90)/\mathrm{T300}(-45*45)/\mathrm{K49}(0*90)_{1/2})_{\mathrm{sym}}/\mathrm{CORE}_{1/2}]_{\mathrm{sym}}$ material, where T300 $(0*90)$ stands for Toray PAN-based carbon-fiber cloth oriented at 0°, T300 $(-45*45)$ is oriented at 45°, K49$(0*90)$ is DuPont aramid-fiber-based Kevlar 49 cloth, and the core is made of aramid honeycomb material.

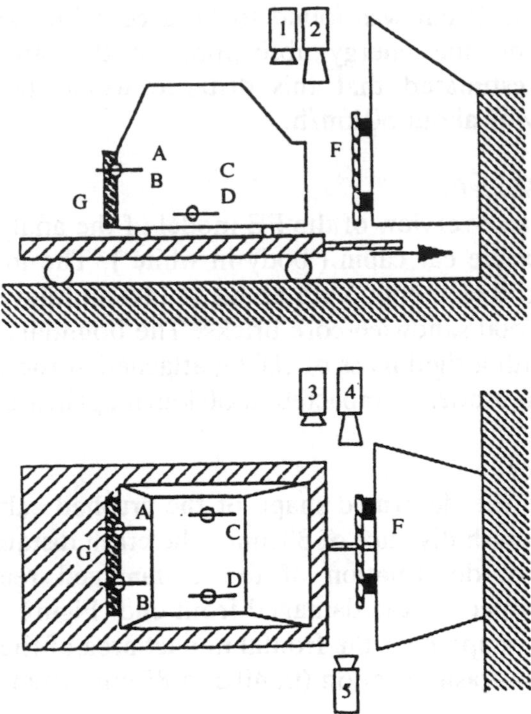

Fig. 24. Frontal crash test setup (INRETS Laboratories). Accelerometers: A, passenger-side back face; B, driver-side back face; C, passenger seat support; D, driver seat support. Load cell: F, on rigid wall. High-speed cameras (16 mm, 500 fps): 1, top general view; 2, top close-up view; 3, left-side general view; 4, left-side close-up view; 5, right-side general view. Additional mass: G, engine and rear frame.

Frontal Crash Test Setup

Figure 24 shows the frontal crash test setup used by INRETS (Lyon–Bron). The cabin is mounted on frictionless rollers on a sled that is stopped short of a rigid barrier. The cabin will then impact on the barrier, and the event is recorded via impact force measurement devices, accelerometers and high-speed cameras.

The purpose of this investigation was to furnish evidence that composite cars can be designed to withstand severe crash events and that state-of-the-art crash simulation codes such as PAM-CRASH can predict the outcome of composite car crash tests as accurately as this is possible today for conventional cars. For that reason, the tests were performed on the cabin alone, with the mass of the engine and the rear subframe (150 kg + 30 kg = 180 kg) attached to the rear end of the cabin via a rigid steel plate.

The frontal impact test was designed to produce noticable deformation of the cabin front, but not to disintegrate the structure. A crash distance of about 10 cm was found to be acceptable, and preliminary studies, based on the energy absorption of the studied composite subassemblies, estimated that this distance would be reached at a forward velocity of about 50 km/h.

Finite Element Model

Figure 25 gives an overview of the FE model of the analyzed half of the prototype composite car cabin ('body in white'). The model comprises about 21 500 finite elements, with about 15 000 multilayered composite thin shells and 6500 sandwich-core bricks. The boundary conditions are as in the test, with a rigid mass of 180 kg attached at the rear wall of the cabin. The cabin model is impacted at 50 km/h against a rigid wall.

Deformed Shapes

Figure 26 shows the deformed shape of the crushed cabin model after 10 ms at a total crash distance of 88 mm. The crash distance is measured from the relative deformation of the undamaged rear part of the structure with respect to the damaged front end. Figure 27 gives details of the deformed shape near the frontal impact area of the central tunnel at 0, 4 and 10 ms crash duration (0, 40 and 85 mm crash distance).

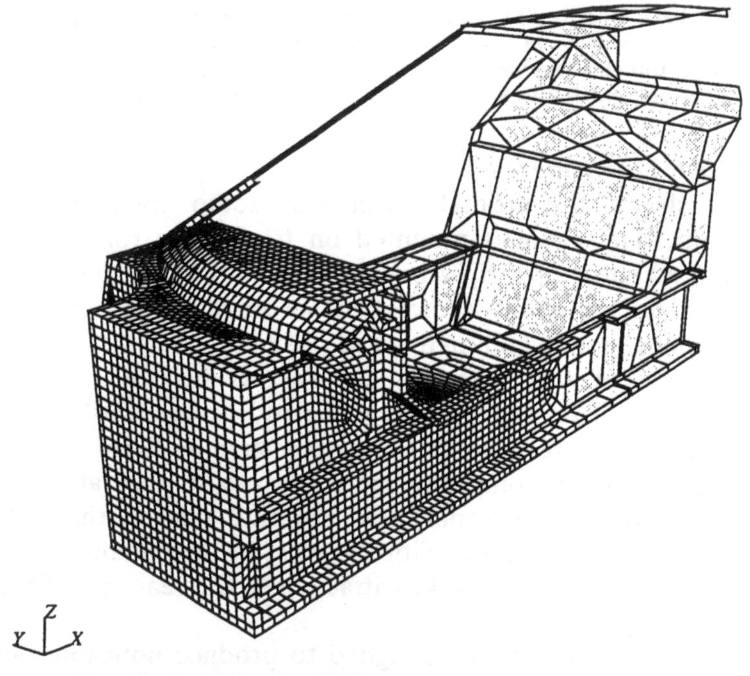

Fig. 25. Frontal crash FE model of composite car cabin.

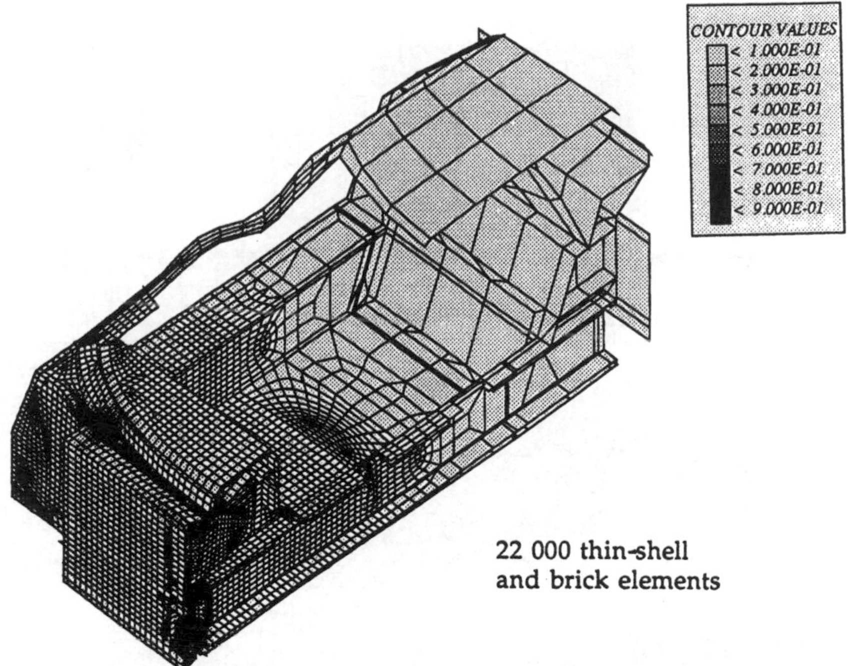

Fig. 26. Frontal crash simulation: deformed shape and damage contours at 88 mm crash distance (10 ms) (PAM-CRASH/DAISY).

The superimposed average damage contours over the multilayered carbon-Kevlar–carbon sandwich facings indicate the local nature of the damage: in a structure made of soft car body steel the damaged (buckled) areas would spread out much further. Black areas indicate values $d = 1{\cdot}0$ of the average damage function, and complete loss of strength. For clarity, the sandwich-core brick elements are not shown in Fig. 27.

Figure 28(a) compares the frontal damage areas of the test (after elastic springback) and the simulation (before springback). Other than, for example, in soft steel structures, there is significant springback in elastic-fracturing composite structures after removal of the impact loads. This is due to the fact that this type of material tends to recover strain after unloading, although with severely damaged secant modulus in the stress–strain curve. This is especially true in the presence of more ductile layers of Kevlar material, which tend to preserve the structural integrity of the damaged wall structure. In elasto-plastic deformation, however, plastic strains are permanent and elastic springback of crushed structures made of ductile material is usually very small.

In the last of the deformed shape plots of Fig. 27 a remote area of

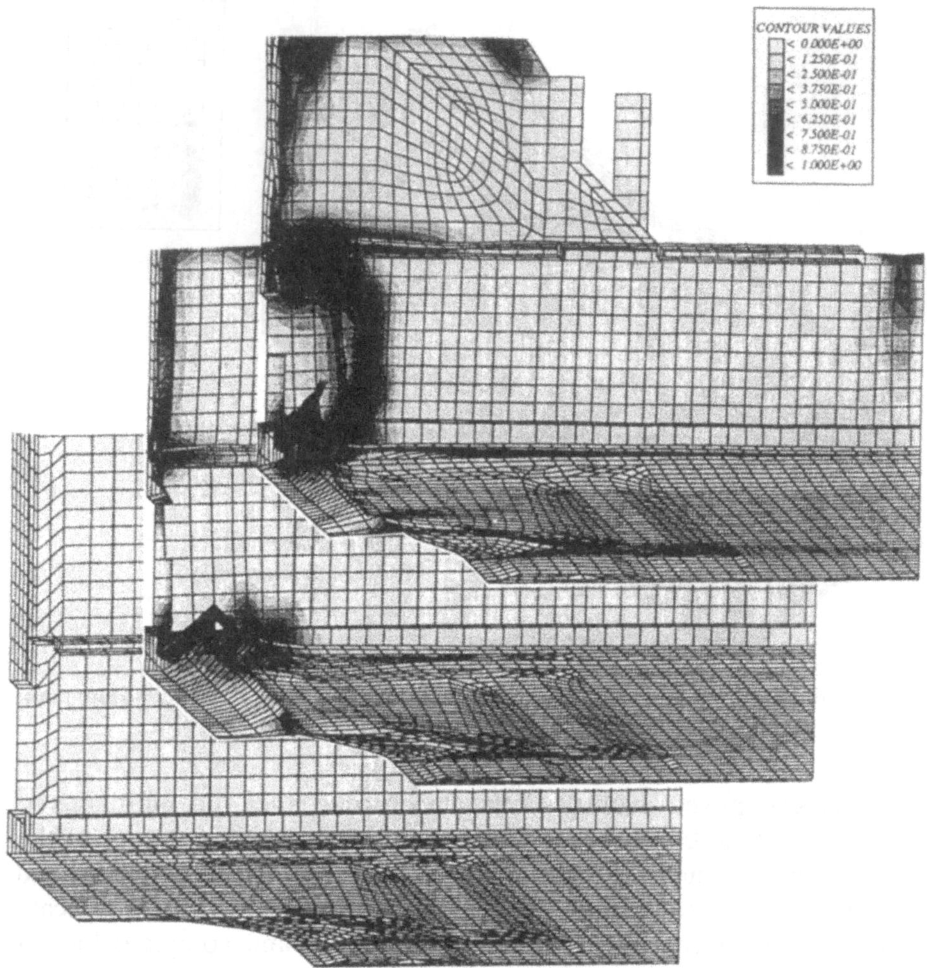

Fig. 27. Frontal crash simulation: deformed shape zooms and damage contours at 0, 4 and 10 ms (0, 39 and 88 ms) (PAM-CRASH/DAISY).

damage in the vertical wall of the central tunnel can be seen near the central hole for the gear stick. Figure 28(b) also contains this damaged area, comparable to the real damage of the crashed structure.

Further Results

The predicted total crash distance of the simulation of the frontal crash test is obtained from the point of zero kinetic energy (Fig. 29). The distance is predicted to be 13·5 cm, which agrees to within a few percent with the total crash distance measured in the frontal crash test. Figure 30 shows plots of impact force versus crash distance for the four

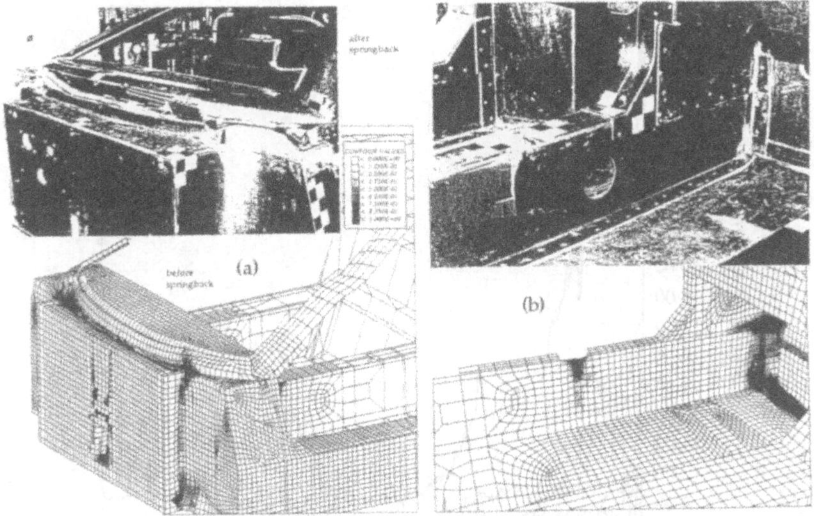

Fig. 28. Frontal crash simulation: deformed shape test versus simulation comparisons at 10 ms (88 ms) (PAM-CRASH/DAISY).

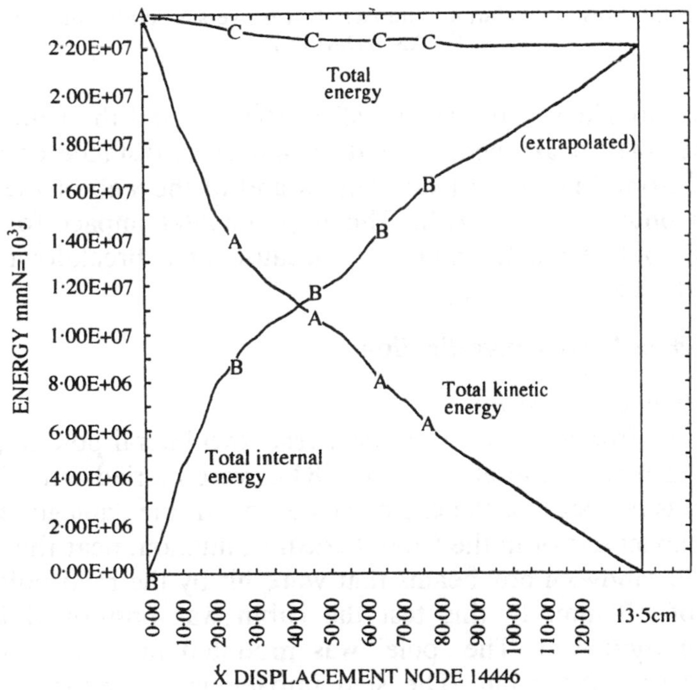

Fig. 29. Frontal crash simulation: energy versus crash distance (PAM-CRASH/DAISY).

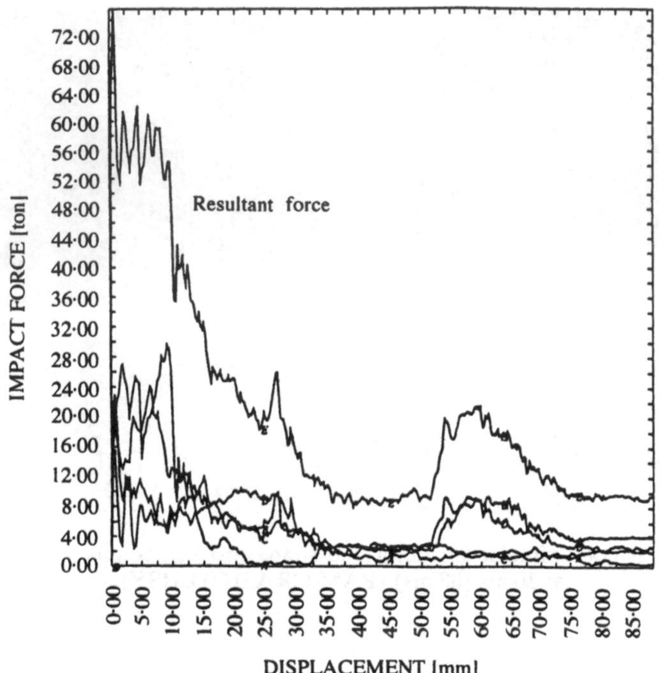

Fig. 30. Frontal crash simulation: impact forces versus crash distance (PAM-CRASH/DAISY).

force plates attached to the rigid wall in the test and the sum over all forces. The small peaks near 27 and 55 mm crash distance correspond to delayed frontal impacts of the flanges and of the web of the central box beam onto the rigid wall. The experimental impact force–time history was not available, however, because of a breakdown of the recording devices.

5.5 Pole-Side Impact Investigation

Pole-Side Impact Test Setup

Following the frontal rigid-wall impact test, two lateral pole impacts at about 20 km/h with a total mass of 288 kg were carried out (Fig. 31). This was feasible because the cabin showed no visible damage, either in the front impact test or in the frontal crash simulation, near the areas of the door sill sandwich box beams that were hit by the rigid pole with a diameter of 270 mm. In this test the cabin was positioned laterally against the rigid wall. The 'pole' was mounted on a sled that was moving towards the cabin. The sled impact was monitored with an accelerometer mounted on the sled, and the impact force of the pole was measured via load cells at the far side of the cabin.

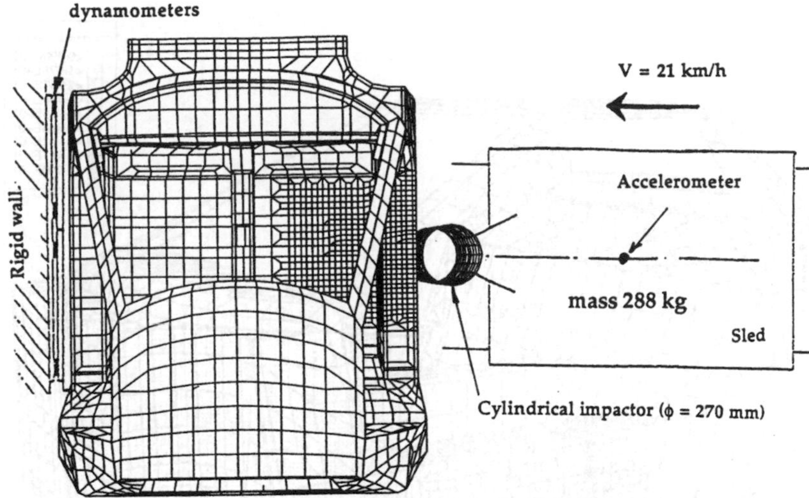

Fig. 31. Pole-side impact test impact (INRETS Laboratories).

Finite Element Model

Figure 31 also shows the finite element model adapted for the lateral pole impact simulation. For this purpose, only the mesh near the impact area around the pole was refined. The mesh contains about 3200 sandwich-core brick finite elements and 8800 multilayered thin shells. The impactor was modelled as a rigid body with mass 288 kg and was restricted to move in the impact direction, with an initial velocity of 21·4 km/h (5944·4 mm/s).

Deformed Shapes

Figure 32 shows the local nature of the deformation after 18 ms (80 mm lateral crash distance) near the sandwich box beam of the door sill that was hit by the pole. At this time the motion of the sled had decreased to about 20% of its initial velocity. For added clarity, the finite elements that had failed were removed from the computer plots. Keeping this in mind, visual agreement of the predicted damage with the photograph from the test is excellent.

Figure 33 contains progressive damage spread in close-up views of the rigid pole impact area at 0, 6, 10 and 14 ms crash duration (0, 35, 52 and 70 mm lateral crash distance). For clarity, the sandwich-core brick elements are not plotted. The contours of the average damage over all plies of the sandwich facings are superimposed on the plots. Black areas indicate a damage function value of $d = 1{\cdot}0$, i.e. elements have lost all strength. The pictures clearly show the zip-like tensile ruptures (tears) due to membrane stresses to both sides of the pole in the outer facing

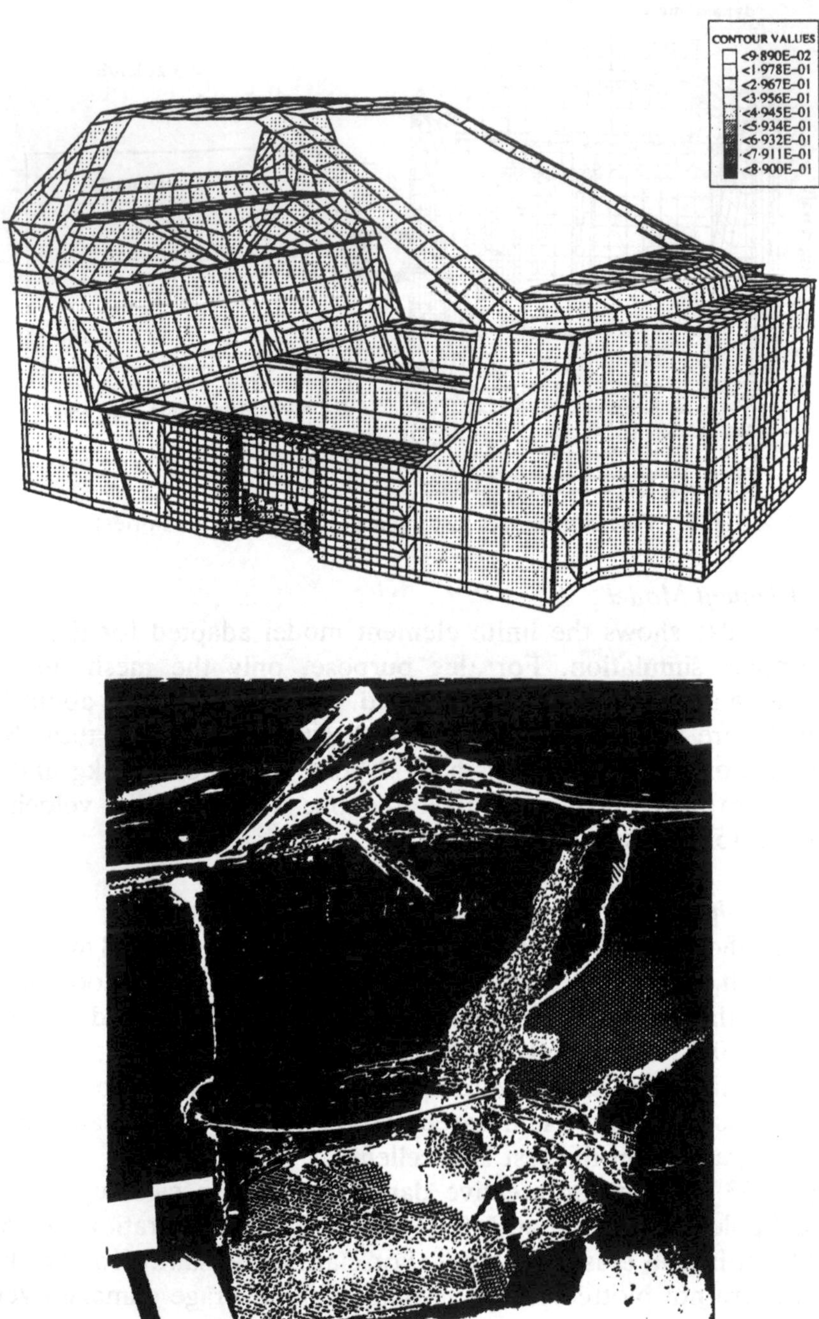

Fig. 32. Pole side impact simulation: deformed shape and damage contours at 14 ms (PAM-CRASH/DAISY).

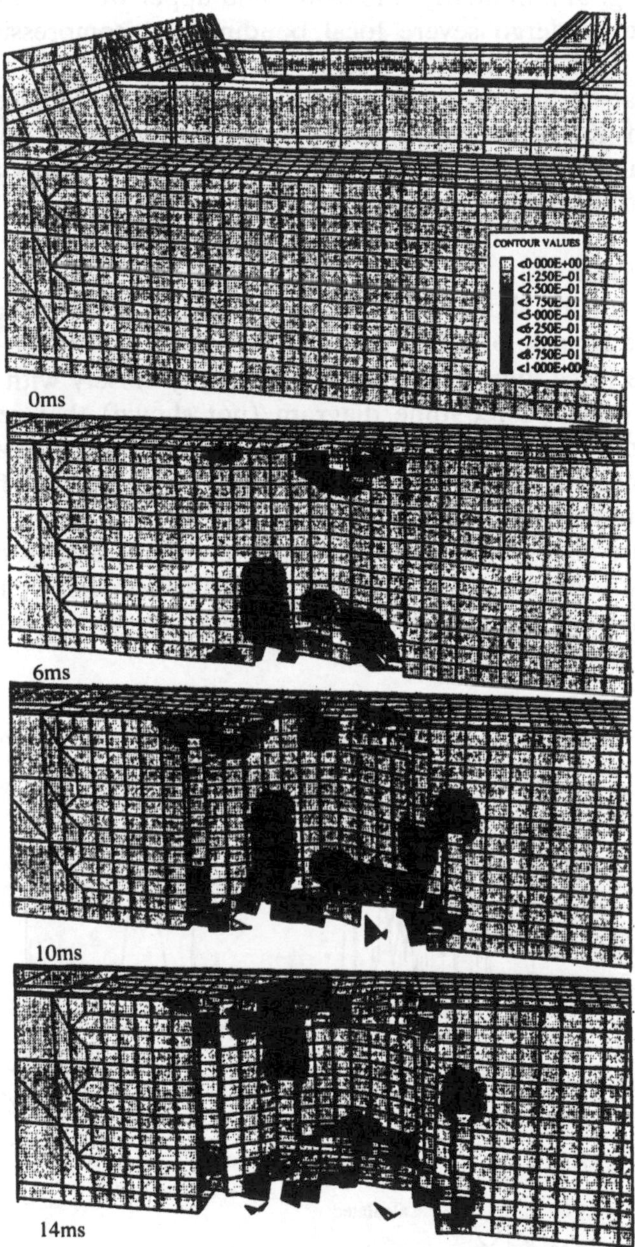

Fig. 33. Pole-side impact simulation: deformed shape close-up views with damage contours (PAM-CRASH/DAISY).

of the sandwich box beam. The same phenomenon has been found in the test (Fig. 32). Similarly, the lower and upper webs of the box beam are seen to undergo severe local bending and compression damage, which has been found in the test.

Note that the test picture contains almost totally damaged shreds of the carbon–Kevlar–carbon facings. Such portions were removed from the computer plots. It is mostly the admixture of the Kevlar plies that preserves this semblance of structural integrity in the tests.

Further Results

Figure 34, finally, shows the calculated total impact force near the rigid wall on the far side of the laterally impacted composite car cabin. As a check, this force–time diagram coincides very closely with the directly measured impact force–time diagram (not shown) that resulted from the accelerations of the impacting mass. This means that the filter effect on the impact signal between the point of impact and the far end rigid wall is quite small.

The figure also shows one of the force–time diagrams from the lateral impact tests. The simulated curve was found to lie well within the scatter range of the two test curves. This is also confirmed by the absorbed energy plots in Fig. 35.

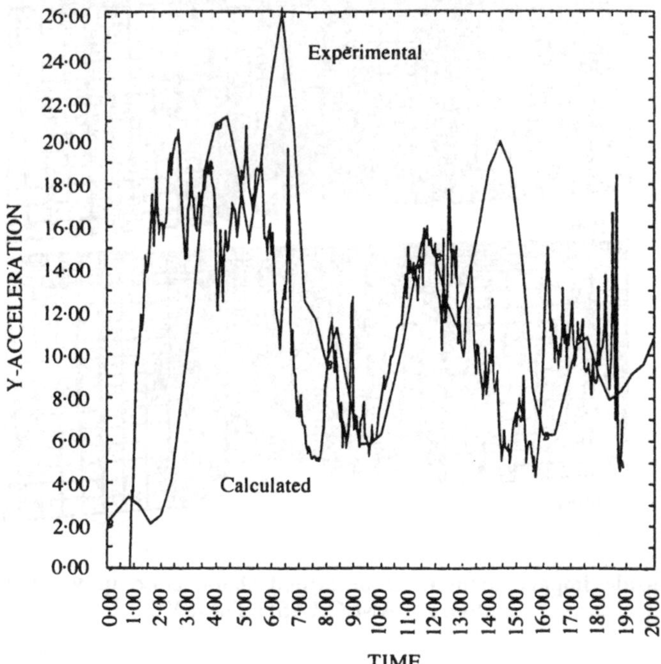

Fig. 34. Pole-side impact simulation: impact force–time histories.

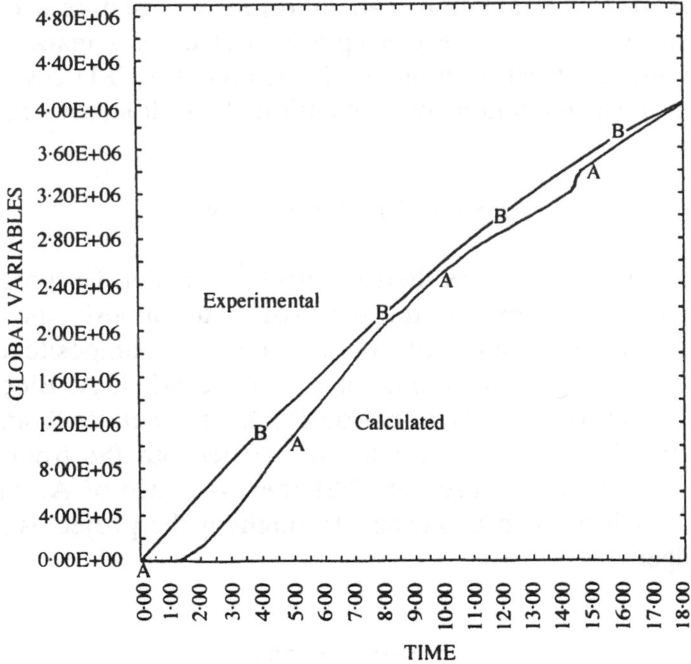

Fig. 35. Pole-side impact simulation: absorbed energy–time histories.

6 CONCLUSIONS

This chapter has reviewed the literature on crashworthiness of composite components. While previous investigations have concentrated on how to trigger and sustain potentially efficient energy absorbing axial crushing modes, where a localized crash frond travels along the component axis, involving splitting and heavy destruction of the composite wall structure, these modes are difficult to obtain in industrial applications, such as in full car structures, made of fiber-reinforced composites. The work reported here has therefore been aimed at more practial composite sandwich wall components and structures, the crushing failure of which involves bending and tearing of the sandwich facings. Such structures can be designed to effectively absorb crash energy without relying on hazardous triggering mechanisms and perfect loading and boundary conditions.

The methodology outlined here and its successful application to full-scale composite sandwich-wall car crash simulations permit the conclusion to be made that the feasibility level of industrial crashworthiness simulation of composite structures has been reached. The crash simulation code PAM-CRASH, augmented by the options for composite crashworthiness simulation, can therefore be used as a design aid and verification code for the conception of structures made from new composite and sandwich wall materials, at no extra CPU cost than that needed for crash simulations of conventionally built structures.

ACKNOWLEDGEMENTS

The authors wish to express their gratitude to Tonen Corporation, Mr Watanabe and Mr Nakada, for permission to publish and for help received during the joint study of the reported composite car crash event. The authors also wish to thank Dr D. Cesari, J. A. Bloch and R. Zac of INRETS (Institut National de Recherche sur les Transports et leur Sécurité, Bron, France) for having carried out the front and side impact tests at their premises. At ESI the assistance of A. Trameçon, O. Fort, G. Milcent and L. Penazzi throughout the project is gratefully acknowledged.

REFERENCES

1. Haug, E., Arnaudeau, F., Dubois, J., de Rouvray, A. & Chedmail, J. F., Static and dynamic finite element analysis of structural crashworthiness in the automotive and aerospace industries. In *Structural Crashworthiness*, ed. N. Jones & T. Wierzbicki. Butterworths, London, 1983, pp. 175–217.
2. de Rouvray, A. & Haug, E., Failure of brittle and composite materials by numerical methods. In *Structural Failure*, ed. T. Wierzbicki & N. Jones. Wiley, New York, 1989, pp. 193–252.
3. Farley, G. L., Bird, R. K. & Modlin, J. T., The role of fiber and matrix in crash energy absorption of composite materials. In *Proceedings of the American Helicopter Society National Meeting: Crashworthy Design of Rotorcraft, 1986.*
4. Schmueser, D. W. & Wickliffe, L. E., Impact energy absorption of continuous fiber composite tubes. *J. Engng Mater. Technol.*, **109** (1987) 72–7.
5. Hull, D., Axial crushing of fiber reinforced composite tubes. In *Structural Crashworthiness*, ed. N. Jones & T. Wierzbicki. Butterworths, London, 1983, pp. 118–35.
6. Berry, J. & Hull, D., Effect of speed on progressive crushing of epoxy–glass cloth tubes. In *Mechanical Properties at High Rates of Strain*, ed. J. Harding. Institute of Physics Conference Series No. 70, Bristol, 1984, pp. 463–70.
7. Fairfull, A. H. & Hull, D., Effect of specimen dimensions on the specific energy absorption of fiber composite tubes. In *Proceedings of ICCM' and ECCM*, Vol. 3, ed. F. L. Mathews *et al.* Pergamon Press, London 1987, pp. 3.36–3.45.

8. Fairfull, A. H. & Hull, D., Energy absorption of polymer matrix composite structures: frictional effects. In *Structural Failure,* ed. T. Wierzbicki & N. Jones. Wiley, New York, 1989, pp. 255–79.
9. Hull, D. & Coppola, J. C., Effect of trigger geometry on crushing of composite tubes. In *Materials and Processing—Move into the 90's,* ed. S. Benson, T. Cook, E. Trewin & R. M. Turner. Elsevier, Amsterdam, 1989, pp. 29–38.
10. Thornton, P. H., Effect of trigger geometry on energy absorption of composite tubes. In *Proceedings of the 5th International Conference on Composite Materials, ICCM-V, San Diego,* ed. Harrington *et al.* The Metallurgical Society, 1985, pp. 1183–99.
11. Thornton, P. H. & Jeryan, R. A., Crash energy management in composite automotive structures. *Int. J. Impact Engng,* **7** (1988) 167–80.
12. Schmueser, D. W., Wickliffe, L. E. & Mase, G. T., Front impact evaluation of primary structural components of a composite space frame. Engineering Mechanics Department, General Motors Research Laboratories, Warren, MI, Paper nb 880890, pp. 67–75.
13. Price, J. N. & Hull, D., The crush performance of composite structures. In *Composite Structures 4,* Vol. 2, ed. I. H. Marshall. Elsevier Applied Science Publishers, London, 1987, pp. 2.32–2.44.
14. Price, J. N. & Hull, D., Axial crushing of glass fibre–polyester composite cones. *Composites Sci. Technol.* **28** (1987) 211–30.
15. Mamalis, A. G., Manolakos, D. E. & Viegelahn, G. L., The axial crushing of thin PVC tubes and frustra of square cross section. *Int. J. Impact Engng,* **8** (1989) 241–64.
16. Price, J. N. & Hull, D., Crush behaviour of square section glass fibre polyester tubes. In *How to Apply Advanced Composites Technology.* ASM International, 1988, pp. 53–61.
17. Hull, D., Impact response of structural composites. *Composite Materials,* January 1985, pp. 35–8.
18. Hull, D., Energy absorbing composite structures. Scientific and Technical Review, University of Wales, No. 3, 1988, pp. 23–30.
19. Pickett, A. K., Haug, E. & Rückert, J., A fracture damaging law suitable for anisotropic short fiber/matrix materials in an explicit finite element code. *J. Composites,* **21** (1990) 297–304.
20. de Rouvray, A. & Haug, E., Industrial calculation of damage tolerance and stress allowables in components made of composite materials using the PAM-FISS/BI-PHASE material model. In *Proceedings of the Workshop: Composites Design for Space Applications, 15–18 October 1985,* ed. W. R. Burke. ESA SP-243, ESA Publications Division, ESTEC, Noordwijk, February 1986, pp. 175–85.
21. Haug, E., Dowlatyari, P. & de Rouvray, A., Numerical calculation of damage tolerance and admissible stress in composite materials using the PAM-FISS/BI-PHASE material model. In *Proceedings of the Conference: Spacecraft Structures, CNES, Toulouse, 3–6 December, 1985,* ed. W. R. Burke. ESA SP-238, ESA Publications Division, ESTEC, Noordwijk, April 1986, pp. 361–71.
22. de Rouvray, A., Haug, E., Dowlatyari, P., Stavrinidis, C. & Kreis, A., Intrinsic damage and strength criteria for advanced composite laminates in space applications. In *Proceedings of the ESA/CNES DFVLR Interna-*

tional Conference on Spacecraft Structures and Mechanical Testing, 19–21 October 1988, ed. W. R. Burke. ESA SP-289, ESA Publications Division, ESTEC, Noordwijk, January 1989, pp. 649–64.

23. de Rouvray, A., Dowlatyari, P. & Haug, E., Validation of the PAM-FISS/BI-PHASE numerical model for damage and strength predictions of LFRP composite laminates, In *Proceedings of MECAMAT 1989, St Etienne, 15–17 November: Mechanics and Mechanisms of Damage in Composites and Multi-Materials, ESIS11,* ed. D. Baptiste. Mechanical Engineering Publications, London, 1991, pp. 183–202.
24. Pickett, A. K., Rückert, J., Ulrich, D. & Haug, E., Material damage law suitable for crashworthiness investigation of random and directional fibre composite materials. In *Proceedings of the 18th International Finite Element Congress, Baden–Baden, 20–21 November 1989.* IKO Software Service GmbH/Richard Wahl, Stuttgart, 1989, pp. 275–94.
25. de Rouvray, A., Haug, E. & Stavrinidis, C., Composite material damage and fracture models for numerical simulations in support of fracture control. In *Proceedings of International Conference on Spacecraft Structures and Mechanical Testing, 24–28 April 1991,* ed. W. R. Burke. ESA SP-321, Vol. 2, ESA Publications Division, ESTEC, Noordwijk, April 1991, pp. 671–8.
26. Haug, E., Dagba, A., Clinckemaillie, J., Aberlenc, F., Pickett, A. K., Hoffmann, R. & Ulrich, D., Industrial crash simulations using the PAM-CRASH code. In *Proceedings of IBM Europe Institute 1988 on Supercomputing in Engineering Structures, Oberlech, Austria, 11–15 July 1988,* ed. P. Melli & C. A. Brebbia. Computational Mechanics Publications/Springer-Verlag, Berlin, 1989, pp. 171–196.
27. Haug, E. & Ulrich, D., The PAM-CRASH code as an efficient tool for crashworthiness simulation and design. In *Automotive Simulation, ASIMUTH, Proceedings of the 2nd European Cars/Trucks Simulation Symposium, Schliersee (Munich), May 1989,* ed. M. Heller. Springer-Verlag, Berlin, 1989, pp. 74–87.
28. Haug, E., Clinckemaillie, J. & Aberlenc, F., Computational mechanics in crashworthiness analysis. Engineering Application of Modern Plasticity (Post-Symposium Short Course of the 2nd International Symposium of Plasticity) at Nagoya Chamber of Commerce and Industry, 4–5 August 1989.
29. Haug, E., Clinckemaille, J. & Aberlenc, F., Contact–impact problems for crash. Engineering Application of Modern Plasticity (Post Symposium Short Course of the 2nd International Symposium of Plasticity) at Nagoya Chamber of Commerce and Industry, 4–5 August 1989, Paper 5.
30. Haug, E., Fort, O., Trameçon, A., Watanabe, M. & Nakada, I., Numerical crashworthiness simulation of automotive structures and composites made of continuous fiber reinforced composite and sandwich assemblies. International Congress and Exposition, Detroit, Michigan, 25 February–1 March 1991, SAE Technical Paper Series 910152.
31. Haug, E., Fort, O., Milcent, G., Trameçon, A., Watanabe, M., Nakada, I. & Kisielewicz, T., Application of new elastic–plastic–brittle material models to composite crash simulation. 13th International Technical Conference on Experimental Safety Vehicles, Paris, 4–7 November 1991.

Chapter 8

DYNAMIC COMPRESSION OF CELLULAR STRUCTURES AND MATERIALS

S. R. Reid, T. Y. Reddy & C. Peng

Department of Mechanical Engineering, UMIST, PO Box 88, Sackville Street, Manchester M60 1QD, UK

ABSTRACT

Cellular structures comprising assemblies of metal tubes or honeycombs are effective impact energy absorbers because they permit gross plastic deformation and have long strokes that enable the designer to limit the decelerating forces. Recent work on the behaviour of such structures is reviewed, with particular emphasis on the localised deformation mechanisms that occur under dynamic loading conditions. Cellular materials comprising open cells deform through a variety of mechanisms governed by the sequential collapse of individual cells and layers of cells, which are reminiscent of those observed in cellular structures. The behaviour of wood is discussed as an example of cellular material behaviour. Experiments have demonstrated that the crushing strengths of wood both along the grain and across the grain change significantly with compression rate. This phenomenon is discussed, and simple models used in the analysis of cellular structures that attribute the stress enhancement to inertial effects are described. The distinctive roles of shock formation and microinertia are discussed.

1 INTRODUCTION

The study of impact energy absorption is an interesting blend of dynamics, materials science and structural mechanics. Structures and materials are required that deform in a predictable manner at controlled force levels if they are to be used in practical systems. The two constraints of limited force and sufficient energy absorption capacity generally lead to the requirement that the energy absorber have a long stroke. Consequently, research has focused on the use of structural elements and materials that can undergo gross deformation. A broad review of this topic, dealing predominantly with metal structures, was produced by Johnson and Reid.[1] In the main, the need for large deformations has led to the use of plastically deforming metal components, especially tubes, whose mode of deformation primarily involves bending in order to accommodate the large changes of shape that are required.[2] Depending upon the mode of loading and geometry of the component, this bending can be accompanied by membrane stretching or compression (as in the progressive buckling of an axially compressed tube[3]) or evolve into a membrane-dominated response at large deflections (as in a clamped tubular beam[4]). However, almost invariably, gross bending is an underlying ingredient in the deformation process. Because it is a dissipative process, plastic deformation has been the dominant mechanism of material deformation within these components, although, especially with the advent of composite materials, other dissipative mechanisms such as fracture, friction and debonding have been utilised.[5]

In the first of this series of conferences, Reid[6] reviewed energy absorbing techniques that used the lateral compression of metal tubes. The deformation mechanisms of a single tube loaded and constrained in various ways were described and analysed, and an introduction was provided to the deformation of systems comprising assemblies of tubes, including the so-called modular crash cushion. The importance of such systems lies in their ability to cope with impact events in which the direction of loading is not controlled. Here systems of this type are termed cellular *structures,* being comprised of macroscopic elements assembled into a cellular array. Recent work on such structures will be reviewed in Section 2. The diameters of the basic components (cells) typically range from 10 mm to 1 m. Examples of such systems include arrays of circular metal tubes, sometimes loosely packed in a container or with adjacent elements mechanically joined (see Figs 1(a,b)). In these systems the geometry of the packing arrangement can be changed, leading to a variety of responses. Alternatively, honeycombs have been tested, these being made up of sheets of material glued or welded together along a pattern of lines and pulled out to form a hexagonal cellular structure.[7] In the latter case, and in much of the work on cellular structures made up of metal tubes, the loads are applied transverse to the cell walls. As such, the work reviewed on this topic in Section 2 provides an update of that given by Reid.[6] However, honeycombs are more usually loaded axially so that the cells buckle progressively.[8] Reference to some work on this topic is also included in Section 2.

Apart from occasional papers such as that by Shaw and Sata,[9] the

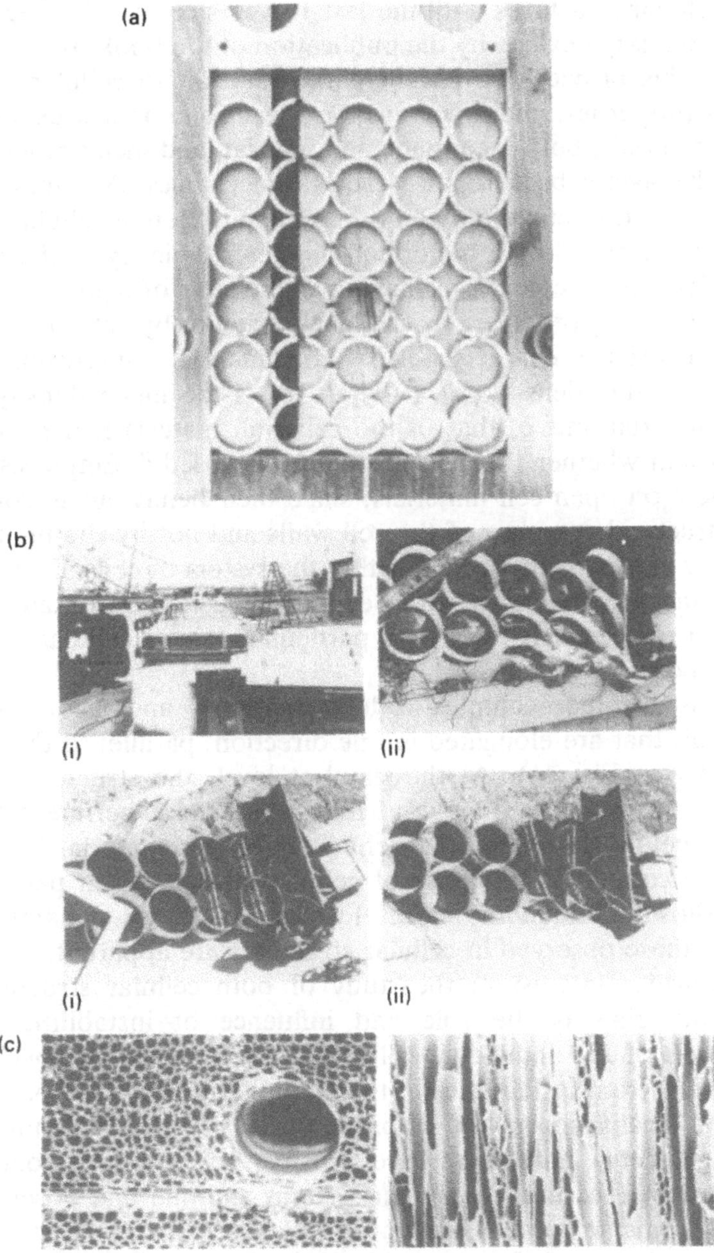

Fig. 1. Cellular structures and materials. (a) Square-packed rectangular array of metal rings. (b) Modular crash cushion before and after impact: plain tubes, (i) and (ii), and selectively reinforced tubes, (iii) and (iv).[24] (c) Cellular structure of balsa wood (magnification ×150). Sections across the grain (left) and along the grain (right).

mechanics of cellular materials had not been explored as extensively as that of cellular structures until the last 10 years or so. This neglect was rectified to a large extent by the publication of the book by Gibson and Ashby.[10] This provides an excellent introduction to cellular materials and their properties, and it surveys the work done on a wide variety of cellular materials, both man-made and natural, and includes models for material behaviour based upon structural mechanics. A comprehensive review is therefore unwarranted in this chapter. Here a cellular *material* will be characterised by a major cell diameter typically of the order of 100 μm–100 mm, the cell walls having a thickness of 3 μm–0·5 mm and the material being produced by chemical action or by natural growth. A cellular material is characterised by the material comprising the cell walls, its relative density ρ_0/ρ_s (where ρ_0 is the mean density of the cellular material and ρ_s that of the cell wall material), the mean cell diameter and whether the cells are open or closed.[10] Emphasis will be placed here on open cell materials, since their behaviour is dominated by the structural response of the cell walls and not by the influence of the flow or compression of fluid within the system of cells. To provide a contrast and a comparison with Section 2, Section 3 contains a brief account of some recent work on a particular class of cellular materials, namely wood.

Wood is a rather complex cellular material, and is distinctive in having cells that are elongated in one direction, parallel to the grain of the wood (see Fig. 1c). At the simplest level, the structure of wood resembles a close-packed array of tubes, and the experimental results discussed below show that the behaviour of this material has certain features in common with those of cellular structures. In particular, it will be shown that under impact loading conditions dynamic effects similar to those observed in cellular structures are apparent.

Of particular interest in the study of both cellular structures and cellular materials is the role and influence of instabilities in the deformation of the individual cells. This is termed microbuckling to distinguish it from instabilities in the whole structure or whole sample. It leads to localisation of the deformation to a degree depending upon the cell geometry and the relative density. Localisation is one of the most distinctive features of cellular systems and one that poses difficult problems from the point of view of theoretical modelling.

The second major influence on the deformation of cellular structures and cellular materials is the inertia of the individual cell walls. The term microinertia will be used for this. It can modify the local quasi-static collapse mechanism within the array, usually leading to less compliant modes, which require higher loads to cause crushing. In material terms, this can result in rate-sensitive material properties that have an inertial origin rather than one dependent on thermal activation or dislocation drag processes. If the cell wall material is sensitive to these latter effects, they too will be present. However, it would appear that *inertial* rate sensitivity can be a dominant effect in cellular systems.

If the loading of individual cells or layers of cells is sufficiently severe, gross deformation leads to a rapidly stiffening phase (usually termed densification), irrespective of whether the deformation mechanism is stable or unstable. Under these circumstances, a different inertial phenomenon, namely shock wave propagation, can be introduced to describe the spread of deformation through the system. This was discussed by Reid[2] for certain (stable) cellular structures, and more recent developments in the use of this model and in the development of models to encompass the wider range of deformation modes are outlined in Section 2, while an application of shock wave theory to wood is described in Section 3. Therefore, bearing in mind the use of cellular structures and cellular materials for impact energy absorption, a review of current knowledge on the mechanics of the dynamic response of such systems will be the main emphasis in this chapter.

2 CELLULAR STRUCTURES

2.1 Applications

Cellular structures constructed from simple engineering components such as bars, plates and tubes have been used effectively for mitigating the effects of impacts.[1] Variants of the basic components can be used, leading to different responses. For example, beams can be of flat, corrugated or tubular cross-section, and the tubes can be circular or non-circular and the loads applied transversely or axially. One example of a cellular/repetitive structure is the familiar Armco-type crash barrier used extensively alongside major roads. These are comprised of tensile elements which are mounted on a series of posts. At least one version of this is designed with frangible posts that snap-off sequentially as a vehicle collides with it.[11]

Cellular arrays manufactured from thin plates or assemblies or circular tubes in a close-packed configuration have been proposed for use in a variety of applications. An example involving lateral compression of metal tubes is a modular crash cushion. This has been used, primarily in the USA, to mitigate the effects of automobile impacts. Portable crash barriers[12] constructed from a line of steel tubes have been used to protect line-striping vehicles on freeways. Variants of this with the tubes in a triangular array (Fig. 1b), sometimes having certain tubes stiffened by the use of diametral bracing members as shown, have been used in the groyne area on freeways to prevent severe impact between automobiles and immovable structures. In the event of an impact, sequential deformation of the steel cylinders gradually dissipates the kinetic energy of the vehicle and usually arrests rather than re-directs the latter. The crushing behaviour of the system depends critically upon the behaviour of individual members that are compressed transversely.[13]

Another example of this type was proposed as a particular form of pipewhip restraint system.[14] This consists of rings (or short tubes) arranged in a close-packed manner in the annulus between the outside (rigid) support ring and the pipe jacket (Fig. 2). A system like this has been examined in detail by Shrive *et al.*[15] These restraints arrest broken high-pressure pipes as they move rapidly under the influence of the reaction force created by the jet of high-pressure fluid that emerges from the break, and they need to be omnidirectionally effective and efficient.

The main application of cellular structures that absorb energy while being crushed axially has been in honeycombs, which have had widespread use, for example as crushable interior padding in automobiles and extensively as core materials for sandwich plates.[16] These have been proposed or developed also for special applications such as in the landing struts of unmanned lunar or planetary landing vehicles[17] and for the protection of nuclear reactors in aircraft[18] and ships.[19]

In most of the above applications the deformation of the elements in the energy absorber is due primarily to bending in the small-

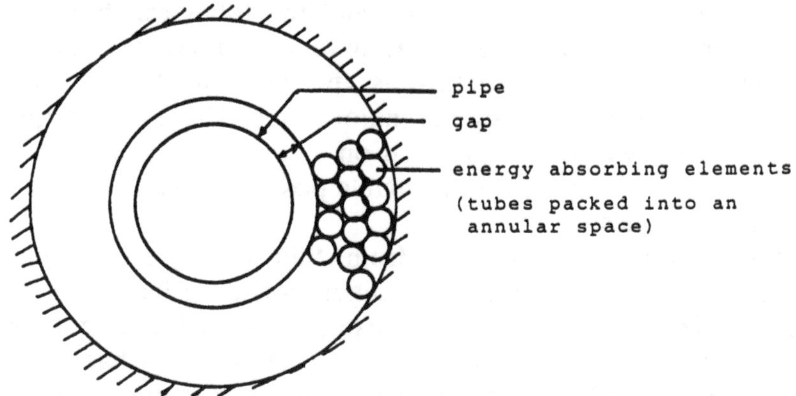

Fig. 2. Pipe whip restraint using laterally loaded tubes.

deformation regime, although membrane extensions and/or twisting can ensue owing to buckling of the tubular elements or the onset of large deflections of the system. The crushing of the individual components leads to the deformation being transmitted through the structure in a manner controlled by the constitutive behaviour of the individual elements and inertia, and special attention will be paid to this phenomenon below.

2.2 Transversely Loaded Systems

(i) One-Dimensional Systems

Earlier work on the deformation of systems made of tubes or rings was reviewed by Reid.[6] The progression of deformation in a portable crash barrier[12] or a modular crash cushion[13] was found to be similar to that through metal ring systems (chains) subject to dynamic compression by a rigid sledge.[20] The propagation of deformation through such systems was shown to be a key factor in the dynamic enhancement of the performance of model systems compared with their behaviour under quasi-static loading.[13] This propagation phenomenon has therefore been studied more systematically, both experimentally and analytically, in the context of single lines of rings undergoing impact along the line of centres of the rings.[21–23]

In chains where the rings are attached directly to each other, wrap-around modes of deformation occur, which lead to rings in alternate positions deforming more than their neighbours.[20] This behaviour was eliminated by inserting plates that ensured that each ring underwent a flat plate compression mode as shown in Fig. 3, and the deformation of the ring elements was affected by incident and reflected structural shock waves.[20] As the lower part of Fig. 3 shows, slowing down the structural shock wave by increasing the mass of the plates can redistribute the deformation in the system. A one-dimensional shock theory based on momentum transfer, analogous to shock wave theory for a bar made from a material with a convex stress–strain curve (Fig. 4a) was constructed.[21] This was later modified to include the effects of elastic waves, strain hardening, strain-rate sensitivity and a modal correction factor by Reid and Bell.[22] The predicted behaviour was found to agree well with experimental observations, as shown in Fig. 4(b). This model has been used recently to analyse the behaviour of ring systems with free distal ends.[23] A similar simple structural shock model for chains where wrap-around modes occur has yet to be produced.

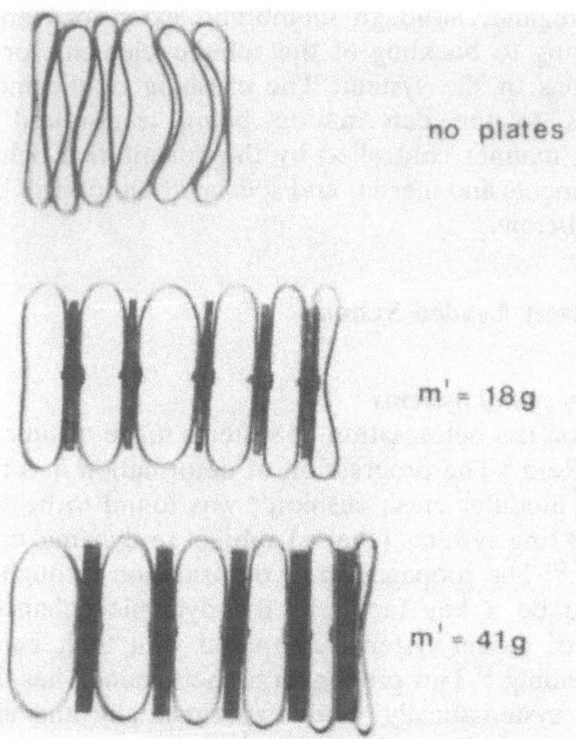

Fig. 3. Deformation of one-dimensional systems of brass rings, showing redistribution of deformation produced by inserting metal plates of mass m'. Brass rings: $D =$ 50·8 mm, $t = 1{\cdot}6$ mm, mass $m = 27{\cdot}5$ g. Striker mass $M = 125\,g$ and velocity $V \approx 34$ m/s. (From Ref. 20.)

To understand the effect of bracings on the behaviour of tubes in modular crash cushions, the behaviour of braced tubes compressed between flat plates has been studied. Experiments on rings with single and asymmetric double braces were carried out by Reid *et al.*,[24] and collapse loads computed using the equivalent structure technique agreed well with the experimentally measured values. The experimental behaviour of rings with symmetrical double braces was reported by Carney and Veillette.[25,26] The load–compression behaviour of doubly braced rings for large deformations was analysed by Reddy *et al.*[27] Here the equivalent structure technique[6,24] was modified to allow for the effects of strain hardening in the plastic hinges which were formed in the braced tubes. The degree of agreement between experimental and predicted load–compression characteristics of doubly braced tubes is shown in Fig. 5 and is reasonably good in the hardening part of the curve up to the maximum load. Thereafter an unstable, softening behaviour ensued. Similar unstable (softening) modes of deformation

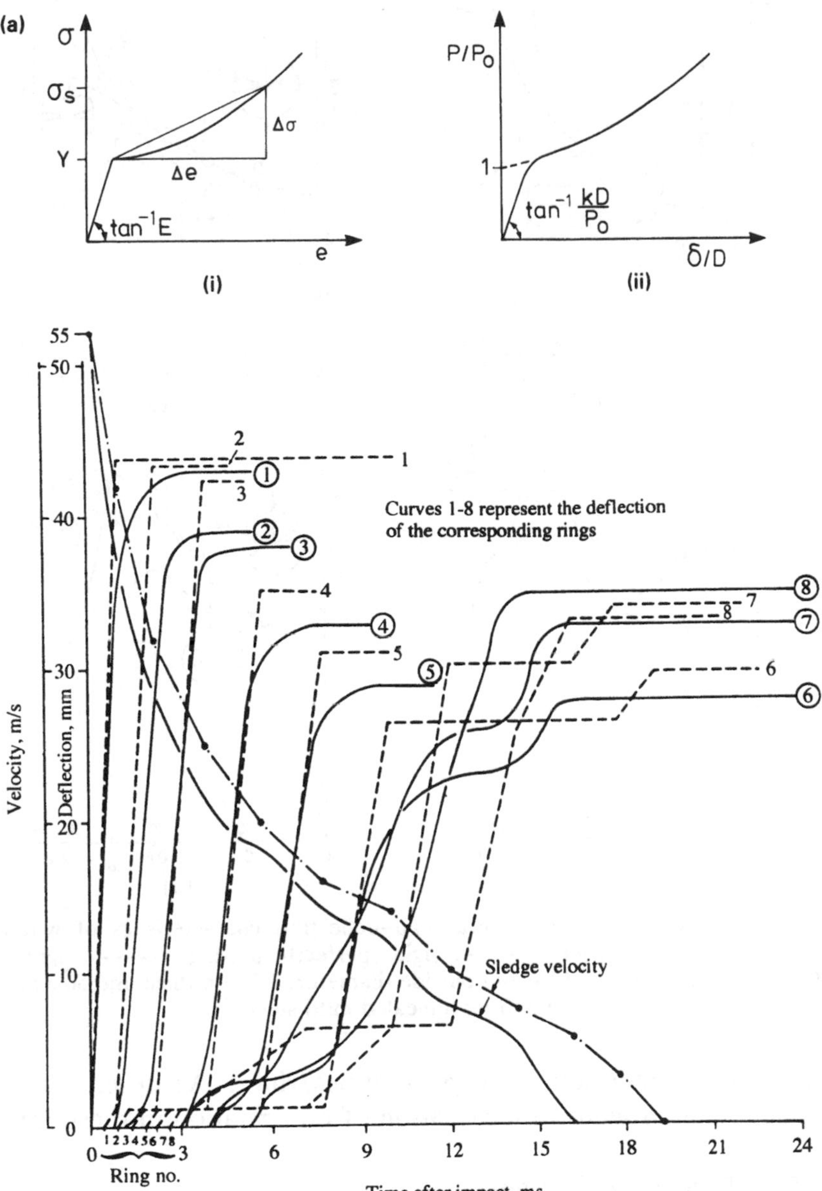

Fig. 4. Shock analysis of one-dimensional ring systems. (a) Comparison between (i) convex nominal stress–strain curve used to examine shock wave propatagion in nickel–chrome steel cylinders and (ii) load–deflection curve for laterally compressed ring/tube.[21] (b) Comparison between results of shock analysis and experimental data.[22]

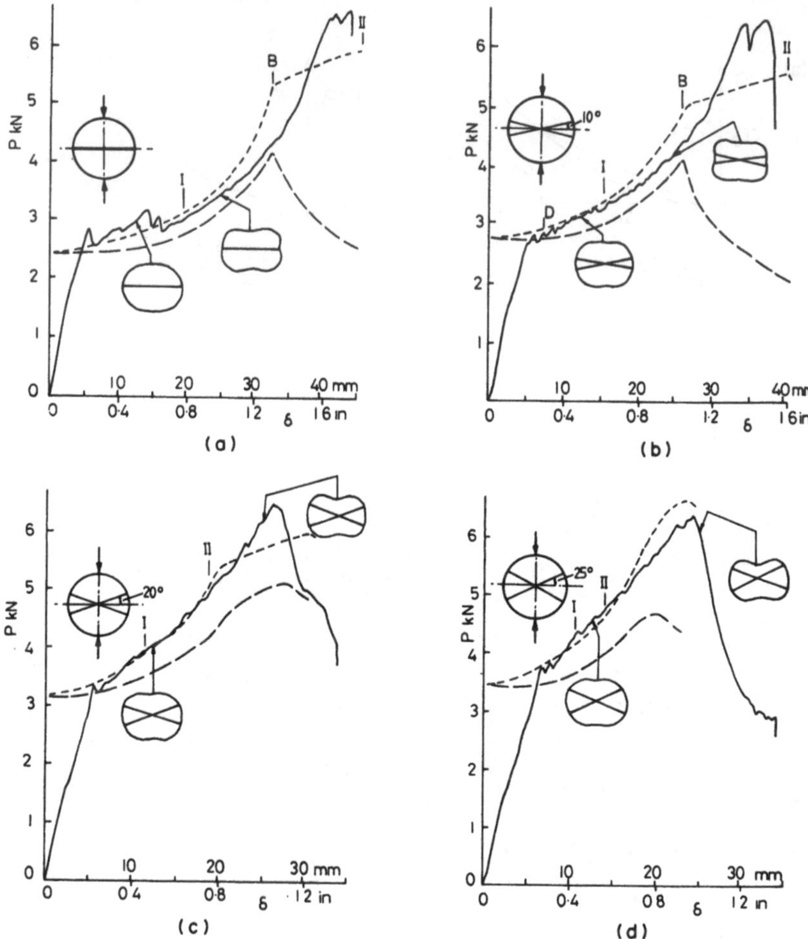

Fig. 5. Experimental and theoretical load–deflection characteristics of wire-braced tubes: ———, experimental; – – – –, rigid, perfectly plastic; - - - - - - rigid, strain-hardening. Tubes were of 101·6 mm outside diameter, 2·2 mm thick and 50·8 mm long, made from annealed mild steel.[27]

occur in two-dimensional arrays of tubes/rings. As described below, this phenomenon leads to localisation of deformation in these arrays.

(*ii*) *Two-Dimensional Systems*

In Ref. 6 the behaviour of laterally compressed open and closed tube systems was discussed, the distinction being that in the latter type the elements exert mutual constraints that lead to unstable behaviour similar to that described above. Only closed systems of tubes with their axes parallel are described in this section. Rings and tubes can be arranged in either square-packed (SP) or hexagonal-close-packed

(HCP) configurations to form two-dimensional arrays. Such arrays can be used as impact energy absorbing systems by loading them in a direction parallel or normal to the cell/tube axes. So also can hexagonal cell structures (honeycombs) be loaded in similar ways. The response to loading in the axial direction is considered in Section 2.3.

Indentation or local loading of cellular structures is not a major theme of this chapter, but some interesting observations can be made from Fig. 6, which shows typical square-packed (SP) and close-packed (HCP) ring systems subjected to crushing and indentation by a flat-faced projectile. Wrap-around modes with consequent alternation of deformation can be observed in the crushed SP systems, and a wedge-shaped deformation front with the sledge face as base can be seen when these systems are indented. HCP systems subjected to indentation are seen to show an enlarging cylindrical deformation front, which is more localised around the indenter than in the SP system. The transmission of the deformation to the sides of the HCP system

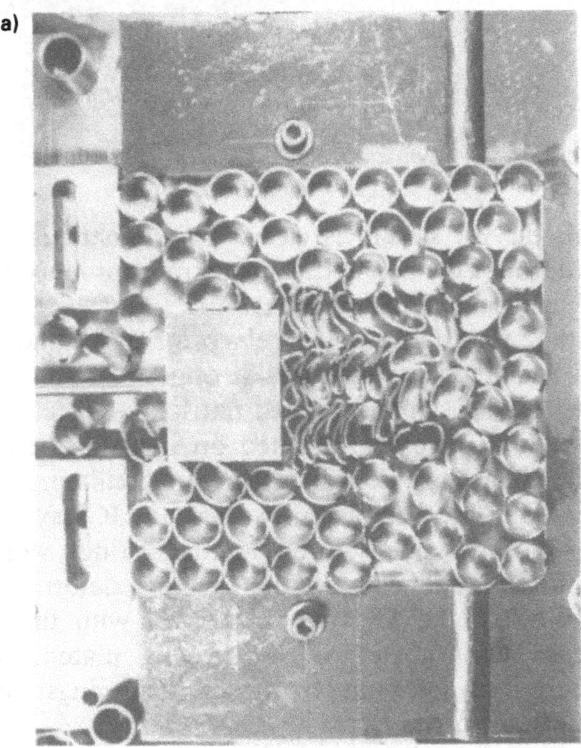

Fig. 6. Dynamic indentation of (a) square-packed (SP) and (b) hexagonal-close-packed (HCP) brass ring systems. Rings have diameter 16 mm, thickness 0·9 mm, length 12·7 mm and mass 4·65 g. They were each struck by a striker of mass 113 g travelling at 79 m/s.

(b)

Fig. 6—(Continued).

compared with the SP system is clear. This illustrates the important effect that the form of the geometrical constraint produced by the initial arrangement has on the dynamic behaviour.

The quasi-static uniaxial crushing behaviour of SP and HCP systems has been studied in detail by Shim and Stronge.[28] As shown in Fig. 7(a), it was found that SP systems exhibited flat load–deflection characteristics, while the HCP systems presented an unstable behaviour in the immediate post-collapse region and then an oscillating characteristic. The experimentally observed mean load for an HCP system was about twice that for an SP system when identical rings were used. The deformation patterns show that HCP systems deform progressively, layer by layer, while the SP systems deform with the formation of V-shaped shear bands ahead of the moving platen, which trigger asymmetric modes of deformation in individual rings. Similar modes are evident in Fig. 6(a).

Shim and Stronge[28] used the equivalent structure technique to obtain collapse loads and load–deflection histories of individual rings subjected to four or six equally spaced radial compressive forces. They identified symmetric and asymmetric modes of deformation, the latter

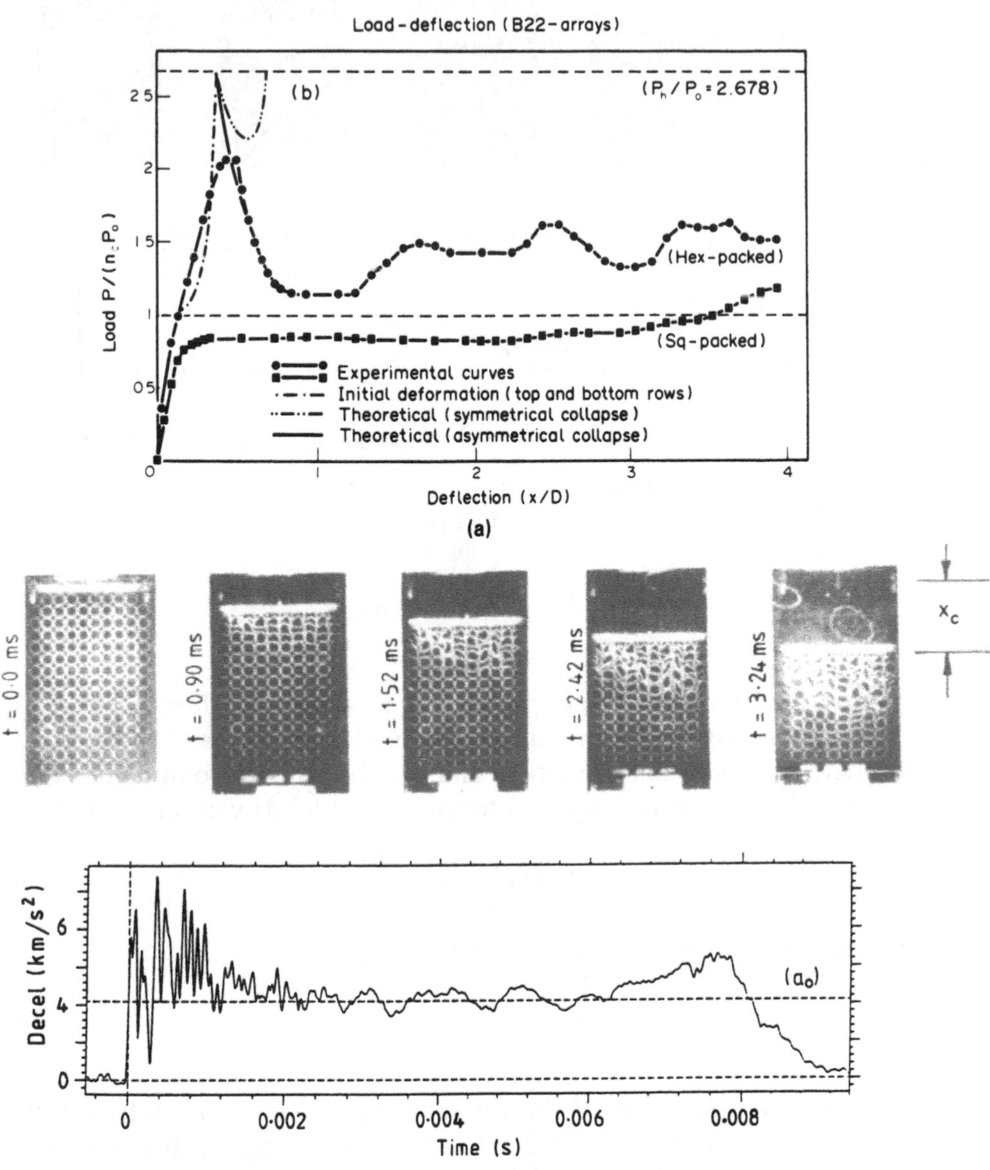

Fig. 7. Dynamic uniaxial crushing of two-dimensional ring systems. (a) Typical quasi-static load–deflection curves for SP and HCP brass ring systems. (b) High-speed films and deceleration of striking mass for (i) a brass tube SP array with $Q = 5{\cdot}2$ showing Phases I, II and IV and (ii) an aluminium tube HCP array with $Q = 7{\cdot}8$ showing Phases I and IV. (From Ref. 29.)

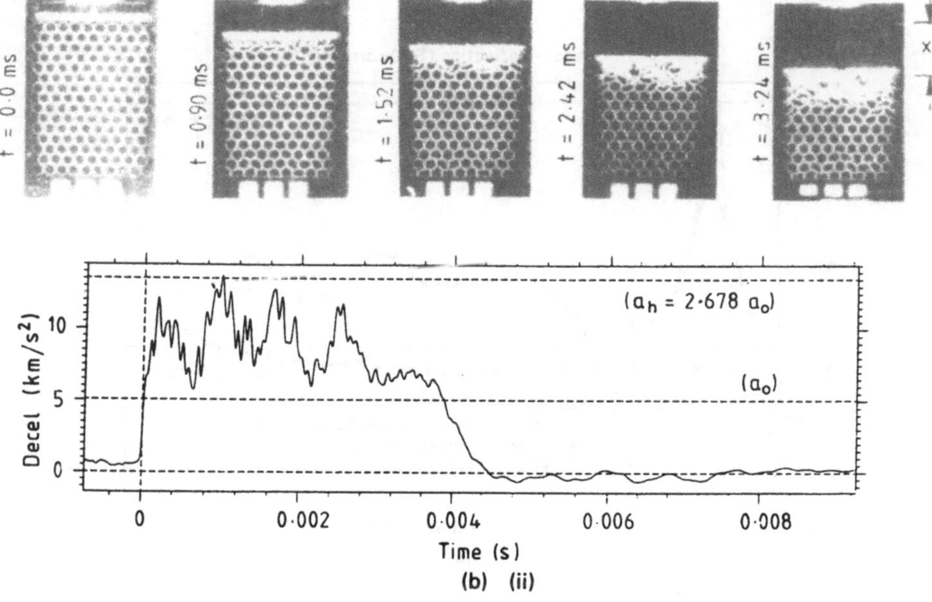

(b) (ii)

Fig. 7—(Continued).

triggered by frictional effects giving rise to tangential forces between the rings. The asymmetric modes were also found to result in unstable load–deformation characteristics and to be more compliant. The collapse load for a ring subjected to four equal loads was shown to be

$$P_0 = 8(1 + \sqrt{2})\frac{M_0}{D} \tag{1}$$

while that for a ring subjected to six equal loads is

$$P_h = 8(3 + 2\sqrt{3})\frac{M_0}{D} \tag{2}$$

where D is the diameter of the ring and $M_0 = \frac{1}{4}Ylt^2$ is the fully plastic bending moment of a longitudinal section of a ring having length l, thickness t and made of a rigid, perfectly plastic material of yield strength Y.

In an interesting and detailed account of the dynamic crushing of tube arrays by a rigid mass M travelling at impact speeds of about 30 m/s, Stronge and Shim[29] found that the uniaxial deformation behaviour of a two-dimensional array of rings with n_c columns and n_r rows was sensitive to a parameter Q, the impact energy ratio, defined as

$$Q = \frac{1}{2}\frac{MV_0^2}{n_c D P_0} \tag{3}$$

This is the ratio of the kinetic energy of the striker to a simplified measure of energy absorption capacity of the row of rings under the indenter, P_0 being given by eqn (1). Figure 7(b) provides examples of high-speed films and of colliding mass deceleration traces for each type of system. The deformation occurred in up to four phases: an initial peak deceleration phase (I) for the striking mass; a crushing phase (II); a densification phase (III); and an unloading phase (IV). Phase I in SP systems involved the surface rings deforming in symmetrical, less compliant modes at loads in excess of $n_c P_0$, which lasted until the wave reflected from the distal end of the system generated by the elastic precursor met the main deformation front. In Phase II rings in SP systems deformed successively in a generally asymmetric mode, resembling quasi-static behaviour, at loads less than $n_c P_0$ until the last row was crushed (see Fig. 7b, i). Phase II was not observed in the HCP systems, and the Phase I mass deceleration pulse consisted of a series of oscillations corresponding to the collapse of successive layers; see Fig. 7(b, ii). The deformation of HCP systems progressed from both ends, and the middle of the system was undeformed if Q was sufficiently low. In Phase III densification of crushed rings occurred during which the residual energy was dissipated. It was noted that Phase III occurred for $Q > 6$ or 7. For small values of Q (~1) only Phases I and IV occurred, and the rings at the distal end support deformed significantly owing to the reflection of the elastic precursor.

The differences in the distribution and the modes of deformation in these ring systems depended to some extent on the effects of side-wall friction, the SP systems being the more sensitive to this. Generally the experimental data suggested that as the rate of crushing (proportional to V_0) increased, 'plastic waves' (the propagation of plastic deformation) became more evident and their effects (e.g. the increase in the duration of Phase I) more predominant.

Klintworth[30] has studied both static and dynamic aspects of the deformation of transversely loaded hexagonal cell honeycombs under uniaxial and biaxial compression. The mechanisms of localisation of deformation resulting from elastic buckling of the cell walls and plastic collapse have been described. Elasto-plastic yield functions and constitutive equations for quasi-static transversely crushed honeycombs have been established by Klintworth and Stronge.[31] With regard to dynamic loading, the effect of compression rate on honeycomb samples

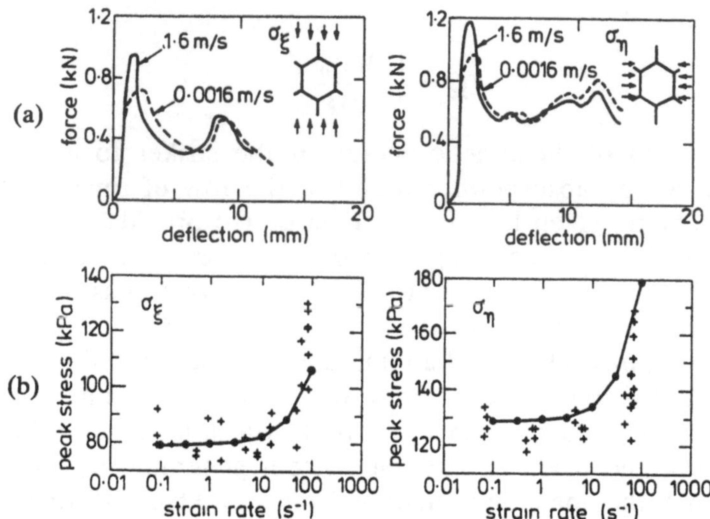

Fig. 8. Microinertial effects in aluminium honeycombs 30 mm thick, cell size about 5 mm. (a) Enhancement of initial crush stress. (b) Comparison between measured peak dynamic stresses with enhancements calculated by Klintworth[30] (solid lines). (From Ref. 32.)

has been examined by Klintworth[30] and discussed recently by Stronge.[32] Figure 8 summarises the results of these studies. It shows that the peak force becomes sensitive to the loading rate at a nominal strain rate of about 10^2. The specimens were only a few cells deep, so that localisation does not interfere with the definition of the nominal strain rate. There are some similarities with the Phase I enhancement of crushing force observed in ring systems and this effect in honeycombs, both of which could be classified as microinertial effects. These are effects that result from a delay in triggering a buckling/collapse mode or from the generation of an alternative less compliant cell collapse mechanism than that occurring in quasi-static compression owing to inertial effects in the cell walls. Klintworth[30] produced a dynamic analysis based on the hexagonal cell geometry that accounted for such effects and showed (see the solid lines in Fig. 8b) that they were indeed responsible for the enhancement in the initial crush stress in aluminium honeycombs. He also investigated the sensitivity of the variations in this peak stress to imperfections in the cell geometry.[30]

The structural shock model for one-dimensional ring systems depicted in Fig. 4 relates to cellular systems composed of elements having a monotonically increasing and steepening load–deflection characteristic. In two-dimensional systems unstable behaviour leading to localisation of deformation introduces a new phenomenon that demands a different treatment. Stronge, Shim and co-workers[33–35] have produced a number of heuristic spring–mass models to explain some of the phenomena observed in two-dimensional ring systems and in open-cell foams. These models are instructive, but there is a need to relate the parameters in them more closely to the parameters defining a particular system before they could yield quantitative results. Figure 9(a) shows the form of the cell deformation characteristic used to simulate a system/material that softens before moving into a densification phase, and Fig. 9(b) shows a set of typical results. There are some similarities with the elastic–plastic shock wave analysis (see Fig. 4b) in terms of the

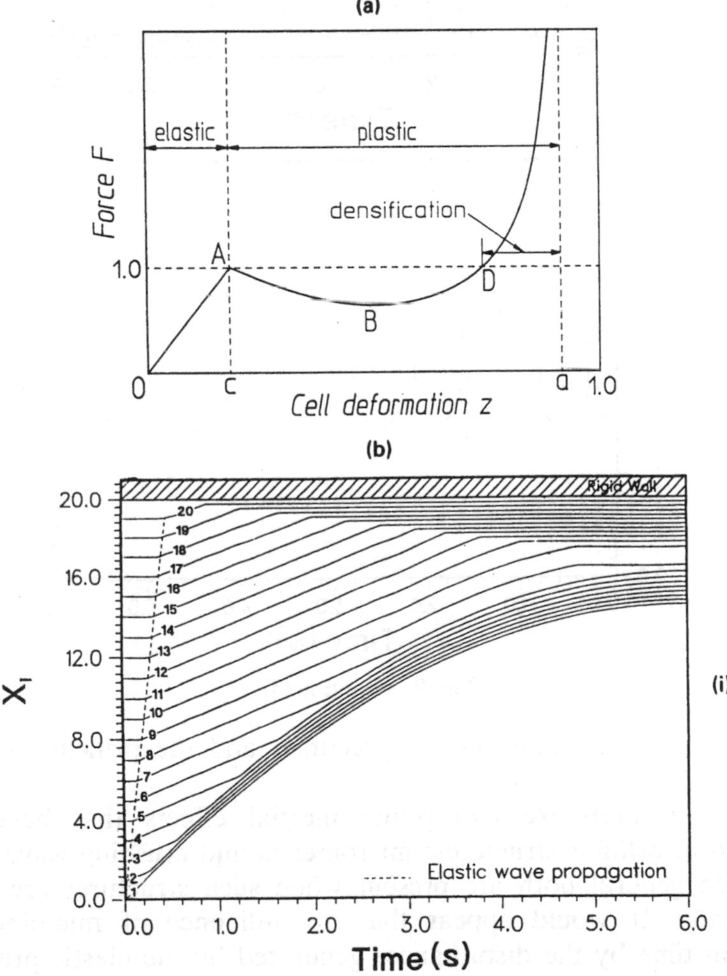

Fig. 9. Dynamic response of strain-softening systems. (a) Cell deformation characteristics. (b) Typical analytical results for a 20-element system: (i) mass positions; (ii) individual cell deformations; (iii) deceleration of colliding body. (From Ref. 35.)

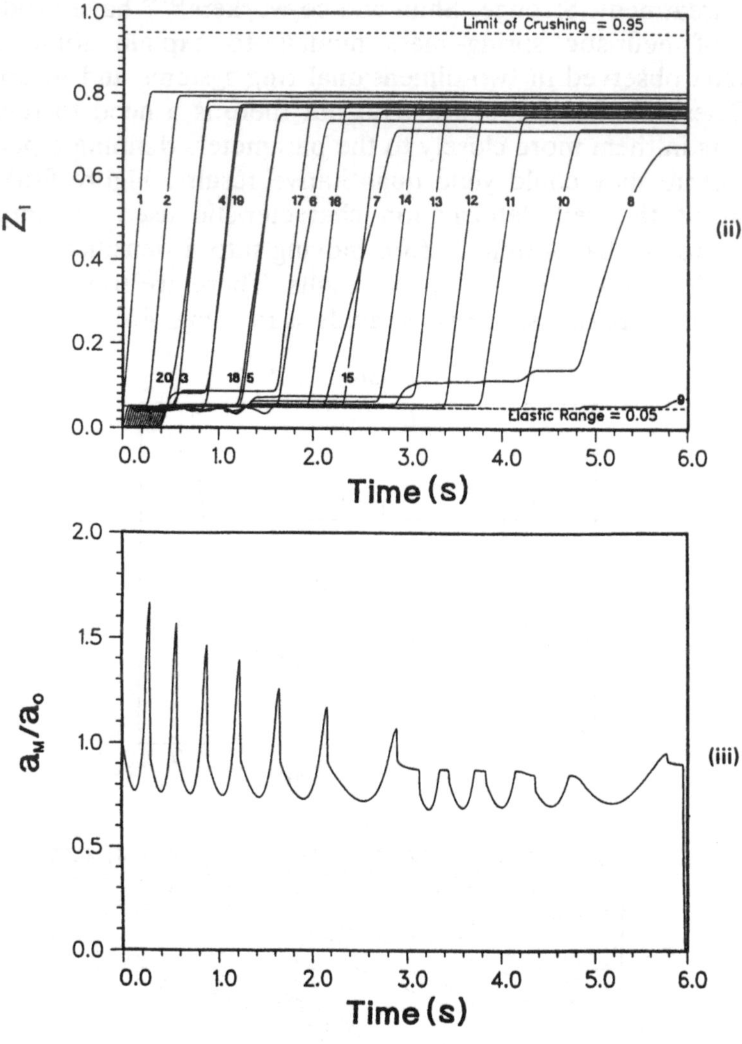

Fig. 9—(Continued).

influence of the reflected elastic precursor and the principal crushing wave.

There are therefore two prime inertial effects that have been identified in cellular structures: microinertia and crushing wave propagation. In general both are present when such structures are loaded dynamically. It would appear that the influence of microinertia is limited in time by the disturbances generated by the elastic precursor, as noted above in relation to the termination of Phase I in SP ring systems.[29] A similar phenomenon was observed in relation to the deformation of single lines of rings.[22] Figure 10 shows that a single ring

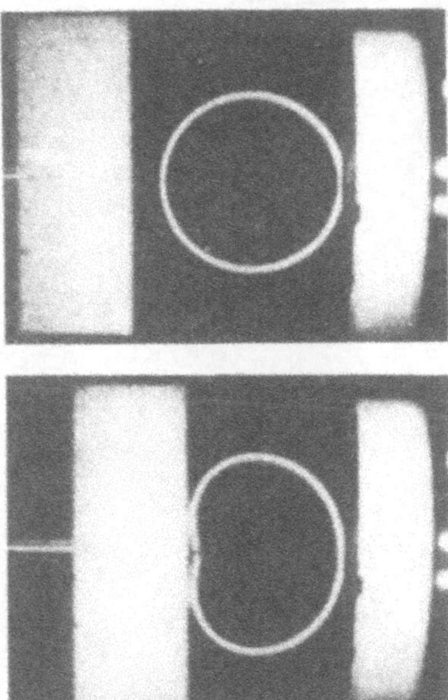

Fig. 10. Non-symmetric mode of deformation of first ring in a one-dimensional ring system (from Ref. 22).

(and indeed the first ring in any system) responds in a non-symmetric mode when struck by a projectile. A crude allowance for this effect was made in calculating the level of deformation in this first ring by using the mode technique.[36] The mode shown in Fig. 10 arises from the inertia of the elements of the ring, and a calculation of the force on the projectile (equivalent to Phase I of the measured pulses shown in Fig. 7) would require a detailed analysis of the problem depicted in Fig. 10. Thus microinertia is a somewhat difficult though important effect to quantify. Similar effects are evident in the behaviour of axially loaded systems, which will now be discussed.

2.3 Axially Loaded Systems

The application of structural mechanics to deduce the basic quasi-static mechanical properties of honeycombs is summarised in Chapter 4 of Ref. 10. Interest here is focused on to the crushing of honeycombs, about which there is relatively little published. McFarland[17] provided a comprehensive set of data on the quasi-static crushing of hexagonal honeycombs and circular tube systems, as well as the results of a limited range of tests on dynamic crushing. Figure 11(a) gives an example of his quasi-static data for aluminium honeycombs of different wall thick-

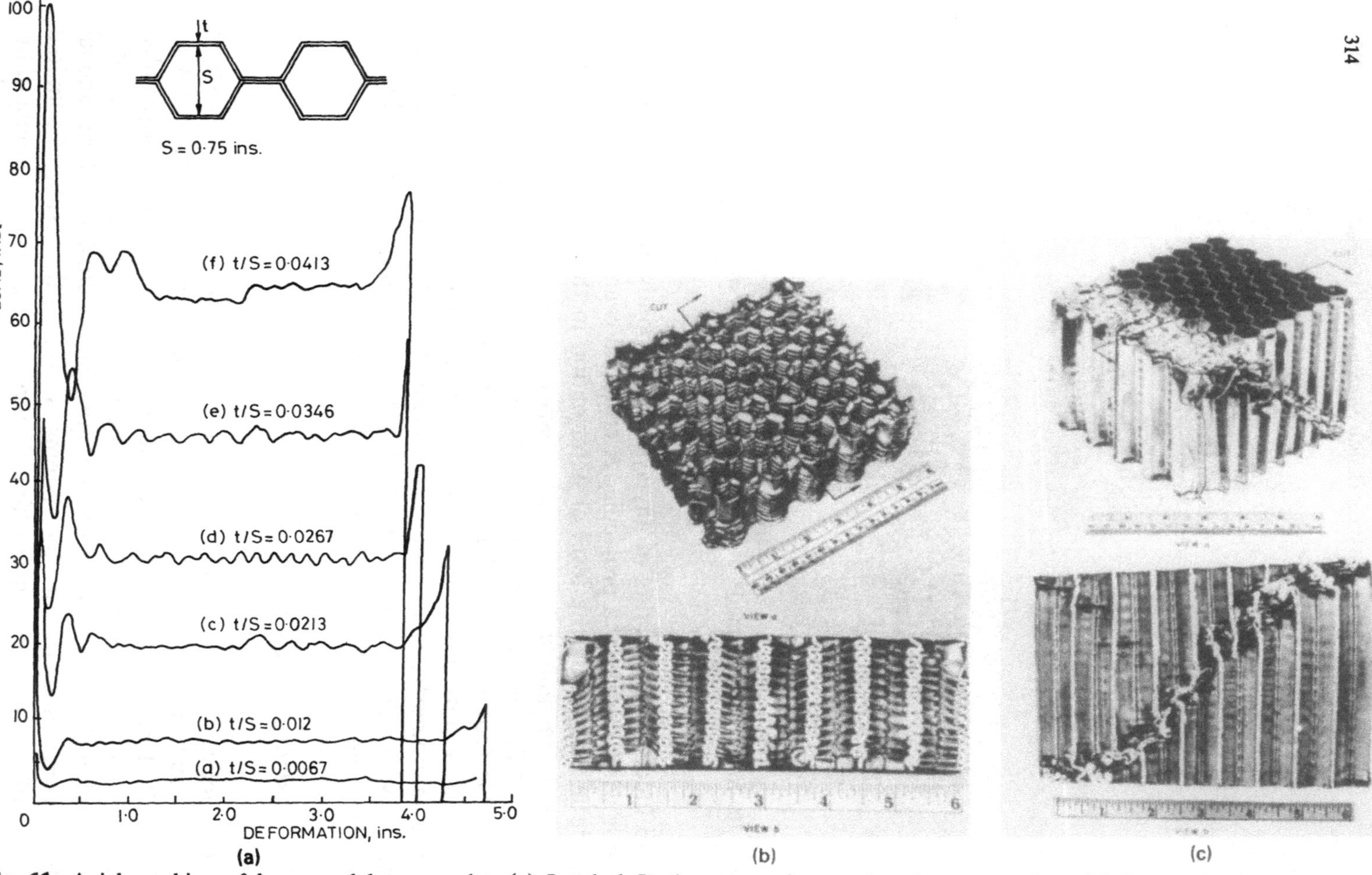

Fig. 11. Axial crushing of hexagonal honeycombs. (a) Load–deflection traces for quasi-static compression. (b) Progressive buckling mechanism. (c) Shear mode of failure for $t/S > 0{\cdot}04$. (From Ref. 17—by courtesy of the Jet Propulsion Laboratory, California Institute of Technology.)

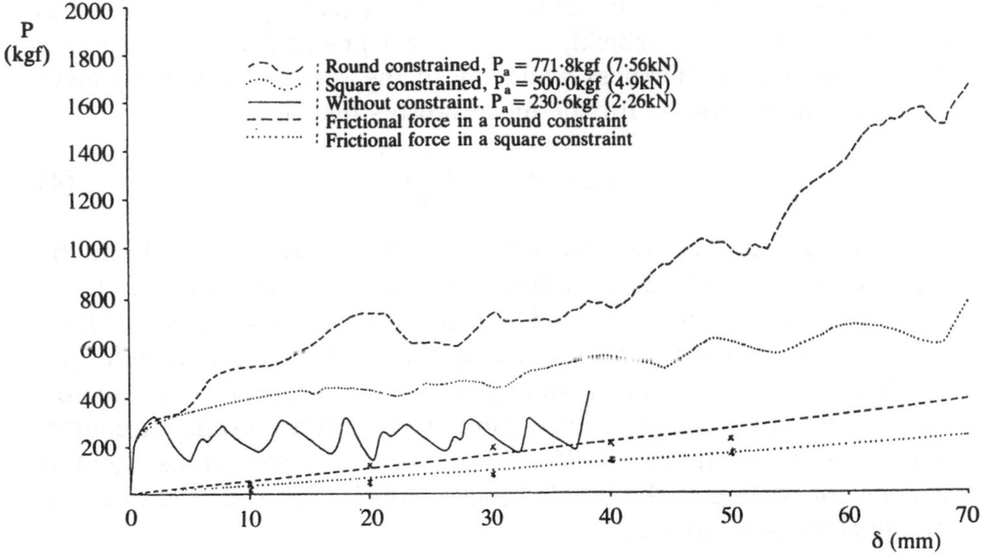

Fig. 12. Load–deflection curves for single aluminium tubes under various external constraints (From Ref. 37).

nesses, and Fig. 11(b) shows the progressive buckling mode that is characteristic of the cell collapse process. This behaviour is very similar to single-tube buckling, which has received a great deal of attention.[3] However, as with lateral tube compression, when arrays of tubes or hexagonal cells deform in this way their modes are modified owing to the mutual constraints exerted by neighbouring elements. Figure 12 shows experimental data due to Lee[37] on single tubes compressed axially with no lateral constraint and when confined externally in a coaxial circular or square tube. As with some of the work on lateral compression of confined tubes, friction between the tube and the confining structure plays a part in the results. When the contribution from friction (which was measured—see Fig. 12) is removed, these results indicate that the mean crushing force for a single tube increases by factors of approximately 1·7 for a square constraint and 2·5 for a circular constraint. A final point to note from Figs 11(a) and 12 is the oscillatory nature of the load–deflection curve. The compression process is inherently unstable. The well-known progressive buckling mode is therefore another example of localisation of deformation produced by softening of the structure, this time by virtue of the buckling mechanism.

McFarland[8] produced a theoretical model to predict the mean crushing stress of honeycombs based on a similar model to that used by

Pugsley and Macaulay[38] for circular tubes. A more precise model has been produced by Wierzbicki,[39] which for a practical honeycomb (i.e. one having double-thickness walls where the sheets are joined) gives the following expression for the mean crush stress $\bar{\sigma}_m$:

$$\bar{\sigma}_m = 16{\cdot}56\sigma_0\left(\frac{t}{S}\right)^{5/3} \tag{4}$$

where σ_0 is the average flow stress, estimated by Wierzbicki to be between 0·66 and 0·77 of the ultimate stress of the material, σ_u. Here t and S are the wall thickness and cell diameter as shown in Fig. 11(a). Equation (4) is based upon the assumption that only plastic deformation contributes to the deformation process. It has recently been noted[16] that this formula tends to underestimate measured crushing strengths, sometimes by as much as 25%. However, in these cases there is evidence of peeling of the bonded joints, which itself contributes to the energy absorbing process.

It is of interest to note that, of the various types of honeycomb investigated by McFarland,[17] HCP systems of circular-section tubes were found to be the most efficient of the systems tested. Of more fundamental interest is the observation[8] that there is an upper limit on the ratio t/S for which progressive buckling occurs (Fig. 11c). The emergence of this shear mode is controlled by the material properties and the ratio t/S. This observation has some bearing on the discussion of the behaviour of wood in Section 3.

McFarland[17] points out that the true test of a honeycomb lies in dynamic loading, and he presents some limited data from such tests. He notes that under dynamic conditions the behaviour is less sensitive to the presence of imperfections in the cell walls and that for aluminium honeycombs there is a 15% increase in the specific energy absorption capacity, which reflects an increase of similar magnitude in the mean crushing force. All his tests were performed in a drop weight apparatus at one impact speed of about 16 m/s.

Goldsmith and Sackman[16] performed dynamic crushing tests on aluminium honeycombs using a slightly rounded projectile with impact speeds up to about 35 m/s and conclude that the plateau crush stress is raised by 20–50% over the range of loading. The use of rounded projectile/indenter appears to remove the initiating peaks in the load–displacement traces that are evident in Fig. 11(a).

No theoretical assessment of crush force enhancement under dynamic loading conditions has been made for honeycombs to date. Abramowicz and Jones[3] have considered the corresponding problem for single steel circular tubes loaded at speeds up to 10 m/s and have predicted enhancements of about 25% for the mean crushing stress stemming from the strain-rate sensitivity of the material. These values agree well with experimental data. However, it should be noted that the experimental mean loads were defined as the ratio of the energy absorbed and the stroke of the tube, and were not measured directly from the tests. It is therefore not clear that the actual mean plateau load increased by this amount.

It is possible that, in addition to strain-rate effects, there may be other deformation mechanisms and inertial effects that play a part in the enhancement of the energy absorption capacity. Such effects have been discussed by Calladine and English,[40] Zhang and Yu[41] and Tam and Calladine[42] in the context of the crushing of slightly bent struts (Fig. 13a). As Fig. 13(b) shows, the quasi-static behaviour of the struts considered by these authors (termed Type II structures) are unstable (or 'soften') when they deform in their simple bending mechanism. However, when struck by a mass G travelling at velocity V, the lateral inertia forces (synonymous with what has been termed microinertia in Section 2.2) on the arms of the strut have the effect of introducing an initial phase of the deformation, which is dominated by plastic axial compression of the strut. In this phase a considerable portion of the kinetic energy (KE) of the striker is absorbed before the simple bending mechanism predominates. Consequently, the deformation produced dynamically can be significantly less than that corresponding to an equal amount of energy dissipated in quasi-static compression, which follows the load–compression characteristic of Fig. 13(b). If SE represents the quasi-static energy absorbed corresponding to the final deflection of the specimen, this effect is revealed in KE/SE ratios greater than unity. Figure 13(c) shows examples of this effect for a series of aluminium specimens.[42] While material strain-rate sensitivity plays a part in this effect, the work of Tam and Calladine[42] clearly demonstrates a significant inertial effect.

The theoretical model due to Zhang and Yu[41] encompassed the energy loss during the compressive phase into a simple plastic impact model, which demonstrated the sensitivity of the results to $\mu = G/m$, the ratio of the mass of the striker to that of the strut. Tam and Calladine[42] produced a more detailed analysis, which showed that the energy loss during this initial phase depends on the impact velocity as well as on μ, as shown in Fig. 13(c). During this phase, the force on the striker is simply equated to the axial yield force of the strut.

It seems reasonable to expect that similar phenomena should occur in structures that buckle, such as single tubes and honeycombs. From an experimental point of view, the measurement of dynamic crushing force

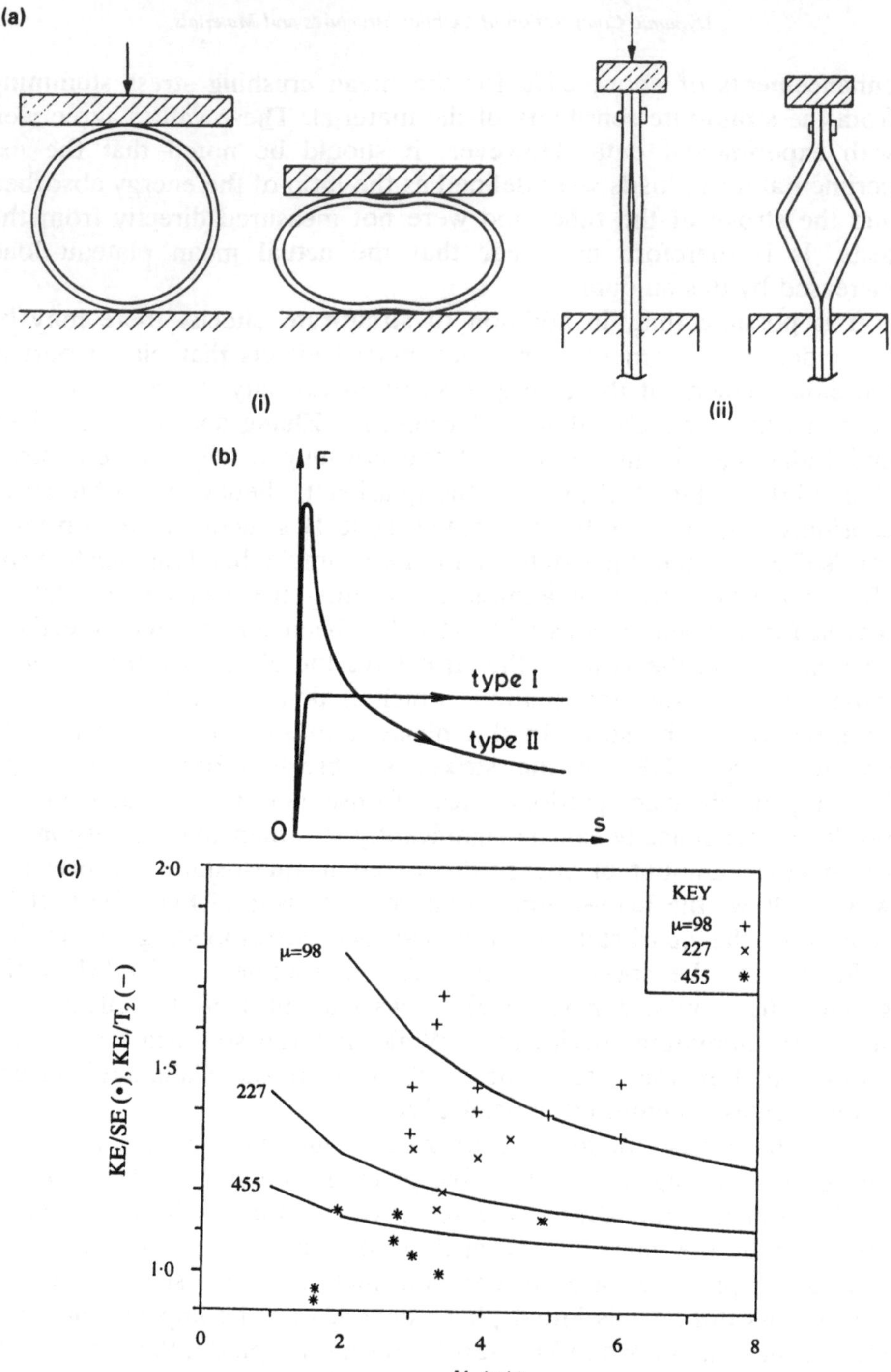

Fig. 13. 'Velocity-sensitive' softening structures. (a) Examples of (i) Type I and (ii) Type II structures. (b) Quasi-static load–deflection curves of each type. (c) Typical set of data showing the sensitivity of the energy absorption capacity of a Type II structure to the mass ratio μ and impact velocity V_0. (From Refs 40 and 42.)

pulses is an important prerequisite to understanding fully the behaviour of such structures.

3 WOOD—A CELLULAR MATERIAL

3.1 Introduction

Interest has grown in recent years in the study of cellular materials such as polymeric foams and wood. One reason for this interest is the extensive use of these materials for packaging and in providing protection from damage under drop conditions; indeed, Gibson and Ashby[10] devote a chapter of their book to the selection of materials for low-speed impact applications. However, wood in particular is also used as a protective material for high-velocity impact events, for example in providing the lining for containment structures surrounding systems that may disintegrate, and as such its response at higher rates of loading is of some interest. Attention will therefore be focused on wood as one example of a rather complex cellular material.

An introduction to the study of wood under dynamic uniaxial crushing conditions and when penetrated by a flat-ended cylindrical projectile was provided by Reid *et al.*,[43] who quoted experimental data for yellow pine and American oak to illustrate the sensitivity of the crush stress of these woods to loading rate. For example, at an impact velocity of around 100 m/s the crushing strength of the two woods increased by factors of between 2 and 4, depending upon the density of the wood and the orientation of the grain with respect to the direction of loading. Dynamic test data are provided below for illustrative purposes on the crush stress of woods of various densities over a wider range of impact velocity up to 300 m/s. The predictions of a simple shock model analogous to that described by Reid *et al.*[21] for ring systems are presented, in which the enhancement of the crush stress is attributed primarily to inertial effects that require that the deformation propagate as a discontinuity surface. The initial strength of this shock wave is directly related to the impact velocity, the density of the wood and the locking strain of the material. As will be seen, despite its simplicity, the model shows good agreement with experimental data for crushing across the grain; but at lower velocities and for loading along the grain it underestimates the enhancement of the crush stress, and the reasons for this are discussed.

To illustrate the range of phenomena involved when wood is crushed, the quasi-static behaviour of the uniaxial crushing of woods is briefly described in Section 3.2. This also provides a reference point for the discussion of the dynamic test data that follows.

3.2 Quasi-Static Compression

Typical stress–strain curves for uniaxial compression tests on a variety of woods are given by Gibson and Ashby,[10] summarising their work with Easterling *et al.*[44] and Maiti *et al.*[45] A detailed account is presented of the deformation mechanisms involved when wood is compressed in its three principal directions, namely axial (along the grain), radial (across the grain transverse to the growth rings) and tangential (across the grain parallel to the growth rings). The essential differences between loading along the grain and transverse to the grain lie in the differences in the collapse mechanisms induced in the cell structure of the wood. The response to compression along the grain is extremely complex, and the initiation of crushing can involve failure of the pyramidal end caps of the cells for low-density woods such as balsa[44] or the onset of various buckling failures such as a progressive concertina-fold microbuckling mode in the crushing cells[10,46] or an Euler-type buckling, which can lead to kink-band formation especially in the denser woods.[46] Compression across the grain (despite the influence of the rays in enhancing the crush stress by approximately 40% when the wood is crushed in the radial as opposed to the tangential direction) is more straightforward. It has been described in terms of plastic bending collapse of a two-dimensional hexagonal cellular array, which provides a good approximation to the structure of many woods.[44] The consequences of these observations on the form of the cell structure and the mechanisms of failure are that simple relationships can be formulated between the crush stresses (i.e. the stresses at which inelastic deformation is initiated in the three principal directions) and the relative density ρ_0/ρ_s of the woods. These are expressed as follows:[10,45,46]

$$\sigma_A = 150\frac{\rho_0}{\rho_s}\ \mathrm{MN/m^2} \tag{5}$$

$$\sigma_R = 1{\cdot}4\sigma_T = 70\left(\frac{\rho_0}{\rho_s}\right)^2\ \mathrm{MN/m^2} \tag{6}$$

where the suffices A, R and T refer to the three principal directions of the wood structure (axial, radial and tangential), $\rho_s = 1500\ \mathrm{kg/m^3}$ is the density of the cell wall material and ρ_0 is the density of the particular wood. ρ_s is virtually the same for all woods. The numerical factors are derived empirically† while the indices arise naturally from the particular collapse mechanisms considered in the structural models. The differences in compressive strength in the radial and tangential directions are relatively small compared with the differences between the strength along the grain and that across the grain. Therefore no distinction is made here between the tangential and radial directions. A single crush stress transverse to the grain defined by $\sigma_{cr}(90°)$ equal to σ_R has been used. The axial crush stress is referred to here as $\sigma_{cr}(0°)$, the angle simply indicating the direction of loading relative to the grain direction.

As noted by Easterling *et al.*[44] and Reid and Peng,[46] the deformation is localised and develops by growth of the regions of cell collapse. This is most noticeable in axial compression, where deformation is affected by the propagation of crush fronts initiated at the ends of the specimen adjacent to the loading platens or the growth of kink bands according to the density of the wood (see Fig. 14). It should be noted that the

(a)

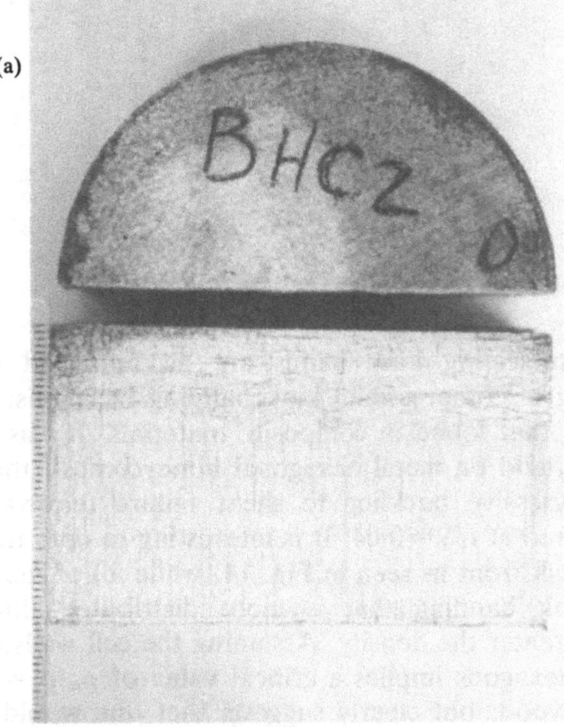

Fig. 14. Split axially loaded wood specimens showing the localised nature of the deformation mechanism: (a) progressive, plane crush front in balsa; (b) kink bands in oak.

† The coefficient of 150 in eqn (5) comes from Ref. 45 and matches the present data (Table 1) and so has been used here. Gibson and Ashby[10] cite a value of 120.

(b)

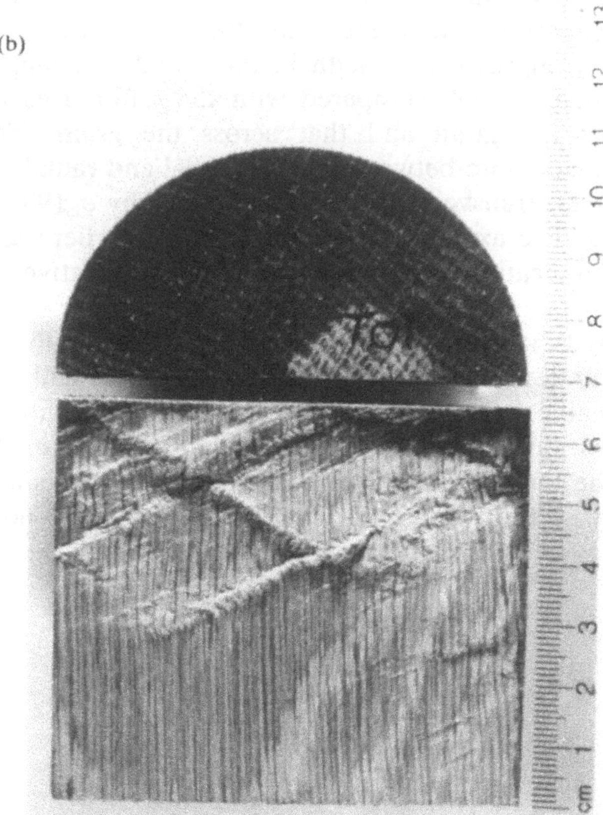

Fig. 14—(Continued).

families of intersecting kink bands are accompanied by cell wall collapse, giving a 'compressible' kink banding mechanism that distinguishes it from that found in composite materials. It was noted in the work by McFarland on metal hexagonal honeycombs[17] that the transition from progressive buckling to shear failure (equivalent to kink banding) occurred at $t/S \approx 0{\cdot}04$. It is interesting to note that only balsa has a plane crush front as seen in Fig. 14, while all of the other woods deform by kink banding that is more distributed throughout the specimen the greater the density. Assuming the cell walls of the woods to be regular hexagons implies a critical value of $\rho_0/\rho_s \approx 0{\cdot}08$. This is low for balsa wood, but clearly suggests that one would expect kink banding as the prevalent failure mechanism in axial compression.

As Fig. 15 shows, following an initial elastic phase, the woods begin to crush at the crush stresses, σ_{cr}, given by eqns (5) and (6). Further crushing occurs in a plateau region. While it is conventional to convert

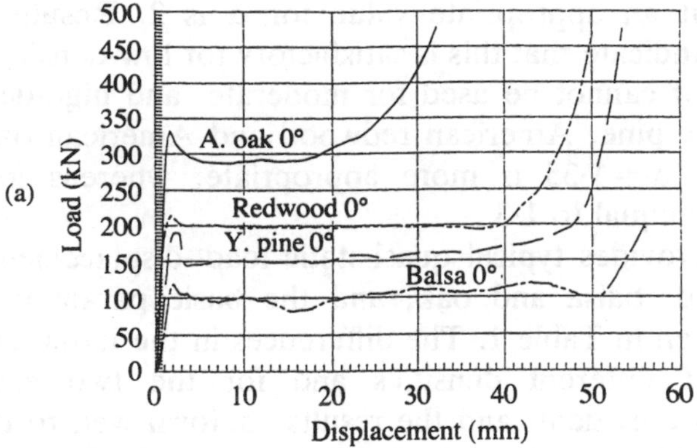

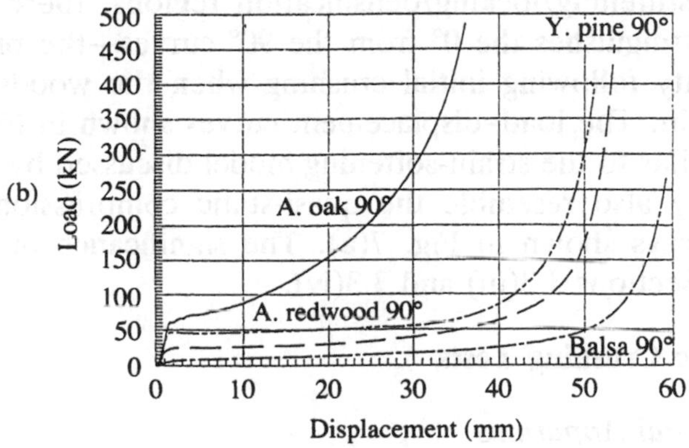

Fig. 15. Quasi-static uniaxial (constrained) load–displacement curves for a variety of woods compressed (a) along (0°) and (b) across (90°) the grain.

load–displacement curves into stress–strain curves in the usual way, it must always be borne in mind that, particularly in the inelastic range, the strain field is very non-uniform throughout the specimen. A rapid increase in the slope of the stress–strain curve occurs when the crushed zones have grown to encompass most of the initial volume of the material. At this point one can reasonably define a critical strain, the locking strain ε_ℓ, which reflects the transition from a cell wall collapse process to solid phase compression of the cell wall material.

Maiti *et al.*[45] discussed the shapes of the stress–strain curves for a variety of cellular solids, including wood. In particular, they produced the following equation to estimate ε_ℓ:

$$\varepsilon_\ell = 1 - \alpha \frac{\rho_0}{\rho_s} \tag{7}$$

indicating that an appropriate value for α is 2. Results from more recent tests[46] indicate that this is satisfactory for low-density woods such as balsa, but it cannot be used for moderate- and high-density woods such as yellow pine, American redwood and American oak. For pine and redwood $\alpha = 1{\cdot}35$ is more appropriate, whereas for oak α is approximately equal to 1·3.

Figure 15 provides typical quasi-static load–displacement curves for pine, redwood, balsa and oak, and the basic parameters for these woods are given in Table 1. The differences in the initial crush stresses for woods of different densities and for the two extreme grain orientations are evident, and the results conform well to the formulae (5) and (6). Also, an important point to note is that, while all the curves in Fig. 15 have similar features, i.e. elastic, yield (initial crush), plateau and stiffening/locking/densification regions, there is one property that distinguishes the 0° from the 90° curves—the presence of a clear instability following initial crushing when the woods are loaded along the grain. The load–displacement curves shown in Fig. 15(a) are therefore similar to the strain-softening model discussed by Shim *et al.*[34] In detail they also resemble the quasi-static compression curves for HCP ring arrays shown in Fig. 7(a). The significance of this will be discussed in Sections 3.3(iii) and 3.3(iv).

3.3 Dynamic Crushing Tests

(i) Experimental Apparatus

Dynamic crushing strengths for a variety of woods have been measured using the apparatus shown schematically in Fig. 16.[47] This consists of a

Table 1

Uniaxial Quasi-Static Parameters for 0° and 90° Wood Specimens Tested under Laterally Constrained Conditions (Uniaxial Compression) (75 mm Diameter × 75 mm Height)

Wood and grain orientation	*Initial density* ρ_0 *(kg/m³)*	*Relative density* ρ_0/ρ_s	*Initial crush stress* σ_{cr} *(N/mm²)*	*Locking strain* ε_ℓ	*Specific locking energy (kJ/kg)*
A. oak 0°	725	0·48	75·0	0·33	30·4
A. oak 90°	695	0·46	12·3	0·37	16·0
Balsa 0°	277	0·18	27·0	0·68	63·4
Balsa 90°	264	0·17	1·6	0·65	10·0
Y. pine 0°	383	0·25	43·3	0·64	62·2
Y. pine 90°	396	0·26	5·1	0·60	15·2
A. redwood 0°	367	0·25	43·0	0·65	59·0
A. redwood 90°	409	0·27	10·0	0·58	16·3

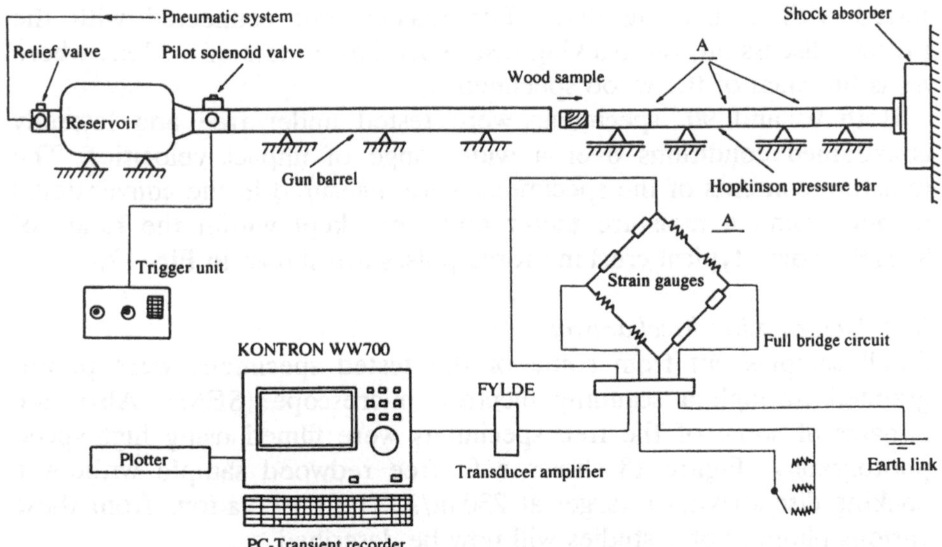

Fig. 16. Schematic of apparatus for measuring the dynamic uniaxial crush stress of wood.

pneumatic launcher comprising of a reservoir that is pressurised from an external compressor, a solenoid valve for releasing the pressure, activated by a trigger unit, and a barrel of internal diameter 45 mm. When the 45 mm diameter specimens were fired from the barrel, their velocities were measured using a simple timing circuit activated by pairs of photocells and light-emitting diodes. The specimens impinged axially on the end of a Hopkinson pressure bar.[48] This method of transient pulse measurement is relatively free from load-cell ringing effects, which tend to distort the signals. The calibration of the bar allows the strain gauge outputs to be transformed into load pulses. By adjusting the reservoir pressure and the position of the specimen in the barrel, impact velocities of between 30 and 300 m/s were attained. A thick-walled tool-steel chamber was specially designed and manufactured to perform impact tests on wood specimens under full lateral constraint (i.e. uniaxial strain). This was mounted between the end of the barrel and the Hopkinson bar.

(*ii*) *Specimens and Results*

Balsa, American oak, American redwood and yellow pine specimens of 45 mm diameter and 45 mm length were cut from the parent blocks and machined to provide a snug fit in the barrel of the launcher. Some of the specimens had an aluminium backing disc attached to them in order to crush them in a manner more akin to a quasi-static test. Others were tested without backing discs. For reasons not connected with the present discussion, the backing discs each has a mass of $4{\cdot}53m_s$, where m_s is the mass of the wood specimen.

Both 0° and 90° specimens were tested under free and laterally constrained conditions over a wide range of impact velocities. The moisture contents of the specimens were measured in the conventional manner using a moisture meter and were kept within the range of 9–12%. Some typical crushing force pulses are shown in Fig. 17.

(iii) Deformation Mechanisms

Small samples cut from some of the tested specimens were photographed through a scanning electron microscope (SEM). Also, the impacts of some of the free specimens were filmed using high-speed photography. Figure 18 shows a 0° free redwood sample without a backing disc striking a target at 230 m/s. The observations from these various photographic studies will now be described.

The dynamic crushing of the wood samples resulted in deformation mechanisms that were even more localised than in the samples tested statically. The main deformation was concentrated at the impact end of the specimens, particularly in those without backing discs (see Fig. 18), the deformation of which resembled bullet impact.[49] The progress of the damaged region in the 0° specimens was through a mechanism reminiscent of projectile erosion, whereas for 90° specimens the deformation was more like mushrooming.[43] The free specimens with backing discs often disintegrated in an explosive manner when impacted at high velocities. The localisation in constrained (i.e. uniaxially compressed) samples is clearly manifested in the 0° woods, where the deformation propagated as a progressive crushing wave. The front of the band of crushed cells progressed from the proximal end towards the distal end, and the width of the band grew as the impact velocity was increased. A band composed of crushed material appeared also at the distal end in some of the specimens with backing discs as a result of the reflection of the compressive elastic stress waves that preceded the main crushing wavefront. The mean engineering compressive strain of the crush band was approximately the same as the uniaxial quasi-static locking strain. There exists an impact velocity for the backed constrained specimens at which the locking strain was attained globally. This can be seen in some of the pulses in Fig. 17.

In the 90° woods the extent of the crushed cells is not as clearly delineated and visible macroscopically as in the 0° woods. This is partly because of the recovery of the samples and partly because the deformation occurs in multiple bands, which appear to be randomly

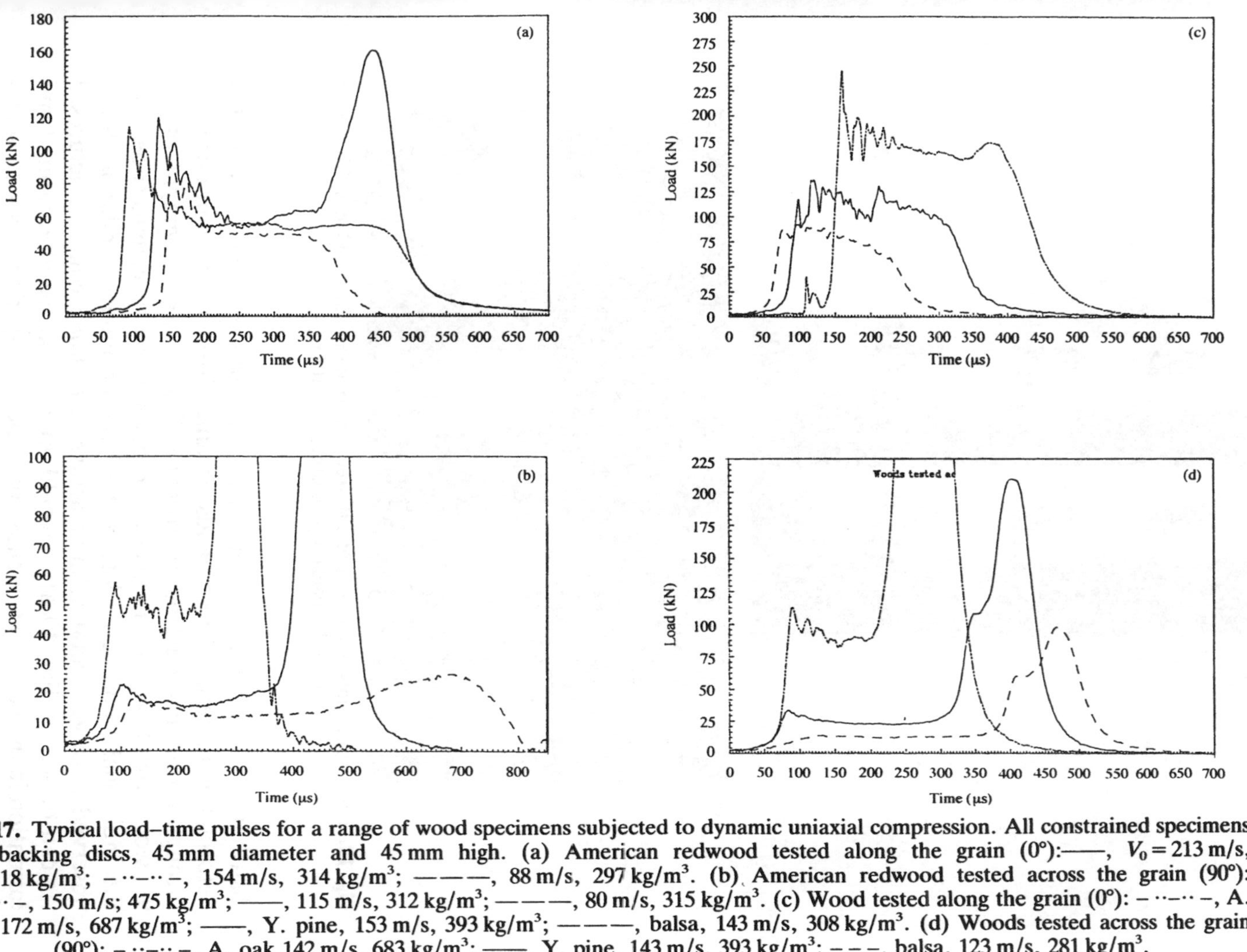

Fig. 17. Typical load–time pulses for a range of wood specimens subjected to dynamic uniaxial compression. All constrained specimens had backing discs, 45 mm diameter and 45 mm high. (a) American redwood tested along the grain (0°):——, $V_0 = 213$ m/s, $\rho = 318$ kg/m^3; – ··–·· –, 154 m/s, 314 kg/m^3; — — —, 88 m/s, 297 kg/m^3. (b) American redwood tested across the grain (90°): – ··–·· –, 150 m/s; 475 kg/m^3; ——, 115 m/s, 312 kg/m^3; — — —, 80 m/s, 315 kg/m^3. (c) Wood tested along the grain (0°): – ··–·· –, A. oak, 172 m/s, 687 kg/m^3; ——, Y. pine, 153 m/s, 393 kg/m^3; — — —, balsa, 143 m/s, 308 kg/m^3. (d) Woods tested across the grain (90°): – ··–·· –, A. oak 142 m/s, 683 kg/m^3; ——, Y. pine, 143 m/s, 393 kg/m^3; – – –, balsa, 123 m/s, 281 kg/m^3.

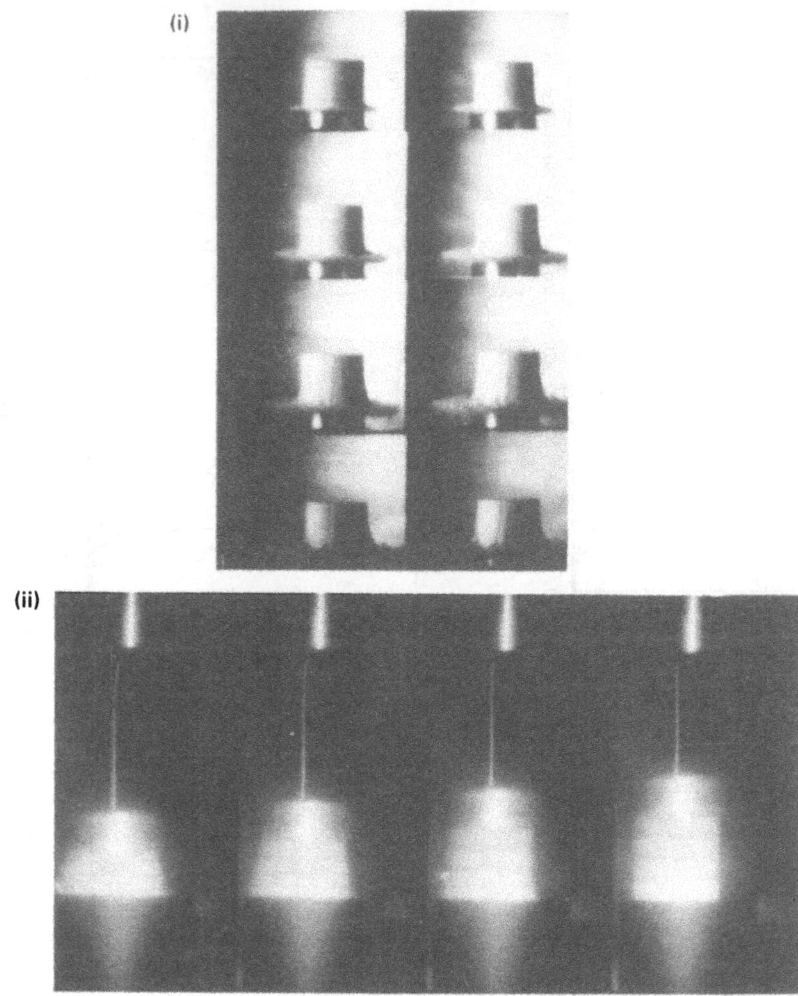

Fig. 18. High-speed films: (i) impact of a 0° free redwood sample at 230 m/s; (ii) impact of a 90° backed pine sample at 108 m/s.

distributed and which lead to more non-uniform yet widely distributed deformation patterns. These bands became visible only when observed through a scanning electron microscope. The microphotographs of the 90° wood sample cut from the crushed parts of the specimens show that the individual wood cells deform by bending modes similar to those observed in larger metal cellular arrays, e.g. honeycomb structures.[31] As can be seen in Fig. 19(a), it would appear that the alternating band formation in the 90° woods results from the propagation of shock waves in a heterogeneous medium having a repetitive structure composed of microstructural layers of earlywood and latewood cells (the latter being

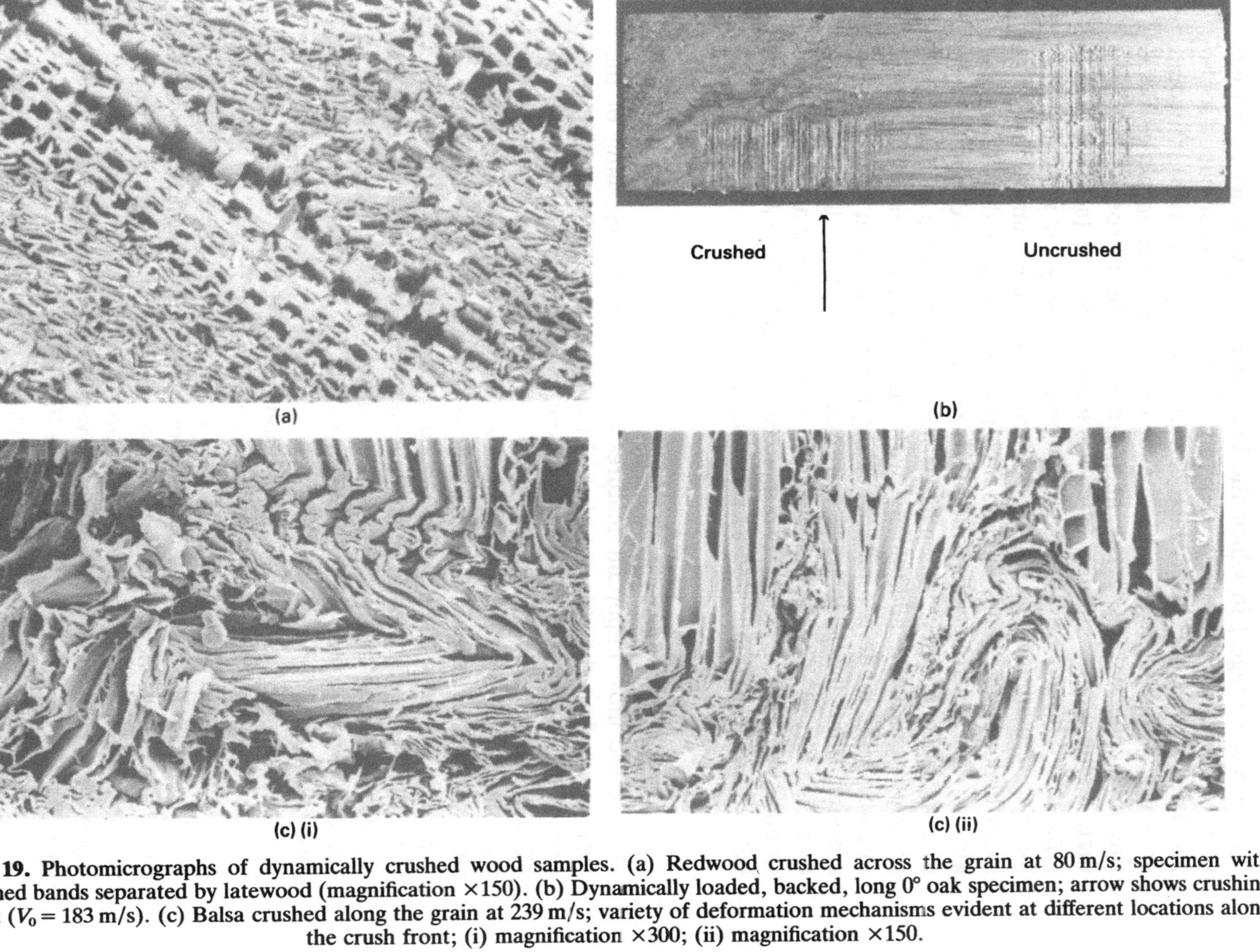

Fig. 19. Photomicrographs of dynamically crushed wood samples. (a) Redwood crushed across the grain at 80 m/s; specimen with crushed bands separated by latewood (magnification ×150). (b) Dynamically loaded, backed, long 0° oak specimen; arrow shows crushing front ($V_0 = 183$ m/s). (c) Balsa crushed along the grain at 239 m/s; variety of deformation mechanisms evident at different locations along the crush front; (i) magnification ×300; (ii) magnification ×150.

thicker and stronger) in which the latewood cell layers act as a rigid rear restraint for the earlywood cells. The same effect can also be produced by the ray cells, which can also act as rigid plates. It would appear that this phenomenon is similar to that found in impact of metal ring and plate systems.[20] The micrographs of the 0° specimens show that, although the crushing front is clearly definable (see Fig. 19b), separating crushed from uncrushed wood, it is irregular in shape. The micro-deformation of the cells is characterised by mechanisms involving shear kinking, microbuckling and flow-microbuckling modes. The shear kinking and microbuckling modes (see Fig. 19c, i) are common in man-made composites, although, as noted earlier, there is clearly cell crushing associated additionally with these mechanisms. The flow-microbuckling mode is a microbuckling deformation that occurs at the crushing front and is sometimes associated with large fibre rotations (see Fig. 19c, ii). The clear delineation between crushed and uncrushed wood is reminiscent of the two-dimensional HCP ring systems, whose behaviour is also dominated by softening behaviour.[29]

(*iv*) *Crush Stress Enhancement*

One of the most important features of the dynamic crushing of the various woods is the enhancement of the crush stress. The initial peak impact loads, plateau loads and maximum stiffening loads (i.e. those when the material is almost fully crushed) all increased to some degree with the increase of the impact velocity. The increase in the initial crush stress can be expressed in terms of the stress ratio S_r. This is defined as the initial dynamic crush stress divided by the quasi-static crush stress, the latter being adjusted for each specimen to allow for the value of its density according to eqns (5) or (6). The variations of S_r with impact velocity for balsa and oak are given in Fig. 20. The enhancement is most significant for loading across the grain (90°), and particularly in the lower-density wood balsa. For instance, for balsa at 90° and at an impact velocity of about 300 m/s the stress ratio S_r rises to 20, whereas this ratio for the 0° specimens is only about 4 at the same velocity. In 90° specimens the stress ratio increases at an increasing rate with the impact velocity, whereas in the 0° woods this relationship is more linear and gradual up to a certain velocity above which it tends also to rise rapidly with increase in the impact velocity. This feature of the 0° specimens is most evident for oak (see Fig. 20a).

Despite the differences in the independent variable, the load–time pulses of the samples with backing discs (Fig. 17) resembled the quasi-static load–deflection curves shown in Fig. 15. There are two interesting similarities with ring systems that are revealed in Fig. 17.

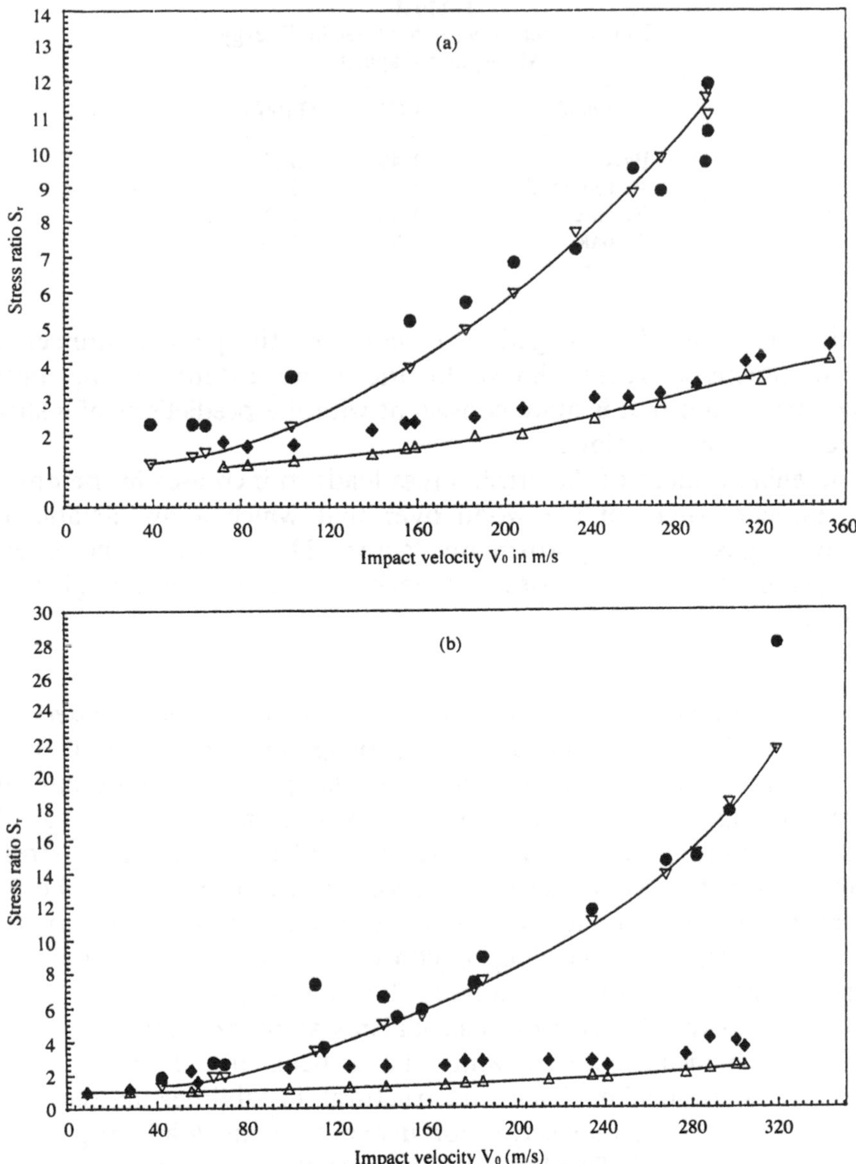

Fig. 20. Comparison between experimental and theoretical crush stress enhancement factors S_r for American oak (a) and balsa (b): ◆, experiment at 0°; △, shock theory at 0°; ●, experiment at 90°; ▽, theory at 90°.

The 0° specimens show a more dramatic drop from the initial peak down to stress levels that are only slightly increased above the quasi-static plateau values. This is a similar characteristic to that detected in aluminium honeycombs[30,32] and in HCP ring systems,[29] and suggests that the enhancement may have a microinertial origin rather than one

Table 2
Enhancement Factor for Specific Energy Absorption Capacity

Wood	Π (0°)	Π (90°)
Balsa	1·49	3·59
A. redwood	1·3	3·5
Y. pine	1·48	2·60
A. oak	1·90	1·64

dominated by shock propagation. In contrast, the plateau stresses for 90° specimens increase almost to the same extent as the initial peak—behaviour that is more consistent with the predictions of a shock model as discussed below.

The enhancement of the crush stress leads to increases in the energy absorption capacity of the wood over that which a quasi-static test would suggest. The approximate factors Π by which the energy absorption capacity increases for each of the woods are given in Table 2.

(*v*) *Simple Shock Theory for Dynamic Crushing of Wood Cylinders*

There are similarities between the propagation of structural shock waves through metal ring systems and the propagation of crushing waves through wood. The basis of this lies in the shapes of the load–deflection curve for a ring (Fig. 4a) and the compressive stress–strain curves for woods such as those shown in Fig. 15. The key common feature is the fact that the curves are generally convex towards the strain axis, and, according to non-linear wave theory, shocks are therefore liable to form. However, it is important to note that the differences in the deformation mechanisms along and across the grain do have significant effects, which have been noted earlier. Axial crushing, in particular, leads to structural instabilities in the wood fibres, which reveal themselves for instance in the load drop at first yield seen in Fig. 15. This load drop implies that, in constitutive terms, we are dealing with a strain-softening material, and this leads to behaviour that, as already noted, is reminiscent of the heuristic model described by Shim *et al.*[34] However, crushing across the grain leads to plastic cell wall collapse, which produces a monotonically increasing constitutive equation reminiscent of the deformation in metal ring systems, for which a shock theory has been shown to be appropriate,[22] provided the impact velocity is sufficiently high.

Shock equations could be derived based on the experimental or theoretical stress–strain relationships for the material. Clearly, in order to bring out the influence of strain-softening, this would need to be done. This approach is not pursued here. Instead, to provide a first-order understanding, a common simplification of the two stress–strain curves is used, as shown in Fig. 21(a). This retains two of the key features of the inelastic stress–strain curves, namely a crushing strength σ_{cr} and a locking strain ε_ℓ. The material model can therefore be described as a rigid, perfectly plastic, locking model. Clearly, from Fig. 15 this model will be more accurate for low-density woods, but it provides useful information for other woods, and it results in analytical

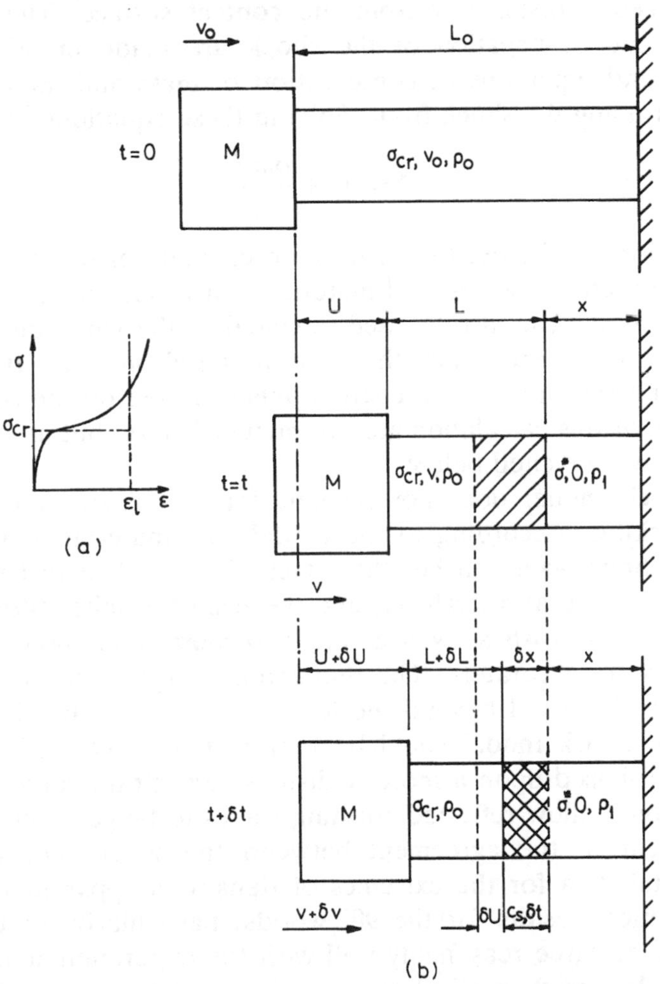

Fig. 21. Shock propagation model for uniaxial compression of a rigid, perfectly plastic, locking material: (a) idealised stress–strain curve; (b) parameters defining propagation of the shock front and the conditions either side of it.

expressions for the dynamic initial crush stress that give a physical feel for the influence of the various parameters on the crushing process.

Figure 21(b) shows a wooden cylinder of initial length L_0, cross-sectional area A_0 and density ρ_0 with an attached mass M striking a rigid target normally with an impact velocity v_0. When impact occurs, a shock wave is initiated that travels along the cylinder. Because of the rigid nature of the initial response, the stress in all the material ahead of the shock is instantaneously raised to σ_{cr}. As the shock passes through the material, it brings it to rest, increases its density by compaction up to the locking strain ε_ℓ and raises the stress to a value σ^*. Figure 21(b) shows the state of the system at time t when the shock has advanced a distance x from the contact surface. The equations governing the propagation of the shock are made up of kinematic equations and equations of conservation of mass and momentum for material crossing the shock front. Solving these equations[47] leads to

$$\sigma^* = \sigma_{cr} + \frac{\rho_0 v^2}{\varepsilon_\ell} \tag{8}$$

This equation can be used to provide an estimate of the initial level of stress generated in the crushed material at a given impact velocity v_0. The shock model can also be used to calculate the variation in σ^* with time.[47] On the assumption that the material unloads rigidly, this function of time represents the stress pulse exerted on the rigid target. The details of this calculation are not included here, but an example of the results will be cited below.

This shock model does not account for the absorption of energy beyond locking. According to the model, the material becomes rigid when the shock wave reaches the attached mass. The real material is not perfectly rigid at a strain ε_ℓ, and the use of a finite stiffness would permit the final crush as well as the maximum loads produced to be evaluated more accurately. The remaining energy would be absorbed by crushing the wood beyond the locking strain ε_ℓ. This limitation of the present shock model could be removed by constructing a shock wave model based upon a more realistic stress–strain curve as in Ref. 22. This would allow reflected crushing waves to be generated.

The extent of the agreement between the shock theory and the experimental data for the extremes of density is apparent in Fig. 20. The theoretical results for the 90° woods, particularly for the lower-density wood, agree reasonably well with the experimental values. For the 0° woods, the theoretical results, while showing reasonable agreement with the slope of the results, are about half of the measured values. However, the convergence between the results for oak and the shock theory as the impact velocity increases should be noted. The load–time pulses obtained from the model agree reasonably well with the experimental pulses, particularly for 90° woods, as shown in Fig. 22.

3.4 Discussion

The level of enhancement of the crush stress of wood across the grain compared with the quasi-static values can be very large at high impact velocities, and this level of increase is well represented by the shock model. This increase is matched by the general level of crushing stress (plateau stress), and leads to an overall increase in the energy absorption capacity of the woods as shown in Table 2 for a set of fully crushed specimens.

The data and observations from the tests described above imply that the dynamic behaviour for compression along the grain is not influenced by shock phenomena but rather by microinertia stabilising the buckling mechanisms that are evident in the quasi-static tests. Comparing the behaviour of wood compressed along the grain with the strut-like elements described at the end of Section 2.3, one would expect the initial peak stress on impact to reflect the axial yield strength of the wood cells σ_{ys}. The applied stress σ_{comp} corresponding to uniaxial yield in the cell wall assuming a hexagonal cell geometry is given by

$$\sigma_{comp} = \sigma_{ys} \frac{\rho_0}{\rho_s} \tag{9}$$

Thus the enhancement over the quasi-static crush strength, eqn (5), is simply

$$\frac{\sigma_{comp}}{\sigma_{cr}} = \frac{\sigma_{ys}}{150} \tag{10}$$

There is some debate about the value of σ_{ys},[10] but Cave[50] gives a value of 350 MN/m^2, which implies a stress ratio of 2·3 from eqn (10) for dynamic axial compression. Given the scatter in the data, a ratio of 2·3 provides a reasonable fit to the 0° data. For balsa this applies throughout the velocity range for which the shock theory predictions are always less than 2·3. However, it is interesting to note that the oak data match the shock theory reasonably well above this value, implying that the stress enhancement mechanism has changed.

For testing across the grain (90°) at low impact velocities it is also clear that the shock theory underestimates the experimental data. This too suggest that microinertia may well be the prevalent strength-

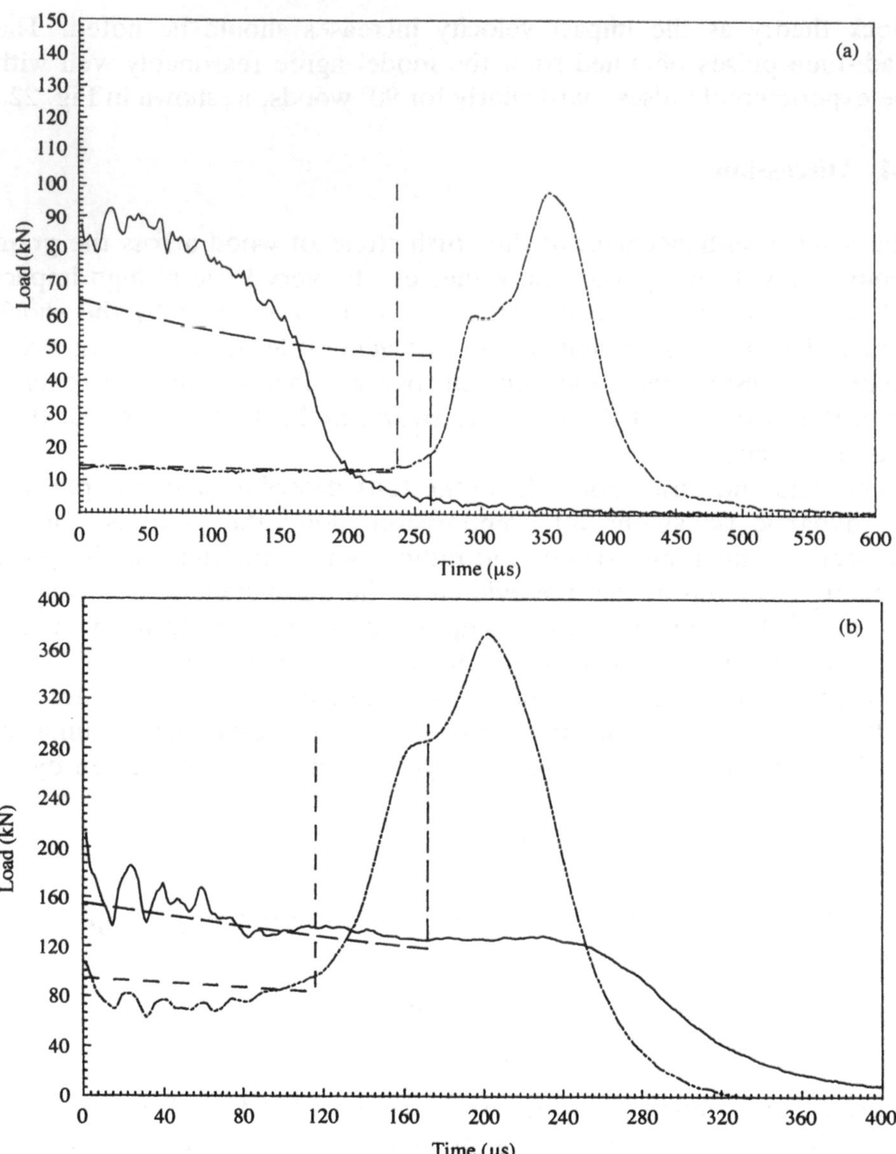

Fig. 22. Comparisons between measured and predicted dynamic force pulses for balsa (a) and American oak (b): ——, experiment at 0°, $V_0 = 143$ m/s (balsa), 142 m/s (oak); — — —, theory at 0°; —·-·—, experiment at 90°, $V_0 = 123$ m/s (balsa), 143 m/s (oak); — — —, theory at 90°.

enhancing mechanism at moderate velocities. Many impact energy absorption applications for wood correspond to impact velocities in the tens of metres per second range, and the discussion above suggests that the modelling of microinertial effects in wood is an important area to pursue.

4 FINAL REMARKS

The main purpose of this chapter has been to review some of the recent work on cellular structures and cellular materials and to draw comparisons between the two. It is clear that there are common features reflecting the common mechanisms that influence their behaviour. Where instabilities occur in the behaviour of the individual cells, localisation of deformation is a dominant feature of the quasi-static behaviour, and this becomes even more so under dynamic loading. This leads to energy absorption mechanisms that result in high loads on the striking mass and the crushing of only part of the structure or material. In contrast, locally stable behaviour leads to a final more uniform distribution of deformation and often a greater enhancement of the energy absorption capacity.

To date, microinertia has been explored in detail in only a few specific problems for which the geometry and the loading conditions are well defined.[30,42] There is, however, evidence of its presence and influence in other problems, such as those discussed above. These would benefit from further study, given the fact that microinertia produces force/stress enhancements that are considerably greater than those resulting from the strain-rate sensitivity of the materials.

Finally, it would appear that engineers could benefit greatly from more careful studies of the mechanics of natural materials such as wood. Biomimetics[51] is proving to be an interesting approach to the design of composite and other advanced materials, and this chapter suggests that a similar philosophy may be useful even in the context of impact energy absorption.

ACKNOWLEDGEMENTS

The authors wish to express their thanks to Dr W. J. Stronge and Dr V. Shim for an advance copy of Ref. 35 and for several helpful discussions during the preparation of this chapter. They are also grateful to British Nuclear Fuels plc, Risley for financial support.

REFERENCES

1. Johnson, W. & Reid, S. R., Metallic energy dissipating systems. In *Applied Mechanics Update,* ed. C. R. Steele. ASME, New York, 1986, pp. 303–19.

2. Reid, S. R., Metal tubes as impact energy absorbers. In *Metal Forming and Impact Mechanics,* ed. S. R. Reid. Pergamon Press, Oxford, 1985, pp. 249–69.
3. Abramowicz, W. & Jones, N., Dynamic axial crushing of circular tubes. *Int. J. Impact Engng,* **2** (1984) 263–81.
4. Reid, S. R. & Goudie, K., Denting and bending of tubular beams under local loads. In *Structural Failure,* ed. T. Wierzbicki & N. Jones. Wiley, New York, 1989, pp. 331–64.
5. Hull, D., Axial crushing of fibre-reinforced composite tubes. In *Structural Crashworthiness,* ed. N. Jones & T. Wierzbicki. Butterworths, London, 1983, pp. 118–35.
6. Reid, S. R., Laterally compressed metal tubes as impact energy absorbers. In *Structural Crashworthiness,* ed. N. Jones & T. Wierzbicki. Butterworths, London, 1983, pp. 1–43.
7. Klintworth, J. W. & Stronge, W. J., Elasto-plastic yield limits and deformation laws for transversely crushed honeycombs. *Int. J. Mech. Sci.,* **30** (1988) 273–92.
8. McFarland, R. K., Hexagonal cell structures under post-buckling axial load. *AIAA J.,* **1** (1963) 1380–5.
9. Shaw, M. C. & Sata, T., The plastic behaviour of cellular materials. *Int. J. Mech. Sci.,* **8** (1966) 469–78.
10. Gibson, L. J. & Ashby, M. F., *Cellular Solids.* Pergamon Press, Oxford, 1988.
11. Highways, *New Civil Engineer* (18 Oct. 1979) 18.
12. Carney III, J. F. & Sazinski, R. J., Portable energy absorbing system for highway service vehicles, *J. Transportation Engng ASCE,* **104** (1978) 407–21.
13. Carney III, J. F., Austin, C. D. & Reid, S. R., Modelling of steel tube vehicular crash cushion, *J. Transportation Engng ASCE,* **109** (1983) 331–46.
14. Johnson, W., Reid, S. R. & Reddy, T. Y., Pipewhip restraint systems. Cambridge University Engineering Department Report, 1979.
15. Shrive, N. G., Andrews, K. R. F. & England, G. L., The impact energy dissipation of cylindrical systems. In *Structural Impact and Crashworthiness,* Vol. 2, ed. J. Morton. Elsevier Applied Science Publishers, Barking, Essex, 1984, pp. 544–54.
16. Goldsmith, W. & Sackman, J. L., An experimental study of energy absorption in impact on sandwich plates. *Int. J. Impact Engng,* **12** (1992) 241–62.
17. McFarland, R. K., The development of metal honeycomb energy absorbing elements. JPL Report TR 32–639, July 1964.
18. Puthoff, R. L. & Gumto, K. H., Parameteric study of a frangible-tube energy absorption system for protection of a nuclear aircraft reactor. NASA TND-5730, 1970.
19. Jones, N., On the collision protection of ships, *Nuclear Engng Des.,* **38** (1976) 229–40.
20. Reid, S. R. & Reddy, T. Y., Experimental investigation of inertia effects in one-dimensional metal rings systems subjected to impact—I. Fixed ended systems. *Int. J. Impact Engng,* **1** (1983) 85–106.

21. Reid, S. R., Bell, W. W. & Barr, R., Structural plastic model for one-dimensional ring systems. *Int. J. Impact Engng,* **1** (1983) 185–91.
22. Reid, S. R. & Bell, W. W., Response of 1-D metal ring systems to end impact. In *Mechanical Properties at High Rates of Strain,* ed. J. Harding. Institute of Physics Conference Series No. 70, Bristol, 1984, pp. 471–8.
23. Reddy, T. Y., Reid, S. R. & Barr, R., Experimental investigation of inertia effects in one-dimensional metal ring systems subjected to end impact—II. Free ended systems. *Int. J. Impact Engng,* **2** (1991) 463–80.
24. Reid, S. R., Drew, S. L. K. & Carney III, J. F., Energy absorbing capacities of braced metal tubes. *Int. J. Mech. Sci.,* **25** (1983) 649–67.
25. Carney III, J. F. & Veillette, J. R., Impact response and energy dissipation characteristics of stiffened metallic tubes. In *Impact and Crashworthiness,* Vol. II, ed. J. Morton. Elsevier Applied Science Publishers, 1984, pp. 546–575.
26. Veillette, J. R. & Carney III, J. F., Collapse of braced tubes under impact loads. *Int. J. Impact Engng,* **7** (1988) 125–38.
27. Reddy, T. Y., Reid, S. R., Carney III, J. F. & Veillette, J. R., Crushing analysis of braced metal rings using the equivalent structure technique. *Int. J. Mech. Sci.,* **29** (1987) 655–68.
28. Shim, V. P. W. & Stronge, W. J., Lateral crushing in tightly packed arrays of thin walled metal tubes. *Int. J. Mech. Sci.,* **28** (1986) 709–28.
29. Stronge, W. J. & Shim, V. P. W., Dynamic crushing of a ductile cellular array. *Int. J. Mech. Sci.,* **29** (1987) 381–406.
30. Klintworth, J. W., Dynamic crushing of cellular solids. PhD Thesis, University of Cambridge, 1988.
31. Klintworth, J. W. & Stronge, W. J., Elasto-plastic yield limits and deformation basis for transversely crushed honeycombs. *Int. J. Mech. Sci.,* **30** (1988) 273–92.
32. Stronge, W. J., Dynamic crushing of elastoplastic cellular solids. In *Mechanical Behaviour of Materials VI (ICM6),* Vol. 1, ed. M. Jono & T. Inoue. Pergamon Press, Oxford 1991, pp. 377–82.
33. Stronge, W. J. & Shim, V. P. W., Microdynamics of crushing in cellular solids. *Trans. ASME, J. Engng Mater. Technol.,* **110** (1988) 185–90.
34. Shim, V. P. W., Tay, B. Y. & Stronge, W. J., Dynamic crushing of strain-softening cellular structures—a one dimensional analysis. *Trans. ASME, J. Engng Mater. Technol.,* **112** (1990) 398–405.
35. Shim, V. P. W., Yap, K. Y. & Stronge, W. J., Effects of nonhomogeneity, cell damage and strain-rate on impact crushing of a strain-softening cellular chain. *Int. J. Impact Engng,* **12** (1992) 585–602.
36. Martin, J. B. & Symonds, P. S., Mode approximations for impulsively loaded rigid–plastic structures. *J. Engng Mech. Div., Proc. ASCE,* **92** (1966) 43–66.
37. Lee, J. W., Behaviour of constrainted tubes and tube bundles under quasi-static and dynamic axial compression. MSc Dissertation, UMIST, 1989.
38. Pugsley, A. & Macaulay, M., The large scale crumpling of thin cylindrical columns. *Q. J. Mech. Applied Maths,* **8** (1960) 1–9.
39. Wierzbicki, T., Crushing analysis of metal honeycombs. *Int. J. Impact Engng,* **1** (1983) 157–74.

40. Calladine, C. R. & English, R. W., Strain-rate and inertia effects in the collapse of two types of energy-absorbing structure. *Int. J. Mech. Sci.*, **26** (1984) 689–701.
41. Zhang, T. G. & Yu, T. X., A note on a 'velocity sensitive' energy absorbing structure. *Int. J. Impact Engng*, **8** (1989) 43–51.
42. Tam, L. L. & Calladine, C. R., Inertia and strain-rate effects in a simple plate-structure under impact loading. *Int. J. Impact Engng*, **11** (1991) 349–77.
43. Reid, S. R., Peng, C. & Reddy, T. Y., Dynamic uniaxial crushing and penetration of wood. In *Mechanical Properties of Materials at High Rates of Strain 1989*, ed. J. Harding. Institute of Physics Conference Series No. 102, Bristol, 1989, pp. 447–55.
44. Easterling, K. E., Harryson, R., Gibson, L. J. & Ashby, M. F., On the mechanics of balsa and other woods. *Proc. R. Soc. Lond.*, **A383** (1982) 31–41.
45. Maiti, S. K., Gibson, L. J. & Ashby, M. F., Deformation and energy absorption diagrams for cellular solids. *Acta Metall.*, **32** (1984) 1963–75.
46. Reid, S. R. & Peng, C., Quasi-static compressive strength and deformation of woods. *Int. J. Mech. Sci.* (1993) (to be published).
47. Peng, C., Crushing and indentation of wood under static and dynamic loading conditions. PhD Thesis, UMIST, 1991.
48. Zukas, J. A., Nicholas, T., Swift, H. F. Greszczuk, L. B. & Curran, D. R., *Impact Dynamics*. Wiley, New York, 1982.
49. Taylor, G. I., The use of flat-ended projectiles for determining dynamic yield stress, I: Theoretical considerations. *Proc. R. Soc. Lond.*, **A194** (1948) 289–99.
50. Cave, I. D., Longitudinal Young's modulus of Pinus Radiata. *Wood Sci. Technol.*, **3** (1969) 40–8.
51. Srinivasan, A. V., Haritos, G. K. & Hedberg, F. L., Biomimetics: advancing man-made materials through guidance from nature. *Appl. Mech. Rev.*, **44** (1991) 463–82.

Chapter 9

ELASTIC EFFECTS IN THE DYNAMIC PLASTIC RESPONSE OF STRUCTURES

T. X. YU†

Department of Mechanics, Peking University, Beijing 100871, China
and
Department of Mechanical Engineering, UMIST, Manchester M60 1QD, UK.

ABSTRACT

Elastic effects in the dynamic response of structures to impact and pulse loading are discussed in the context of the assessment of structural crashworthiness and structural failure under dynamic loading. Previous studies of single- and two-DoF (degrees of freedom) models are first reviewed, and the validity of the widely adopted rigid–plastic idealization is thoroughly examined. By taking an impulsively loaded cantilever beam as a typical example, various aspects of the differences between dynamic elastic–plastic and rigid–plastic behaviours are revealed; while the results obtained from a Timoshenko beam finite element model, a mass–spring finite difference model and a simplified beam–spring model are compared with each other. Numerical results confirm that the energy ratio R (input energy divided by the maximum elastic energy that can be stored in the structure) is not a unique index in judging whether or not elastic effects are negligible. In impact problems the mass ratio (colliding mass divided by the mass of the structure itself) significantly affects the deformation mechanism and energy dissipation pattern. A general discussion indicates that more complex response patterns are likely to appear if an elastic–plastic system has many DoF or deforms in a mode of combined flexure and extension.

NOTATION

b	Width of rectangular cross-section
C	Elastic spring coefficient

d_p	D_p/M_p, non-dimensional energy dissipation
D_p	Total energy dissipation due to plastic deformation
$\dot{D}_p$	Energy dissipation rate due to plastic deformation
e_0	K_0/M_p, impact energy ratio
E	Young's modulus
E_{in}	Input energy
G	Mass of colliding particle
$\bar{G}$	Lumped mass in two-DoF model
h	Depth of doubly symmetric cross-section
I	Second moment of area about transverse axis through centroid
k	κL, non-dimensional curvature
k_Y	$\kappa_Y L$, non-dimensional curvature at yield
K	Kinetic energy
K_0	$\frac{1}{2}GV_0^2$, impact energy
L	Length of beam
m	M/M_p, fully plastic bending moment ratio
M	Bending moment
M_p	Fully plastic bending moment
M_Y	Yield moment
N	Force applied on spring in a single-DoF model; membrane force in a spherical shell
N_Y	Yield force for spring in a single-DoF model
p	Internal loading pulse for a spherical shell
P	Loading pulse for a single-DoF model
Q	Shear force
R	Energy ratio E_{in}/U_e^{max}
R'	Ratio D_p/U_e^{max}
R_s	Radius of a spherical shell
t	Time
t_d	Duration of a pulse
T_0	Characteristic time, $\rho\sqrt{\rho L/M_p}$
T_1	Fundamental period of elastic vibration
u	Displacement of mass in a single-DoF model
U_e	Elastic deformation energy
U_e^{max}	Maximum elastic deformation energy for a structure
v	VT_0/L, non-dimensional transverse velocity
v_0	V_0T_0/L, non-dimensional initial transverse velocity of colliding mass
V	Transverse velocity of colliding mass
V_0	Initial transverse velocity of colliding mass
w	W/L, non-dimensional transverse deflection; radial displacement of a spherical shell

w_e	Amplitude of the final elastic vibration
W	Transverse deflection
x	X/L, non-dimensional coordinate
X	Axial coordinate in undeformed configuration
Y	Yield stress
α	E_t/E, coefficient of strain-hardening
β	ratio of peak force to yield force
γ	$G/\rho L$, mass ratio
ε	Strain
ε_Y	Yield strain
ζ	Velocity ratio in a two DoF system
η	Ratio w_f/w_e
θ	Rotation angle, inclination
$\dot{\theta}$	Rotation rate
θ_0	Non-dimensional initial release angle in the double-tier spring model
κ	Curvature
κ_Y	Maximum elastic curvature
λ	Λ/L, non-dimensional coordinate of hinge
Λ	Coordinate of hinge location H
μ	Mass per unit surface of a shell
ν	Poisson's ratio
ρ	$\rho_v A$, mass per unit length of cantilever
σ	Normal stress on cross-section
τ	t/T_0, non-dimensional time; shear stress on cross-section
ϕ	Angular displacement in the Shanley-type model
ϕ_0	Initial release angle in the Shanley-type model
ψ	Relative rotation angle
ψ_p	Relative rotation angle at yield in the MS-FD model
ω	Non-dimensional initial angular velocity in the double-tier spring model

Subscripts

a	Average
A	Tip of cantilever
B	Root of cantilever
f	Final
H	Plastic hinge
in	Input
Y	Yield
0	Initial

Superscripts

′ (prime)	Differentiation with respect to coordinate x
· (dot)	Differentiation with respect to time variable
e	Elastic
ep	Elastic–plastic
p	Plastic
rp	Rigid–plastic

1 INTRODUCTION

1.1 Motivation for Studying Elastic Effects in Structural Dynamics

In order to assess the crashworthiness and energy absorbing capacity of a structure under quasi-static or dynamic loading, it is often necessary to know the largest deflection, deformed shape and partition of energy dissipation in the structure. From the theoretical point of view, these deformation and energy absorbing quantities are closely related to a certain *deformation mechanism* or *deformation mode,* which may eventually lead to a certain *failure mode.* As shown by many previous investigations, the dynamic deformation mechanism of a structure can be quite different from that of the same structure in the quasi-static cases, owing to inertia and the differences of the dynamic properties of material from the static ones. The dynamic deformation mechanism of a structure also depends very much on the *material modelling*. Most previous analyses of the dynamic plastic response of structures neglected elastic strains in the material model in order to simplify the formulation and solutions. However, will the analyses based on this simplification still give reasonable and reliable estimates for the deformation and energy absorbing properties over the whole spectrum of loading parameters? Will the neglect of elastic deformations significantly alter the dynamic deformation mechanism or the failure mode of structures? It is evident that a further study of elastic effects will be not only beneficial to the modelling of the impact dynamics of structures, but also useful for the practical assessment of engineering structures.

1.2 Early Studies of Dynamic Behaviour of Elastic–Plastic Structures

Ductile structural metals generally display elastic–plastic behaviour. Hence, when an engineering structure such as a beam, plate or shell composed of such materials is subjected to an intense dynamic loading, the initial response is *elastic*. After the yield stress is reached at some point, a more complex motion follows in which *plastic flow* occurs in certain regions while the rest of the structure behaves elastically. The interfaces between elastic and plastic regions move during the response, and the plastic regions may disappear and reappear. Plastic work is done during a certain time interval, after which the structure continues to vibrate elastically until the kinetic energy is gradually damped out, leaving the structure in a state of permanent deflection and an associated field of residual stresses.

Owing to the complex intermingling of elastic and plastic behaviour in the structural response, general *analytical* methods are not available; there is only one 'exact' theoretical solution for a beam composed of elastic–plastic material. This theory was developed by Duwez, Clark and Bohnenblust,[1] and was based on a result obtained much earlier by Boussinesq.[2] By introducing a new independent variable cx^2/t, with c being a suitable constant, they reduced the equation of the beam motion from a partial differential equation to an ordinary differential equation, so that in principle a complete solution for an elastic–plastic material can be obtained. However, their solution is only applicable to an *infinite* beam subjected to a concentrated transverse impact of constant velocity. Although this type of solutions was later extended to semi-infinite beams by Conroy,[3] up to now no analytical solution has been published for *finite* elastic–plastic beams under impact or pulse loading.

In an early attempt to develop a general *numerical* method for structural impact problems, Witmer *et al.*[4] included not only elastic deformations but also finite changes of geometry, strain hardening and strain-rate sensitivity in the analyses. Their finite difference scheme employed a step-by-step explicit method, while the structure was replaced by a finite number of masses and connecting links; each connecting link contained two or more layers carrying tension or compression stresses. The axisymmetric responses of shells, plates, rings and beams to impulsive or blast loading producing large elastic–plastic deformations were given as examples and compared with experimental data. A typical example is shown in Fig. 1, in which the response of an explosively loaded clamped 6061-T6 beam predicted by their numerical scheme compares well with experimental results. Experience with various numerical codes shows that difficulties exist in dealing with the floating boundaries between elastic and plastic regions, and the computation of dynamic problems is expensive.

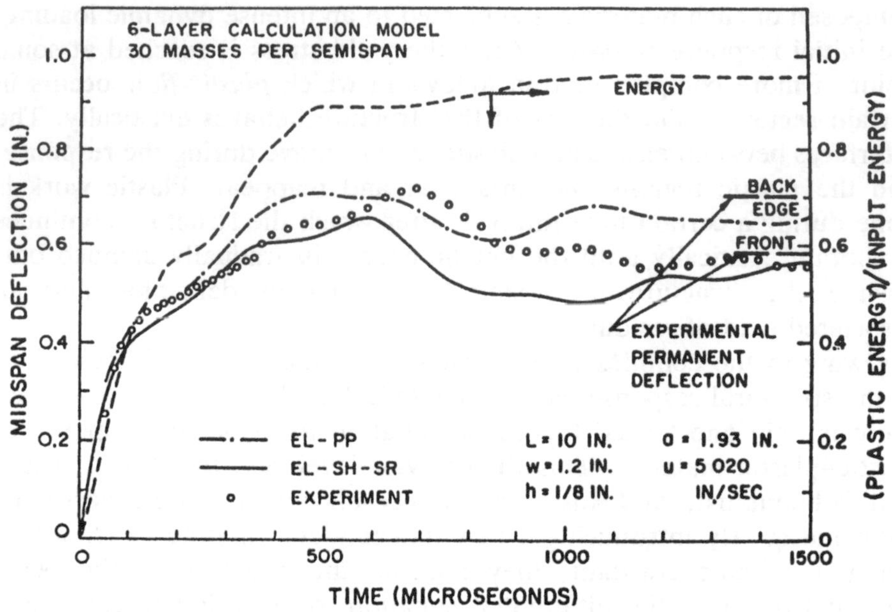

Fig. 1. Response of an explosively loaded clamped 6061-T6 beam, predicted by a numerical scheme, compared with experimental results.[4]

1.3 Validity of Rigid–Plastic Idealization

The difficulty of obtaining a complete elastic–plastic solution by either analytical or numerical means has called for consideration of simplifications in the material modelling. In plasticity theory the neglect of elastic strains in comparison with plastic ones (i.e. effectively taking the elastic moduli as infinite) leads to a powerful simplification in the analyses. After pioneering works by Lee and Symonds[5] for a finite free beam and Conroy[6] for an infinite beam, the *rigid–plastic idealization* has been widely accepted and extensively used in the structural dynamics over the past four decades. In these rigid–plastic analyses deformation occurs only in regions (in fact, only at discrete *plastic hinges* in beams, if both material strain-rate sensitivity and strain hardening are neglected) where a yield condition is satisfied; everywhere else, only rigid-body motion occurs.

This rigid–plastic idealization for dynamically loaded structures is based upon the fact that the plastic deformation of a structure is much larger than the associated elastic deformation when the dynamic loading on the structure is intense, while material elasticity is not likely to play an important role in the overall response of the structure. However, the elastic deformation is unknown in the rigid–plastic analyses, so that the validity of the rigid–plastic idealization has to invoke an *energy hypothesis*. That is, if the total input energy imparted to a structure by the dynamic loading is much larger than the corresponding maximum elastic strain energy that the structure is able to store then almost all the input energy has to be transformed into plastic deformation energy, i.e. dissipated by the plastic deformation of the structure; accordingly, the effect of the elastic deformation can be neglected. Thus we may introduce an *energy ratio*

$$R \equiv \frac{E_{in}}{U_e^{max}} \tag{1}$$

where E_{in} is the total input energy imparted to a structure by dynamic loading and U_e^{max} is the *maximum* elastic energy that can be stored in the structure. Previous investigations have confirmed that rigid–plastic analysis provides a good approximation when the structure is subjected to intense short pulse loading while the energy ratio R is large (say $R > 5$). In the limiting case of impulsive loading (i.e. initial velocities being specified) the 'error' in the final deflection caused by the rigid–plastic idealization is estimated to be of order $1/R$.[7] In these cases the error is easily seen to be positive, since the specified initial energy must all go into plastic deformations rather than being divided between plastic work and elastic strain energy.

However, as will be seen later, $R \gg 1$ is a necessary but not sufficient condition for the validity of a rigid–plastic idealization, and the elasticity could affect many aspects of the dynamic behaviour of plastic structures. Hence the implication of the 'elastic effects on structural response' could be far beyond that represented by a simple comparison of the final deflection. Our intention here is not to provide a universally applicable technique for solving the dynamic response of elastic–plastic structures; instead, we shall try to illustrate (i) how elasticity affects the basic feature of dynamic plastic response of structures, and (ii) the conditions under which the elastic effects become insignificant.

2 SINGLE- AND TWO-DOF SYSTEMS

2.1 Approach by a Single-DoF System

In the case of a pulse loading, Symonds[8,9] first noticed that if the duration t_d of a loading pulse is not small in comparison with the fundamental period T_1 of the elastic vibration of the structure then the errors caused by the rigid–plastic idealization may be large (e.g.

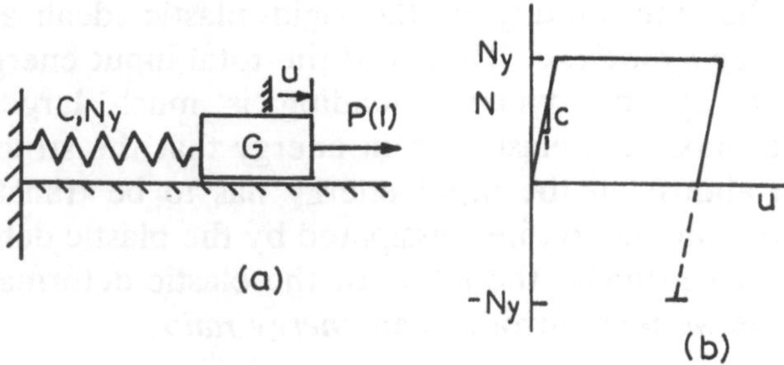

Fig. 2. (a) A single-DoF mass–spring model loaded by pulse $P(t)$, proposed by Symonds and Frye.[10] (b) Relation between force $N(t)$ and displacement $u(t)$ for the spring.

30–60%) and may become negative, i.e. on the unsafe side, even when the actual plastic deformations are large (e.g. 10–20 times elastic strain magnitudes). In general, therefore, rigid–plastic predictions are applicable only if the conditions $R \gg 1$ and $t_d < T_1$ both are satisfied.

To fully examine the error caused by the rigid–plastic idealization in the case of a pulse loading, Symonds and Frye[10] studied the dynamic behaviour of a single-degree-of-freedom (DoF) mass–spring model (Fig. 2a). The equation of motion for the mass G in the model is

$$G\frac{d^2u}{dt^2} + N = P(t) \tag{2}$$

where $P(t)$ is the loading pulse. The force $N(t)$ provided by the spring is related to the displacement u by an elastic–plastic relation shown in Fig. 2(b) or by a rigid–plastic relation (taking the spring constant $C \to \infty$). After calculating the response of this simple model due to six different pulse shapes (rectangular, half-sine and triangulars with four kinds of rise time), Symonds and Frye[10] concluded that a large energy ratio R is indeed a necessary, but not sufficient, condition for the rigid–plastic idealization to give a good estimate of an elastic–plastic solution. For example, in the case of a linear decreasing pulse, the dependence of the ratio u_f/u_y on the ratio t_d/T_1 is shown in Fig. 3(a), where u_f is the final displacement given by the elastic–plastic solution, $u_y = N_Y/C$ and N_Y are the displacement and force respectively of the spring at the yield state, the parameter $\beta \equiv P_0/N_Y$ and P_0 is the peak force applied on the mass. Figure 3(b) shows the error of the rigid–plastic prediction relative to the elastic–plastic solution for the final displacement as a function of the ratio t_d/T_1. In the figure $R' \equiv D_p/U_e^{max}$, with D_p being the plastic dissipation; for this single-DoF model $R' = R - 1 = 2u_f/u_y$, where R is defined by eqn (1). It can be

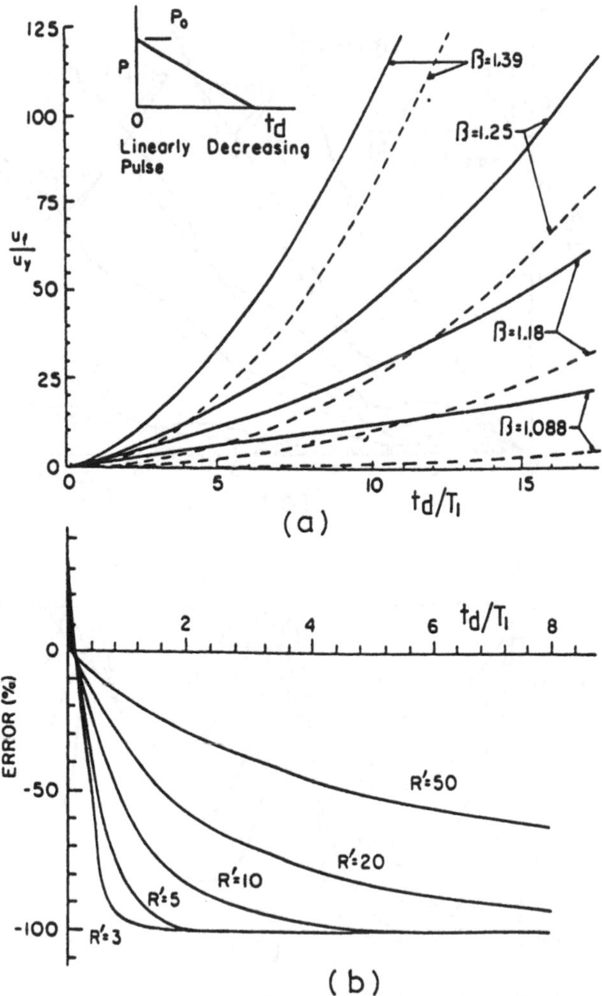

Fig. 3. Response of the single-DoF model to a linearly decreasing pulse: (a) dependence of displacement ratio on loading parameters, obtained from elastic–plastic (solid lines) and rigid–plastic (dashed lines) solutions; (b) error of rigid–plastic prediction relative to elastic–plastic solution on final displacement.

seen that, while the error of a rigid–plastic prediction decreases with increasing R, the magnitude of the error increases as the pulse duration t_d increases.

A notable fact explored by Symonds and Frye[10] is that when the loading pulse has a *non-zero rise time,* the error of the rigid–plastic prediction displays a wavy character over a large range of the ratio t_d/T_1, and the discrepancies between the rigid–plastic and the elastic–plastic solutions sometimes become quite significant. For example, Fig.

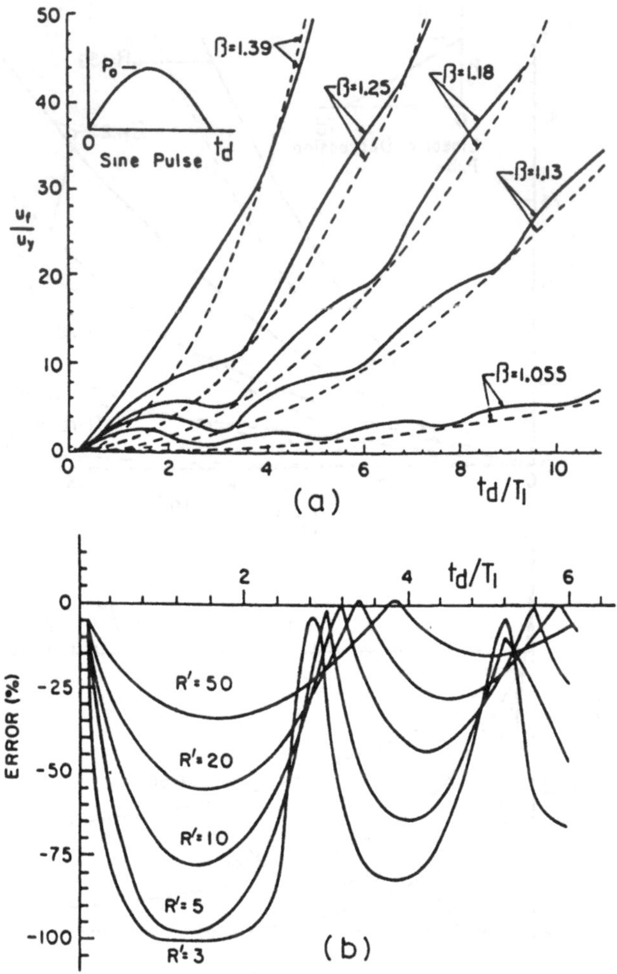

Fig. 4. Response of the single-DoF model to a half-sine pulse: (a) dependence of displacement ratio on loading parameters, obtained from elastic–plastic (solid lines) and rigid–plastic (dashed lines) solutions; (b) error of rigid–plastic prediction relative to elastic–plastic solution on final displacement.

4 is obtained from the single-DoF model when it is subjected to a half-sine pulse with a duration t_d and the peak force $P_0 = \beta N_Y$. The error of the rigid–plastic prediction is shown in Fig. 4(b), which indicates that the error is smaller for the larger value of R, and the peaks in the error curves decrease with increasing pulse duration. Triangular pulses with non-zero rise time result in similar error curves. Obviously, in these cases no simple formula based solely on the energy ratio R could provide an estimate of the error for a rigid–plastic solution.

It is worthwhile to point out that if the transient phase in the dynamic response of a structure is neglected and only the *modal* phase of the motion is retained then the motion of the structure is described by a single-DoF model. Therefore, although the single-DoF mass–spring model is extremely simple, the conclusions drawn from the analysis of this model may serve as a source of insight into the dynamic response of elastic–plastic structures.

2.2 Spherically Symmetric Response of a Complete Spherical Shell

The elastic–plastic response of a complete thin spherical shell to internal blast loading is essentially a single-DoF problem, since the response is spherically symmetric while the shell undergoes a biaxial stretching and a uniform expansion, which depend simply on a unique spatial variable, namely the radial displacement w. In fact, the equation of motion for the spherical shell is

$$\mu \frac{d^2 w}{dt^2} + \frac{2N}{R_s} = p(t) \qquad (3)$$

where μ is the mass per unit surface area, N is the membrane force, R_s is the radius of the spherical shell and $p(t)$ is the internal loading pulse. Equation (3) is obviously similar to eqn (2), so that analytical solutions can easily be found for both elastic–plastic and rigid–plastic responses of the shell to a pulse loading. The elastic–plastic solution was first given by Baker.[11] In his numerical calculation he also took account of strain hardening, shell thinning and the increase in radius during deformation. By considering impulsive loading only, Duffey[12] later incorporated the strain-rate effect into Baker's elastic–plastic analysis. Jones[13] then presented a rigid–plastic solution for this problem and made a comprehensive comparison between the elastic–plastic and rigid–plastic predictions for the final displacements and the partition of energy. Figure 5 shows a typical radial displacement–time history for an elastic–perfectly plastic spherical shell subjected to a rectangular pressure pulse. The final radial displacement predicted by the rigid–perfectly plastic solution for this example is $w_f^{rp} = 6{\cdot}940 w_y$, with w_y being the displacement of the shell when the shell material first yields plastically. The value of w_f^{rp} almost coincides with the average radial displacement w_a^{ep} given by the elastic–perfectly plastic solution.

In the case of impulsive loading, based on the above-mentioned analyses,[11–13] we can draw the following conclusions.

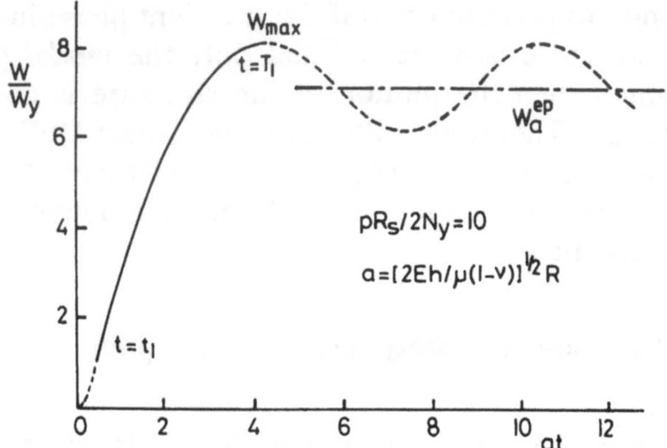

Fig. 5. Radial displacement versus time response for an elastic–perfectly plastic symmetric shell subjected to a rectangular internal pressure pulse $p(t)$, given by Jones.[13]

(i) If $V_0 > [2(1-\nu)Y^2h/\mu E]^{1/2}$, the response is elastic–plastic; otherwise, it is purely elastic. Here V_0 is the initial radial outward velocity, h is the thickness of the shell, and E, ν and Y are respectively the Young's modulus, Poisson's ratio and yield stress of the material.

(ii) The error in the final radical displacement caused by the rigid–plastic idealization is

$$\text{error} \equiv \frac{w_f^{rp} - w_a^{ep}}{w_a^{ep}} = \frac{1}{R-1} = \frac{1}{R'} \tag{4}$$

where R and R' are as defined before. It is not surprising that eqn (4) lead to exactly the same conclusion as that given by Symonds[9] for a single-DoF mass–spring system subjected to impulsive loading, since the two problems are essentially identical.

2.3 Approach by a Two-DoF System

Simple lumped mass models of impulsively loaded beams can greatly simplify the beam dynamics by limiting the number of DoF; likewise, they limit the number of dynamic modes. To study a cantilever subjected to general impulsive loading by a two-DoF model, Stronge and Hua[14] considered a cantilever of length L with two equally spaced masses $\bar{G}$, as shown in Fig. 6(a). These masses are located at the midpoint and the tip, and they have initial velocities V_0 and ζV_0 respectively. Assume that deformations occur only at discrete hinges

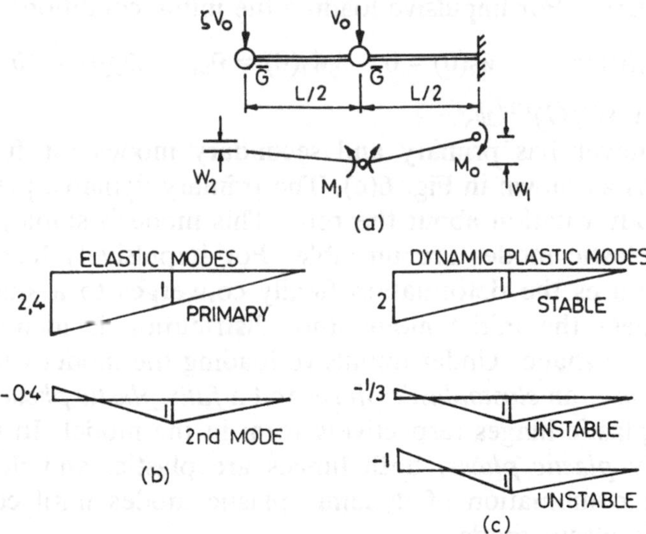

Fig. 6. (a) A two-DoF model of an impulsively loaded elastic–plastic cantilever, proposed by Stronge and Hua.[14] (b) Dynamic elastic modes for cantilever. (c) Dynamic fully plastic modes for cantilever.

located at the root and the midpoint, which are assumed to have elastic–perfectly plastic behaviour; that is, if the relative rotation at hinge i ($i = 1, 2$) is ψ_i then the moment M_i during loading is given by

$$\left.\begin{aligned} M_i &= C\psi_i \quad (|\psi_i| < \psi_Y) \\ |M_i| &= M_p \quad (|\psi_i| \geqslant \psi_Y, \quad \dot{\psi}_i > 0) \end{aligned}\right\} \quad (i = 1, 2)$$

where the parameter ψ_Y is the yield rotation angle at the hinges. The deformation of this model is always initially elastic while both hinges have rotations $|\psi_i| < \psi_Y$; then it can be elastoplastic after one of the hinges has a rotation magnitude $|\psi_i| \geqslant \psi_Y$; and finally it can become fully plastic if both hinges have rotation magnitudes $|\psi_i| \geqslant \psi_Y$. The analysis given by Stronge and Hua[14] neglects elastic unloading, so the final deformation configuration is unambiguous.

The equation of motion of this two-DoF model in non-dimensional form is

$$\begin{Bmatrix} \ddot{\bar{w}}_1 \\ \ddot{\bar{w}}_2 \end{Bmatrix} + \begin{bmatrix} 1 & -2 \\ 0 & 1 \end{bmatrix} \begin{Bmatrix} m_1 \\ m_2 \end{Bmatrix} = \begin{Bmatrix} 0 \\ 0 \end{Bmatrix}$$

where $\bar{w}_i \equiv 2W_i/L\psi_Y$ is the non-dimensional transverse nodal displacement, $m_i \equiv M_i/M_p$ is the moment at node $i-1$, and $(\dot{\ }) \equiv \mathrm{d}(\ )/\mathrm{d}\bar{\tau}$, with

$\bar{\tau} \equiv 2t(C/\bar{G}L^2)^{1/2}$. For impulsive loading the initial conditions are

$$\bar{w}_1(0) = 0, \qquad \bar{w}_2(0) = 0, \qquad \dot{\bar{w}}_1(0) = \bar{v}_0, \qquad \dot{\bar{w}}_2(0) = \zeta\bar{v}_0$$

where $\bar{v}_0 \equiv V_0(C/\bar{G})^{1/2}/\psi_Y$.

The cantilever has primary and secondary modes for fully plastic deformations as shown in Fig. 6(c). The primary dynamic plastic mode is a rigid-body rotation about the root. This mode is stable, while the secondary plastic modes are unstable. For impulsively loaded rigid–plastic structures the deformation finally converges to a stable modal solution unless the initial momentum distribution is identical to an unstable mode shape. Under impulsive loading the model will undergo an *elastic phase,* an *elastoplastic phase* and a *fully plastic phase,* in which 0, 1 and 2 plastic hinges respectively form in the model. In particular, for the *fully plastic phase,* both hinges are plastic, and deformation occurs in a combination of dynamic plastic modes until coalescence with a stable plastic mode.

The energy ratio for the model is $R = \frac{1}{2}\bar{v}_0^2(1 + \zeta^2)$. For $\frac{1}{2}\bar{v}_0^2 = 5{\cdot}0$ and $0 \leqslant \zeta \leqslant 3$ the velocity trajectories for elastoplastic deformations of the impulsively loaded cantilever are illustrated in Fig. 7(a). The stable dynamic plastic mode corresponds to an initial momentum distribution

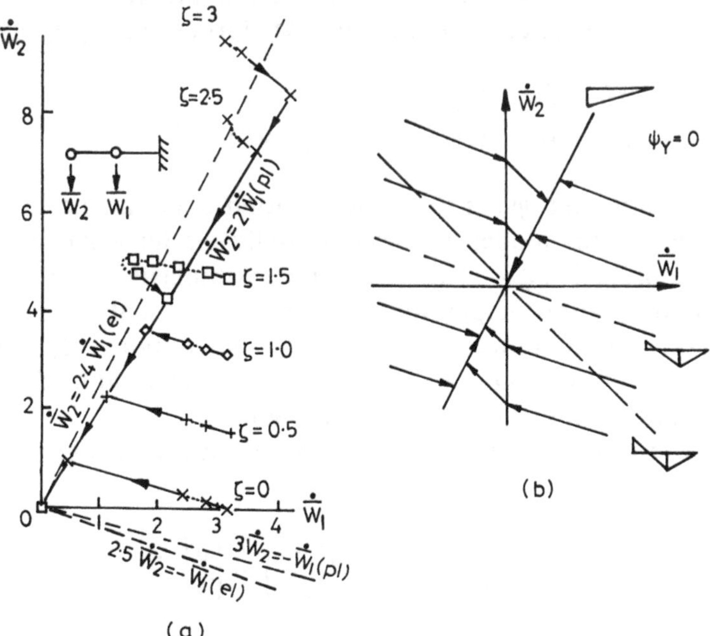

Fig. 7. Velocity trajectories for the two-DoF model: (a) elastic–plastic case; (b) rigid–perfectly plastic case.

$\zeta = 0$. For initial conditions in the range $1{\cdot}1 < \zeta < 2{\cdot}0$ the deformation at the central hinge is still elastic when motion is coincident with the stable plastic mode, so the trajectory crosses this mode line. Velocity trajectories corresponding to the rigid–perfectly plastic approximation for the same model are shown in Fig. 7(b). These solutions converge to the stable plastic mode along paths that parallel to the unstable plastic modes; the discontinuities in these trajectories at $\dot{\bar{w}}_1 = 0$ are caused by a change in the sense of the moment at the root. For a specified energy ratio R, in comparison with the elastoplastic response, the error of the rigid–plastic approximation is largest for an initial momentum distribution that is not very different from the stable plastic mode shape.

Stronge and Hua[14] also studied a two-DoF model of a simply supported elastoplastic beam, which has two stable dynamic plastic modes. Their analysis shows that elastic effects during an intermediate phase of elastoplastic deformations can significantly affect subsequent widespread plastic deformations when multiple stable dynamic plastic modes exist. As a result, the accuracy of the rigid–plastic approximation is very sensitive to the initial momentum distribution, especially when the latter is close to the stable plastic mode.

3 FINITE ELEMENT APPROACHES USING A TIMOSHENKO BEAM (TB-FE) MODEL

3.1 General Description

In order to examine the effects of material elasticity in structures subjected to impact loading, in Sections 3–6 we shall focus on the analysis of an elastic–plastic cantilever beam struck by a rigid particle mass at its tip. A cantilever is a very simple and special structure, but much more complicated than a single- or two-DoF system. The reasons for choosing it are

(i) there is an analytical complete rigid–plastic solution for comparison;[15]
(ii) there are some experimental results available for comparison;
(iii) besides the energy ratio, the mass ratio is a parameter that can be varied to change the response characteristics;
(iv) no axial force is induced by large deflections, so the flexural deformation remains the main concern.

In this particular impact problem, the energy ratio defined by eqn (1) becomes

$$R = \frac{K_0}{U_e^{max}} = \frac{GV_0^2EI}{M_p^2L} \tag{5}$$

where G is the mass attached at the tip, V_0 is its initial velocity, M_p is the fully plastic bending moment, I is the second moment of the cross-section and L is the length of the cantilever. Note that the value $U_e^{max} = M_p^2L/2EI$ taken here overestimates the elastic capacity of a cantilever since it comes from a simplification that all the cross-sections of the latter are supposed to be in the fully plastic state *simultaneously*, which is unrealistic for the flexural deformation of a cantilever.

Two important finite element approaches to impulsively loaded cantilevers are due to Symonds and Fleming[16] and Reid and Gui.[17] Both employed the finite element code ABAQUS and assumed an elastic–perfectly plastic material behaviour. The model used in their approaches describes the large deflection dynamic response of a Timoshenko beam of rectangular cross-section; so in the following we denote it as the TB-FE model. The parameters in these calculations were taken to be as listed in Table 1.

3.2 Symonds and Fleming's Work

Bending moment distributions at various times were presented in Ref. 16 for $\gamma = 1{\cdot}64$ and $R = 2$; this case is related to a moderately large mass ratio and a rather large proportion of elastic deformation. It was reported that those corresponding to early times bore little resemblance to the bending moment distribution corresponding to the travelling hinge phase in the rigid–plastic analysis.[15] However, at later times the

Table 1
Parameters Used in the Finite Element (TB-FE Model) Calculations[16,17]

Parameters	*Ref.* 16 $\gamma = 1{\cdot}64$	*Example* 1 *in Ref.* 17 $\gamma = 1{\cdot}64$	*Example* 2 *in Ref.* 17 $\gamma = 0{\cdot}0228$
M_p (N m)	16·5	16·5	24·8
L (mm)	355·6	355·6	304·8
G (kg)	0·336	0·336	0·0023
V_0 (m s^{-1})	4·7–12·9	12·9	251·5
$e_0 = K_0/M_p = GV_0^2/2M_p$	0·228–1·69	1·69	2·93
$\gamma = G/\rho L = G/\rho_v bhL$	1·64	1·64	0·0228
$R = 2K_0EI/M_p^2L$	2·0–14·8	14·8	51·7
R^{-1}	0·5–0·0676	0·0676	0·0193
Number of elements	28	28	24

finite element solution did show a clear modal type of moment distribution with a plastic hinge at the root of the cantilever, as found in the modal phase of the rigid–plastic solution. The results also indicate that the plastic work dissipated in the interior of the cantilever in the elastic–plastic response is significantly smaller than that predicted by the rigid–plastic theory; and this discrepancy increases as R increases over the range shown in Table 1.

From these numerical results, Symonds and Fleming[16] concluded that the transient phase of motion, which notably appears in the rigid–perfectly plastic solution, is not in general observed in their finite element solutions; plastic deformation at the root may take place prior to or simultaneously with plastic deformation in the beam interior; the presence of elastic action is seen to eliminate or substantially reduce the plastic work done in the interior of the beam, even at very large values of the energy ratio R. As noted later by other researchers, however, these conclusions are only valid for the cases of large or moderately large mass ratios; see below.

3.3 Reid and Gui's Work

With the principal aim of exploring how the presence of elastic deformation changes the response history and the distribution of internal plastic deformation, Reid and Gui[17] have also used the TB-FE model to examine the response of impulsively loaded cantilevers. From the bending moment distributions at various times obtained for Example 1, they described the elastic–plastic response of the cantilever as the following sequence of events.

Phase 1: *Elastic–plastic bending wave.* The moment distributions in this phase show many of the characteristics of the wave propagation in Timoshenko beams. At the tail of the disturbed region there exists a maximum bending moment M_p, which may be regarded as a plastic hinge. The position of this hinge moves along the cantilever as time progresses, in a manner that is described well by the rigid–plastic solution.

Phase 2: *Reversed hinge at root—arrest of travelling hinge.* After the elastic bending wave is reflected at the root, the progress of the travelling hinge is arrested and its position tends to move back towards the tip, as a result of the interaction between the primary bending wave and the reflected wave. Meanwhile the bending moment oscillates and increases in the magnitude at the root until a reversed hinge is formed at the root.

Phase 3: *Initiation of positive root rotation.* While a certain amount of plastic work has been consumed in reversed bending at the root, the position of the internal hinge sweeps back and forth, until the moment at the root reduces in magnitude and then a positive root hinge forms.

Phase 4: *Root rotation and elastic vibration.* The plastic deforming region shrinks quickly to a root hinge; and finally only elastic vibrations occur as the cantilever oscillates about the root.

Redrawn from a figure in Ref. 17, Fig. 8(a) provides a summary of the evolution of the plastic regions. Unlike the rigid–plastic moment distribution in which full plastic moment is reached in the whole region between the travelling hinge and the root, here full plasticity is only achieved over small regions of the cantilever. The position of the travelling hinge given by the rigid–plastic solution is also shown in Fig. 8(a) for comparison. It can be seen that the rigid–plastic analysis accounts quite accurately for the movement of the internal hinge until the formation of the reversed hinge at the root, which signals the arrest of the travelling hinge and causes it to spread and move back towards the tip. Figure 8(b) shows the increase in plastic work, which is distributed along the cantilever. The peak values of plastic work are found around the middle of the cantilever; this is a result of the oscillation in the position of the plastic hinge in *phases 2 and 3.* However, the vast majority of the energy is dissipated in the root rotation and this dominates the overall behaviour. For Example 1, both Refs 16 and 17 found that almost 87% of the available kinetic energy is absorbed in the element adjacent to the root compared with 72% as predicted by the rigid–plastic solution.

Example 2 listed in Table 1 pertains to a bullet impact on a cantilever, with a much smaller mass ratio $\gamma = 0{\cdot}0228$ and much higher initial velocity. The response history follows broadly the same phases as in Example 1. The evolution of the plastic regions for Example 2 are plotted in Fig. 9(a). Again the rigid–plastic analysis shows good agreement with the elastic–plastic solution until the arrest of the travelling hinge following the formation of the reversed hinge at the root. The root rotation phase in the elastic–plastic solution is initiated later than that in the rigid–plastic solution. For Example 2 the majority of the energy (91·75%) is absorbed internally. As shown in Fig. 9(b), the distribution of plastic work approaches more closely that predicted by the rigid–plastic analysis; however, a major discrepancy exists around the central part of the cantilever, where a higher curvature region (or a 'kink' in the final deformed shape) appears as a result of the arrest of the travelling hinge.

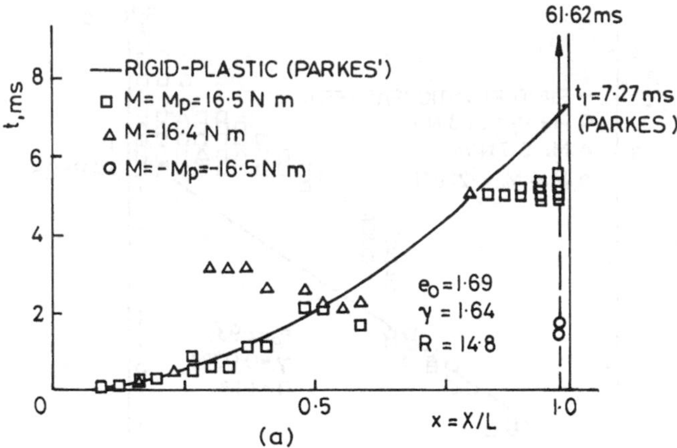

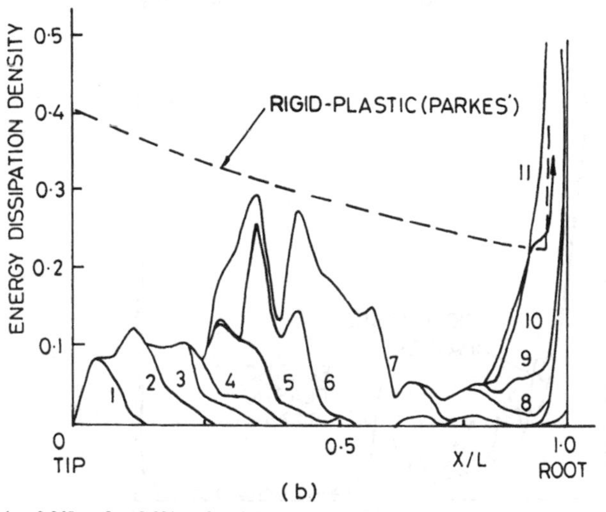

1 = 0.065ms 2 = 0.281ms 3 = 0.506ms 4 = 0.638ms 5 = 0.891ms 6 = 1.158ms
7 = 3.590ms 8 = 4.464ms 9 = 5.014ms 10 = 18.88ms 11 = 61.01ms

Fig. 8. Results for Example 1 ($\gamma = 1{\cdot}64$, $e_0 = 1{\cdot}69$ and $R = 14{\cdot}8$), related to the TB-FE model:[17] (a) evolution with time of plastic regions along the cantilever; dots represent the results obtained using the TB-FE model; the solid curve pertains to the rigid–plastic (Parkes) solution; (b) energy dissipation density along cantilever; solid curves are obtained from the TB-FE model at various instants during response; the dashed curve pertains to the rigid–plastic (Parkes) solution.

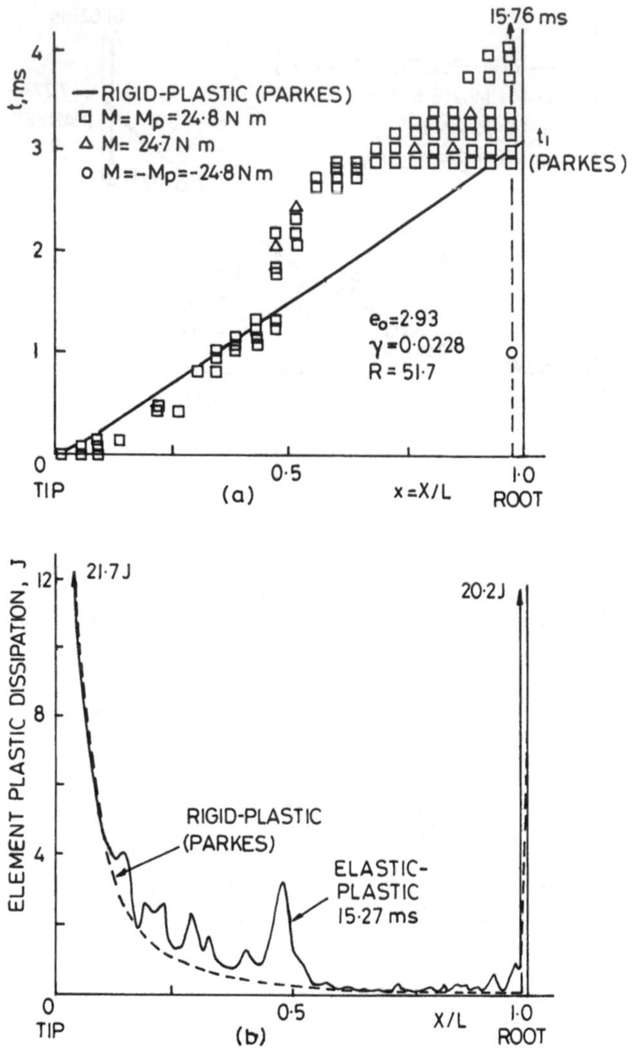

Fig. 9. Results for Example 2 ($\gamma = 0{\cdot}0228$, $e_0 = 2{\cdot}93$ and $R = 51{\cdot}7$), related to the TB-FE model:[17] (a) evolution with time of plastic regions along the cantilever; dots represent the results obtained using the TB-FE model; the solid curve pertains to the rigid–plastic (Parkes) solution; (b) energy dissipation density along cantilever; solid curves are obtained using the TB-FE model with 75 elements at $t = 15{\cdot}27$ ms; the dashed curve pertains to the rigid–plastic (Parkes) solution.

4 A FINITE DIFFERENCE APPROACH USING A LUMPED MASS–SPRING (MS-FD) MODEL

4.1 Major Feature of MS-FD Model

There are various ways to model elastic–plastic cantilevers subjected to impact loading; here we shall illustrate another kind of discrete modelling, namely a lumped mass–spring model in connection with a finite difference procedure (denoted as the MS-FD model), which has been studied by Hou, Yu and Su.[18] As sketched in Fig. 10(a), the MS-FD model discretizes the original cantilever into n elements of equal length, while the mass of each element is assumed to be concentrated at its two ends. The nodes are numbered sequentially, beginning from the tip as the 0th node, where the colliding mass G is added. Hence the model consists of $n + 1$ lumped masses connected by n massless rigid links of length L/n. Since both shear and axial deformations are neglected, only flexural deformation is allowed; and

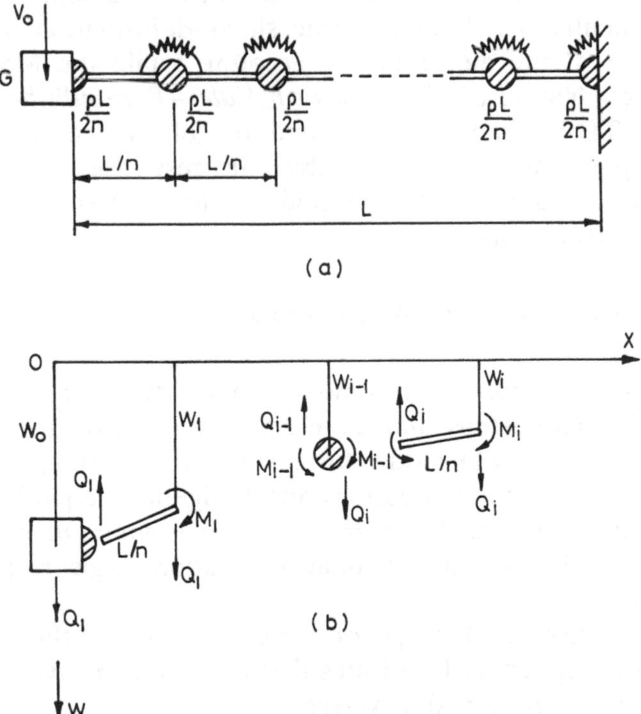

Fig. 10. Mass–spring finite difference (MS-FD) structural model, proposed by Hou, Yu and Su:[18] (a) discretization of a cantilever; (b) positive sense of deflection, shear force and bending moment.

this is represented by the relative rotations between adjacent rigid links. This relative rotation between each pair of links is resisted by an elastic–plastic rotational spring, which reflects the flexural rigidity of the cantilever. This structurally based modelling is convenient for carrying out a semi-analytical study and naturally brings forth non-dimensional variables that represent the system. More elaborate multilayered beam discretizations suggested by Witmer *et al.*[4] and Hashmi *et al.*[19] are more versatile but less easy to interpret for comparisons with analytical results.

The major differences between the TB-FE model and the MS-FD model are as follows:

(i) The TB-FE model employs an *elastic–perfectly plastic* constitutive relation between *stress* σ and *strain* ε (not between bending moment M and curvature κ); while the MS-FD model employs an *elastic–linear hardening* constitutive relation between *moment* M and *relative rotation angle* ψ.

(ii) The TB-FE model essentially represents *Timoshenko beams of rectangular cross-section,* since the beam elements B21 adopted in calculations[16,17] take elastic shear deformation and rotatory inertia of the elements into account; while the MS-FD model reflects feature of *ideal sandwich Euler–Bernoulli beams* only.

(iii) The TB-FE model allows one to follow the *large geometry changes* that occur in the dynamic response of cantilevers to impact; while the MS-FD model is formulated on the basis of *small deflections*.

4.2 Formulation with the MS-FD Model

As given in Appendix I, a combination of dynamic equations with geometrical relationships and an elastic–strain-hardening constitutive relation for the MS-FD model formulates $2n$ governing equations (n second-order differential equations and n algebraic equations) with $2n$ unknown functions ψ_i and m_i ($i = 1, 2, \ldots, n$), where m_i and ψ_i denote non-dimensional moment and relative rotation angle respectively, at the ith node.

An examination of these governing equations and the initial conditions given in Appendix I indicates that there are *five non-dimensional parameters* γ, e_0, R, α and n, where

$$\gamma \equiv \frac{G}{\rho L}, \qquad e_0 \equiv \frac{K_0}{M_p} = \frac{GV_0^2}{2M_p}, \qquad R \equiv \frac{K_0}{U_e^{max}} \tag{6}$$

with α being the coefficient of strain hardening and n the number of links in the model. With initial conditions, a solution can be obtained by numerical integration of the initial value problem using a Runge–Kutta procedure, provided that these five parameters are specified. Calculations confirm that when n is sufficiently large, say $n \geq 20$, then the qualitative features of the solution are insensitive to n.

4.3 A Comparison with the TB-FE Model Using a Numerical Example

In order to compare the results with the solutions obtained for perfectly plastic cantilevers, a small value of $\alpha = 0{\cdot}001$ is taken, which in fact refers to a case of very weak strain hardening. Apart from this, all the parameters of Example 1 listed in Table 1 are used in the calculation of the MS-FD model.

Figures 11(a) and (b) show the evolution of plastic regions and a distribution of the energy dissipation respectively along the cantilever. While they differ from the rigid–plastic (Parkes) solution, a clear similarity is found between the results obtained using the MS-FD model (Fig. 11) and those using the TB-FE model (Fig. 8). It may be concluded that for this particular example both TB-FE and MS-FD models give almost identical results, in spite of some notable differences in modelling, as illustrated in Section 4.1.

4.4 The Influences of Mass Ratio γ and Energy Ratio R

The MS-FD model can be employed to easily examine the influences of the non-dimensional parameters, such as mass ratio γ and energy ratio R, on the dynamic behaviour of impulsively loaded cantilevers.

Figures 12(a) and (b) shows the evolution of plastic regions during the response for the case of heavy colliding mass ($\gamma = 5{\cdot}0$) and that of very light colliding mass ($\gamma = 0{\cdot}025$) respectively. It can be seen from Fig. 12(a) that when $\gamma = 5{\cdot}0$ and $R = 5$, a plastic region first appears at the middle part of the cantilever for a very short time, and then reappears at a region close to the root at a time later than that predicted by the rigid–plastic solution; this is more like a modal solution, and differs from the rigid–plastic solution. When $\gamma = 5{\cdot}0$ but $R = 15$, a plastic region also first appears at the middle part of the cantilever, but there are indications that the plastic region will continue to move (backwards then forwards) along the cantilever until it finally reaches the root at a time similar to that predicted by the rigid–plastic solution. It is noted that both reverse yielding at the root and the

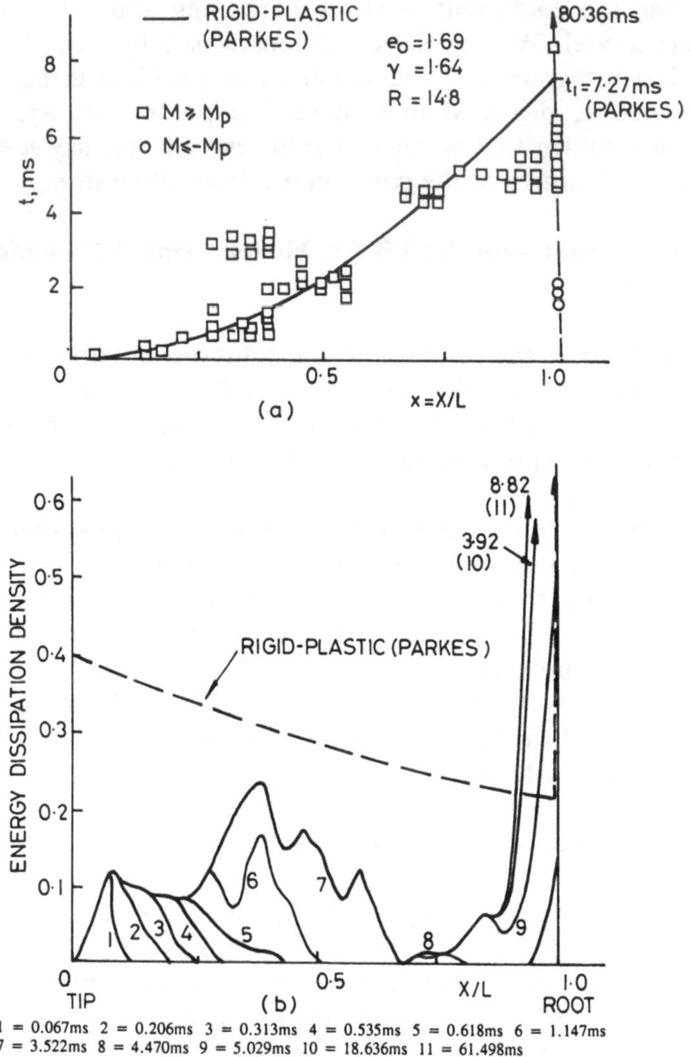

Fig. 11. Results for Example 1 ($\gamma = 1{\cdot}64$, $e_0 = 1{\cdot}69$ and $R = 14{\cdot}8$), related to the MS-FD model:[18] (a) evolution with time of plastic regions along the cantilever; dots represent the results obtained using the MS-FD model; the solid curve pertains to the rigid–plastic (Parkes) solution; (b) energy dissipation density along the cantilever; solid curves are obtained from the MS-FD model at various instants during response; the dashed curve pertains to the rigid–plastic (Parkes) solution.

arresting of the travelling hinge occur only in the case of $R = 15$; in case of smaller R ($R = 5$) the root section directly enters into a modal-type yielding without prior reverse yielding. The latter case was not reported in the previous finite element approaches, since they only examined large values of R ($R = 14{\cdot}8$ and $51{\cdot}7$).

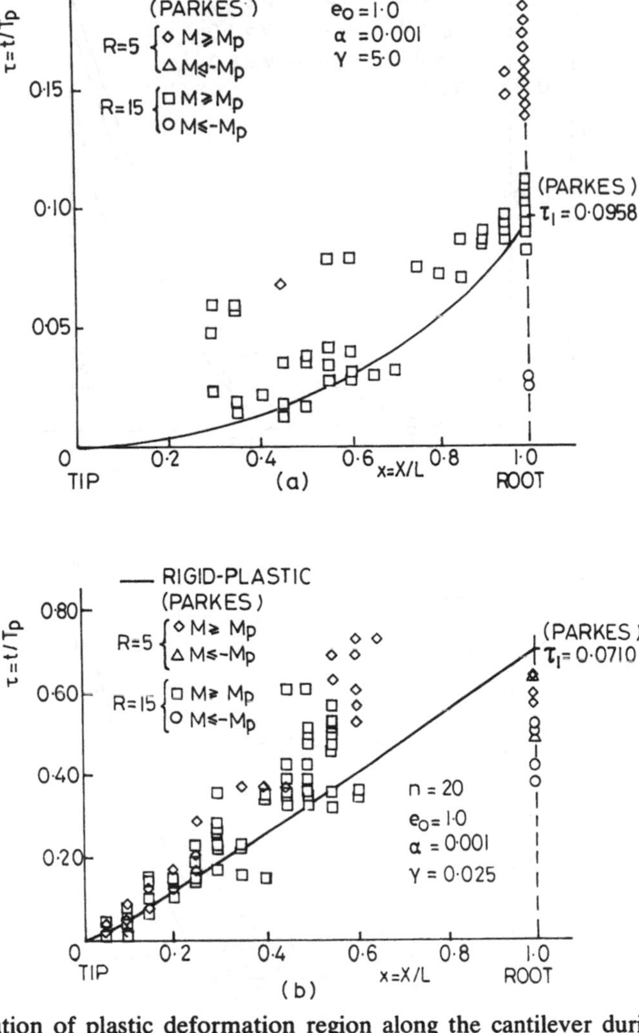

Fig. 12. Evolution of plastic deformation region along the cantilever during dynamic response, obtained from the MS-FD model;[18] $e_0 = 1$, $\alpha = 0{\cdot}001$, $R = 5$ and 15: (a) $\gamma = 5$; (b) $\gamma = 0{\cdot}025$.

As shown in Fig. 12(b), when the mass ratio γ is very small, e.g. $\gamma = 0{\cdot}025$ (which is very close to the value $\gamma = 0{\cdot}0228$ adopted in Example 2), the travelling hinge reaches about 0·5–0·6 of the whole length, and disappears afterwards. This implies that the response pattern is quite similar to *Pattern* II_b in Section 5 below; that is, only a part of the cantilever undergoes plastic deformation when the colliding mass is very light or the cantilever itself is very long (both result in a very small mass ratio γ). Also, in this case it is expected that most of

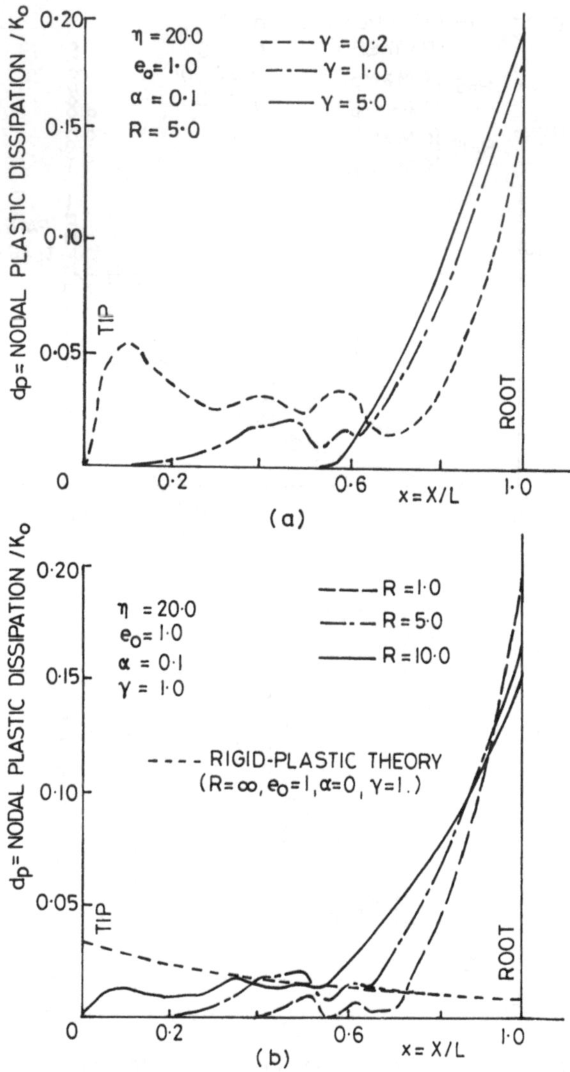

Fig. 13. Distribution of energy dissipation along cantilever, obtained from the MS-FD model:[18] (a) influence of mass ratio γ; (b) influence of energy ratio R.

input energy is dissipated in the interior sections, while a small fraction of energy is dissipated at the root only by its reverse yielding.

Figures 13(a) and (b) demonstrate the influences of mass ratio γ and energy ratio R respectively on the distribution of the nodal plastic dissipation d_p, which is proportional to the final curvature remaining in the cantilever. It can be seen that more energy is dissipated in the outer half of the cantilever (near the tip) when the mass ratio γ is smaller or the elastic energy ratio R is larger.

5 EFFECT OF ELASTIC DEFORMATION AT THE ROOT OF A CANTILEVER

5.1 A Rigid–Plastic Cantilever with an Elastic–Plastic Spring at the Root

In order to study elastic effects in the impulsively loaded cantilever problem, we must provide the system with a certain capacity for storing elastic deformation energy. The single-DoF mass–spring system studied by Symonds *et al.* (see Section 2.1) merely refers to the modal motion of a structure. For a dynamically loaded cantilever this single-DoF system is in fact equivalent to setting an elastic–plastic rotational spring at the root, while the cantilever itself is regarded as a rigid body. Combining this elastic–plastic rotational spring at the root and the rigid–perfectly plastic beam in classical analyses, Wang and Yu[20] proposed a simplified structural model that has a certain capacity for storing elastic deformation energy in addition to its essential rigid–plastic property. This model is certainly not as good as the previous MS-FD model in representing a real elastic–plastic cantilever beam, but it could lead to much simpler analyses and reflect some basic feature of the dynamic response of elastic–plastic cantilevers.

The simplified structural model proposed by Wang and Yu[20] is shown in Fig. 14(a); it consists of a *rigid*-perfectly plastic cantilever and an *elastic*–perfectly plastic rotational spring at the root. The material is assumed to be rate-independent; the influence of shear on yielding is neglected; the deflection is assumed to be small, and the colliding body is treated as a particle with mass G. Under these assumptions, the dynamic problem of this structural model is almost the same as that analysed by Parkes[15] except that the bending moment at the root is related to the root rotation by a moment–rotation-angle relationship for an elastic–perfectly plastic spring; see Fig. 14(b), where C is the elastic constant of the spring and ψ is the root rotation angle, composed of an elastic part ψ_e and a plastic part ψ_p. During loading, the moment at the root M_B is elastic for rotations $\psi \leqslant M_p/C$ and fully plastic (M_p) for $\psi \geqslant M_p/C$. Here M_p is the fully plastic bending moment for any cross-section of the cantilever.

Based on the idea that the entire elastic capacity of the cantilever $U_e^{max} = LM_p^2/2EI$ is represented by the elastic capacity of the root spring $U_e^{max} = M_p^2/2C$, the spring constant C can be determined as $C = EI/L$.

According to the assumptions made above, the impact at the tip causes plastic bending of the cantilever, as well as elastic–plastic deformation of the rotational spring at the root. This leads to a

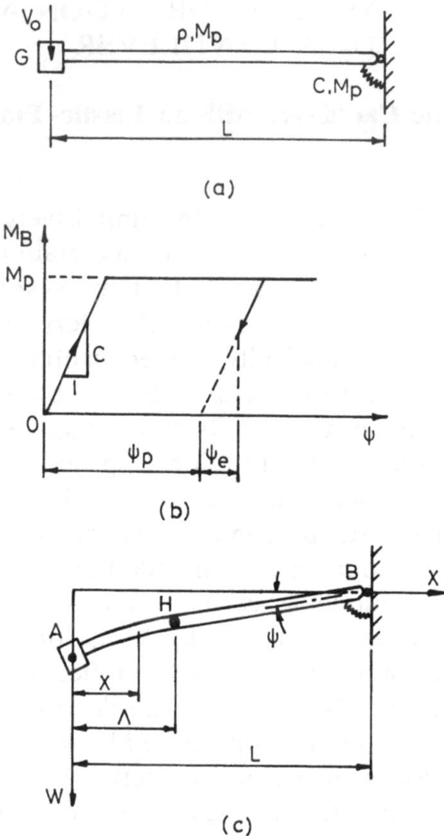

Fig. 14. (a) A rigid–plastic cantilever with a rotational spring at the root—a model proposed by Wang and Yu.[20] (b) The elastic–plastic characteristic of the rotational spring. (c) Deformation mechanism with a travelling hinge at H and rotation at root B.

deformation mechanism with a travelling hinge and a root rotation, as shown in Fig. 14(c). As given in Appendix II, a formulation with this deformation mechanism results in a set of *three* second-order differential equations, together with a constitutive relation for the rotational spring at the root. The governing equations and initial conditions only consist of *three non-dimensional parameters,* namely γ, e_0 and R.

5.2 Response Patterns

The appearance of an elastic deformation at the root notably changes the dynamic behaviour from that of the rigid–perfectly plastic idealization. The latter always has a travelling hinge that propagates from the impact point to the root of the cantilever; then it is followed by a modal phase of motion. In contrast, the present model exhibits two distinct patterns of response.

Pattern I. A travelling plastic hinge moves from the tip towards the root, but before it reaches the root, the rotation at the root exceeds the elastic range $\psi > M_p/C$. In this case *two hinges* occur simultaneously in the cantilever. With this pattern, the travelling hinge always reaches the root of the cantilever; thereafter the deformation enters a modal phase where the remaining kinetic energy is partly dissipated and partly transformed into an elastic vibration by deformation at the root.

Pattern II. A travelling plastic hinge moves from the tip towards the root, but at a certain *interior* section the hinge motion halts and the hinge disappears; meanwhile the root rotation remains elastic. After the disappearance of the travelling hinge, the root *may* or *may not* enter into a plastic state, depending on the remaining kinetic energy. Thus *Pattern II* can be further separated into Pattern II_b and *Pattern* II_a, depending on the largest root rotation.

Numerical examples have established that the inverse energy ratio R^{-1} has the largest effect on the pattern of response. A small value of R^{-1} results in a *Pattern I* response that is somewhat similar to the rigid–perfectly plastic solution. When the energy ratio R^{-1} is very small, the root enters the plastic state soon after the hinge begins to move from the tip. In this case the shear force at the hinge vanishes, so there is no angular acceleration of the segment between the hinge and the root; this is just Parkes' solution. On the other hand, a large value of R^{-1} results in *Pattern II.* Hence, if the energy stored in elastic deformation at the root is relatively large, only a part of the cantilever undergoes plastic deformation.

The variation of the hinge location with time is plotted in Fig. 15, which shows that the value of R^{-1} has only a minor effect on the speed of the hinge before it reaches $\lambda \equiv \Lambda/L \approx 0{\cdot}4$; thus the elastic–plastic response of the present model is not very different from the rigid–perfectly plastic (Parkes) solution during an initial period. According to the rigid–perfectly plastic solution, the travelling hinge passes through the *entire* cantilever, whatever the value of γ. The present results indicate, however, that if the input energy is relatively small (compared with the elastic capacity of the cantilever) then the travelling hinge passes through only *a part* of the cantilever before it disappears (*Pattern II*). Thereafter, the entire cantilever simply rotates about the root. If the energy is small at the beginning of the modal phase, the root

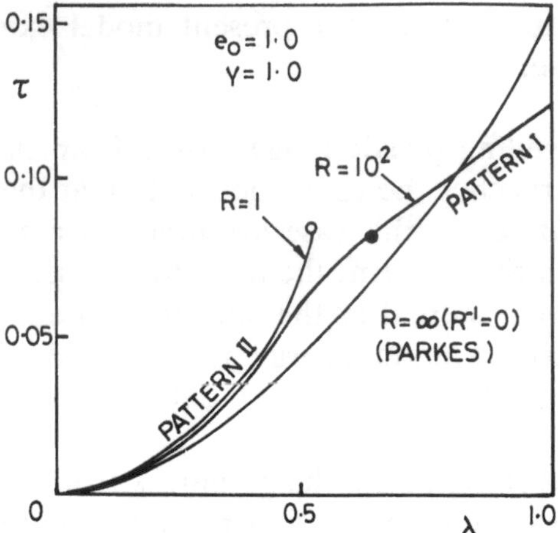

Fig. 15. Travelling of plastic hinge along cantilever; $e_0=1$, $\gamma=1$ and $R^{-1}=0$, 10^{-2} and 1·0.

deforms in the elastic range (*Pattern* II_a); if this energy is large, the root deforms plastically (*Pattern* II_b). Figure 15 also shows that for either *Pattern I* or *Pattern II* the hinge slows down near the middle of the cantilever. This is similar to the results obtained using finite element approaches; it results in a high-curvature region (or a 'kink') around the middle part of the cantilever in its final configuration.

For $e_0=1$, Fig. 16 shows a map in the R^{-1}–γ plane, which illustrates the regions of occurrence of these patterns. When $R^{-1}>1$, the elastic

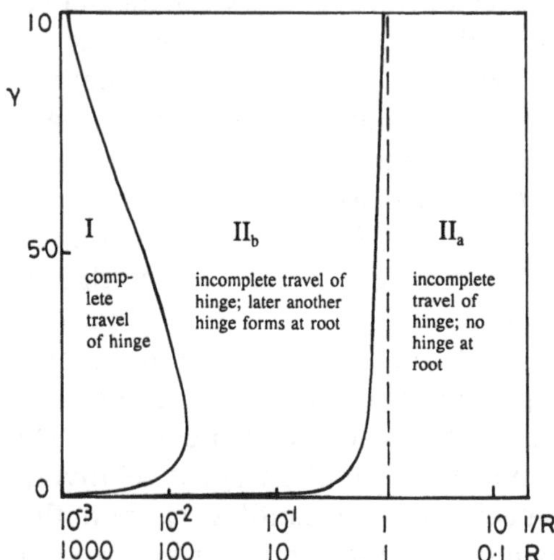

Fig. 16. A map in the R^{-1}–γ plane, showing the parameter regions in which various response patterns occur; $e_0=1$.

capacity is beyond the input energy, so the response is always *Pattern* II_a. If $R^{-1} < 1$ but the mass ratio $\gamma \gg 1$, the response is always *Pattern* II_b. In this case most of the input energy is dissipated plastically at the root in a modal phase of deformation. Finally, if R^{-1} is very small (or R is very large) while the mass ratio γ is moderate, the response is *Pattern I*.

It is interesting to compare this conclusion with the previous results from finite element analyses. For instance, the parameters adopted for Example 1 in Table 1 seem to fall in the region of modal response, *Pattern* II_b. Reid and Gui[17] pointed out that the final modal response is delayed by the interaction between the reflected elastic flexural wave and the travelling hinge. This feature is clearly described by the present Pattern II_b.

In most of the results shown here we consider $e_0 = 1$, i.e. $K_0 = M_p$. Wang and Yu[20] has also reported that the effect of e_0 on the pattern of response is somewhat similar to that of R^{-1}. For specified values of γ and R^{-1}, the response tends to *Pattern I* or *Pattern II* if e_0 is very small or very large respectively. Also, they found that e_0 has little influence on the total energy dissipation d_p.

6 REMARKS ON A CANTILEVER UNDER IMPACT

(1) Unlike a single-DoF systems as discussed in Section 2.1, a structure, even as simple as a cantilever, has an infinite (or by approximation, a large) number of DoF, so that a mere comparison on the final deformation at a particular point (here it is the final deflection at the tip) is *insufficient* for assessing different theories. We have shown that the elastic–plastic behaviour of a cantilever (or a simplified structural model of it) may differ significantly from the corresponding rigid–perfectly plastic solution in many aspects, such as the distribution of bending moment, the evolution of plastic regions, the energy dissipation pattern, the spatial distribution of plastic dissipation and the final deformed shape.

(2) For a cantilever struck by a particle mass at its tip, the MS-FD model (Section 4) and the simplified structural model with a root spring (Section 5) indicate that, besides the energy ratio R, the mass ratio $\gamma = G/\rho L$ and the input energy ratio $e_0 = K_0/M_p$ both play significant roles in governing the dynamic elastic–plastic response of cantilevers. The combined effect of the energy ratio R and the mass ratio γ on the characteristics of the dynamic behaviour of impulsively loaded elastic–

Table 2
Features of Dynamic Response of Impulsively Loaded Cantilevers, Affected by Energy Ratio R and Mass Ratio γ

	Large γ	*Small* γ
Very large R (e.g. $R > 100$)	Similar to the rigid–plastic (Parkes) solution;[15] a travelling hinge moves along the whole cantilever	
Large R (e.g. $10 < R < 100$)	As Example 1 in Table 1, described in detail in Section 3.2; the root dissipates most energy; similar to *Pattern* II_b in Section 5	As Example 2 in Table 1; incomplete travelling of hinge; the root dissipates little energy; similar to *Pattern* II_a in Section 5
Small R (e.g. $2 < R < 5$)	Modal-like response; the root dissipates almost all the energy	Similar to Example 2 in Table 1; hinge travelling is delayed.
Very small R (e.g. $0{\cdot}5 < R < 1$)	Almost entirely elastic response, with a little plasticity at the root	Very localized plastic deformation at the region near the tip

plastic cantilevers may be summarized as in Table 2, which clearly confirms the essential influence of the mass ratio γ on the deformation mechanism.

(3) A comparison between the results obtained from the MS-FD model (Section 4) and the TB-FE model (Section 3) indicates that the basic feature of the dynamic response of elastic–plastic cantilevers is *not* significantly altered by some minor differences in modelling, such as the difference between a Timoshenko beam of rectangular cross-section and an ideal sandwich Euler–Bernoulli beam, and that between small deflection and large deflection formulations, etc. Hence simplified structural models do provide useful tools for studying the dynamic behaviour of elastic–plastic structures. Compared with finite element approaches, simplified models have some notable advantages, including the following.

(i) A few non-dimensional parameter groups governing the dynamic behaviour of structural models have naturally emerged from the formulation, so one can explore how each non-dimensional group affects the response feature.
(ii) By means of semi-analytical approaches to the models, various response patterns can be distinguished; this contributes a deeper insight into the intrinsic mechanism in which the elastic deformation of the models is involved.

(iii) Compared with finite element approaches, semi-analytical approaches to the simplified models require much less time for computation and data processing; this is particularly beneficial to engineering applications during the preliminary design stage.

7 MORE GENERAL DISCUSSIONS

A straight cantilever is a very special structure in which axial forces are not developed during its large flexural deformation. However, for most practical structures (e.g. clamped beams, arches, plates and shells) the large deflection caused by intense dynamic loading does induce *axial* or *membrane forces* because inextensional deformation modes are not kinematically admissible. In these cases elastoplasticity appears in both flexural and extensional deformations, so the interaction makes problems extremely complicated, and the application of analytical methods becomes more difficult. Some numerical studies (see Refs 4 and 19) have given examples of large elastic–plastic deflections of dynamically loaded clamped beams, circular plates and hemispherical shells, since the multilayer structural discretization adopted in these numerical schemes allows the incorporation of both extensional and flexural deformations.

Yankelevsky[21] proposed a new approximate model to analyse the large deformation dynamic response of axially constrained elastic–plastic beams. In contrast with conventional finite difference models, his model beam is composed of two rigid parts connected by a gap of zero width built of fibres having an imaginary (or *equivalent*) length, as shown in Fig. 17. This equivalent length governs the strains and stresses

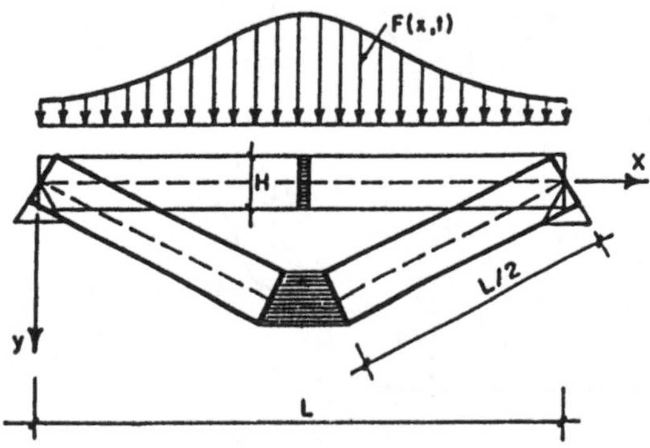

Fig. 17. An approximate model of an elastic–plastic beam in large deflection, proposed by Yankelevsky.[21]

in the beam, for equal deflections in the real and model beams, and is found to be almost constant in the elastic and elastic–plastic domains. The final deflections predicted by this model showed a good correspondence with test results, while it may also be used to calculate a complete response time history. Yankelevsky[22] later extended his model to the analysis of the elastic–plastic response of rectangular plates under blast loading. The plate is modelled with rigid segments interconnected by hinge lines that yield a roof-shaped displacement field. It is still possible to find the 'equivalent' length of the fibres at the hinge lines by equating deflections in the real and model plates, but this requires a rather complicated assessment.

In recent years some large general purpose finite element numerical codes have become available for calculating the large-deflection dynamic response of elastic–plastic structures subjected to arbitrary loading. Many numerical examples have been reported, but it is well known that in some cases the calculated results are sensitive to minor differences in the numerical schemes. A typical example was given by Symonds and Yu,[23] who examined a pin-ended elastic–plastic beam subjected to a uniformly distributed rectangular pressure pulse over the whole span (Fig. 18a). Using several well-established numerical codes, they identified *three elastic–plastic response patterns* according to the variation of the midpoint deflection w with time. If the loading pulse is beyond the static collapse load but remains moderate, *Type I* response occurs, i.e. the beam vibrates between both sides of its original straight configuration, but the final deflection at the midpoint is positive (i.e. in the same

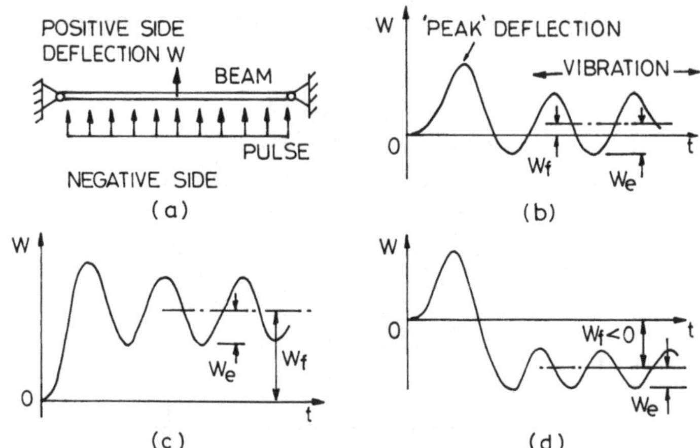

Fig. 18. (a) A pin-ended elastic–plastic beam subjected to a uniformity distributed pulse, studied by Symonds and Yu.[23] (b) Type I response. (c) Type II response. (d) Type III (anomalous) response.

direction as that of the external loading), as sketched in Fig. 18(b). If the pulse is intense, then a *Type II* response occurs; that is, the beam vibrates within the positive side, as sketched in Fig. 18(c); the later part of the response is similar to the vibration of a shallow arch. A quite unexpected result was reported by Symonds and Yu,[23] who found in addition to these two response patterns that some intermediate loading parameters can result in an *anomalous* (*Type III*) response pattern, as sketched in Fig. 18(d); that is, the beam first deforms in the positive side then rebounds to the negative side and continues to vibrate within the negative side. After the elastic vibration is eventually damped out, the final defection of the beam is expected to be negative (i.e. in the direction opposite to the external loading), which confounds intuition. This counter-intuitive response pattern is a result of a complex interaction between the flexural and axial elastic–plastic deformations in a beam.

To explain this anomalous phenomenon, Symonds and Yu[23] employed a simple Shanley-type model of a beam (Fig. 19a), which is essentially a single-DoF system but retains the elastic–plastic properties of a material and the interaction between the bending moment and the axial force. They found that the response pattern is strongly dependent on the 'peak' deflection or the 'release' angle ϕ_0 at which reverse motion starts, and the counter-intuitive response pattern only occurs

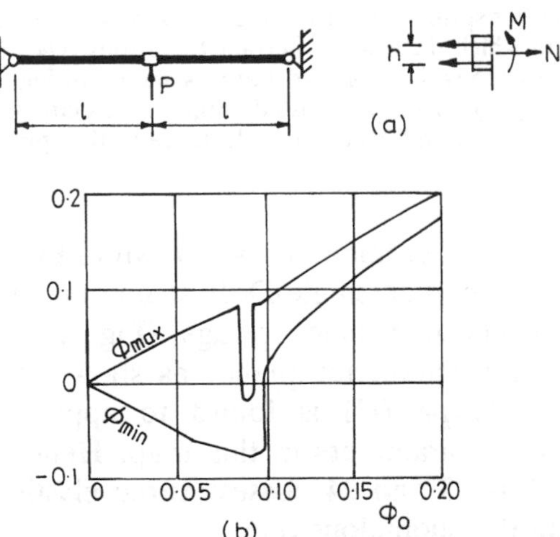

Fig. 19. (a) A single-DoF Shanley-type model proposed by Symonds and Yu.[23] (b) The maximum and minimum angular displacements in the final elastic vibration phase of the Shanley-type model, as functions of the release angle ϕ_0, for $1/\varepsilon_Y = 400$ and $h/l = 0{\cdot}0271$; note that the narrow 'slot' in the ϕ_{max} curve that represents the anomalous response pattern.

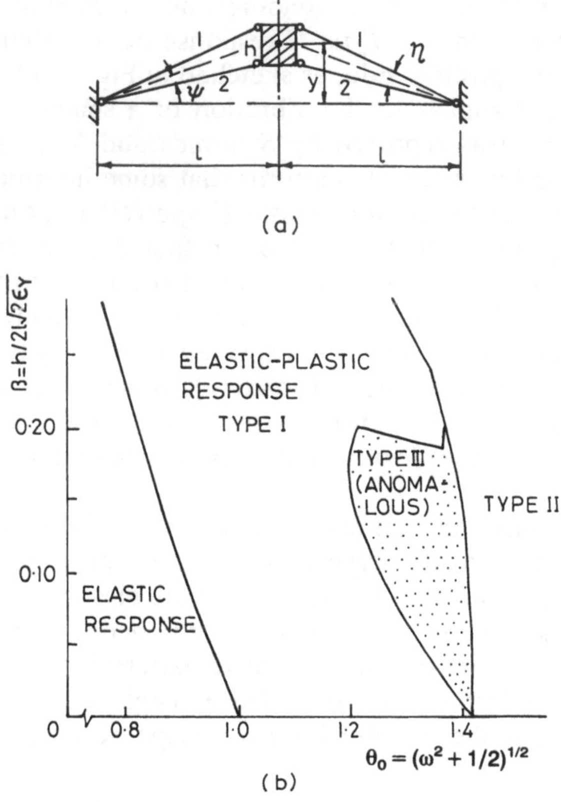

Fig. 20. (a) A double-tier spring structural model proposed by Yu and Xu.[24] (b) A map in parameter plane, in which the regions pertain to various types of response patterns; ε_Y is the yield strain of the springs, ψ and l are as shown in Fig. 20(a), $\theta = \psi/\sqrt{2\varepsilon_Y}$, $\omega = d\theta(0)/d\tau$ is the non-dimensional initial angular velocity of the mass m, $\tau = t\sqrt{4N_Y/ml}$, and N_Y is the yield force of the springs.

within a narrow range of parameters, as shown in Fig. 19(b). Later, Yu and Xu[24] suggested another single-DoF structural model, which contained double-tier elastic–plastic springs (Fig. 20a). Their analysis results in a map in a parameter plane, as shown in Fig. 20(b). The anomalous pattern (*Type III*) is found to appear only in a well-determined region of parameters in this map. Hence both single-DoF models given in Refs 23 and 24 possess three elastic–plastic response patterns, including the anomalous one.

A possible way to summarize these response patterns is to make a comparison between the final deflection w_f and the amplitude w_e of the final elastic vibration by introducing an index $\eta \equiv w_f/w_e$. Obviously, the responses of *Types I, II and III* sketched in Fig. 18(b), (c) and (d) pertain $0<\eta<1$, $\eta>1$ and $\eta<0$ respectively; while $\eta=0$ represents a pure elastic response. *Types I and II* are quite normal; for instance, the results shown in Figs 1 and 5 both are *Type II* ($\eta>1$). In general, η is expected to increase with increasing R, but the existence of the anomalous (*Type III*) response pattern with $\eta<0$ does violate this general observation. For instance, a reconsideration of the example shown in Fig. 19(b) produces a rather complex relationship between $\eta \equiv w_f/w_e$ and R, as depicted in Fig. 21. Besides showing the appearance of negative η (*Type III*), this figure also demonstrates a jump in the value of η at the transition from *Type I* to *Type II*.

Lee and Symonds[25] recently reported that a two-DoF model of a fixed-ended beam subjected to impulsive loading can display a *chaotic response*. It is evident that the more DoF a dynamic system possesses, the more complex is the response pattern. Therefore we have reason to believe that continuous structures could display a great deal of variety in their dynamic elastic–plastic response patterns. Symonds and his collaborators also noticed that, along with the extreme sensitivity of the anomalous phenomenon to structural and loading parameters, damping

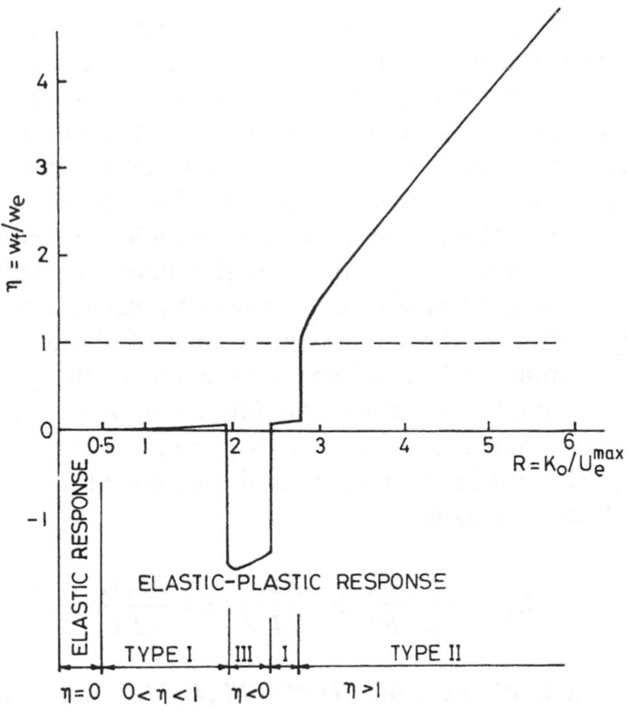

Fig. 21. The variation of $\eta = w_f/w_e$ with the energy ratio $R = K_0/U_e^{max}$ for the example given in Fig. 19(b).

plays a vital role in determining the response pattern of the single-DoF model. Since damping is a complex factor in reality, the response could indeed be unpredictable in certain ranges of the parameters. Recently two groups of researchers[26,27] have utilized different loading techniques and have independently reported their experimental observations related to the anomalous response of beams. In a couple of specimens the midsection undergoes a *Type III* response history, while the remainder of the cantilevers undergo a *Type II* response.

8 CONCLUDING REMARKS

(1) It has been pointed out in Section 1 that the energy ratio R defined by eqn (1) is an important index in judging the validity of the rigid–plastic idealization. Although $R \gg 1$ seems to be a reasonable requirement from the energy point of view, the simple use of this condition could be misleading. The maximum elastic capacity U_e^{max} expression in eqn (1) is usually estimated by assuming that all parts of the structure reach the yield or fully plastic state *simultaneously*; for example, see eqn (5) for a cantilever. This simultaneous yielding state is unrealistic when a structure mainly deforms in a *flexural* mode. The overestimation of U_e^{max} results in a underestimation of the value of R. This becomes severe for a large structure subjected to a local loading. For instance, for a very long cantilever subjected to an impact at the free end, the value of R estimated by eqn (5) would be very small, although the rigid–plastic solution still provides a reasonable approximation for the deformation in the region near the impact point. A better estimate of U_e^{max} could be obtained by calculating the elastic strain energy related to the fundamental mode of elastic vibration, for which the maximum strain is taken to be equal to the yield strain of material. For example, by taking the fundamental elastic mode of a simply supported beam of length L as $y = y_0 \sin(\pi x/L)$ and $\kappa_{max} = \kappa(L/2) = M_Y/EI$, with M_Y being the yield moment of the cross-section, the related elastic energy is

$$U_e^{max} = \int_0^L \frac{M_Y^2}{EI} \sin^2\left(\frac{\pi x}{L}\right) dx = \frac{M_Y^2 L}{4EI}$$

This is just a half of the value $U_e^{max} = M_Y^2 L/2EI$ that is obtained by assuming that $\kappa_Y = M_Y/EI$ simultaneously appears at every cross-section of the beam.

(2) It has been emphasized that $R \gg 1$ is not a sufficient condition for entirely neglecting elastic effects. The pulse shape, rise time and pulse duration all significantly influence the final deformation of pulse-loaded structures. In cases of impact loading the ratio of impinging mass to the mass of the structure itself plays an important role, as brought to light by Table 2 for a typical cantilever impact problem.

(3) A comparison between the elastic–plastic solution and the rigid–plastic solution for the same structure indicates that the involvement of elastic deformations changes not only the magnitude of the final deformation, but also the deformation mode and energy dissipation pattern. For a many-DoF system or a continuous structure the elastic–plastic response generally displays much more complex deformation modes and energy dissipation patterns in comparison with the rigid–plastic response. For example, the anomalous response pattern illustrated in Section 7 cannot be found for a rigid–plastic beam or a rigid–plastic model, although the other conditions remain unchanged. Strictly speaking, however, a rigid–plastic analysis should not be used for relatively small values of R when this anomalous behaviour occurs.

(4) For designers who are concerned with structural crashworthiness, energy absorbing systems or structural failure under dynamic loading, the rigid–plastic idealization may serve as a *first-order approximation*, and it usually provides a *good prediction* of the final deformations for relatively large value of R; for example, the error could be less than 20% if $R > 5$. The rigid–plastic methods therefore still remain valuable in design for most structures under intense dynamic loading. However, the analyses in this chapter do suggest that the following checks should be made before directly applying the rigid–plastic results to structural problems.

(i) Is the magnitude of R large enough (say $R > 5$) in the problem? Do the values of other important parameters (rise time of the pulse, pulse duration, mass ratio etc.) fall within reasonable ranges?

(ii) Has engineering experience or experimental results shown any particular deformation mode that is disregarded in the rigid–plastic analysis? If so, examine to see if these modes occur in the elastic–plastic response.

(iii) Although a single- or two-DoF system can be taken as a preliminary model for an actual structure (e.g. a modal solution), it should be borne in mind that an increase in the numbers of DoF in a system is likely to lead to more complex deformation patterns.

ACKNOWLEDGEMENTS

The author wishes to thank Dr W. J. Stronge of Cambridge University and Dr X. D. Wang of Peking University for many helpful discussions with them during the preparation of the first draft of this chapter. The author is indebted to Professor T. Wierzbicki and Professor N. Jones for their stimulatiang comments and valuable suggestions. The author would like to thank the Royal Society of London for its support during the period of his stay in Cambridge, when the first draft was prepared. The assistance offered by Mr J. J. Batty of UMIST in preparing tracings is gratefully acknowledged.

REFERENCES

1. Duwez, P. E., Clark, D. S. & Bohnenblust, H. F., The behavior of long beams under impact loading. *Trans. ASME, J. Appl. Mech.*, **72** (1950) 27–34.
2. Boussinesq, M. J., *Application des potentiels a l'étude de l'équilibre et du mouvement des solides élastiques*. Gauthier-Villars, Paris, 1885, p. 444.
3. Conroy, M. F., Plastic deformation of semi-infinite beams due to transverse impact. *J. Appl. Mech.*, **23** (1956) 239–43.
4. Witmer, E. A., Balmer, H. A., Leech, J. W. & Pian, T. H. H., Large dynamic deformations of beams, rings, plates and shells. *AIAA J.*, **1** (1963) 1848–57.
5. Lee, E. H. & Symonds, P. S., Large plastic deformations of beams under transverse impact. *J. Appl. Mech.*, **19** (1952) 308–14.
6. Conroy, M. F., Plastic–rigid analysis of long beams under transverse impact loading, *J. Appl. Mech.*, **19** (1952) 465–70.
7. Symonds, P. S., Survey of method of analysis for plastic deformation of structures under dynamic loading. Division of Engineering Report BU/NSRDC/1-67, Brown University, Providence, RI, 1967.
8. Symonds, P. S., Elastic–plastic deflections due to pulse loading. In *Dynamic Response of Structures*, ed. G. Hart. ASCE, New York, 1981, pp. 887–901.
9. Symonds, P. S., A review of elementary approximation techniques for plastic deformation of pulse-loaded structures. In *Metal Forming and Impact Mechanics*, ed. S. R. Reid. Pergamon Press, Oxford, 1985, pp. 175–94.
10. Symonds, P. S. & Frye, C. W. G., On the relation between rigid–plastic and elastic–plastic predictions of response to pulse loading. *Int. J. Impact Engng*, **7** (1988) 139–49.
11. Baker, W. E., The elastic–plastic response of thin spherical shell to internal blast loading. *J. Appl. Mech.*, **27** (1960) 139–44.
12. Duffey, T., Significance of strain-hardening and strain-rate effects on the transient response of elastic–plastic spherical shells. *Int. J. Mech. Sci.*, **12** (1970) 811–25.

13. Jones, N., *Structural Impact.* Cambridge University Press, 1989, Sect. 5.6.
14. Stronge, W. J. & Hua Yunlong, Elastic effects on deformation of impulsively loaded elastoplastic beams. *Int. J. Impact Engng,* **9** (1990) 253–62.
15. Parkes, E. W., The permanent deformation of a cantilever struck transversely at its tip. *Proc. R. Soc. Lond.,* **A228** (1955) 462–76.
16. Symonds, P. S. & Fleming, Jr, W. T., Parkes revisited: on rigid–plastic and elastic–plastic dynamic structural analysis. *Int. J. Impact Engng,* **2** (1984) 1–36.
17. Reid, S. R. & Gui, X. G., On the elastic–plastic deformation of cantilever beams subjected to tip impact. *Int. J. Impact Engng,* **6** (1987) 109–27.
18. Hou, W. J., Yu, T. X. & Su, X. Y., A study of the dynamic elastic–plastic response of impulsively loaded cantilever beam by a lumped masses-springs model. (In Chinese.) Submitted to *Acta Mech. Solida Sinica.*
19. Hashmi, S. J., Al-Hassani, S. T. S. & Johnson, W., Large deflexion elastic–plastic response of certain structures to impulsive load: numerical solutions and experimental results. *Int. J. Mech. Sci.,* **14** (1972) 843–60.
20. Wang, X. D. & Yu, T. X., Parkes revisited: effect of elastic deformation at the root of a cantilever beam. *Int. J. Impact Engng,* **11** (1991) 197–209.
21. Yankelevsky, D. Z. & Boymel, A., Dynamic elasto-plastic response of beams—a new model. *Int. J. Impact Engng,* **2** (1984) 285–98.
22. Yankelevsky, D. Z., Elasto-plastic blast response of rectangular plates. *Int. J. Impact Engng,* **3** (1985) 107–19.
23. Symonds, P. S. & Yu, T. X., Counterintuitive behavior in a problem of elastic–plastic beam dynamics. *J. Appl. Mech.,* **52** (1985) 517–22.
24. Yu, T. X. & Xu, Y., The anomalous response of an elastic–plastic structural model to impulsive loading. *J. Appl. Mech.,* **56** (1989) 868–73.
25. Lee, J.-Y. & Symonds, P. S., Extended energy approach to chaotic elastic–plastic response to impulsive loading. *J. Appl. Mech.,* **59** (1992) 139–157.
26. Li, Q. M., Zhao, L. M. & Yang, G. T., Experimental results on the counterintuitive behaviour of thin clamped beams subjected to projectile impact. *Int. J. Impact Engng,* **11** (1991) 341–8.
27. Kolsky, H., Rush, P. & Symonds, P. S., Some experimental observations of anomalous response of fully clamped beams. *Int. J. Impact Engng,* **11** (1991) 445–56.

APPENDIX I

Referring to Fig. 10(b), the equations for nodal shear force Q_i and bending moment M_i give

$$Q_1 = \left(G + \frac{\rho L}{2n}\right)\frac{\mathrm{d}^2}{\mathrm{d}t^2}W_0, \qquad Q_i - Q_{i-1} = \frac{\rho L}{n}\frac{\mathrm{d}^2}{\mathrm{d}t^2}W_{i-1} \quad (i = 2, 3, \ldots, n) \tag{I.1}$$

$$M_i - M_{i-1} = -\frac{L}{n}Q_i \quad (i = 1, 2, \ldots, n) \tag{I.2}$$

Under the assumption of small deflections, the transverse nodal displacement W_i is related to the rotation angle θ_i of each link by

$$\theta_i = n(W_{i-1} - W_i)/L \quad (i = 1, 2, \ldots, n) \tag{I.3}$$

Thus at the ith joint the *relative* rotation ψ_i between the adjacent links i and $i+1$ is

$$\psi_i \equiv \theta_i - \theta_{i+1} = \frac{n(W_{i+1} - 2W_i + W_{i-1})}{L} \quad (i = 1, 2, \ldots, n-1), \qquad \psi_n = \theta_n \tag{I.4}$$

Assume that the rotational springs between every pair of adjacent links have an elastic–linearly strain-hardening relation between the moment and the relative rotation angle; that is,

$$M_i(\psi_i) = \begin{cases} C\psi_i & (\psi_i \leqslant \psi_p) \\ M_p + \alpha C(\psi_i - \psi_p) & (\psi_i \geqslant \psi_p,\ \dot{\psi}_i > 0) \end{cases} \tag{I.5a}$$

After a joint has a largest relative rotation $\psi_i^* > \psi_p$, the moment–rotation relation for unloading is

$$M_i(\psi_i) = M_i(\psi_i^*) - C(\psi_i^* - \psi_i) \quad (\psi_i^* > \psi_i > \psi_i^* - 2\psi_p) \tag{I.5b}$$

where the range of applicability is limited to $\psi_i > \psi_i^* - 2\psi_p$ by the Bauschinger effect for kinematic hardening.

For a cantilever the boundary conditions are

$$Q_0 = M_0 = 0, \qquad W_n = 0 \tag{I.6}$$

In the original problem the concentrated mass G acquires a transverse velocity V_0 at time $t = 0$ owing to impulsive loading. To retain the same kinetic energy, for the present MS-FD model this results in initial conditions

$$\left.\begin{aligned} W_0 = 0, \quad & \frac{dW_0}{dt} = \frac{V_0}{\sqrt{1 + \rho L/2nG}} \\ W_i = \frac{dW_i}{dt} = 0 \quad & (i = 1, 2, \ldots, n) \end{aligned}\right\} \tag{I.7}$$

Using matrix notation and introducing

$$\{Q\} \equiv (Q_1, Q_2, \ldots, Q_n)^{\mathrm{T}}, \quad \{M\} \equiv (M_1, M_2, \ldots, M_n)^{\mathrm{T}},$$
$$\{W\} \equiv (W_0, W_1, \ldots, W_{n-1})^{\mathrm{T}}$$
$$\{\theta\} \equiv (\theta_1, \theta_2, \ldots, \theta_n)^{\mathrm{T}}, \quad \{\psi\} \equiv (\psi_1, \psi_2, \ldots, \psi_n)^{\mathrm{T}}$$

$$\mathbf{A} \equiv \begin{bmatrix} 1 & & & & 0 \\ -1 & 1 & & & \\ & \ddots & \ddots & & \\ & & & 1 & \\ 0 & & & -1 & 1 \end{bmatrix}, \qquad \mathbf{B} \equiv \begin{bmatrix} n\gamma + \frac{1}{2} & & & 0 \\ & 1 & & \\ & & \ddots & \\ 0 & & & 1 \end{bmatrix}$$

and

$$T_0 \equiv L\sqrt{\frac{\rho L}{M_p}}, \quad \tau \equiv \frac{t}{T_0}, \quad (\dot{\ }) = \frac{d}{d\tau}(\ ), \quad \{w\} \equiv \frac{1}{L}\{W\}$$

$$\{q\} \equiv \frac{L}{M_p}\{Q\}, \quad \{m\} \equiv \frac{1}{M_p}\{M\}, \quad \psi_p \equiv \frac{M_p L}{nEI}$$

$$\bar{\psi} \equiv \frac{\psi}{\psi_p}, \quad \bar{\theta} \equiv \frac{\theta}{\psi_p}, \quad \gamma \equiv \frac{G}{\rho L}, \quad e_0 \equiv \frac{K_0}{M_p} = \frac{GV_0^2}{2M_p}, \quad R \equiv \frac{K_0}{U_e^{max}}$$

eqns (I.1)–(I.4) can be condensed in non-dimensional form

$$\mathbf{A}\{q\} = \frac{1}{n}\mathbf{B}\{\ddot{w}\}, \quad \mathbf{A}\{m\} = -\frac{1}{n}\{q\}$$

$$\{\bar{\theta}\} = \frac{n}{\psi_p}\mathbf{A}^T\{w\}, \quad \{\bar{\psi}\} = \mathbf{A}^T\{\bar{\theta}\}$$

Combining these equations gives

$$\{\ddot{\bar{\psi}}\} = -\frac{n^3}{\psi_p}\mathbf{F}\{m\} \tag{I.8}$$

where

$$\mathbf{F} \equiv (\mathbf{A}^T)^2\mathbf{B}^{-1}(\mathbf{A})^2 = \begin{bmatrix} (n\gamma+\frac{1}{2})^{-1} & +5 & -4 & 1 & & & & \\ & -4 & 6 & -4 & \ddots & & 0 & \\ & 1 & -4 & 6 & \ddots & & & \\ & & \ddots & \ddots & \ddots & & & \\ & & & & 6 & -4 & 1 \\ & 0 & & & -4 & 5 & -2 \\ & & & & 1 & -2 & 1 \end{bmatrix}$$

The constitutive relation (I.5) can also be recast in non-dimensional form:

$$m_i = \begin{cases} \bar{\psi}_i & (\bar{\psi}_i \leqslant 1) \\ 1 + \alpha(\bar{\psi}_i - 1) & (\bar{\psi}_i \geqslant 1,\ \dot{\bar{\psi}}_i \geqslant 0) \end{cases} \tag{I.9a}$$

$$m_i = \bar{\psi}_i - (1-\alpha)(\bar{\psi}_i^* - 1) \quad (\bar{\psi}_i^* - 2 \leqslant \bar{\psi}_i \leqslant \bar{\psi}_i^*,\ \dot{\bar{\psi}}_i < 0) \tag{I.9b}$$

with $i = 1, 2, \ldots, n$.

Taking the elastic capacity of the present model, $U_e^{max} = \frac{1}{2}nM_p\psi_p$, to be equal to the elastic capacity of an elastic–plastic cantilever, $U_e^{max} = M_p^2L/2EI$, we find

$$\psi_p = \frac{M_p L}{nEI} \tag{I.10}$$

Accordingly, the constant C appearing in (I.5a) is $C = M_p/\psi_p = nEI/L$. Since $\psi_p = 2U_e^{max}/n\,M_p = 2e_0/n\,R$, eqn (I.8) can be rewritten as

$$\{\ddot{\psi}\} = -\frac{n^4 R}{2e_0}\mathbf{F}\{m\} \tag{I.8'}$$

The combination of (I.8′) and (I.9) gives the $2n$ governing equations with $2n$ unknown functions $\bar{\psi}_i$ and m_i $(i = 1, 2, \ldots, n)$. As the hardening coefficient α tends to zero, this structural representation approaches that of an elastic–perfectly plastic ideal sandwich beam.

Referring to (I.6), the second-order differential equation (I.8′) has the following initial conditions:

$$\begin{aligned} &\bar{\psi}_1 = 0, \quad \dot{\bar{\psi}}_1 = n^2 R\sqrt{\frac{1}{e_0(2\gamma + 1/n)}} \\ &\bar{\psi}_i = 0, \quad \dot{\bar{\psi}}_i = 0 \quad (i = 2, 3, \ldots, n) \end{aligned} \tag{I.13}$$

APPENDIX II

Suppose that at any time t a plastic hinge is located at an interior section H, a distance $\Lambda(t)$ from the tip (see Fig. 14c). The equations of motion in non-dimensional form are (for a detailed derivation see Ref. 20)

$$m_B - 1 + \tfrac{1}{2}(1-\lambda)^3\ddot{\psi} = 0 \tag{II.1}$$

$$(\gamma + \tfrac{1}{2}\lambda)\dot{v} + [\gamma + \lambda(1 - \tfrac{1}{2}\lambda)]\ddot{\psi} + \tfrac{1}{2}v\dot{\lambda} = 0 \tag{II.2}$$

$$\tfrac{1}{6}\lambda^2\dot{v} + (\tfrac{1}{2} - \tfrac{1}{3}\lambda)\lambda^2\ddot{\psi} + \tfrac{1}{2}v\lambda\dot{\lambda} - 1 = 0 \tag{II.3}$$

where

$$x \equiv \frac{X}{L}, \quad w \equiv \frac{W}{L}, \quad \lambda \equiv \frac{\Lambda}{L}, \quad m_B \equiv \frac{M_B}{M_p}, \quad \gamma \equiv \frac{G}{\rho L}$$

$$T_0 \equiv L\sqrt{\frac{\rho L}{M_p}}, \quad v \equiv \frac{V}{L/T_0}, \quad \tau \equiv \frac{t}{T_0}, \quad (\dot{\ }) \equiv \frac{d}{d\tau}(\ )$$

The non-dimensional bending moment provided by the rotational spring at the root is given by

$$m_B = \begin{cases} C\psi_e/M_p & \text{(elastic or unloading)} \\ 1 \text{ or } -1 & \text{(otherwise)} \end{cases} \tag{II.4}$$

The three unknown functions $\psi(\tau)$, $v(\tau)$ and $\lambda(\tau)$ can be solved from eqns (II.1)–(II.3) with the initial conditions

$$\psi(0) = 0, \quad \dot{\psi}(0) = 0, \quad v(0) = v_0, \quad \lambda(0) = 0 \tag{II.5}$$

Chapter 10

IMPACT PERFORMANCE OF ALUMINIUM STRUCTURES

I. J. McGregor

Alcan International Limited, Banbury Laboratory, Southam Road, Banbury, Oxfordshire OX16 7SP, UK

D. J. Meadows, C. E. Scott & A. D. Seeds

Alcan Automotive Structures (UK) Limited, Southam Road, Banbury, Oxfordshire OX16 7SA, UK

ABSTRACT

The weight saving achieved through the use of aluminium automotive structures in comparison with equivalent steel structures is discussed, with particular reference to the crashworthiness performance. The differences between aluminium material properties and those of steel are highlighted and the implications for the design of crashworthy structures discussed. Experimental results for the performance of aluminium structural components subjected to axial and bending collapse are presented, and the important design variables for each of these collapse modes identified. Design tools for assessing the energy absorption, stability and collapse modes for axial and bending collapse are also presented and compared. Finally, the use of the design tools and the solving of the issues raised in the crashworthiness design of vehicle structure are illustrated with reference to a project conducted with a major automotive vehicle manufacturer.

1 INTRODUCTION

Interest in the use of aluminium for automotive structures is increasing as a result of the need for automotive manufacturers to address the issues related to the environment. Pressures being brought to bear on manufacturers include the need to reduce fuel consumption to conserve resources, reduce CO_2 and other emissions and maximise material recycling.

Aluminium is a material that is able to address all of these issues by offering significant weight savings, without compromising strength or structural performance, as well as providing outstanding corrosion resistance and demonstrated recyclability. These major benefits have

Table 1
Relative Weight of Aluminium and Steel Automotive Structures

Vehicle	*Steel weight (kg)*	*Aluminium weight (kg)*	*Weight saving (%)*	*Proportion of steel torsional stiffness (%)*
Pontiac Fiero	444	303[a]	32	—
Ferrari 408	99	68	31	115
BL Metro	137	74[a]	46	76
Bertone X1/9	192	130[a]	33	101
Honda NSX	350	210	40	—
Medium-volume saloon[b]	262	138	48	113
Medium–high-volume saloon[b]	170	101	41	107
Low-volume sports soft top[b]	340	213	37	288

[a] Weight of aluminium replica of production vehicle.
[b] Current development projects.

encouraged many of the world's largest car companies to explore actively the increased use of aluminium for skins, structure, power train and chassis components.

The weight saving that results from the substitution of aluminium for steel may be used to provide reduced fuel consumption, increased space and improved vehicle performance. The weight savings that may be achieved are demonstrated in Table 1, which compares the relative weight of the load bearing structure of various vehicles in steel and aluminium. Weight savings of between 40 and 50% are possible. For the whole vehicle, indications are such that a weight saving of as much as 25% compared with conventional steel structures may be possible.

It is important, however, that this weight saving does not compromise the safety or structural performance of the vehicle. Crashworthiness is one aspect of structural performance that manufacturers are required to address in order to meet regulatory performance requirements. The objectives of a good crashworthiness design are to

- dissipate in a controlled manner the kinetic energy of the impact
- retain a survival space for the occupants
- minimise the forces and accelerations experienced by the occupants

Extensive testing of steel vehicles for crashworthiness has been conducted to validate the designs. However, with the need to optimise for weight saving, the importance of understanding the mechanisms controlling performance has become vital. This is especially the case for aluminium, since only a limited amount of experimental data have been generated on the crashworthiness performance of aluminium structures.[1] Most of the information available concentrates on the performance of box sections.[2–5] As part of the development of an Aluminium Vehicle Technology,[6] Alcan International Ltd has published several papers on the performance of aluminium structured vehicles.[7,8] In Ref. 9, the test track evaluation of an aluminium structured Bertone X1/9 replica vehicle is described. This included a front impact test, in which the replica vehicle demonstrated the required crashworthiness performance. This was achieved with a weight saving of 31% compared with the equivalent steel body.

One of the objectives of this paper is to demonstrate the weight savings that may be achieved using aluminium impact structure, while retaining the required structural performance. The important aspects that must be considered for designing aluminium structure for crashworthiness will be discussed. It will also be demonstrated that existing design tools, developed for steel, may be used for aluminium, although some consideration must be given to the original assumptions used in the development of the tools, which may not be suitable for aluminium. The most important objective is to demonstrate that a satisfactory aluminium crashworthy structure may be designed without requiring radical changes from conventional steel designs.

2 ALUMINIUM MATERIAL PROPERTIES

It is useful to compare the material properties of aluminium and steel and to consider the implications of these properties for the design of impact structure. Table 2 and Fig. 1 compare the material properties of aluminium alloys AA5754-O and AA6111-T4 to mild steel. AA5754-O is typical of the medium-strength ductile aluminium alloys that may be used for the load-bearing structure in automotive applications. AA6111-T4 is an aluminium alloy used as a skin material or as a structural material where high strength is required.

Table 2
Typical Material Properties of Aluminium and Steel

Material	*Elastic modulus (MPa)*	*Poisson's ratio*	*Density (kg/m^3)*	*0·2% proof stress (MPa)*	*UTS (MPa)*	*Elongation (50 mm gauge) (%)*
AA5754-O	70 000	0·33	2 700	100	220	23
AA6111-T4 (+30 min at 180°C)	70 000	0·33	2 700	180	320	25
Mild steel	205 000	0·30	7 850	220	370	39

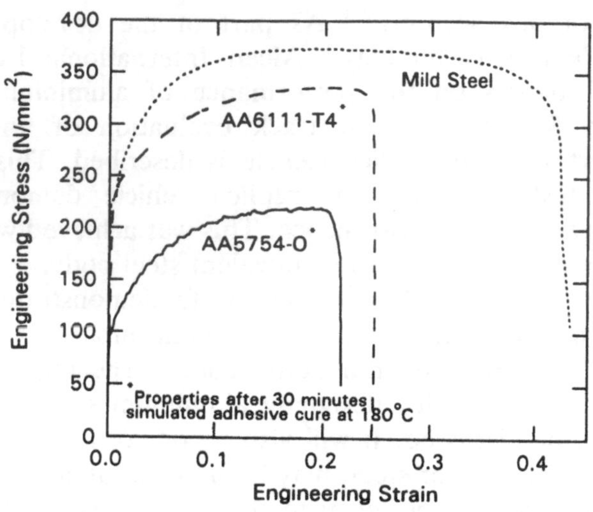

Fig. 1. Comparison of stress–strain curves for aluminium and steel.

From Table 2, it may be seen that the density of aluminium is about one-third that of steel. This results in an aluminium structure at 50% the weight of an equivalent steel structure when using 50% thicker gauges of material. Since, in the collapse of structures, the main deformation mode and consequent energy absorption is through bending of the sheet, aluminium structures, with their thicker gauge, have an advantage over steel. Although this is somewhat offset by the higher strength of steel, it will be demonstrated that the impact performance of aluminium structures is such that the same performance as a steel structure may be achieved with weight savings of around 45%.

It is generally accepted that the strain rate sensitivity of aluminium is low in comparison to steel. In Ref. 10, a comparison is made between the strain-rate sensitivity of steel and aluminium alloys used in an experimental programme looking at the failure of clamped beams under impact. It was reported that the strain-rate sensitivity of the aluminium alloy was very small in comparison with the steel. However, the material used in that programme of work was much thicker (5 mm) that the sheet material thicknesses used in the construction of automotive structures (typically 1·0–3·0 mm).

In order to check the strain-rate sensitivity of AA5754-O and AA6111-T4 automotive sheet, tensile tests were conducted on specimens at different strain rates. To establish the appropriate strain rates, impact tests were conducted on hexagonal box sections (Fig. 2) on a drop hammer rig at the University of Liverpool Impact Research

Fig. 2. Aluminium hexagonal box section before and after impact.

Centre. Strain gauges were placed around the section in order to measure the strain rates during the development of the folds within the box. The total energy input was 10 kJ at a velocity of around 11 m/s, and the results indicated that strain rates in the section were in the range from 3 to 64 s^{-1}.

Figures 3 and 4 show the influence of strain rate on the 0·2% proof stress and the UTS of AA5754-O and AA6111-T4 aluminium sheet. The sheets, as supplied, were pre-conditioned by being heated to 180°C for 30 min to simulate an adhesive cure cycle. It is clear that the influence of strain rate on the proof stress and the UTS for both materials is small. This is especially true for the AA6111 alloy, which showed almost no change in properties as a function of strain rate. For the AA5754 alloy, there was no change in the UTS, but a 25% increase in proof stress was observed. From Ref. 10, the lower yield strength of mild steel increased by 60% and the UTS by 20% over the same range of strain rates. This has an important implication for aluminium impact structure, in that the strength increases that may be obtained from steel at high impact velocities will not occur for aluminium. Consequently, the designer of aluminium impact structures will have to compensate for this through efficient design.

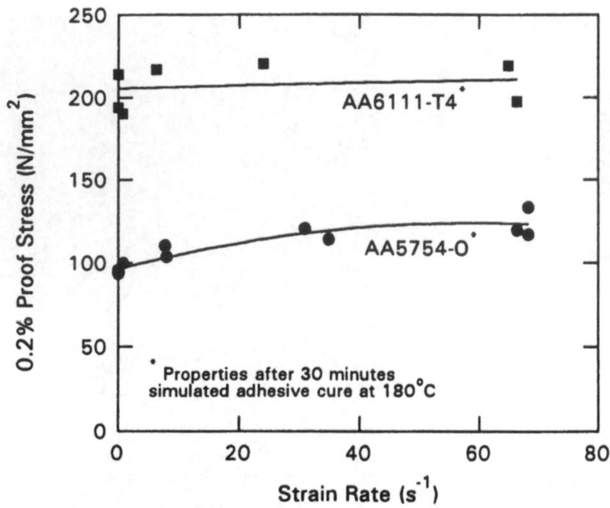

Fig. 3. Influence of strain rate on the 0·2% proof stress of aluminium.

3 DESIGN OF ALUMINIUM IMPACT MEMBERS

In the crash performance of full-scale structures, there is a wide range of elastic and plastic deformation modes, which include axial crush, deep bending collapse and lateral crush. The crash event often starts with the simple deformation of a few components, but, especially in the case of high-energy impacts, concludes with a complex interaction of the components.

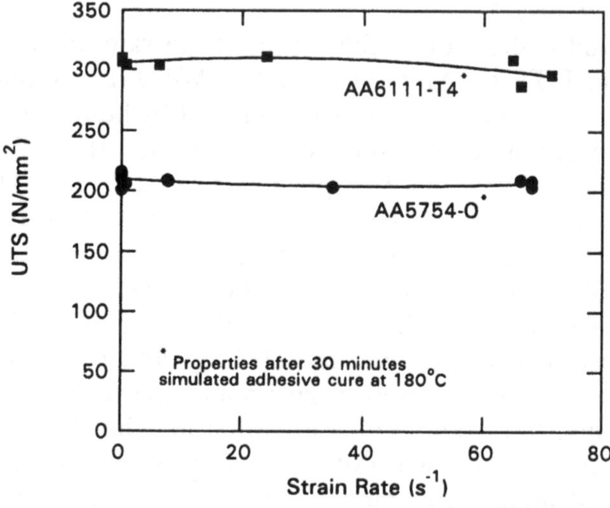

Fig. 4. Influence of strain rate on the UTS of aluminium.

Despite this apparent complexity, much of the research and development of both material suppliers and car companies is concentrated on the failure of components under axial and bending collapse. The performance characteristics of the components under this type of loading are then used in non-linear finite element models of the full vehicle structure. Often the axial and bending collapse modes account for the majority of the energy absorbed, and it is therefore desirable to identify and understand the variables that influence the performance of components in these failure modes. The next sections of this chapter will therefore discuss the performance of aluminium components under axial and bending collapse, and make a comparison with the performance of equivalent steel components.

3.1 Axial Collapse

(i) Important Design Variables

There has been a good deal of work reported on the dynamic testing and analysis of steel sections under axial collapse.[11–13] A good design of axial impact member will demonstrate a controlled and progressive collapse as illustrated in Fig. 2, as well as achieving the required energy absorption. Figure 5 shows the force–displacement curve for a slow crush test on an aluminium hexagonal impact member. The important design loads are indicated, and include the load to initiate collapse and the average crush force.

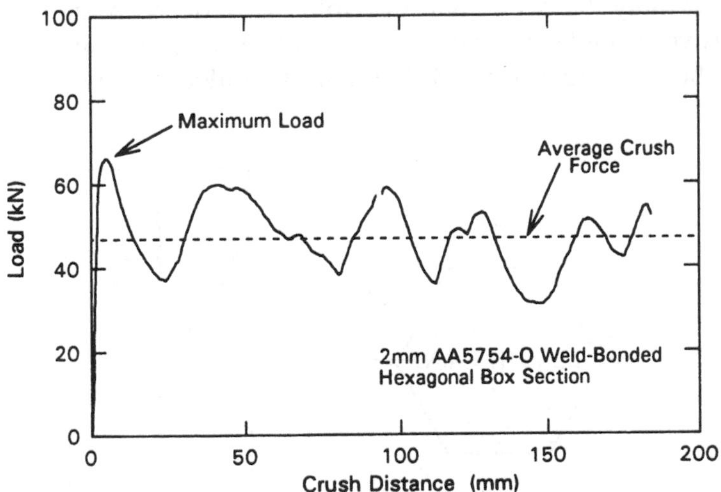

Fig. 5. Force–displacement curve for slow crush testing of an aluminium hexagonal box section.

Table 3
Comparison of Impact Properties of Aluminium and Steel Hexagonal Box Sections

Material and joint type	*Average crush force (kN)*	*Mass relative to steel section mass (%)*	*Specific energy absorption (kJ/kg)*
1·2 mm mild steel spot-welded (25 mm pitch)	52	100	14·5
2·0 mm AA5754-O aluminium spot-welded (25 mm pitch)	46	57	22·4
2·0 mm AA5754-O aluminium weld-bonded (75 mm pitch)	53	57	25·8

Table 3 compares the impact performance of a spot-welded 1·2 mm mild steel hexagonal section with those of 2·0 mm spot-welded and a 2·0 mm weld-bonded AA5754-O aluminium hexagonal section. The geometry of the hexagonal section tested is shown in Figure 6. Weld-bonding is a joining technique that uses adhesive in combination with spot-welds. The adhesive used in these tests was a single-part heat curing epoxy adhesive developed for Alcan International Ltd as part of its aluminium vehicle technology. Figure 7 compares the folding pattern in the spot-welded steel section and the weld-bonded aluminium section. It may be seen that the folding pattern in the aluminium section is very similar to that obtained in the steel section.

It can be seen from Table 3 that a spot-welded aluminium section at

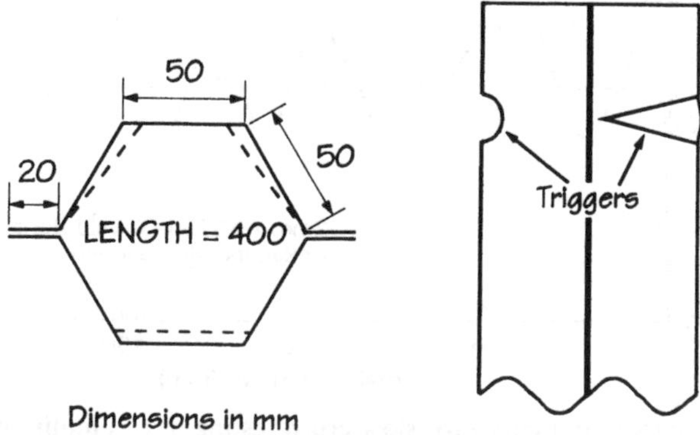

Fig. 6. Geometry of hexagonal box test specimen.

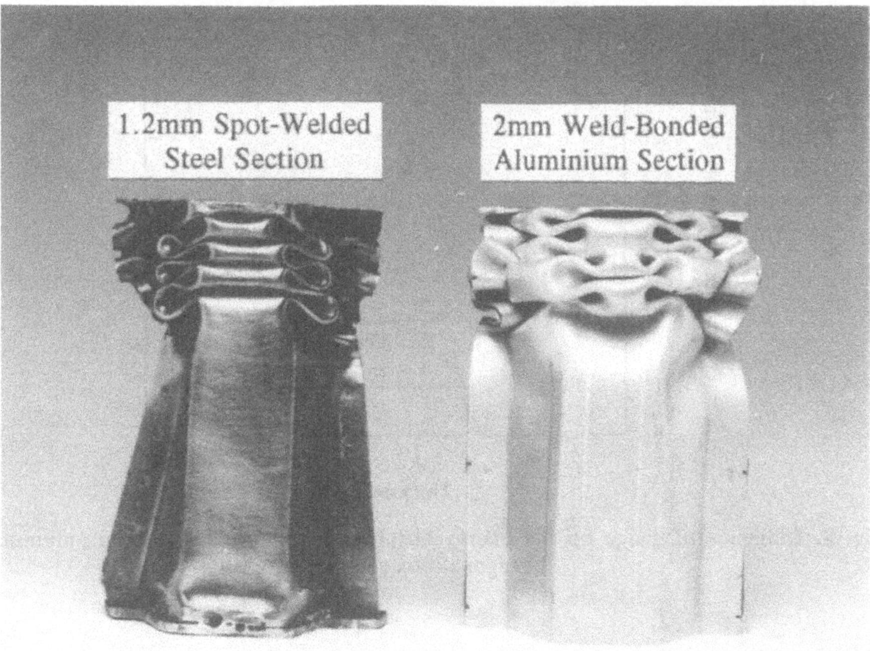

Fig. 7. Comparison of folding pattern for aluminium and steel hexagonal box sections.

57% the weight of an equivalent steel section has a much higher specific energy absorption. It may be demonstrated that the spot-welded aluminium section would have to be 65% of the weight of its steel counterpart in order to give the same average collapse force. The reason for the improved energy absorption of aluminium structures per unit mass is in the use of thicker gauges, compared with equivalent steel structures, and the lower density of aluminium compared with steel.

Figure 8 shows the influence of gauge on the energy absorption per unit mass of aluminium hexagonal impact members. As can be seen, the gauge has a very significant influence. Figure 8 also shows the prediction of the influence of gauge using the CRASH-CAD design software.[14,15] This is a PC-based computer package that uses the theoretical work of Wierzbicki and Abramowicz.[16,17] Alcan has contracted the developers of this package, which was originally developed for materials with a stress–strain behaviour similar to that of steel, to develop a version specifically adapted for aluminium. As may be seen, there is a very good correlation between predicted and measured values.

CRASH-CAD is based on the theoretical developments summarised in Ref. 18, which gives the average crush force P_{av} for a 120° angle

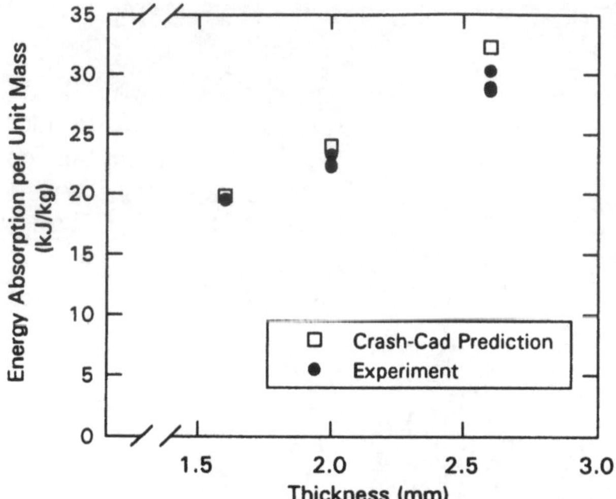

Fig. 8. Influence of gauge on the energy absorption per unit mass for an aluminium hexagonal box section.

element in the form

$$P_{av} = 3{\cdot}37\sigma_0 C^{0{\cdot}4} t^{1{\cdot}6} \tag{1}$$

where C is the length of the corner element, σ_0 the flow stress and t the sheet thickness. As may be seen from eqn (1), the average crush force is proportional to the gauge of the material to the power of 1·6. This gives the energy per unit mass proportional to the gauge to the power of 0·6. Using these factors, the designer is able easily to estimate the influence of a change of gauge on the predicted energy absorption of a component.

Table 3 also shows the influence of the joining method on the performance of impact members. A comparison is made between an aluminium section that has been spot-welded at a pitch of 25 mm with a section that has been weld-bonded. Figure 9 compares the fold patterns of weld-bonded and spot-welded aluminium reinforced double top-hat sections. Both show very similar fold patterns. However, the weld-bonded section typically gives an improvement of 15% in energy absorption compared with the spot-welded section. As a result of this improvement, weld-bonding in aluminium sections results in comparable performance to a spot-welded steel section at around 55% of the weight.

Impact testing has been conducted on hexagonal box sections at various temperatures. The box sections were the same as those shown

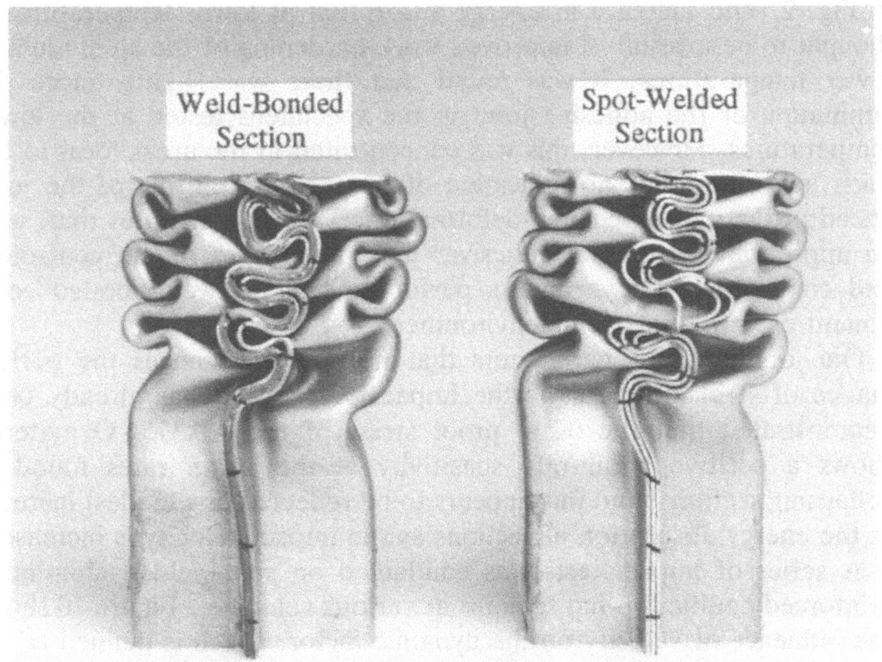

Fig. 9. Comparison of folding pattern for weld-bonded and spot-welded aluminium reinforced double top-hat sections.

in Figs 2 and 6, and were joined by weld-bonding with the spot-weld pitch at 75 mm. The objective of the test programme was to check the performance of the joint as a function of temperature. The impact energy was 10 kJ and the tests were conducted at three temperatures, −40°C, −10°C and +18·5°C. Table 4 shows the average crush force obtained as a function of temperature. It can be seen that the average crush force at −40°C was around 10% higher than at +18·5°C. All of the box sections collapsed in a controlled manner similar to that shown

Table 4
Influence of Temperature on the Average Crush Force of Weld-Bonded Aluminium Hexagonal Box Sections

Test temperature (°C)	*Average crush force (kN)*
−40	58·7
−10	57·4
+18·5	53·0

in Fig. 2. The increase in energy absorption at lower temperatures is thought to be a result of improved work-hardening of the aluminium at lower temperatures. It was found that there was slightly more delamination of the adhesive joint in the specimens tested at the lower temperatures. However, this was concentrated in the areas local to the folds, and there was no indication of fracture or splitting of the joint ahead of the main specimen deformation. This demonstrates that, with an appropriate structural adhesive, it is possible to obtain controlled and consistent crashworthiness performance of a weld-bonded component even in an extreme environment.

One of the other parameters that normally influences the performance of steel structures is the impact velocity. It has already been demonstrated that the 0·2% proof stress of the AA5754-O material shows a positive strain-rate sensitivity at the strain rates found in collapsing sections, and this appears to be reflected in a modest increase in the energy absorption of sections as the impact velocity is increased.

A series of impact tests was conducted on spot-welded aluminium reinforced double top-hat sections at various velocities. Figure 10 shows the influence of velocity on the dynamic factor, which is defined as the ratio of the average crush force under impact to that under slow crush loading. The box sections were made from 2 mm AA5754-O aluminium alloy and the impact energy was maintained constant at 5 kJ. For an impact velocity of 12 m/s, the dynamic factor is around 1·1. This is small in comparison with mild steel, as demonstrated in Ref. 19, which

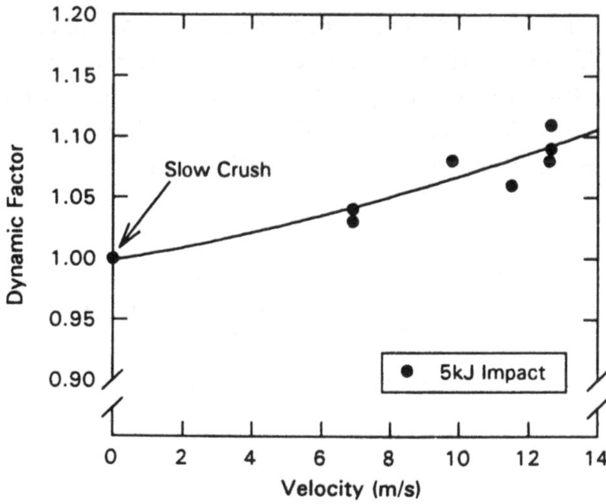

Fig. 10. Influence of impact velocity on the dynamic factor for an aluminium reinforced double top-hat section.

Table 5
Relative Axial Crush Performance of Various Aluminium Impact Section Geometries

Material	2·6 mm AA5754-O sheet					AA6063	
Geometry	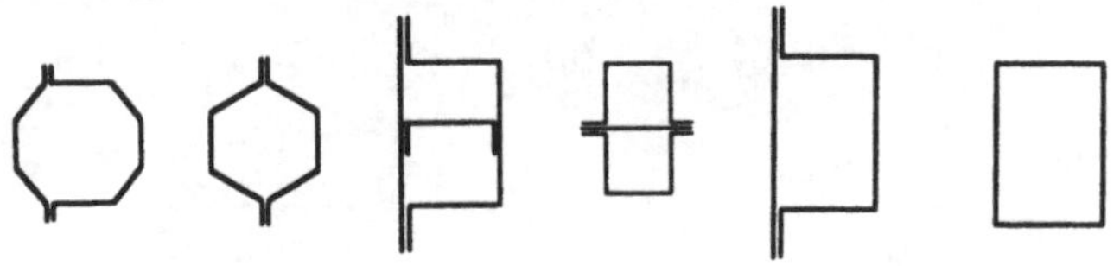						
						Extrusion	
						T4	T6
Average crush force (kN)	85	80	95	82	60	34	58
Mass-specific energy absorption (kJ/kg)	29·9	29·5	28·8	24·8	22·5	16	28

gives dynamic factors for mild steel, for an impact velocity of 13·4 m/s, in the range of 1·3–1·7, depending on the material's yield and tensile strength.

One of the most important design aspects that controls component performance is not a material property, however, but is related to section geometry. In general, most section geometries that work for steel will also work for aluminium. Table 5 compares the relative axial crush performance of various aluminium sections. It can be seen that the mass-specific energy absorption is strongly influenced by the geometry of the section. In particular, the more corners in the section, the higher the mass-specific energy absorption.

This has important implications in the choice of the manufacturing route for components. Aluminium extrusions are attractive for impact members owing to their low cost of manufacture and the similarity to seam-welded steel sections. However, an extruded section will normally contain fewer corners than an equivalent folded sheet section, since the latter requires flanges for assembly and these automatically introduce extra corners into the section. This can, in principle, be offset by using the extruded material in a higher strength condition. However, the balance of material properties against section efficiency is important. Figure 11 compares the performance of an aluminium extrusion in AA6082 material in both the T4 and T6 (heat-treated) conditions. The T6 condition has a much higher strength, and this provides the

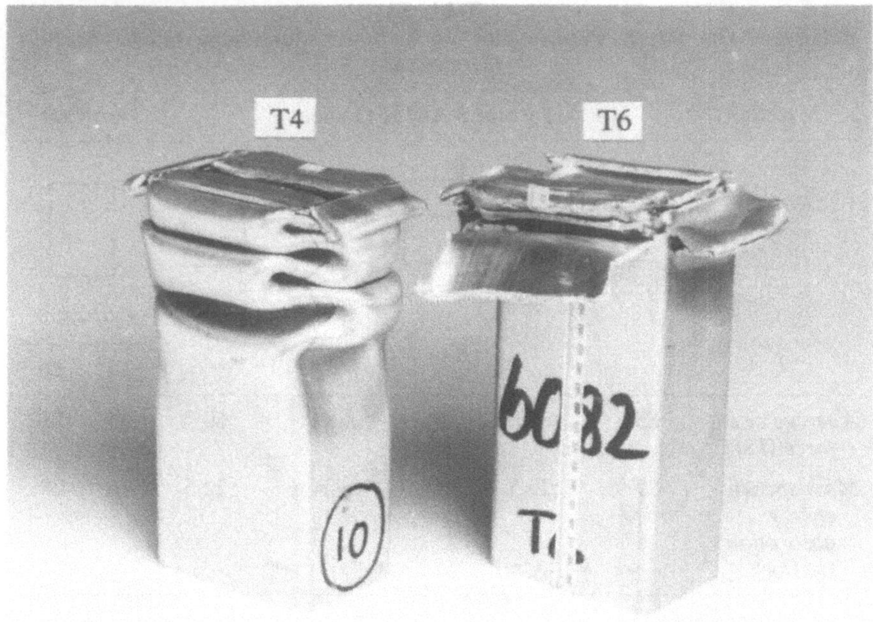

Fig. 11. Comparison of impact performance of AA6082 aluminium extrusions in T4 and T6 conditions.

opportunity to produce a more efficient section, as is shown in Table 5 for AA6063 alloy. However, from Fig. 11, it can be seen that there is a greater possibility of fracture in the T6 heat-treated condition compared with the lower-strength T4 condition.

The designer therefore has to carefully balance material form against section form in order to achieve the most mass efficient solution. In order to assist in this, design tools are very important. There have been many different design tools assessed and developed for steel, and it is important to have similar tools available for aluminium.

(ii) Designing for Axial Collapse

The design procedure for aluminium impact structure is, in most respects, no different from that of steel. In Refs 20–22, design approaches for steel automotive components subjected to axial collapse are presented. Figure 12 illustrates the principal stages involved in the design process of an impact member. As has already been discussed, the selection of geometry is one of the most important in determining the mass efficiency of the section. However, in practical design, the geometry of the section is often compromised by space constraints and

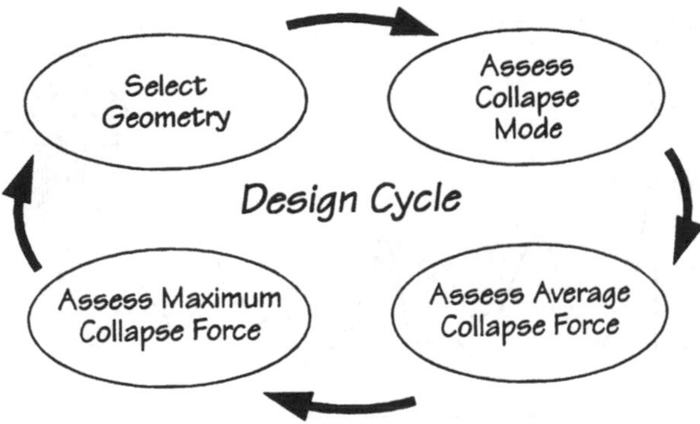

Fig. 12. Design cycle for impact members.

other functional requirements from the component. This normally results in the component being specifically designed for the particular application.

Once a preliminary geometry is selected, the three performance aspects that should be considered are collapse mode, average collapse force and maximum collapse force. In order to satisfy these design requirements, the design cycle shown in Fig. 12 should be used several times in an iterative process. This requires the ability to assess the performance of the proposed design quickly.

The various design tools available to the designer include experimental testing, finite element modelling, PC-based modelling and hand calculations. Experimental testing is expensive and time-consuming, and is usually only used in the stages of design when the design is close to being finalised. Finite element modelling may be used, using codes such as LS-DYNA3D,[23] but normally involves many hours of computation on a powerful computer. As a result of this, the use of a PC-based software package such as CRASH-CAD, as described earlier, is very attractive. This package may be used on a desktop personal computer to assess a proposed design, with only a few minutes of computation required.

An example of the use of LS-DYNA3D in the prediction of the axial crush performance of a 2 mm AA5754-O aluminium hexagonal box section is shown in Fig. 13. Full non-linear properties of the aluminium were used in the analysis, with the section modelled using shell elements. Although the load to initiate collapse is over-predicted, there is a good correlation between the experimental and predicted average crush force. The predicted deformation pattern is also close to that observed in practice. However, as already discussed, the problem in

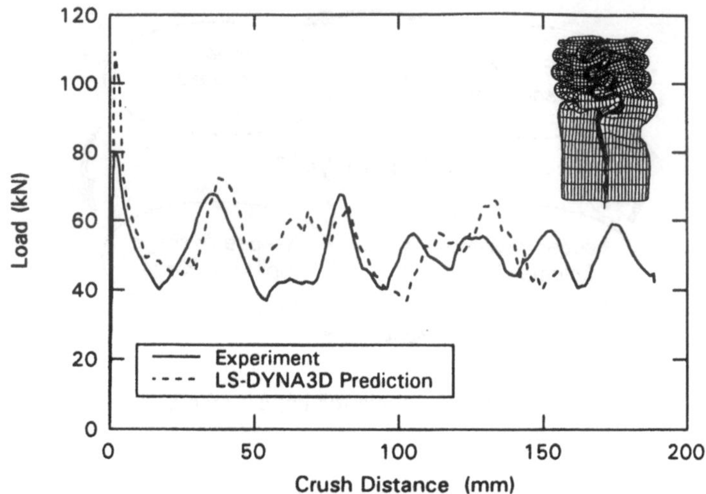

Fig. 13. Comparison of predicted and measured force–displacement curves using LS-DYNA3D.

using this type of finite element analysis as part of the design cycle is the large amount of computational time required.

The CRASH-CAD package is based on the modelling of the corners of sections using superfolding elements.[18] It is therefore possible to model complex section geometries with only a few elements, which results in very short analysis times. The use of this package for designing steel components is described in Ref. 24. Although originally developed for materials having a stress–strain curve similar to that of steel, the package is being continually developed to extend its application to other materials. In its commercial form, the package does not use experimental factors in its prediction, and is based on a theoretical approach. However, Alcan International Ltd has worked closely with the developers of the package to produce a calibrated version for aluminium. The objective of the development was to use Alcan's extensive experimental database on the performance of aluminium components to produce a design tool with a high accuracy for aluminium.

Table 6 shows a comparison between the CRASH-CAD prediction and the experimental results of average crush force and fold half-wavelength for several different aluminium section geometries. It can be seen that high accuracy is achieved for a wide range of component geometries. The prediction of the fold half-wavelength is important in determining the collapse mode of the component. The geometry of the component is such that it must allow complete development of the folds

Table 6
Comparison of Predicted and Measured Crush Performance Using CRASH-CAD for Various Aluminium Section Geometries

Geometry (2 *mm AA5754-O spot-welded*)					
Average crush force (*kN*)	*CRASH-CAD*	28·1	53·2	41·9	66·6
	Experiment	26·0	49·6	39·8	70·1
Fold half-wavelength (*mm*)	*CRASH-CAD*	35·9	32·4	36·0	46·1
	Experiment	37·5	33·5	36·5	41·5

within the component in order to achieve consistent progressive collapse. The fold half-wavelength is also important for the design of the joint in the component. By placing the spot-welds in the flange at the same pitch as the fold half-wavelength, it is possible to improve the overall performance of the component both in terms of a stable collapse mode and average collapse force.

Figure 14 compares the CRASH-CAD predicted force–crush-distance curve with that measured from slow crush axial testing of a 2 mm AA5754-O aluminium hexagonal section. It can be seen that,

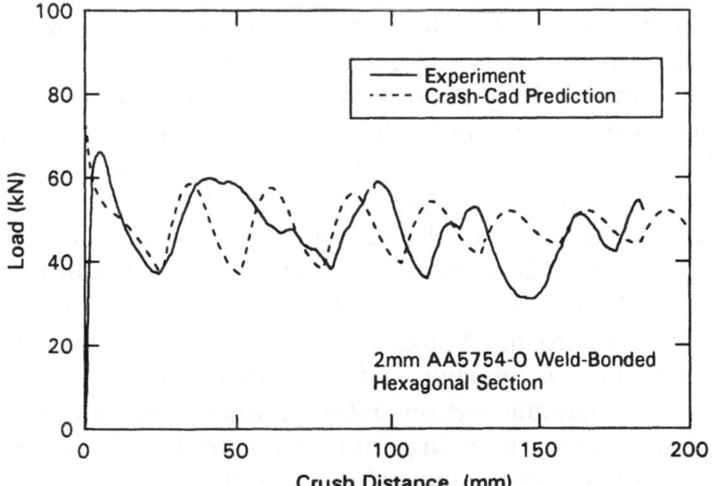

Fig. 14. Comparison of predicted and measured force–displacement curves using CRASH-CAD.

although the mean level of the crush force is in very good agreement, the shapes of the curves are not identical. This was also true for the LS-DYNA3D prediction in Fig. 13. In fact, the shape of the curve is influenced by imperfections, and is generally different for each replica test specimen. The important point is that these imperfections do not have a large influence on the average crush force; therefore the average crush force is largely independent of the exact shape of the force–displacement curve.

The prediction of the maximum collapse force is also important, since this will influence the design of the supporting structure of the main impact components, which must withstand the impact collapse loads without itself collapsing. Often the load to initiate collapse is controlled by collapse initiators. These are local indentations incorporated into the part during manufacture, as shown in Fig. 6. The collapse initiators also control the starting deformation pattern for the component, and the CRASH-CAD package will predict the most suitable initiator design for progressive collapse. However, the prediction of the maximum collapse force with initiators is very difficult, since the extent of the deformation and the work-hardening of the material local to the initiators is not easily determined. As a result, experimental testing is often used to develop the optimum collapse initiator depth.

For stability of the section, however, it is often important to assess the load to initiate collapse in the rear end of the component. This load is compared with the collapse initiation load for the front of the section and the local peak loads obtained during progressive collapse to ensure overall stability of the section. Various buckling analysis methods have been developed for steel.[25] However, as can be seen from Fig. 1, the stress–strain curves of steel and aluminium are very different, especially in the area around the yield or proof stress. Steel usually exhibits a linear stress–strain curve up to the yield stress. Aluminium, however, only exhibits linearity up to the limit of proportionality, with subsequent strain hardening all the way to the UTS. For aluminium, a proof stress is defined instead of a yield stress, normally the stress at 0·2% plastic strain. The stability of a section is dependent on the slope of the material stress–strain curve. For material in the elastic range, this slope is equal to the Young's modulus; in the plastic range, it is equal to the tangent modulus of the stress–strain curve. For the prediction of aluminium section stability, where buckling often occurs in the plastic range, it is important to consider the shape of the aluminium curve above the limit of proportionality.

Figure 15 shows experimental results for a series of tests on square aluminium extrusions. The thickness of the extrusions was maintained

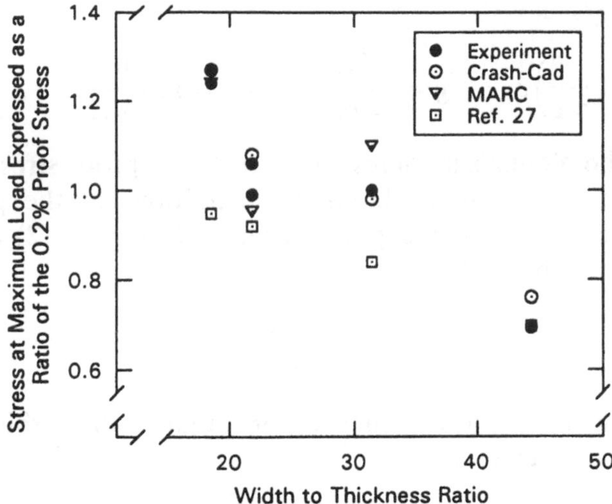

Fig. 15. Influence of wall width-to-thickness ratio on the static maximum load of square aluminium extrusions.

constant at 1·6 mm and the influence of the wall width on the maximum load measured. The extrusions were in AA6063 alloy aged to give properties similar to AA5754-O. In Fig. 15, a comparison is made between the experimental results and the predictions from hand calculations, MARC finite element analysis[26] and CRASH-CAD for the maximum load expressed as a ratio of the 0·2% proof stress versus the section width-to-thickness ratio. The important point to note is that for a width-to-thickness ratio of less than 30, the experimental results give the stress at maximum load greater than the 0·2% proof stress for the aluminium alloy. Accurate prediction for aluminium collapse loads therefore requires the ability to predict instability well into the plastic range of the material. Indeed, for aluminium, the stress–strain curve becomes non-linear at the limit of proportionality, which is typically around 80% of the 0·2% proof stress.

In Ref. 27, an approach is given for the calculation of the maximum load of an aluminium plate subjected to axial compression. The approach takes into account the shape of the stress–strain curve between the limit of proportionality and the 0·2% proof stress in the calculation of the buckling stress. The maximum load of a plate (P_{max}) subjected to axial compression is given by

$$P_{max} = bt(B - D\lambda) \tag{2}$$

where b is the plate width and t the plate thickness. B, D and λ are defined for heat-treatable alloys in T4 and lower tempers and non-heat-treatable alloys as

$$B = \left(1 + \frac{3}{g}\right)\sigma_y, \quad g = \sqrt{\frac{E}{\sigma_y}}, \quad D = 0{\cdot}123\frac{B^{1{\cdot}5}}{\sqrt{E}}, \quad \lambda = \frac{mb}{t} \tag{3}$$

where E is the Young's modulus, σ_y is the 0·2% proof stress, and m is a factor that depends on the boundary conditions of the plate, and is taken as 1·63 for a simply supported plate. Equation (2) applies for a plate with a width less than

$$b = \frac{Bt}{2mD} \tag{4}$$

For plates with a width equal to or greater than this value, the maximum load is given by

$$P_{\max} = \frac{2g}{m}\sqrt{1 + \frac{3}{g}}\,\sigma_y t^2 \tag{5}$$

These equations were used in the prediction for the square aluminium extrusions shown in Fig. 15, and, as can be seen, the theory generally under-predicts the experimental results, especially for small width-to-thickness ratios. This is a result of the theory limiting the maximum collapse stress to the 0·2% proof stress. This is a typical limitation of closed-form solutions, and the alternative approach is to use non-linear finite element analysis or an approach that considers the shape of the stress–strain curve above the 0·2% proof stress.

Figure 15 also gives the results from MARC finite element analysis. A shell element model was used, with full non-linear material properties of the aluminium included. In order to model the natural imperfections in the sections, it was found necessary to introduce local displacements into the model. These were obtained from the buckled shape of the component calculated from an eigenvalue buckling analysis. For aluminium, the maximum magnitude of the displacements were chosen as 10% of the aluminium gauge. It may be seen from Fig. 15 that the prediction is good, with the technique able to predict collapse loads above the 0·2% proof stress.

The problem with using MARC analysis for predicting collapse loads is that, since the analysis is non-linear, it takes a lot of computing time. As has been discussed for the analysis of the energy absorption of components, it is desirable to have the capability to perform the analysis quickly as part of the design cycle shown in Fig. 12.

The approach that has been developed for aluminium in the CRASH-CAD programme is to use an analytical approach in combination with the non-linear stress–strain curve. The approach is based on using the von Kármán effective width concept[28] for plates that buckle in the elastic range, and the Stowell theory[29] for plates that buckle in the plastic range.

In the von Kármán theory, it is assumed that at the point of ultimate strength, the stress in the compressed plate is carried by two strips adjacent to the supported edges. In the case of a plate supported on only one edge, the stress is assumed to be carried by only one strip. The effective width b_{eff} is given by

$$b_{eff} = Kt\sqrt{\frac{\pi^2}{12(1-\nu^2)}\frac{E}{\sigma_{lp}}} \tag{6}$$

where σ_{lp} is the limit of proportionality, ν is Poisson's ratio and K is an experimentally determined factor normally referred to as the buckling coefficient. For plates with a width larger than the effective width, the stress at maximum load will be below the limit of proportionality and

$$P_{max} = b_{eff} t \sigma_{lp} \tag{7}$$

For the plates that have a width smaller than the effective width, the stress at maximum load will be above the limit of proportionality. It is assumed that, for plates buckling plastically, this buckling stress is the stress at the maximum load of the plate. The Stowell theory is used to determine the critical strain ε_{cr} at buckling, which is given by

$$\varepsilon_{cr} = K\frac{\pi^2}{9}\left(\frac{t}{b}\right)^2\left(2+\sqrt{1+3\frac{E_t}{E_s}}\right) \tag{8}$$

where E_t is the tangent modulus, E_s is the secant modulus and K is an experimentally determined coefficient depending on the plate boundary conditions. Once ε_{cr} has been determined from eqn (8), the critical stress is obtained directly from the non-linear stress–strain curve. For plastic buckling, the critical buckling stress σ_{cr} is taken as the stress at maximum load, and therefore the maximum load is determined from

$$P_{max} = bt\sigma_{cr} \tag{9}$$

Figure 15 shows the correlation obtained using this approach for aluminium square extrusions. Good correlation is obtained for all the width-to-thickness ratios tested, with a maximum stress above the 0·2% proof stress being predicted for several of the sections.

Table 7 shows the maximum load predictions of CRASH-CAD using the theory described above for several different cross-sections. The correlation with experiment is very good and typically within 11%. One

Table 7
Comparison of Predicted and Measured Maximum Load Using von Kármán and Stowell Theories for Various Aluminium Section Geometries

Geometry				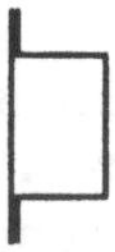	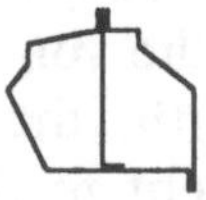
Maximum load (kN)	*CRASH-CAD*	26·5	99·0	57·0	86·5
	Experiment	25·1	103·1	55·0	98·0

of the most important aspects in predicting the maximum load of the section is the determination of the boundary conditions at the unloaded edges of the plate. In the predictions shown in Table 7, a lower-bound approach was used where, for the majority of the plates, the boundary conditions were assumed to be simply supported. The exception to this was for plates where spot-welding was used to join the section during fabrication. In the aluminium version of CRASH-CAD, the boundary conditions for spot-welded plates has been determined from the results of an experimental programme to obtain appropriate calibration factors.

With this theory implemented into the CRASH-CAD programme, it is possible for the designer to assess very quickly all of the aspects highlighted in the design cycle in Fig. 12 for the design of aluminium components subjected to axial crush. The work for the future is to develop a similar capability for the design of members subjected to bending collapse.

3.2 Bending Collapse

(i) Important Design Variables

Bending collapse of components is the other main deformation mode considered by designers. Figure 16 shows the experimental test arrangement that has been used to determine the bending collapse response of aluminium and steel members tested as cantilever beams. The three important aspects of performance for bending collapse are the maximum moment to initiate collapse, the energy absorbed and the mode of collapse.

Figure 17 shows the moment–rotation curves for three beams demonstrating three different modes of collapse. The first curve shows the performance of a beam that fails by local buckling. This is typical of

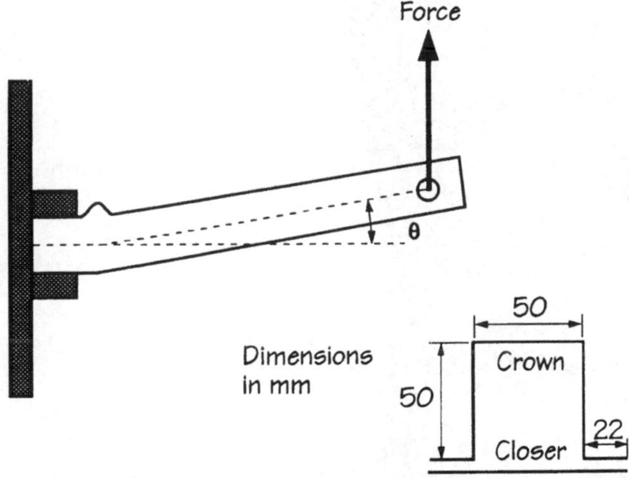

Fig. 16. Experimental test arrangement for bending collapse testing.

a beam with a high width-to-thickness ratio. Although the maximum moment and energy absorbed are low, the mode of collapse is stable and reliable. Figure 18 shows this type of failure in both an aluminium and a steel single top-hat section. It is seen that the mode of collapse is very similar in both sections.

The second curve in Fig. 17 shows the response of a beam that fails by tensile tearing. In this failure, the beam deforms predominantly plastically until local tensile strains are exceeded, with subsequent

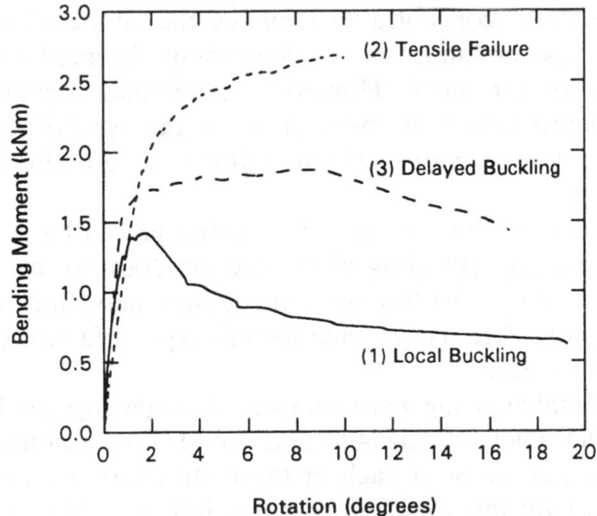

Fig. 17. Moment–rotation curves for aluminium beams under bending.

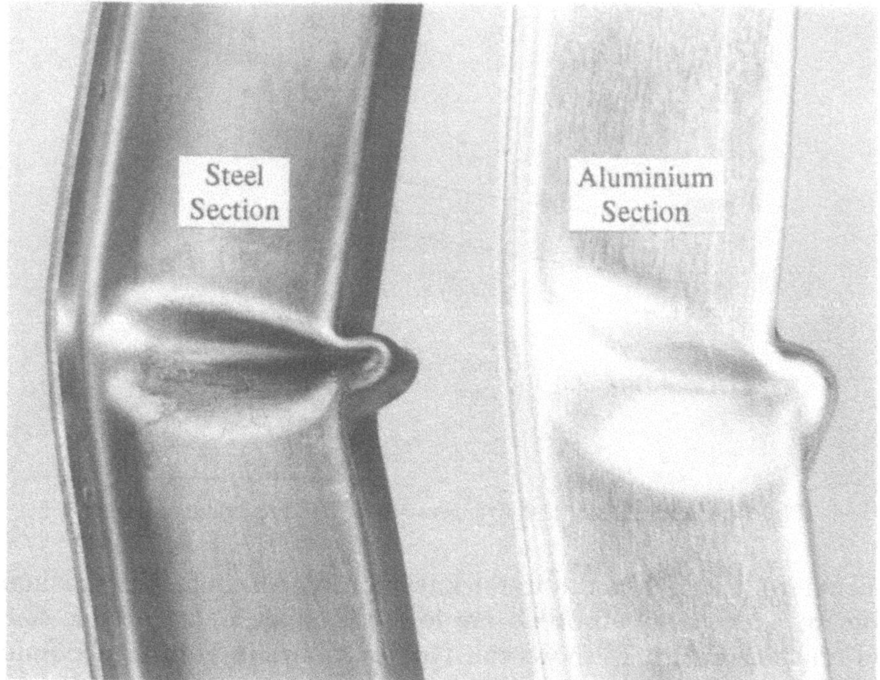

Fig. 18. Comparison of local buckling failure of aluminium and steel sections under bending.

failure of the beam. After failure, there is no residual load-carrying capacity. Although this failure mode gives the highest maximum moment and energy absorption, the abrupt loss of bending strength and sensitivity to stress concentrations makes this mode of failure undesirable. This type of failure is not commonly observed for aluminium and very rarely for steel. However, aluminium beams with small width-to-thickness ratios are more prone to this type of failure. Figure 19 shows the tensile and buckling failures in an aluminium single top-hat section.

The third curve in Fig. 17 shows the performance of a beam that fails by delayed buckling. Buckling of the section does not occur until well into the plastic range, with a subsequent high maximum moment and energy absorption. Beams designed for this type of failure give the most weight-efficient design.

Figure 20 compares the performance of single top-hat beams made from aluminium alloys AA5754-O and AA6111-T4 and mild steel. The cross-sectional geometry of each of the beams was the same, with the beams loaded with the crown in compression (Fig. 16). It can be seen that the weld-bonded 2 mm AA5754-O and the 1·2 mm spot-welded

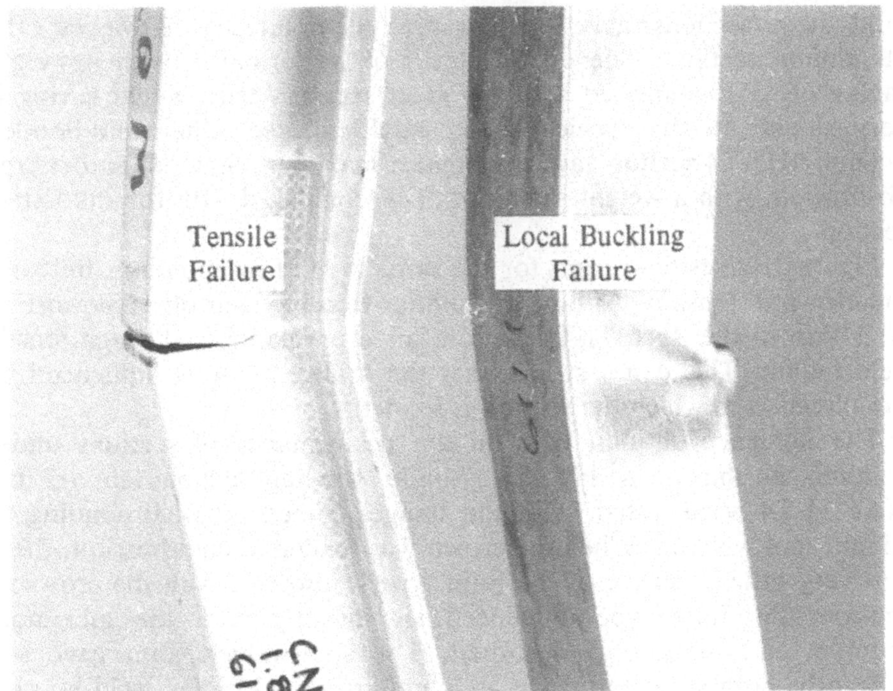

Fig. 19. Comparison of tensile and local buckling failure in an aluminium section under bending.

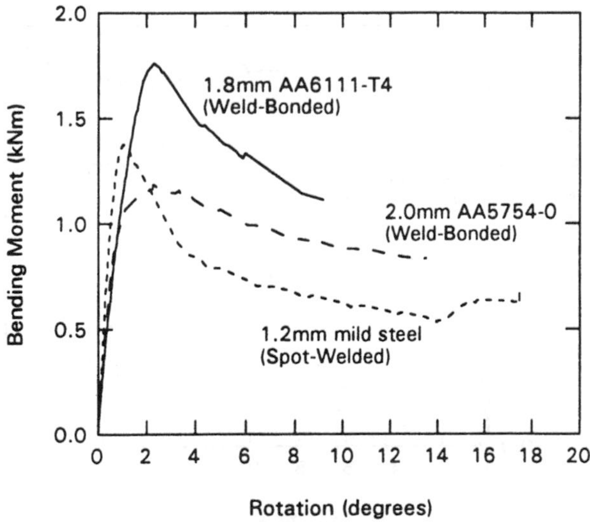

Fig. 20. Comparison of aluminium and steel sections under bending with crown in compression.

mild steel sections gave comparable performances. However, the aluminium section achieved this level of performance with a weight saving of 45% compared with the steel section. This weight saving is very similar to that achieved for axial collapse. The weld-bonded 1·8 mm 6111-T4 section gave the highest maximum moment and energy absorption, with a weight saving of 50% compared with the mild steel section.

Figure 21 gives the results for the same type of components, but with the crown in tension. Although a similar trend to that observed in Fig. 20 is shown, the AA6111-T4 section failed prematurely through tensile field failure. This demonstrates that the failure mode is influenced by the direction in which the section is loaded.

The influence of joint type on the performance of sections under bending is shown in Fig. 22. Single top-hat sections in 1·2 mm AA6111-T4 were tested, with the flanges joined by weld-bonding or 25 mm spot-welding. When the crown was loaded in compression, there was very little influence of the joint type. However, with the crown in tension, the joint type influenced significantly both the maximum moment and the energy absorbed. The spot-welded joint gave significantly poorer performance in comparison with the weld-bonded joint, with failure occurring through buckling between the spot-welds.

As has been mentioned, certain aluminium sections fail by tensile tearing. Figure 23 shows the moment–rotation curves for single top-hat sections of various width-to-thickness ratios, tested with the crown in

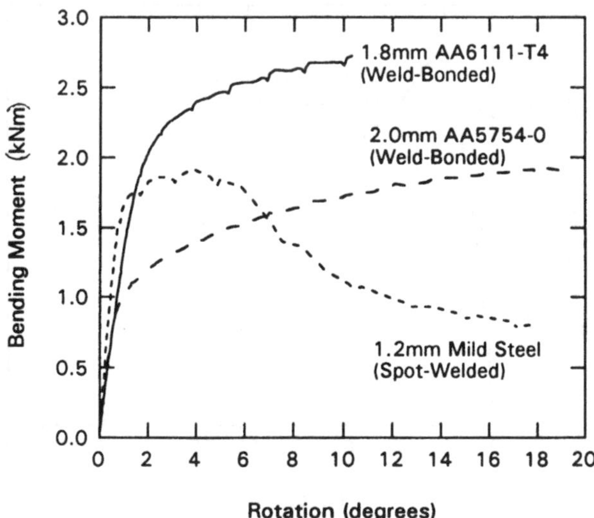

Fig. 21. Comparison of aluminium and steel sections under bending with crown in tension.

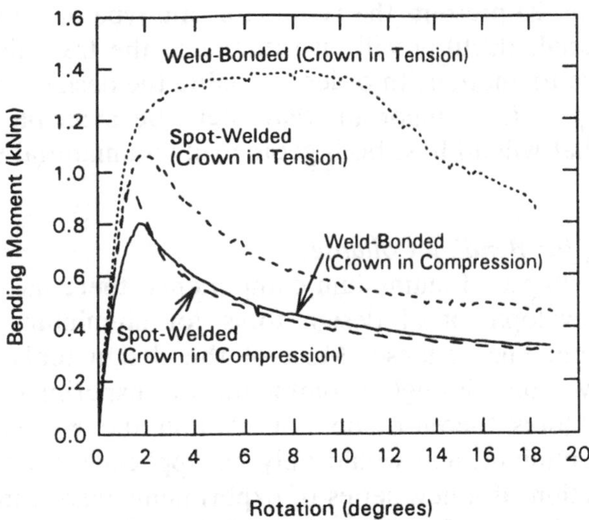

Fig. 22. Influence of joint type on the bending performance of aluminium single top-hat box sections.

tension. The single top-hat was a weld-bonded AA5754-O 60 mm × 60 mm section tested as a cantilever beam with the material thickness as the only variable. It is seen from Fig. 23 that, with width-to-thickness ratios less than 30, thc failure of this type of section is by tensile tearing. With width-to-thickness ratios of 30 or greater, the failure is by local buckling of the closer plate and sidewall.

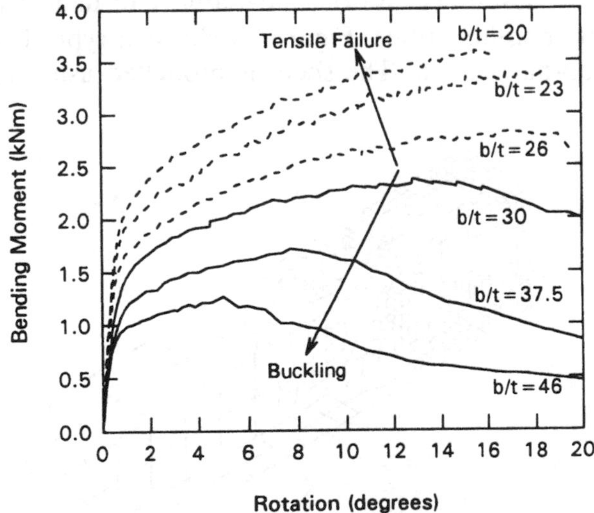

Fig. 23. Influence of wall width-to-thickness ratio on moment–rotation curves of aluminium box sections.

Although Fig. 23 presents the results for one type of geometry with a constant material, ductility will also influence the transition from one mode of failure to another. In order to enable the designer to design for bending collapse, it is important that there be available appropriate design tools that will address both geometric and material variables.

(*ii*) *Designing for Bending Collapse*

As for axial collapse of aluminium components, there has been almost no specific development of design tools for aluminium components subjected to bending collapse. Most of the design tools for steel[30–32] have been developed through a combination of experiment and theory. These design tools therefore do not contain the correct calibration factors for aluminium, and would only be applicable for the design of aluminium sections if a new series of experiments were conducted.

For axial collapse, the CRASH-CAD programme has been developed for aluminium using a limited number of experimental results. The superfolding elements used in that package are now being developed and extended for use in bending collapse.[33] Work is also in progress to develop the package for the bending collapse of aluminium components. However, for the present, finite element modelling has been found to be the most useful analysis technique for predicting the performance of aluminium beams subjected to bending.

The finite element package MARC has been used to predict the performance of steel and aluminium sections under bending collapse. This code was chosen because of its capability to handle large material and geometric non-linearities. Figure 24 shows a typical MARC model of a single top-hat section. The sheet is modelled using shell elements,

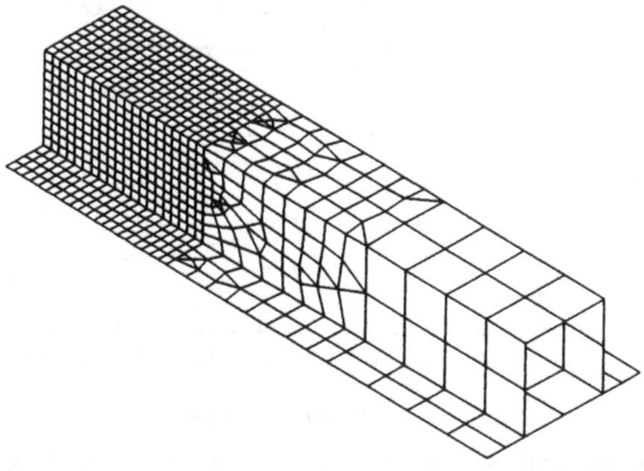

Fig. 24. MARC model of a single top-hat section.

with the flanges modelled as shells with double thickness. It is not possible to obtain accurate prediction of the maximum moment without introducing imperfections into the model to account for the natural imperfections present in the real component. This calibration is introduced in the form of local displacements obtained from the buckled shape of the component calculated from an eigenvalue buckling analysis. For aluminium, the magnitude of the maximum initial displacement was chosen as 10% of the aluminium gauge.

Figure 25 compares the prediction of MARC analysis with experimental results for weld-bonded AA6111-T4 and spot-welded mild steel 50 mm × 50 mm single top-hat sections. It can be seen that very good prediction of the shape of the curves for both the aluminium and the steel beams was obtained. This prediction of the shape of the curve may also be used to assess the potential for failure of the section by tensile failure. A predicted response such as that in Fig. 25 for the 6111-T4 section with crown in tension would indicate that failure by tensile tearing was possible, as indeed was observed in the experimental test of this component.

However, for general prediction of tensile failure, accurate predictions of the strain distributions within the component are required, as well as a good failure criterion, such as a fracture strain limit diagram similar to the forming limit diagram used in material forming predictions.[34] This is an area of active research that will benefit both material suppliers and end users of the material. It will enable material

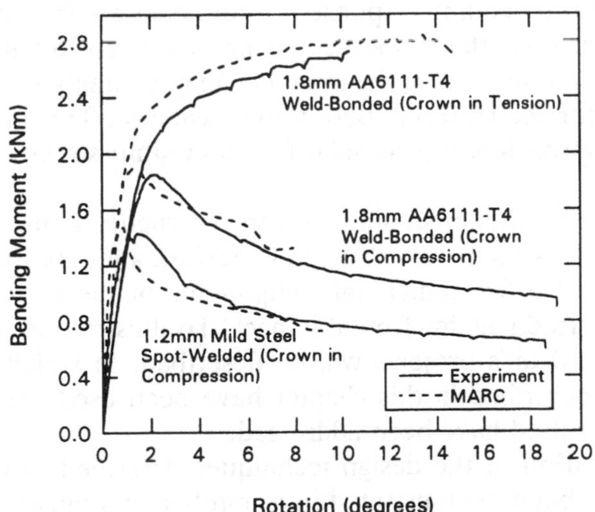

Fig. 25. Comparison of predicted and measured moment–rotation curves using MARC.

suppliers to establish the material properties required for typical applications, and it will enable end users to establish more effectively the best design forms for avoiding tensile failure.

With the development of a better understanding of the performance of aluminium members under bending collapse and the development of appropriate design tools, it will be possible for the designer to design components that fail in modes, such as delayed buckling, that give the most weight-efficient solutions. This will assist in the development of weight-efficient aluminium automotive structures.

4 DESIGN OF ALUMINIUM STRUCTURES

The preceding sections have discussed the design parameters and the design tools that may be used to design aluminium components for use in aluminium impact structures. It has been demonstrated that the parameters that influence steel performance also influence the performance of aluminium components. Similarly, it has been shown that, with calibration, the same design tools developed for steel may also be used very effectively for aluminium. This is also true when it comes to the design of aluminium structures.

In the analysis of steel impact absorbing structures, the structure is often modelled as a collection of beams that behave as non-linear springs.[35,36] The non-linear spring responses are obtained either from analysis or testing of the actual beams in the structure. Normally both the axial and the bending responses of the beam are included. Hence in this approach, once the properties of the beams are obtained from the component analysis or test, the structure may be analysed without need for concern for the material used in the structure. This means that this modelling approach is just as valid for aluminium structures as it is for steel.

However, for the modelling of some structures and impact load cases, the assessment of structural performance is based on the performance of a few individual components of the structure, such as the front longitudinals for frontal impact. To illustrate this, an example from one of Alcan's projects will be described, in which many of the design tools described in this chapter have been used and where some of the issues raised have been addressed.

The application of the design techniques described in the preceding sections have been demonstrated in a number of vehicle projects. One such project was the replacement of the complete steel front structure of a Volvo 960 with aluminium. The objective of this project was to demonstrate that the same performance targets could be achieved in aluminium as with steel, but with significant weight savings.

In the design of the structure, the issues of energy absorption, stable progressive collapse, and the strength and stability of the support structure had to be addressed. The main front impact structure of the Volvo 960 vehicle comprises a pair of front longitudinal members running from the bumper down, at a slight angle, to below the front floor of the passenger compartment. The primary energy absorption is by the front parts of these longitudinals, which are of a rectangular closed section. The rear parts of the longitudinals provide the support for these impact members, but are also expected to deform later in the impact event.

A constraint on the design of the aluminium impact energy absorbing components was to achieve the original steel performance, but without increasing the external dimensions of the components for reasons of assembly of the final vehicles. The initial calculations for the impact members were performed using the CRASH-CAD design software. This provided an initial aluminium cross-section design, which was subsequently slow crush tested to establish the progressive collapse of the component and to indicate maximum and average collapse forces.

The strength and stability of the rear support structure was calculated with simple hand techniques. Engineers' theory of bending was used in combination with the Engineering Sciences Data Unit design sheets for buckling stability[37] and the von Kármán theory for the maximum load. The ESDU sheets use a very similar analysis approach for buckling as that in Ref. 27, and generally give a conservative estimate of buckling strength.

Further slow crush testing was conducted to include the influences of bending and the effects of a bumper shock absorber that was installed inside the impact member and which was known to influence the folding of the section and hence the collapse forces. Figure 26 shows the test specimen after testing and Fig. 27 the force–displacement curve. It can be seen that a good progressive collapse was achieved. However, it should be noted that Fig. 27 shows an increase in the collapse force during the collapse as a result of the shock absorber and the folds in the collapsing section interacting.

Testing then progressed to a series of dynamic impact sled tests, which involved a pair of front longitudinals complete with the full bumper system. This resulted in minor modifications to increase the suppport structure strength, and to reduce the maximum forces that occurred later in the collapse process owing to the interaction of the internal damper with the developing folds of the impact member.

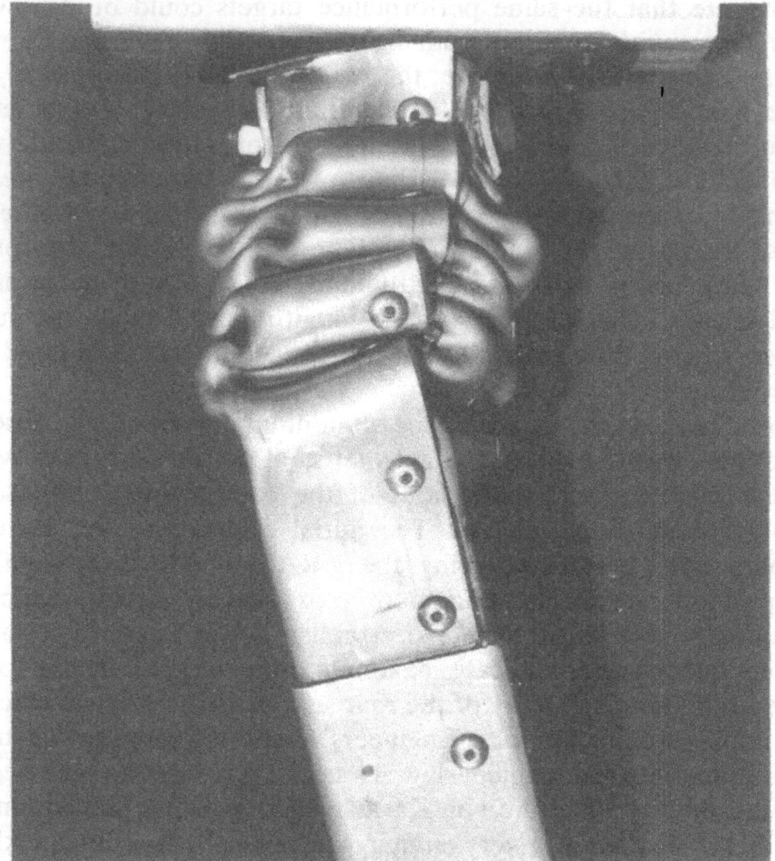

Fig. 26. Volvo 960 aluminium front longitudinal after slow crush testing.

The final design of the impact member, which successfully withstood the full vehicle crash test at 35 mph (55 km/h), is shown in Fig. 28. This was a fabricated, rivetted and bonded rectangular section in AA5754-O aluminium alloy, with angle stiffeners in the longer vertical side walls. The original steel section is also shown in Fig. 28 for comparison. The weight saving of the aluminium impact member over the original steel design was 40%. The weight saving for the complete front structure, which successfully met the other structural performance criteria of stiffness, strength and fatigue durability, was approximately 40%. This is consistent with the weight savings achieved in many of Alcan's projects as demonstrated in Table 1.

5 CONCLUSIONS

It has been demonstrated that it is possible to obtain weight savings of around 45% in the impact structure of aluminium automotive structures

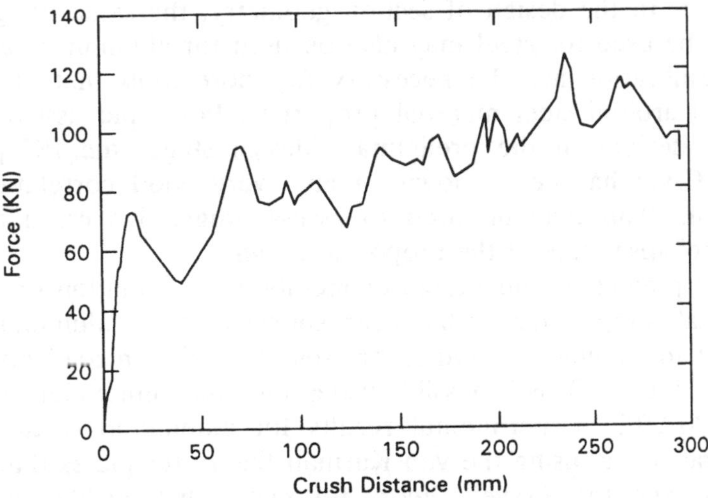

Fig. 27. Force–displacement curve for slow crush test on a Volvo 960 aluminium front longitudinal.

in comparison with an equivalent mild steel structure. This may be achieved while still meeting the required structural performance in terms of crashworthiness and the other structural performance criteria such as stiffness, strength and fatigue durability. Although the strength of aluminium is generally less than that of steel, the low density and the thicker gauges of material that can be used in aluminium structures results in high energy absorption per unit mass.

Testing has demonstrated that section geometries that work for steel generally work for aluminium, and that the geometry of the section is important in determining the mass-specific energy absorption of the

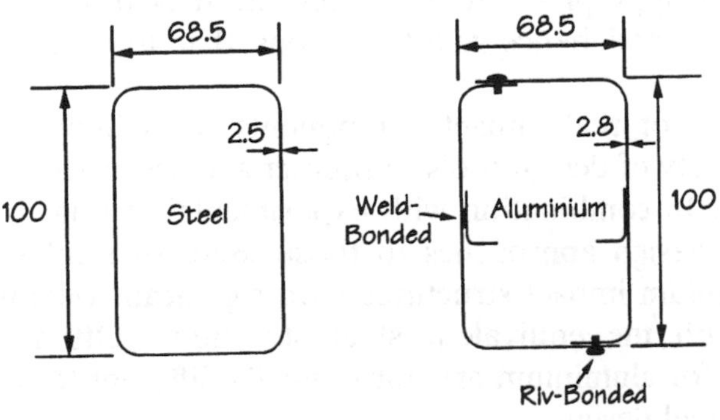

Fig. 28. Comparison of aluminium and steel designs of a Volvo 960 front longitudinal.

component. In the design of section geometry, the same design tools that may be used for steel may also be used for aluminium, although some recalibration may be necessary for those tools that have been developed around steel material properties. For rapid assessment of proposed designs in the preliminary design stage, the PC package CRASH-CAD has been shown to give very good correlation with experiment. This may be used to assess progressive crush and the average collapse force of the proposed design.

For the prediction and design of members for initiation of collapse under axial compression, it has been shown that, for aluminium, it is important to be able to predict stresses at maximum load above the 0·2% proof stress. This is possible using non-linear finite element codes such as MARC. Experimental results for aluminium sections have demonstrated that using the von Kármán theory for plates that buckle elastically, and the Stowell theory for plates that buckle plastically, gives very good correlation between prediction and experiment. This approach has been incorporated into a version of CRASH-CAD for aluminium, and may be used to predict the load to initiate collapse with high accuracy, even in components with complex cross-sections.

In bending collapse of aluminium components, weight savings of around 45% were achieved compared with equivalent steel sections with the same performance. This weight saving is similar to that achieved for axial collapse. The joint type was also shown to be important for bending collapse, with weld-bonding giving a better performance for certain loading directions compared with spot-welding.

Occasionally, aluminium sections fail by tensile failure, and experiments have demonstrated that aluminium beams with small wall width-to-thickness ratios are more prone to this type of failure. For the prediction of the performance of aluminium beams in bending collapse, a finite element analysis code such as MARC has been identified as being the most appropriate tool at present. It is possible, using this tool, to identify and hence avoid sections that are likely to fail by tensile failure.

In the design of real impact components, it is important to have available a variety of design tools to predict and assess the performance of the designs. In combination with experimental testing, it is possible to use similar design approaches to those used for steel structures to produce aluminium impact structures with significant weight savings in comparison with the equivalent steel structures. Often, the design solutions used for aluminium are not radically different to those used in conventional steel design.

ACKNOWLEDGEMENTS

The authors would like to thank Volvo for their permission to publish the results of the design project. They would also like to express their thanks to Steve Buckley and the rest of their colleagues for their contributions, and the managements of Alcan International Limited and Alcan Automotive Structures (UK) Limited for their permission to publish this paper.

REFERENCES

1. Sadeghi, M. M. & Tidbury, G. H., Bus roll over accident simulation. In *Proceedings of the 3rd International Conference on Vehicle Structural Mechanics, Troy, MI,* SAE, 1979, pp. 255–65.
2. Rivet, R. M. & Riches, S. T., Drop weight impact testing of spot-welded, weld-bonded and adhesively bonded box sections. Welding Institute Report 9431.01/85/478.3, 1986.
3. Seeds, A. & Sheasby, P. G., Evaluation of adhesive joining systems in aluminium box beams. SAE Technical Paper Series 870152, 1987.
4. Nardini, D. & Seeds, A., Structural design considerations for bonded aluminium structured vehicles. SAE Technical Paper Series 890716, 1989.
5. McGregor, I. J., Seeds, A. D. & Nardini, D., The design of impact absorbing members for aluminium structured vehicles. SAE Technical Paper Series 900796, 1990.
6. Wheeler, M. J., Sheasby, P. G. & Kewley, D., Aluminium structured vehicle technology—a comprehensive approach to vehicle design and manufacturing in aluminium. SAE Technical Paper Series 870146, 1987.
7. Selwood, P. G., Law, F. J., Sheasby, P. G. & Wheeler, M. J., The evaluation of an adhesively bonded aluminium structure in an Austin–Rover Metro vehicle. SAE Technical Paper Series 870149, 1987.
8. Seeds, A. D., Nardini, D. & Cassese, F., The development of a centre cell structure in bonded aluminium for the Ferrari 408 research vehicle. SAE Technical Paper Series 890717, 1989.
9. Warren, A. S., Wheatley, J. E., Marwick, W. F. & Meadows, D. J., The building and test-track evaluation of an aluminium structured Bertone X1/9 replica vehicle. SAE Technical Paper Series 890718, 1989.
10. Yu, J. & Jones, N., Further experimental investigations on the failure of clamped beams under impact loads. *Int. J. Solids Structures,* **27** (1991) 1113–37.
11. Abramowicz, W. & Jones, N., Dynamic axial crushing of square tubes. *Int. J. Impact Engng,* **2** (1984) 179–208.
12. Abramowicz, W. & Jones, N., Dynamic axial crushing of circular tubes. *Int. J. Impact Engng,* **2** (1984) 263–81.
13. Abramowicz, W. & Jones, N., Dynamic progressive buckling of circular and square tubes. *Int. J. Impact Engng,* **4** (1986) 243–70.
14. CRASH-CAD, Licensed by Impact Design Inc., Winchester, MA.

15. Wierzbicki, T. & Abramowicz, W., CRASH-CAD—a computer program for design of columns for optimum crash. SAE Technical Paper Series 900462, 1990.
16. Wierzbicki, T. & Abramowicz, W., On the crushing mechanics of thin-walled structures. *J. Appl. Mech.*, **50** (1983) 727–34.
17. Abramowicz, W. & Wierzbicki, T., Axial crushing of multicorner sheet metal columns. *J. Appl. Mech.*, **56** (1989) 113–20.
18. Wierzbicki, T. & Abramowicz, W., The mechanics of deep plastic collapse of thin walled structures. In *Structural Failure*, ed. T. Wierzbicki & N. Jones. Wiley, New York, 1989, pp. 281–329.
19. Crash energy management. Automotive Steel Design Manual, Rev. 2, American Iron and Steel Institute, 1988.
20. Magee, C. L. & Thornton, P. H., Design considerations in energy absorption by structural collapse. SAE Technical Paper Series 780434, 1978.
21. Mahmood, H. F. & Paluszny, A., Design of thin walled columns for crash energy management—their strength and mode of collapse. 4th International Conference on Vehicle Structural Mechanics, Detroit, 1981, SAE Paper 811302.
22. Shirasawa, K., Akamatsu, T. & Ohkubo, Y., Mean crushing strength of closed hat section members. *Bull. JSAE,* **7** (1976) 107–14.
23. LS-DYNA3D, Licensed by Livermore Software Technology Corporation, Livermore, CA.
24. Wierzbicki, T. ed., *Impact Design* No. 1. Impact Design Inc., Winchester, MA, 1990.
25. Bulson, P. S., *The Stability of Flat Plates*. Chatto & Windus, London, 1970.
26. MARC. MARC Analysis Research Corporation, USA.
27. Marsh, C., *Strength of aluminium,* 5th edn. Alcan Canada Products Ltd, 1983.
28. von Kármán, T., Sechler, E. & Donnell, H. L., The strength of thin plates in compression. *Trans. ASME,* **54** (1932) 53–7.
29. Iyengar, N. G. R., *Structural Stability of Columns and Plates*. Wiley, New York, 1988.
30. Kecman, D., Bending collapse of rectangular section tubes and square section tubes. *Int. J. Mech. Sci.,* **25** (1983) 623–36.
31. Kecman, D., Program WEST for optimisation of rectangular square section tubes from the safety point of view. 4th International Conference on Vehicle Structural Mechanics, Detroit, 1981, SAE Paper 811312.
32. Kecman, D., Deep bending collapse of thin walled beams and joints. *Manual of Crashworthiness Engineering,* Vol. 5. Centre for Transportation Studies, MIT, Cambridge, MA, 1989.
33. Abramowicz, W., On the development and implementation of superbeam element. PAM'91, ESI International Seminar, Paris, La Villette, November 1991.
34. Leroy, G. & Embury, J. D., The utilisation of failure maps to compare the fracture modes occuring in aluminium alloys. In *Formability Analysis, Modelling and Experimentation,* ed. S. S. Hecker, A. K. Ghosh & H. L. Gegel. The Metallurgical Society of AIME, Chicago, 1977, pp. 183–207.

35. Sadeghi, M., Test analysis interaction in crash simulation of automotive structures. *Modern Vehicle Design Analysis: International Journal of Vehicle Design Conference, London, 21–24 June 1983.*
36. Wardill, G. A. & Kecman, D., Theoretical prediction of the overall collapse mode and maximum strength of bus structure in a roll over situation. 18th International Congress, FISITA, Hamburg, 1980, Transport Systems, Paper 1.7.4, p. 281.
37. ESDU, Vol. 11, Data Items 010108, 010119; Vol. 1, Data Item 76016.

Chapter 11

MOTORWAY IMPACT ATTENUATION DEVICES: PAST, PRESENT AND FUTURE

J. F. CARNEY III

Department of Civil Engineering, Vanderbilt University, Nashville, Tennessee, 37235, USA

ABSTRACT

The energy dissipation characteristics of impact attenuation systems employed in motorway safety applications are examined. Such devices have a life-saving potential during an impact event, where kinetic energy must be dissipated in a controlled manner such that decelerations remain within acceptable limits. The first part of this chapter deals with the mathematical modeling and performance characteristics of various energy dissipating safety devices that have been proven effective in saving lives and reducing injury severities during crash events. Both guardrail end treatments and crash cushions are considered. Attention in directed in the second part of the chapter to several motorway safety issues that must be addressed as the 21st century approaches. These issues include the prospects for developing maintenance-free, reusable safety hardware, the need for developing more accurate analytical occupant risk tools and more powerful and user-friendly crash simulation software, and the importance of considering non-tracking, side impacts when developing motorway safety hardware.

1 INTRODUCTION

Motor vehicle related accidents are a major worldwide health problem, and constitute a great economic loss to society. For example, vehicular crashes kill more Americans between the ages of 1 and 34 than any other source of injury or disease.[1] Put another way, for almost half the average life span, people are at greater risk of dying in a roadway crash than in any other way. In the USA, more than 95% of all transportation deaths are motorway-related, compared with 2% for rail and 2% for air.[2] The yearly worldwide societal costs of motorways deaths and injuries runs into hundreds of billions of pounds. Indeed, the productive or potential years of life that are lost prior to age 65 as a result of motor vehicle related injuries or death are greater than those lost to cancer or heart disease.[3]

Measures are being taken to reduce the billions of pounds lost in medical expenses, earnings, insurance claims, and litigation, as well as the intangible costs associated with human suffering. One important contribution to improved motorway safety has been the development of impact attenuation devices that prevent errant vehicles from crashing into fixed object hazards that cannot be removed, relocated, or made breakaway. These devices have existed since the 1960s, and many technical improvements and innovative designs have been developed in the last 25 years. A motorway impact attenuation system is usually designed to gradually decelerate a vehicle to a safe stop under head-on impacts, or to redirect the vehicle away from a hazard under side impact conditions. Their prudent use has saved numerous lives through the reduction of accident severities. It is not the aim of this chapter to discuss the energy dissipation characteristics and relative effectiveness of each and every one of these devices. Instead, attention is directed at describing the response characteristics of certain systems that, in the opinion of the author, have proven to be or show great promise of becoming particularly effective energy dissipation hardware. Special attention is given to these devices which lend themselves to the mathematical modeling of their impact responses. Finally, several new design concepts and safety issues that will become important in the near future are discussed.

2 INJURY MECHANISMS AND OCCUPANT RISK CRITERIA IN CRASH EVENTS

The occupant of a vehicle can be injured in a crash as a result of the occurrence of one or more of the following four events:

- crushing of the occupant compartment
- impact with part of the vehicle interior
- unacceptably high decelerations
- ejection

Research is ongoing on the biomechanics of injury and the associated occupant risk criteria that should be employed to determine the severity of crash events. Many of the advances in this general area have been made in the research laboratories of automobile manufacturers and in research supported by government departments of transportation. In the 1960s the reality of US traffic fatalities occurring at a rate of 1000 per week prompted the Federal Highway Administration to initiate a research and development program to provide rapid improvement in highway safety. The development of roadside safety appurtenances was an important part of this safety program, and a variety of devices have evolved during the last 25 years. The installation of these devices on the roadway system has substantially reduced the severity of many accidents.

The first recommended procedures for performing full-scale crash tests in the United States were contained in the single-page Highway Research Board Circular 482 published in 1962.[4] This document specified a 17·8 kN test vehicle, two impact angles (7° and 25°), and an impact velocity of 96·6 km/h for testing guardrails. In 1974, an expanded set of procedures and guidelines were published as NCHRP Report 153.[5] This report was the first comprehensive specification that addressed a broad range of roadside hardware, including longitudinal barriers, terminals, transitions, crash cushions, and breakaway supports. Specific evaluation criteria were presented, as were specific procedures for performing tests and reducing test data. In the years following the publication of Report 153, a wealth of additional information regarding crash-testing procedures and evaluation criteria became available, and in 1976 Transportation Research Board Committee A2A04 was given the task of reviewing Report 153 and providing recommendations. The result of this effort was Transportation Research Circular No. 191.[6] As this was being published, a new NCHRP project was initiated to update and revise Report 153. The result of this NCHRP project was Report 230,[7] published in 1981 and updated in 1992.

Report 230 covers the test procedures and evaluation criteria to be followed in the United States in evaluating the effectiveness of roadside safety hardware. A full-scale crash testing program is specified, and the performance of a device is judged on the basis of three factors: (i) structural adequacy, (ii) occupant risk, and (iii) vehicle trajectory after collision. The structural adequacy of an appurtenance is evaluated by its ability to interact with a selected range of vehicle sizes and impact conditions in a predictable and acceptable manner. The unit should remain intact during impact so that detached debris will not present a hazard to traffic. The occupant risk evaluation of a highway appurtenance is a surrogate measure of the response of a hypothetical vehicle occupant during the vehicle attenuator impact. The vehicle kinematics are used to estimate the impact velocity and ridedown accelerations of the occupant, and limiting values are recommended. Another essential crash test requirement of NCHRP Report 230 is that the impacting vehicle remain upright during and after collision and that the integrity of the passenger compartment be maintained. The vehicle trajectory after collision is of concern because of the potential risk to other traffic. An acceptable vehicle trajectory after impact is characterized by minimal intrusion into adjacent traffic lanes.

The Report 230 procedures for acquiring occupant risk data involve the numerical integration of acceleration–time data. Of primary concern are the magnitudes of the hypothetical occupant impact velocity with the interior of the vehicle and the maximum 10 ms average deceleration of the occupant following this impact. The recommended threshold values for occupant impact velocity and ridedown deceleration are 12 m/s and 20*g* respectively.

The value of the occupant impact velocity depends on the 'flail' distance available before impact occurs. Report 230 assumes the occupant to be a rigid body whose acceleration $\mathbf{a}_o$ can be expressed as

$$\mathbf{a}_o = \mathbf{a}_v + \mathbf{a}_{o/v} \tag{1}$$

where $\mathbf{a}_v$ is the acceleration of the reference frame moving with the vehicle and $\mathbf{a}_{o/v}$ is the acceleration of the occupant relative to the moving reference frame. Assuming an unrestrained occupant has no acceleration until colliding with the vehicle interior, eqn (1) becomes

$$|\mathbf{a}_{o/v}| = |\mathbf{a}_v| \tag{2}$$

Furthermore, the velocity of the occupant $\mathbf{v}_o$ is

$$\mathbf{v}_o = \mathbf{v}_v + \mathbf{v}_{o/v} \tag{3}$$

where $\mathbf{v}_v$ is the velocity of the reference frame moving with the vehicle and $\mathbf{v}_{o/v}$ is the velocity of the occupant relative to the moving reference frame. Since $\mathbf{v}_0$ is a constant vector, $\mathbf{C}$, prior to impact with the vehicle interior, eqn (3) beomces

$$\mathbf{C} = \mathbf{C} - \int \mathbf{a}_v \, \mathrm{d}t + \mathbf{v}_{o/v} \tag{4}$$

which yields

$$\mathbf{v}_{o/v} = \int \mathbf{a}_v \, \mathrm{d}t \tag{5}$$

Equations (2) and (5) form the basis for calculating the occupant/compartment impact velocities and the ridedown accelerations given in Report 230.

Another important measure of occupant risk deals with injuries to organs within the thoracic cage. Damage to the liver, kidneys, and/or spleen can be life-threatening. Extensive lateral impact tests have been performed on human cadavers and surrogate specimens to determine physiological responses and develop an injury index. The result is the thoracic trauma index (TTI),[8] which can be expressed in the form

$$\mathrm{TTI} = 0{\cdot}5(G_\mathrm{r} + G_\mathrm{ls}) \tag{6}$$

where G_r is the greater of the peak of either the upper or lower rib acceleration in *g*s and G_ls is the lower spine peak acceleration in *g*s. Life-threatening injuries are unlikely when the TTI is less than 100.

Head injuries have been recognized for many years as being among the most debilitating types of trauma experienced in accidents. Injuries sustained by the head and brain are difficult to treat, and often result in long-term disfunction. Such injuries often involve great cost to society either because of losses due to an early death or the costs of long-term treatment and loss of productivity. In the United States, it is estimated that approximately 135 000 persons are hospitalized each year for brain injuries as a result of motor vehicle accidents. The in-hospital costs of these injuries is of the order of 370 million dollars. In a recent examination of 49 143 trauma patients in 95 trauma centers, Gennarelli *et al.*[9] discovered that, although patients with head injuries accounted for one-third of the trauma cases, they represented two-thirds of the trauma deaths.

Tests on embalmed cadaver heads have led to the following head injury criteria (HIC):[10]

$$\mathrm{HIC} = \left[\frac{1}{t_2 - t_1}\int_{t_1}^{t_2} a\,\mathrm{d}t\right]^{2{\cdot}5}(t_2 - t_1) \tag{11.7}$$

where a is the resultant head acceleration, and t_1 and t_2 are two points in time during any interval in which the head is in continuous contact with a part of the vehicle other than the belt system. Values of HIC greater than 1000 are to be avoided.

3 GUARDRAIL END TREATMENTS

The most commonly employed guardrail system in much of the world consists of a galvanized steel beam shaped in the form of a W (W-beam) and supported on steel or wood posts. Early installations of these W-beam longitudinal barriers were constructed with untreated, blunt ends. This design resulted in many severe accidents, which were sometimes characterized by a piercing of the occupant compartment of the errant vehicle by the sharp end of the W-beam section, as illustrated in Fig. 1. An inexpensive design modification has been extensively employed to reduce the impact severity associated with an untreated, blunt end collision. The end of the W-beam guardrail is twisted through 90° and sloped to the ground.[11,12] This eliminated the 'spearing' problem, but vehicles were often launched into the air by the sloped W-beam, resulting in potentially serious rollover accidents.

Fig. 1. Guardrail spearing accidents.

Fig. 2. Breakaway cable terminal.

In the early 1970s, a breakaway cable terminal (BCT) guardrail end treatment was developed to minimize the spearing and rollover problem.[13] The BCT design, shown in Fig. 2, consists of two wood posts that are designed to fracture on impact, allowing the rail element to bend away from the vehicle, which then passes behind the terminal. For redirectional hits away from the end of the terminal, it is necessary to develop tensile strength in the rail. This is provided by means of a cable that transfers the tensile forces from the W-beam to the base of the end post.

The impact performance of the BCT is greatly affected by the specific impact conditions and the way in which the barrier end is flared away from the roadway service. Furthermore, even correctly installed BCT systems develop unacceptably high ridedown decelerations when impacted with light vehicles, and side impacts with non-tracking vehicles continue to be a serious problem.

3.1 The Guardrail Extruder

An ingenious new end treatment for guardrails was developed in the late 1980s that employs the guardrail itself to dissipate kinetic energy. The concept involves placing an extruder over the end of a straight W-beam guardrail.[14,15] Energy is dissipated in the extrusion process by deforming and deflecting the W-beam. Consider the squeezer–deflector system shown in Fig. 3(a). During the impact event, the rail is first forced through the squeezer, where its cross-sectional height is reduced from 82·6 mm to 25·4 mm. The deflector then curves the rail away from the path of the the impacting vehicle. The analytical model for this complicated mechanical process was developed by Quersky[14] and involved the modeling of the following subprocesses:

- transverse and longitudinal deformations and frictional losses in the squeezer;
- deformation of the flattened W-beam in the deflector;
- frictional losses due to static and dynamic forces in the deflector;
- mass acceleration effects in the longitudinal direction.

In the transverse direction, the squeezer section transforms the cross-sectional shape. An idealized free-body diagram of the assumed collapse mechanism is presented in Fig. 3(b), in which f_1 and f_2 are friction forces. Applying the equations of equilibrium yields the following expression for the force P of a reference cross-section of W-beam:

$$P = M_p \frac{b_1 + 2b_2 - \mu h}{(b_1 - \mu h)(b_2 - \mu h)} \tag{8}$$

where b_1, b_2, and h are defined in Fig. 3(b), μ is the coefficient of friction, and

$$M_p = \tfrac{1}{4}\sigma_y b t^2 \tag{9}$$

where σ_y is the yield stress, b is the depth of the W-beam along the y-axis, and t is the thickness of the W-beam. The process is discretized to yield the total force over the guardrail–squeezer interface. It is of interest to note that, although the squeezer section terminates with a 25·4 mm opening, the W-beam becomes completely straightened during the squeezing process.

For a given differential element, summation of forces in the longitudinal direction yields the force required to sustain the squeezing process, F_s, which can be expressed in the form

$$F_s = 4M_p \frac{\sin\theta + \mu\cos\theta}{\cos\theta - \mu\sin\theta} \frac{b_1 + 2b_2 - \mu h}{(b_1 - \mu h)(b_2 - \mu h)} \tag{10}$$

where $\tan\theta$ is the slope of the squeezer section boundary.

Once the W-beam is flattened in the squeezer section, the deflector causes it to curl away from the impacting vehicle. Energy is dissipated in plastically deforming the flattened W-beam and through friction over the guardrail–deflection interface. Furthermore, as the extruder is forced over the guardrail by the impacting vehicle, the guardrail is

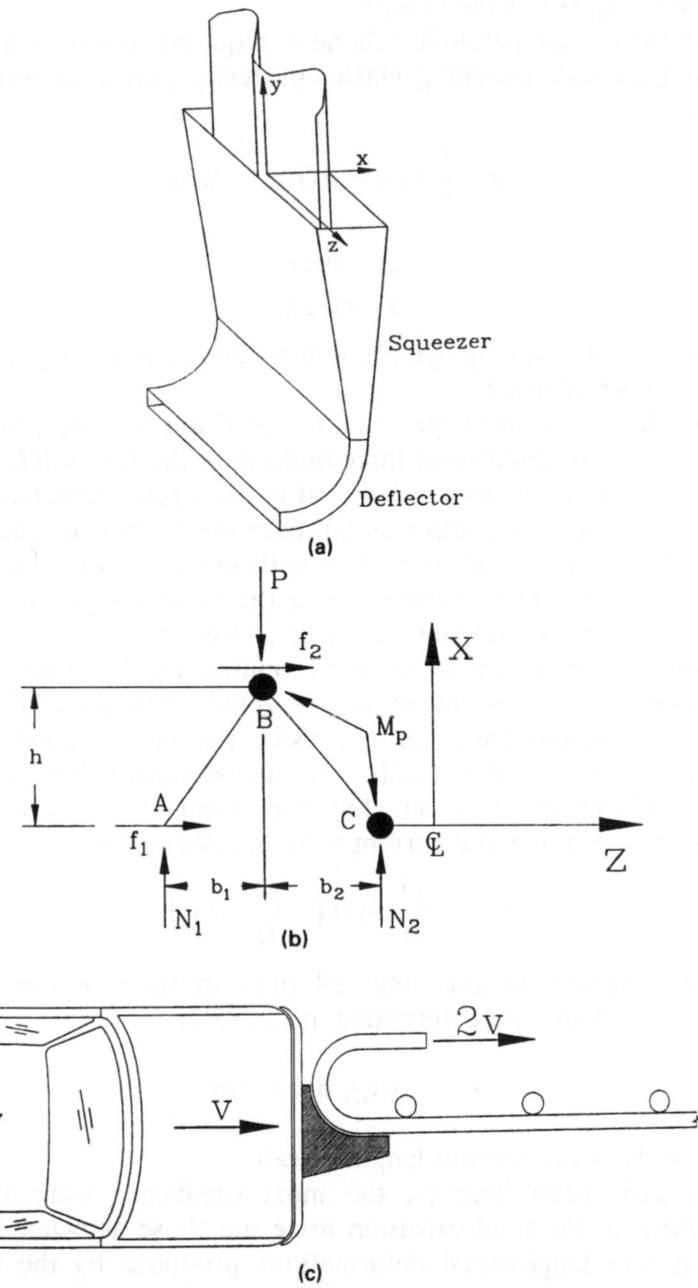

Fig. 3. The guardrail extruder.[14] (a) squeezer–deflector device; (b) idealized collapse mechanism; (c) mass-acceleration effect.

curled 180° and pushed ahead of the impacting vehicle at twice the instantaneous speed of the vehicle.

The strain energy per unit volume e required to curl a length l of flattened guardrail, assuming elastic–perfectly plastic constitutive behavior, is

$$e = \int_e \sigma \, d\varepsilon = E(\varepsilon_y \varepsilon_1 - \tfrac{1}{2}\varepsilon_y^2) \tag{11}$$

where

$$\varepsilon_y = \sigma_y / E \tag{12}$$

$$\varepsilon_1 = t/2R \tag{13}$$

Equation (11) can be employed to determine the force required to curl a unit length of W-beam.

Friction forces are developed as a result of the bending process itself and because of the centrifugal force buildup as the rail is deformed into a circular arc. Both effects are handled in an approximate manner, and the force distribution extends for 180° because the rail continues to curl in a circular arc after it loses contact with the deflector. These friction force contributions play minor roles in the complete process, and the details of their derivation may be found in Ref. 14.

The mass acceleration effect is accounted for by considering the acceleration of an element of rail traveling through 180° while its velocity is increased from zero to twice the vehicle's instantaneous velocity, as shown in Fig. 11.3(c). It can be shown that, assuming a constant vehicle deceleration, the rail element acceleration as a function of its position and current vehicle velocity $\dot{y}$ is

$$\mathbf{a}_p = -\frac{\dot{y}^2}{R}\sin\theta\,\mathbf{j} - \frac{\dot{y}^2}{R}\cos\theta\,\mathbf{i} \tag{14}$$

The force required to push the rail back in front of the impacting vehicle can therefore be determined in the form

$$F_y = \int (a_p)_y \, dm = 2\rho\dot{y}^2 \tag{15}$$

where ρ is the mass per unit length of rail.

Under high-speed impacts, the most significant elements in the development of the total extrusion force are those associated with the transverse and longitudinal deformations produced by the squeezing process and the mass acceleration effect caused when the curled rail is pushed ahead of the impacting vehicle. Calculated force components for an assumed 96·6 km/h impact speed and a coefficient of kinetic friction of 0·39 are shown in Table 1.

Table 1
Dynamic Extrusion Forces (1b)[14]

Squeezer force	*Deflection flattening force*	*Deflector friction force*	*Mass-acceleration force*	*Total force of extrusion*
5 629	1 036	838	3 256	10 759

Note: 1 lb = 4·448 N.

The commercial version of the guardrail extruder is called the ET 2000,[16] and is shown in Fig. 4.

3.2 The Brakemaster

This new device, presented in Fig. 5, dissipates energy by developing friction between brakes and a wire rope cable.[17] The cable/brake assembly consists of two brakes positioned on a 25·4 mm diameter, 6 × 25 IWRC galvanized wire rope cable and supported within the brake/tension support. One end of the cable is attached to the embedded anchor. The cable then passes through holes in diaphragms and the down-stream guardrail posts, and is secured at the last post with a large plate washer and nut.

The brakes use aluminum sleeves as friction wear elements. These sleeves are pre-loaded against the cable at the factory by high-strength

Fig. 4. ET 2000 guardrail extruder.

Fig. 5. The brakemaster.

steel spring plates. The spring plates and aluminium sleeves provide a friction resistance, which brings an impacting vehicle to a complete stop. The brake/tension support assembly allows the brakes to slide within horizontal steel channels for 127 cm before they move on the cable. This reduces vehicle deceleration during the first part of an impact into the system.

3.3 The Crash Cushion/Attenuating Terminal (C-A-T)

This device, called the C-A-T, is shown in Fig. 6. Note that the steel rails are slotted. In fact, although only three rows of slots are visible in Fig. 6, there are four such rows in each rail. The slots are 88·9 mm long and 22·2 mm high, and are separated by 7·9 mm long steel lands. The rails nearest the impact end of the C-A-T are made of 12-gauge steel and are 3·8 m long. They envelope the succeeding 10-gauge rails of the same length. During an end-on impact event, some energy is dissipated during the fracturing of the wooden posts that occurs as the system telescopes back upon itself. A significant percentage of the energy is dissipated, however, by the shearing off of the multitude of steel land sections between the slots in the rails along the length of the system. The C-A-T can be transitional to median and shoulder guardrail, concrete safety shape median barriers, and vertical walls or piers. This effective energy dissipation device was developed through an extensive full-scale crash testing program.[18] No mathematical modeling of the energy dissipating mechanisms has been performed.

Fig. 6. The crash cushion/attenuating terminal.

3.4 Hex-Foam Cartridges (GREAT)

Several very effective impact attenuation devices have been developed that usc a matrix of hex-shaped cardboard honeycomb filled with polyurethane foam to dissipate energy. The carboard is stacked in 25 mm layers in a cross-ply orientation, as shown in Fig. 7.

When the hex-foam material is crushed, the walls of one honeycomb layer shear into the walls of the adjoining honeycomb layer. The

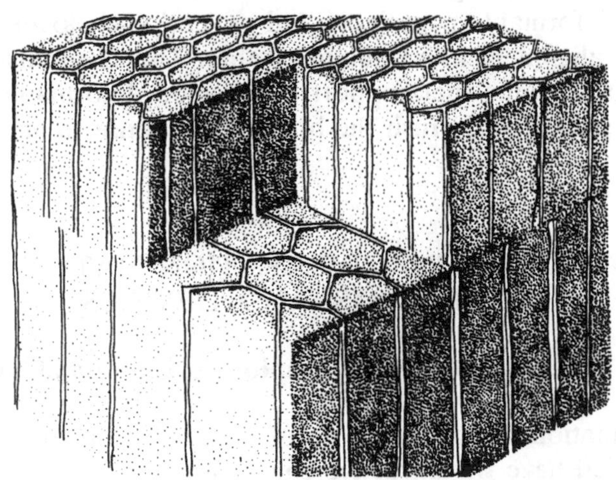

Fig. 7. Hex-foam cartridge.

Fig. 8. Guardrail energy absorbing terminal.

polyurethane foam provides both gussetting and stabilization for all the honeycomb cells, as well as additional shearing resistance when the honeycomb walls shear into the foam. Crushing of foam and honeycomb walls also occurs, and adds to the overall crushing force. During crushing, the compressive force level continually increases until full crush is achieved, at approximately 90% compression. Because of the interlocking effect of the honeycomb shear matrix, the rebound after impact is virtually eliminated. In addition, the material provides a certain degree of rate sensitivity during crush.

Hex-foam is employed in the Guardrail Energy Absorbing Terminal (GREAT)[19] shown in Fig. 8. Note that the stacked hex-foam sheets are placed in a deformable container, and several such loaded boxes are supported in alignment in a retaining frame. This frame has a plurality of nest retaining units that support the loaded deformable boxes and telescope to successively crush the boxes in response to an axial impact force.

4 OTHER ENERGY DISSIPATION MATERIALS, MECHANISMS, AND DEVICES

4.1 Impact Attenuation Devices Employing Clusters of Metal Tubes

Impact attenuation devices made up of connected metal tubes that are laterally loaded have the attractive characteristic of a collapsing stroke that can approach 95% of the undeformed length of the system. Reid[20] has presented a thorough survery of the load–deformation characteristics of unbraced metal tubes and tubular arrays. Systems loaded both quasi-statically and dynamically are treated, the effects of system inertia and strain-rate sensitivity are discussed, and 38 important technical references are identified. Additional work on the response of metal clusters has centered on lateral crushing characteristics of tightly packed arrays of tubes.[21–23] The crushing force is a function of the properties and packing arrangement of the array.

The Connecticut Impact Attenuation System

In certain impact attenuation system applications, it is advantageous to employ metallic tubes that are stiffened across their diameters.[24–27] An example of such a design is shown in Fig. 9. This device, known as the Connecticut Impact Attenuation System (CIAS),[28] is composed of 14

Fig. 9. Connecticut impact attenuation system.

mild steel tubes of 0·91 or 1·22 m diameters, formed from straight mild steel plate sections. These tubes are bolted together, rest on a concrete pad, and are attached to an appropriate backup structure. In order to cope with the redirectional crash test case involving an impact near the rear of the system, steel 'tension' straps (ineffective under compressive loading) and 'compression' pipes (ineffective in tension) are employed. This bracing system ensures that the crash cushion will respond in a stiff manner when subjected to an oblique impact near the rear of the unit, providing the necessary lateral force to redirect the errant vehicle. On the other hand, the braced tubes retain their unstiffened response when the attenuation system is crushed by impacts away from the back of the device. In a head-on impact, for example, the tension bracing is loaded in compression, and buckles. The compression bracing, being welded to the tube at one end only, carries no load during the collapse process because its free end separates from the tube wall when collapse occurs. The internal bracing system is only activated under side impact conditions. The energy dissipation characteristics of braced tubes are summarized in the following section.

Quasi-Static Response of Braced Tubes

Consider a tube that is symmetrically braced across its diameter as shown in Fig. 10 and laterally loaded at its top and bottom between flat

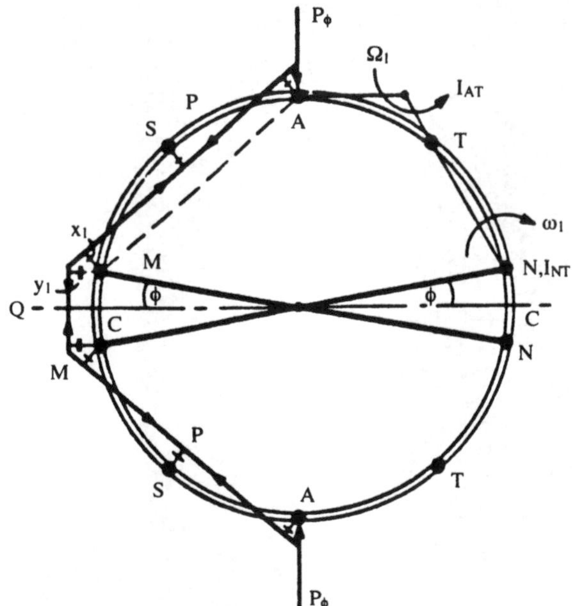

Fig. 10. Braced circular tube and possible collapse mechanism.

plates. The initial collapsing mode, associated with a loading of magnitude P_ϕ, will be a function of the angle of double bracing ϕ, and can be determined using upper- and lower-bound methods, assuming that the material is rigid, perfectly plastic. In the upper-bound approach, a kinematically admissible collapse mechanism is employed, and the work rate of the system of applied loads is equated to the corresponding internal energy dissipation rate. The lower-bound technique employs the equivalent structure method in which the yield condition is not violated and equilibrium is satisfied.

It can be shown[24] that one of the three possible collapse mechanisms will occur, depending on the value of ϕ. Consider, for example, the candidate collapse mechanism shown in Fig. 10. The left-hand half of the figure shows the lines of action of the resultant forces in the various sections of the tube, while the kinematics of the collapse mechanisms are shown in the upper right-hand quadrant.

Lower bound. The condition imposed on the solution shown is that the line of action of P lies equidistant from the three hinges at A, S, and M. By symmetry, the line of action of Q is vertical and must be concurrent with that of the force(s) P and the lines MN produced, in order to ensure equilibrium at M. From the figure,

$$x_1 = \tfrac{1}{2}R[1 - \cos(\tfrac{1}{4}\pi - \tfrac{1}{2}\phi)], \qquad y_1 = \frac{x_1 \cos\phi}{\sin(\tfrac{1}{4}\pi + \tfrac{1}{2}\phi)} \tag{16}$$

The yield condition gives

$$P = \frac{M_p}{x_1}, \qquad Q = \frac{M_p}{y_1} \tag{17}$$

while equilibrium leads to

$$P^l_\phi = 2P\cos(\tfrac{1}{4}\pi + \tfrac{1}{2}\phi) = \frac{4M_p \cos(\tfrac{1}{4}\pi + \tfrac{1}{2}\phi)}{R[1 - \sin(\tfrac{1}{4}\pi + \tfrac{1}{2}\phi)]} \tag{18}$$

The limit on the validity of this bound is imposed by the condition that the distance between C and the line of action of Q must not exceed y_1. As ϕ increases, the line of action of Q crosses C, and the condition for the validity of the lower bound given by eqn (18) is that $R - R\cos\phi \leqslant 2y_1$. Using the above equations, this inequality reduces to $\phi < \tfrac{1}{6}\pi$.

Upper bound. I_{AT} and I_{NT} are instantaneous centers of rotation for the sections AT and NT respectively. Assuming that T is at the midpoint of the arc AN (which is required by the lower bound above) leads to

$$I_{AT}A = R[\cos\phi - (1-\sin\phi)\tan(\tfrac{1}{8}\pi + \tfrac{3}{4}\phi)], \quad I_{NT}T = 2R\sin(\tfrac{1}{8}\pi - \tfrac{1}{4}\phi) \tag{19}$$

and

$$I_{AT}I_{NT} = \frac{R(1-\sin\phi)}{\cos(\tfrac{1}{8}\pi + \tfrac{3}{4}\phi)} \tag{20}$$

Continuity of velocity at T implies that $I_{AT}T\Omega_1 = I_{NT}T\omega_1$. The principle of virtual work leads to

$$2P^{u}_{\phi}I_{AT}A\Omega_1 = 8M_p(\omega_1 + \Omega_1). \tag{21}$$

which finally yields

$$P^{u}_{\phi} = \frac{4M_p\cos(\tfrac{1}{4}\pi + \tfrac{1}{2}\phi)}{R[1-\sin(\tfrac{1}{4}\pi + \tfrac{1}{2}\phi)]} \tag{22}$$

Thus, in the range $0 \leqslant \phi \leqslant \tfrac{1}{6}\pi$, we have coincident upper and lower bounds.

Similar analyses for $\phi > \tfrac{1}{6}\pi$ can be performed, and the results are summarized as follows:

$$P_{\phi} = \frac{4M_p}{R} \times \begin{cases} \dfrac{\cos(\tfrac{1}{4}\pi + \tfrac{1}{2}\phi)}{1-\sin(\tfrac{1}{4}\pi + \tfrac{1}{2}\phi)} & (0 \leqslant \phi \leqslant \tfrac{1}{6}\pi) \\ \dfrac{1-\sin\phi}{1-\cos\phi} & (\tfrac{1}{6}\pi \leqslant \phi \leqslant \tfrac{1}{4}\pi) \\ 1 & (\phi \geqslant \tfrac{1}{4}\pi) \end{cases} \tag{23}$$

where ϕ is the bracing angle with respect to the horizontal, R is the radius of the cylinder, and M_p is the plastic moment of the cross section.

These results are illustrated in Fig. 11, where the dimensionless collapse load ratio $P_{\phi}R/4M_p$ is the ratio of the braced to unbraced collapse loads. The corresponding collapse modes for three regimes have 10, 8, or 4 plastic hinges, depending on the value of ϕ. Note that for $\phi \geqslant \tfrac{1}{4}\pi$, the diametrical bracing does not affect the collapse of the tube.

The equivalent structure technique has been effectively applied to problems involving large deformations of rigid–perfectly plastic structures produced by plastic bending.[21,22,25,29–31] The magnitudes of the forces can be obtained from the geometry of the structure and its

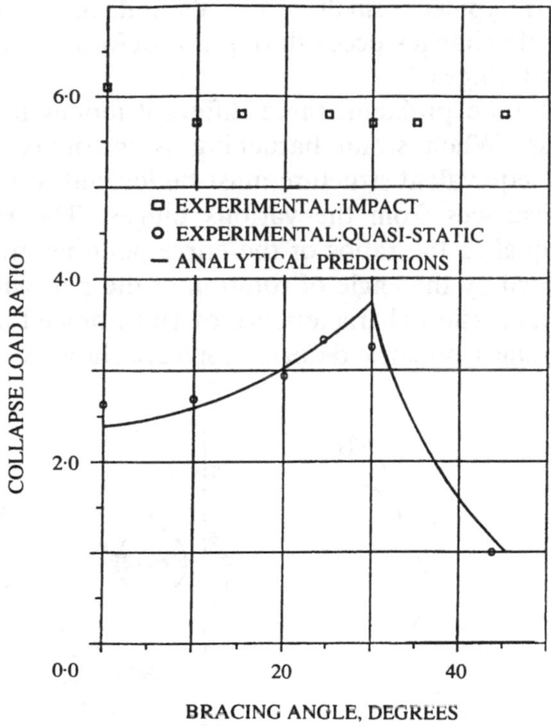

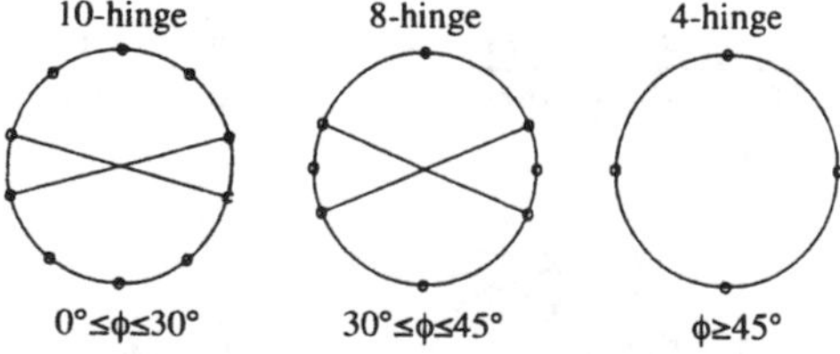

Fig. 11. Dimensionless initial collapse loads versus bracing angle for quasi-static and impact loading.

material properties. The change in geometry can be followed with a convenient kinematic model, and the load–deformation relationship for the structure can be derived.

In applying an equivalent structure approach, it is assumed that concentrated hinges are developed during the collapse process. Nevertheless, it is possible to account for the effects of strain hardening.[25] The inclusion of strain hardening is inconsistent with the assumption that point plastic hinges are formed, since strain-hardening spreads a point hinge into a plastic region. However, the length of this plastic region is typically small,[32] and its influence on deformation minor when mode changes occur during the deformation as a result of the movement of hinges.[30]

In the braced tube problem, three different modes form during the collapse process. When strain hardening is introduced, the various segments of the equivalent structure must be located so that they are at the requisite distances from the various hinges. The ratios of these distances are equal to the ratios of the corresponding moments. These in turn are defined by the angle of rotation at the particular hinge.

The load–compression characteristics of rings braced at 0°, 10°, 20°, and 25° undergoing symmetric deformations are shown in Fig. 12. Their

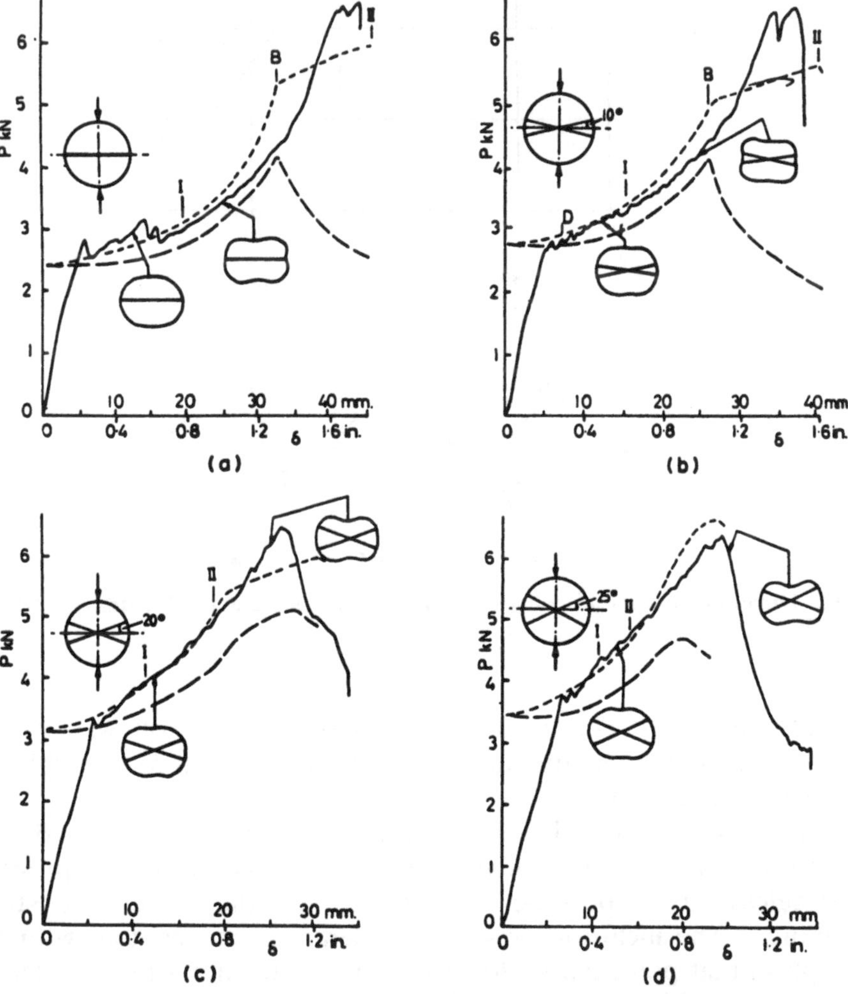

Fig. 12. Experimental and theoretical load–deflection characteristics of wire-braced rings: ——, experimental; — —, rigid–perfectly plastic; - - -, rigid–strain-hardening.

deformations are also shown schematically in the insets in this figure. Further details of the tests can be found in Ref. 25.

Impact Response of Braced Tubes

Metallic tubes that are employed in crash cushions to dissipate kinetic energy are subjected to impact loading conditions. Strain-rate effects play a significant role in increasing the energy dissipation capacities of such impulsively loaded tubes compared with equivalent tubes loaded quasi-statically. The influence of strain-rate effects, however, does not account for the total increase in initial collapse load and the associated energy dissipation that typically occurs under impact loading. In addition to the influence of the strain rate, tubes subjected to lateral impact loading conditions exhibit dynamic collapse configurations during the early phases of the impact event that are quite distinct from those of their quasi-static counterparts. This behavior is of great importance in the design of impact attenuation devices, since the deceleration levels to which the occupant of an errant vehicle is subjected must remain within allowable limits.

In Ref. 27, eight braced-tube experiments were conducted involving impact velocities that varied over the narrow range of 49·6–53·6 km/h. In view of the weak dependence of flow stress on strain rate, one would expect similar strain-rate effects to exist in the eight cases. Figure 11 shows the variation in dimensionless initial collapse load with the double bracing angle under both quasi-static and impact loading conditions. Although quasi-static collapse loads are clearly sensitive functions of the bracing angle, the impact collapse loads are essentially independent of the bracing orientation. This anomalous behavior can be explained by considering the deformation patterns that occur in the early stages of the impact event. Under quasi-static conditions, the early collapse mechanisms are symmetric with respect to the horizontal diameter of the tube. This is not the case under impact loading conditions. In fact, a localized region of plastic deformation adjacent to the impacted area dominates the early-time response of the tube. At later times (and deformations), the impact and quasi-static mode shapes coalesce. Figure 13 shows the early stages of the collapse mode for an impact loading. The left-hand part shows results obtained from ABAQUS and the right-hand part illustrates an idealized model. It is clear that, at the beginning of the event, neither the bracing nor the majority of the tube experiences any strain. The response is localized, and independent of the bracing orientation.

Figure 14 illustrates the importance of accounting for strain-rate effects in the numerical modeling of impact problems. The software

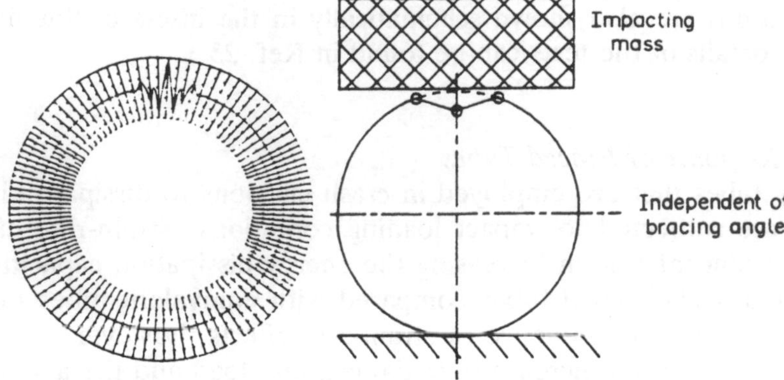

Fig. 13. Collapse mode during early stage of impact loading.

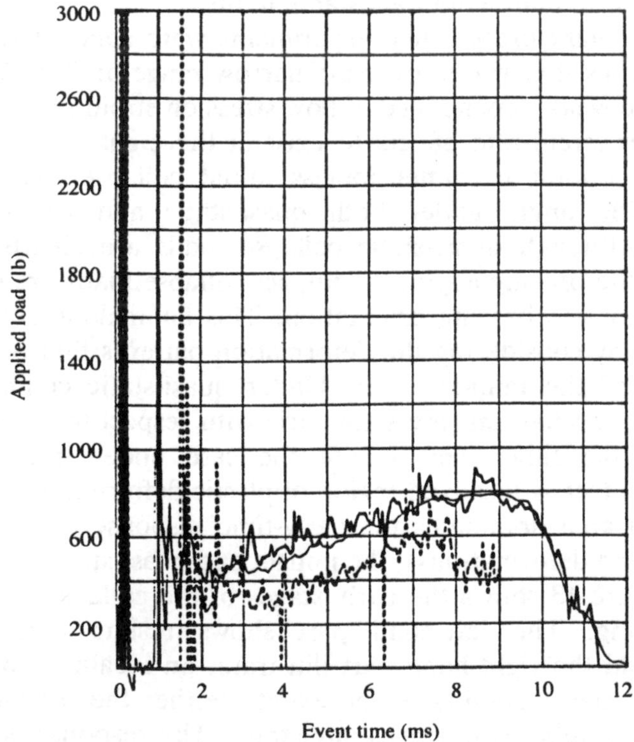

Fig. 14. Tube force history with strain rate (——), without strain rate (– – –), and experimental (——) (1 lb = 4·448 N).

package ABAQUS does an excellent job of predicting the experimental results when strain-rate effects are included.

In addition to the Connecticut Impact Attenuation System shown in Fig. 9, several other devices employing mild steel cylinders to dissipate kinetic energy have been developed and are described below.

4.2 Truck Mounted Attenuator

A truck-mounted attenuator is a portable crash cushion that is employed in short-term stationary or slow-moving maintenance and repair operations (e.g. line-striping and pavement overlay) to provide protection for both the motoring public and maintenance personnel. The Connecticut Crash Cushion,[33,34] shown in Fig. 15, is a truck-mounted attenuator that uses four 0·61 m diameter steel cylindrical sections to dissipate energy. The energy absorbing system involves three components:

- the service vehicle guidance frame
- the energy absorbing pipes
- the impacting plate assembly

The impacting plate is constructed of 6061-T6 aluminum. The alumi-

Fig. 15. Truck-mounted attenuator.

num tubing in the impacting plate assembly slides inside the steel structural tubing upon collapse of the system.

4.3 Narrow Hazard Crash Cushion

The Narrow Connecticut Impact Attenuation System[35] is employed at such narrow hazard locations as concrete barrier ends and bridge pillars. The crash cushion design is shown in Fig. 16. The system is composed of a single row of eight 0·91 m diameter mild steel cylinders of different thicknesses, which are formed from flat plate stock. All cylinders are 1·22 m high, and a total of four 25·4 mm diameter cables (two on each side of the system) provide lateral stability and assist in redirecting errant vehicles under side impact conditions. The 7·32 m length was chosen as the probable minimum acceptable length for the crash cushion if NCHRP Report 230 crash test requirements were to be met. The 0·91 m width was selected because most narrow highway hazards are approximately 0·61 m wide, and the crash cushion should be slightly wider than this dimension.

Fig. 16. Narrow Connecticut Impact Attenuation System.

A generalized CIAS design has been developed[36] in the form of an expert system computer program to optimize the design of the crash cushion when given the unique characteristics of a proposed site. These conditions include the available site dimensions and the speed limit of the roadway. This expert system (called CADS) can be used to optimally design crash cushions in multiple service level applications. The individual cylindrical wall thicknesses are determined so that the occupant impact velocities and ridedown accelerations are minimized, subject to the dual constraints of system length and the required energy dissipation capability. This computer-based design system allows the non-expert to optimally design site-specific versions of the Connecticut Impact Attenuation System.

Two additional effective attenuation devices deserve mention. The first is the Hi-Dro Sandwich System.[37] This is composed of any array of polyvinyl plastic tubes filled with water. The kinetic energy of the impacting vehicle is dissipated by forcing the water through orifices in the top of each cartridge. This total process is complicated, and an appropriate mathematical model has not been developed. However, the system has been extensively crash-tested, and a series of standard units has been developed to suit most crash cushion requirements. The second system is the sand-filled plastic barrel crash cushion.[38] The primary energy dissipation mechanism in this very popular system is the transfer of the vehicle's momentum to the variable weights of sand in the barrels that are impacted. The initial cost of sand-filled plastic barrel cushions is low, primarily because no backup structure is required, and they have been effectively employed for many years. However, they cannot redirect errant vehicles, and extreme care should be taken in the placement of those sand-filled barrels near the rear of an array. Otherwise, impalement on the corner of the hazard being shielded may occur.

5 HIGHWAY SAFETY HARDWARE IN THE 21st CENTURY

The day will probably arrive when fully automated vehicle control systems will exist. Vehicles will be capable of operating in any traffic environment and traveling portal to portal without driver intervention. For the first quarter of the 21st century, however, a continuing need will exist for effective roadside safety hardware. Truck-mounted attenuators, crash cushions, terminals, longitudinal barriers, and other appurtenances designed to enhance the safety of roadways have proven their cost-effectiveness. The crashworthiness of these devices continues to be improved as new systems are developed under more sophisticated crash testing guidelines. Almost without exception, however, these highway safety appurtenances have high maintenance costs associated with their use. Following a vehicular impact, the energy dissipating material and other system components must be discarded and replaced. Many devices in widespread use are proprietary, and the maintenance and replacement costs associated with restoration of these systems over their service lives are extremely high. Furthermore, because of this cost and the fact that the human resources of transportation agencies are usually spread rather thinly, impacted safety devices often remain unrefurbished for long periods of time. This unsafe situation represents a danger to the motoring public and an increased liability exposure to the managing transportation entity involved.

The root of the problem lies in the sacrificial nature of the energy dissipating mediums employed in current impact attenuation devices. Is is possible to develop an energy dissipating medium that can dissipate large amounts of kinetic energy, undergo large deformations and strains without fracturing, and, most importantly, restore itself to its original size, shape, and energy dissipation potential when the forcing function is removed? Put another way, can a reusable impact attenuation device be designed and constructed?

The potential financial, legal, and safety payoffs for highway operations associated with developing highway safety devices that are essentially maintenance-free are enormous. Maintenance costs associated with the repair of impact safety devices would be greatly reduced or eliminated. Tort liability exposure related to damaged or collapsed hardware would be significantly decreased. Finally, the safety of the motoring public and the maintenance personnel involved in maintaining and repairing damaged hardware would be greatly enhanced.

5.1 High-Molecular-Weight/High-Density Polyethylene

A promising candidate material is high-molecular-weight/high-density polyethylene (HMW HDPE). This 'smart' energy dissipating thermoplastic has the properties of self-restoration and reusability, possessing a memory of its undeformed shape. A HMW HDPE tube, for example, when crushed laterally between two plates to complete collapse, will restore itself to approximately 90% of its original shape upon removal of the load. It can be reloaded and unloaded repeatedly, exhibiting almost identical load–deformation/energy dissipation characteristics. It remains ductile at temperatures well below 0°F (−18°C) and its energy dissipation potential is still significant at temperatures above 100°F (38°C).

High-density polyethylene is a thermoplastic material, which is solid in its natural state. This polymer is characterized by its opacity, chemical inertness, toughness at both low and high temperatures, and chemical and moisture resistance. High density can be achieved because of the linear polymer shape, which permits the tight packing of polymer chains. The physical properties of high-density polyethylene are also affected by the weight-average molecular weight of the polymer. When this high-density polymer is used with a high-molecular-weight resin (in the 200 000–500 000 range), a high-molecular-weight/high-density polyethylene is produced that exhibits the following favorable material characteristics:

- high stiffness
- high abrasion resistance
- high chemical corrosion resistance
- high moisture resistance
- high ductility
- high toughness
- high tensile strength
- high impact resistance over a wide temperature range

When loaded laterally between two plates, an HMW HDPE tube deforms in the same 'dog-bone' configuration as would a metal tube. Of primary interest is the energy dissipation capacity of a collapsed tube. The geometrical and material nonlinearities associated with the large deformations of interest make an exact solution difficult to obtain. However, this problem is simplified here by making the following crucial assumption: the ring curvature during the post-buckling process is given by

$$\kappa = \frac{1}{R}\left[1 - B\cos\left(\frac{2s}{R}\right)\right] \tag{24}$$

where $R = \frac{1}{2}D$ is the radius of the ring, s is the arc length measured along the circumferential direction of the ring, and B is a dimensionless parameter. Equation (24) can be numerically integrated to obtain ring configurations as functions of the parameter B, and four of the infinitely large number of such configurations are reported in Fig. 17, for $B = 0$, 0·2, 1·15, and 2·33. Because this problem involves double symmetry, only a quarter of the ring is shown. The parameter B may be interpreted as a generalized coordinate that controls the buckling behavior of the ring, and, as seen in Fig. 17, total collapse of the ring is

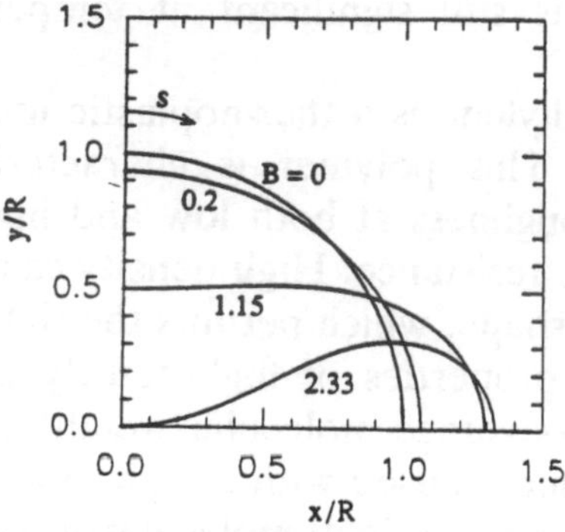

Fig. 17. Ring configurations for different values of B.

achieved for $B \approx 2{\cdot}33$. Equation (24) gives the exact curvature variation associated with the small-deflection first buckling mode for the ring. Therefore the assumption of eqn (24) implies that the ring locks itself into that curvature variation until it totally collapses. This assumption leads to a final collapse configuration for the ring ($B = 2{\cdot}33$) that is quite similar to the well-known 'dog-bone' configuration. It is noted that this curvature variation has been previously considered as potentially useful for the collapse analysis of circular rings under external pressure, and has been used in experimental investigations of the buckle geometry in submarine pipelines.

With the ring curvature defined by eqn (24), it is possible to calculate the bending strain energy U as a function of B. Extensive calculations yield

$$U = \begin{cases} \dfrac{\pi}{24} B^2 E_1 \dfrac{t^3}{R} & (B \leqslant B_0) \qquad (25) \\[2ex] \dfrac{4\varepsilon_y^3}{3B} R^2 (E_1 - E_2) \ln \dfrac{1+\lambda}{\varepsilon_0} & \\ \quad + 4\varepsilon_y^2 (E_2 - E_1) S_1 t + \varepsilon_y \lambda B (E_1 - E_2) t^2 & \\ \quad + \dfrac{B^2}{6} \left[E_1 \left(\dfrac{\pi}{4} - \dfrac{S_1}{R} - \dfrac{\varepsilon_0 \lambda}{2} \right) + E_2 \left(\dfrac{S_1}{R} + \dfrac{\varepsilon_0 \lambda}{2} \right) \right] \dfrac{t^3}{R} & (B_0 \leqslant B \leqslant 2{\cdot}33) \end{cases} \qquad (26)$$

where $B_0 = \varepsilon_y D/t$, $\varepsilon_0 = B_0/B$, $\lambda = (1 - \varepsilon_0^2)^{1/2}$, and $S_1 = 0{\cdot}5R \cos^{-1} \varepsilon_0$. In evaluating U, it has been assumed that the bending strains vary linearly through the thickness of the ring. Equation (26) has been shown to give accurate results for energy dissipation capacities of both metallic and non-metallic tubes under both plate and pressure loading conditions.

Some Experimental Results

The energy dissipation and self-restoration characteristics of HMW HDPE tubes are functions of

- temperature
- diameter-to-thickness ratio D/t
- strain
- deformation level
- repeated/cyclic loading

An extensive series of quasi-static and impact experiments have been performed on a wide variety of HMW HDPE tubes over a wide range of test temperatures.

Typical load versus deformation plots are presented in Fig. 18, where IPS is the nominal tube diameter and $SDR = D/t$. This data were employed to determine the relationship between energy dissipation

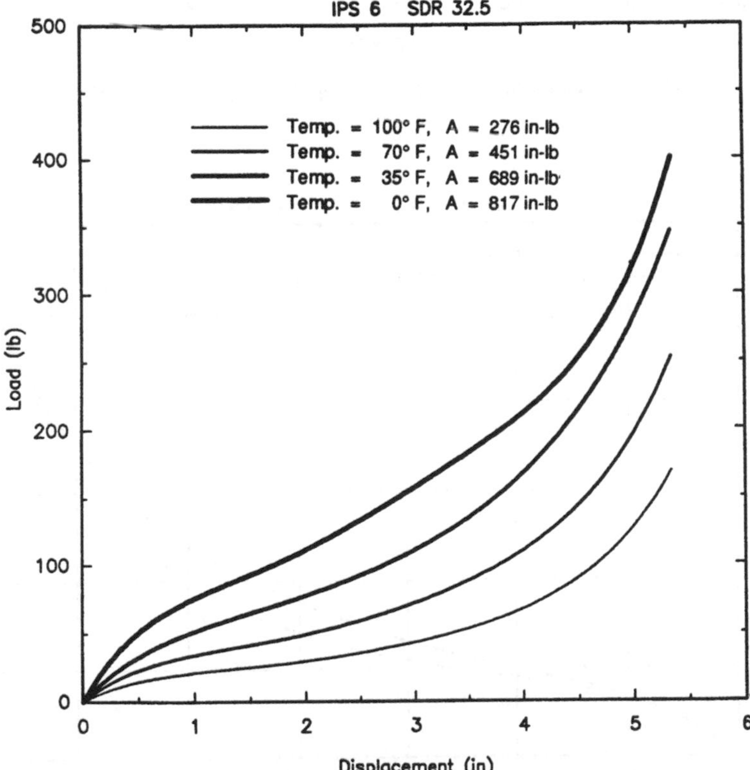

Fig. 18. Quasi-static load–displacement curves and areas for IPS 6 SDR 32·5 for temperatures 0, 35, 70, and 100°F (approx. −18, 2, 21, and 38°C) (1 mph = 1·61 km/h, 1 lb = 4·448 N, 1 in = 25·4 mm).

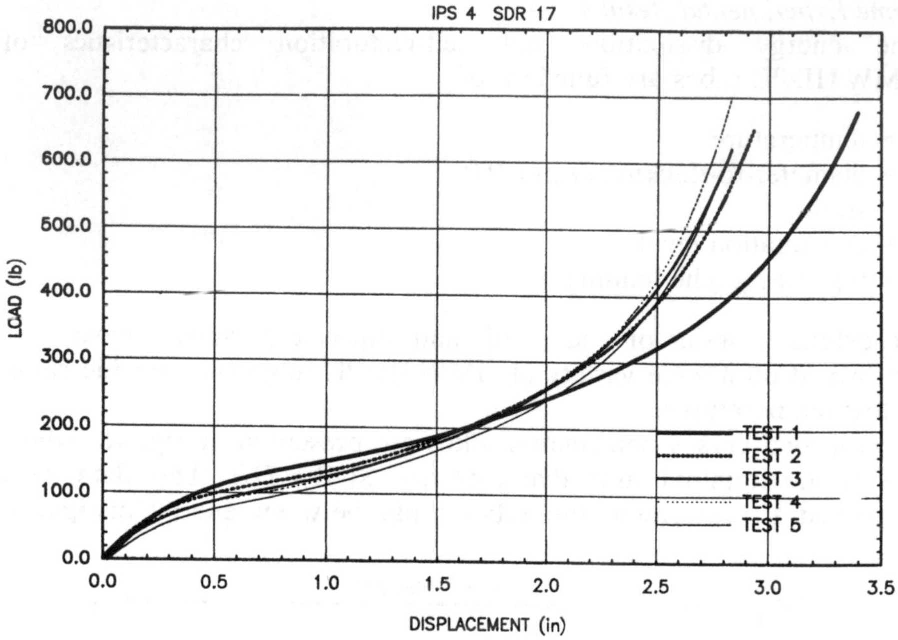

Fig. 19. Reloading tests (1 lb = 4·448 N, 1 in = 25·4 mm).

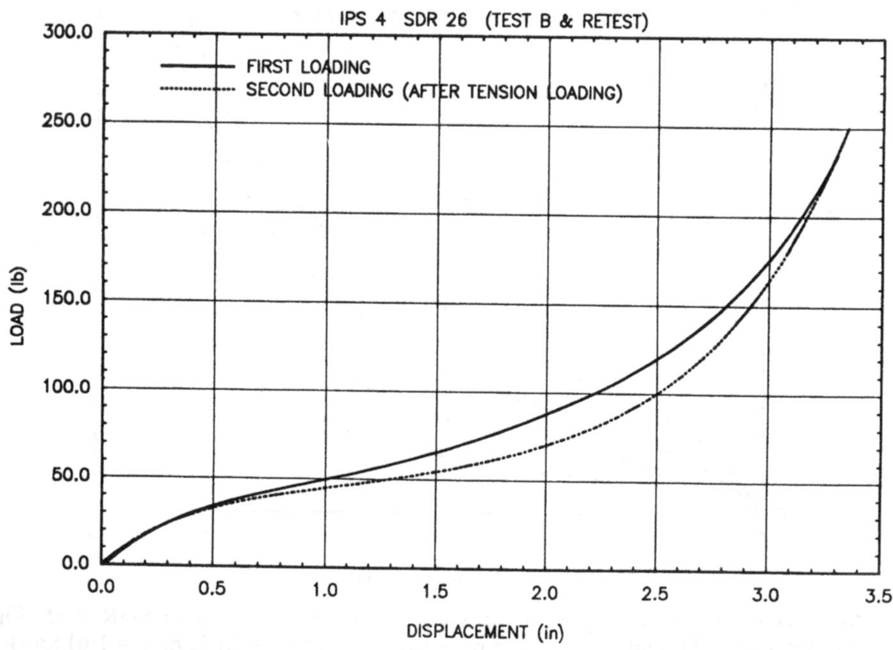

Fig. 20. Reloading of a restored HMW HDPE tube (1 lb = 4·448 N, 1 in = 25·4 mm).

(area under the load–displacement curve), tube length, radius, thickness, and test temperature. The effects of reloading previously crushed tubes are illustrated in Fig. 19. Five different load–deformation tests were conducted at 24 h intervals on the same 102 mm diameter specimen. Note that, after the first cycle, self-restoration approached 100% and the energy dissipation potential did not deteriorate. A variation of this experiment is illustrated in Fig. 20. After loading the tube to complete collapse, the specimen was allowed to restore itself naturally to approximately 88% of its original circular shape. The tube was then loaded in tension until, upon load removal, the original circular shape was restored. The subsequent second loading of the tube yielded results quite similar to those of the virgin test. Under quasi-static loading conditions, the areas under the load–displacement curves are sensitive functions of temperature. Consider, for example, the ratio of areas for the two temperature extremes in Fig. 18:

$$\left(\frac{A_{0^\circ}}{A_{100^\circ}}\right)_{\text{static}} = \frac{817}{276} = 2{\cdot}96 \tag{28}$$

An impact loading program has demonstrated that this extreme temperature sensitivity that is present under quasi-static conditions is much reduced under impact test conditions. This very significant and heretofore unknown fact is illustrated by the impact test results shown in Fig. 21. The sensitivity of the energy dissipation potential of an HMW HDPE tube to temperature under impact loading conditions is clearly significantly less than under quasi-static ones. Consider the following result from Fig. 21:

$$\left(\frac{A_{0^\circ}}{A_{100^\circ}}\right)_{\text{impact}} = \frac{976}{534} = 1{\cdot}83 \tag{29}$$

If the strain-rate sensitivity factor (SRS) is defined as the ratio of the impact to quasi-static energy dissipation capacities of a tube, the influence of strain rate as a function of temperature can be effectively illustrated. Strain-rate sensitivity factors are presented in Fig. 22 for two sets of impact velocities. Note that the rate of loading is of little import at low temperatures and very significant at high temperatures.

It is expected that impact attenuation devices employing HMW HDPE tubes will be crash-tested in 1993.

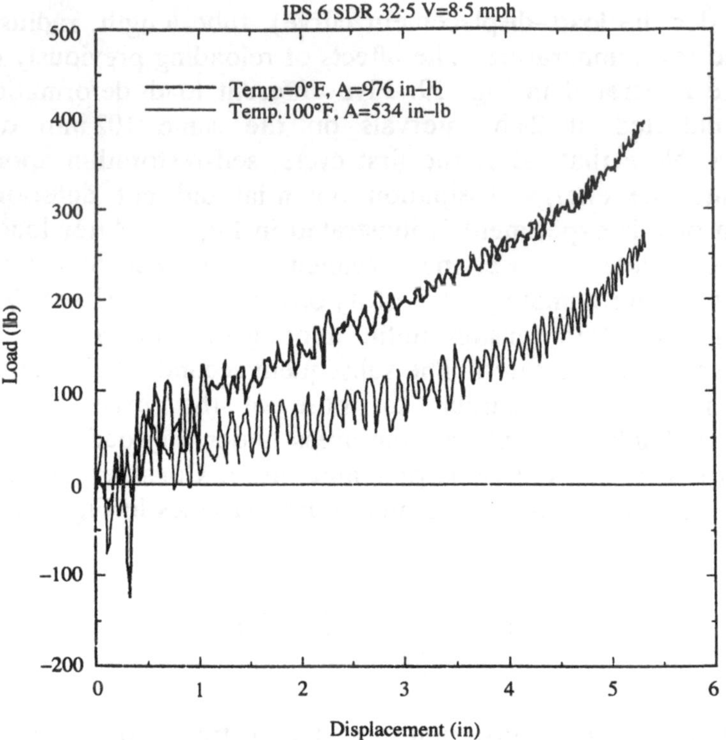

Fig. 21. Impact load–displacement curves and areas for IPS 6 SDR 32·5 tor temperatures of 0 and 100°F (approx. −18 and 38°C) (1 mph = 1·61 km/h, 1 lb = 4·448 N, 1 in = 25·4 mm).

6 OTHER ISSUES

6.1 Improved Flail Space Model

The commonly employed Report 230 procedure for calculating the occupant impact velocity described earlier is valid only for rectilinear motion. It should not be employed for cases when the impact event involves curvilinear motion. For example, consider a 25°, 96·6 km/h impact into a longitudinal barrier.

Let an (r, s, t) coordinate system represent a reference frame attached to the test vehicle and the (x, y, z) axes represent a global system attached to the Earth. If **i**, **j**, and **k** are unit vectors in the (r, s, t) reference frame, it can be shown[39] that

$$\dot{v}_r = a_r - v_t\dot{\theta}_{\mathrm{p}} + v_s\dot{\theta}_{\mathrm{y}} \tag{30}$$

$$\dot{v}_r = a_s - v_r\dot{\theta}_{\mathrm{y}} + v_t\dot{\theta}_{\mathrm{r}} \tag{31}$$

$$\dot{v}_t = a_t - v_s\dot{\theta}_{\mathrm{r}} + v_r\theta_{\mathrm{p}} \tag{32}$$

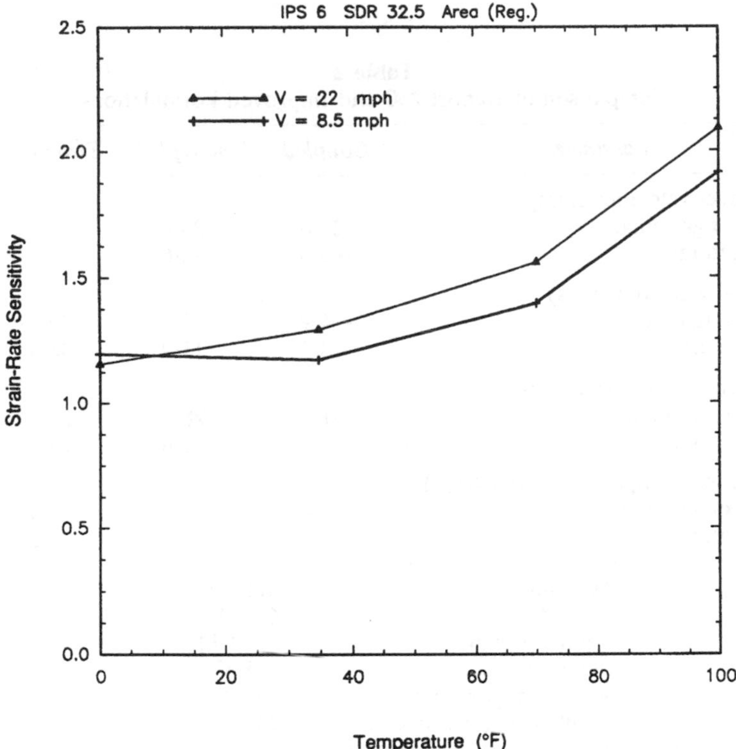

Fig. 22. Strain-rate sensitivity of IPS 6 SDR 32·5 as a function of temperature for two different impact velocities (1 mph = 1·61 km/h).

where $\dot{\theta}_r$, $\dot{\theta}_p$, and $\dot{\theta}_y$ are the roll, pitch, and yaw angular velocities of the vehicle, and a_r, a_s, and a_t are accelerations measured by transducers fixed to the vehicle. It is clear from eqns (30)–(32) that the vehicle velocity components are coupled.

To accurately calculate occupant impact velocities, researchers must make use of the angular velocity data in evaluating test data. The error associated with neglecting these vector relations increases as the angular velocity increases. Table 2 shows the significant errors that are associated with neglecting the coupled terms in the equations of motion. Both cases described are associated with Report 230 test 10 conditions involving high-speed impacts into longitudinal barriers at 25° impact angles. Employing the uncoupled procedures of Report 230 leads to errors in lateral velocity values that can be in excess of 200%. Furthermore, if the test matrix is expanded in the future to include non-tracking or side impact tests, neglecting the vector nature of the individual acceleration quantities would result in even larger errors,

Table 2
Comparison of Report 230 and Improved Formulations

Parameter	*Coupled*	*Uncoupled*	*Error (%)*
Impact velocities (m/s)			
Longitudinal	26·8	26·8	—
Lateral	−0·6	−0·6	—
Final velocities (m/s)			
Longitudinal	−4·9	−1·4	70·7
Lateral	−8·7	−15·9	82·9
Change in velocity (m/s)			
Longitudinal	31·7	28·3	10·8
Lateral	9·3	1·6	76·3
Occupant impact velocities (m/s)			
Longitudinal	8·4	8·2	1·6
Lateral	5·7	6·1	6·1

Test number	WE4-1
Appurtenance	G4(1S)
Vehicle weight	17·8 kN
Impact angle	25·5°
Maximum yaw rate	165·8°/s
Number of 1 ms time steps	740

Parameter	*Coupled*	*Uncoupled*	*Error (%)*
Impact velocities (m/s)			
Longitudinal	26·3	26·3	—
Lateral	−0·2	−0·2	—
Final velocities (m/s)			
Longitudinal	16·9	18·8	11·2
Lateral	5·3	16·5	209·4
Change in velocity (m/s)			
Longitudinal	9·4	7·5	20·2
Lateral	5·7	16·7	193·9
Occupant impact velocities (m/s)			
Longitudinal	—	—	—
Lateral	4·7	4·9	3·9

Test number	SPI-1
Appurtenance	G4(1S)
Vehicle weight	20·0 kN
Impact angle	25·3°
Maximum yaw rate	198·3°/s
Number of 1 ms time steps	577

because in such tests the angular velocities and angular accelerations can be very large.

6.2 Effects of Modified Structural Shapes on Load–Deflection Response

The effectiveness of employing clusters of laterally loaded circular tubes to dissipate kinetic energy in a controlled manner has been demonstrated. In certain applications (e.g. truck-mounted attenuators), it might be necessary to design an energy dissipation system that is essentially rectangular in plan view form, with a length-to-width ratio greater than one. Because of practical considerations, the required length of a truck-mounted attenuator, which is a function of impact speed and vehicle mass, should be minimized subject to satisfaction of occupant risk constraints. Given the conditions described above, it can be advantageous to modify the structural shapes of the components of the system. More specifically, the use of elliptical tubes shows great potential in such applications. When loaded parallel to their major axes, elliptical tubes possesses a long collapsing stroke and an almost constant force level that exists for much of the collapse event. Figure 23 presents load–deformation curves for five different elliptical aluminum specimens that were all formed from the same 50·8 mm diameter stock.

Braced elliptical tubes also exhibit interesting characteristics. The initial collapse loads for braced circular tubes are given in eqn (23) and

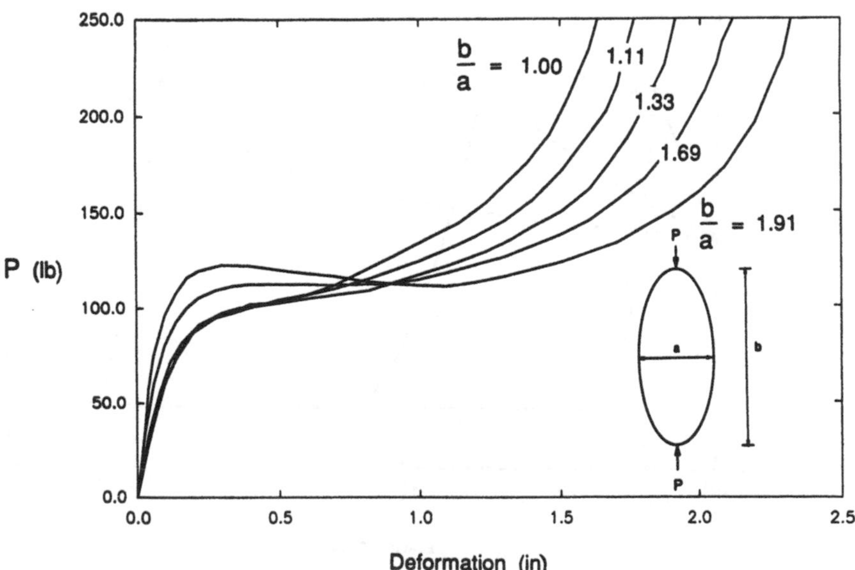

Fig. 23. Load–deflection curves for elliptical aluminum tubes (1 lb = 4·448 N, 1 in = 25·4 mm).

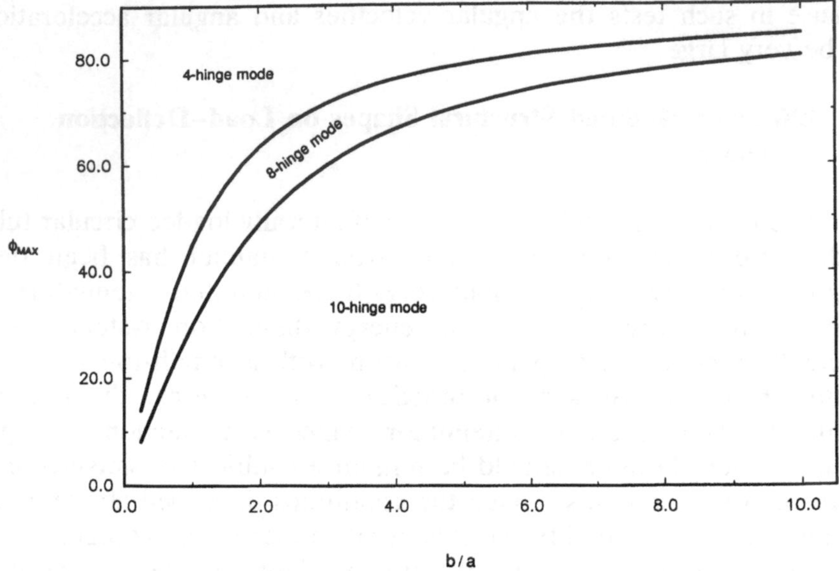

Fig. 24. Collapse load regimes for braced elliptical tubes.

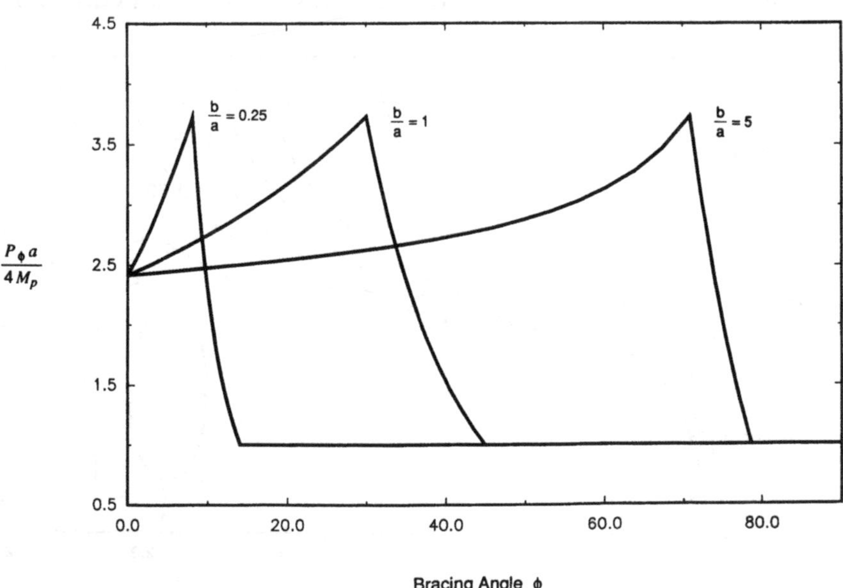

Fig. 25. Effect of bracing angle on collapse load of elliptical tubes.

illustrated in Fig. 11. In the elliptical tube case, although the same three general collapse mechanisms exist, both the magnitude of the collapse load and the critical collapse mode are sensitive functions of the ratio of the lengths of the two axes, $\alpha = b/a$. A mathematical analysis yields the following results:

$$P_\phi = \frac{4M_p}{a} \times \begin{cases} \dfrac{(\alpha^2 + \tan^2\phi)^{1/2} - \tan\phi}{\sqrt{2}\,[\alpha^2 + \tan^2\phi - \tan\phi\,(\alpha^2 + \tan^2\phi)^{1/2}]^{1/2} - \alpha} & \left(0 \leqslant \phi \leqslant \tan^{-1}\dfrac{\alpha}{\sqrt{3}}\right) \\ \dfrac{(\alpha^2 + \tan^2\phi)^{1/2} - \tan\phi}{(\alpha^2 + \tan^2\phi)^{1/2} - \alpha} & \left(\tan^{-1}\dfrac{\alpha}{\sqrt{3}} \leqslant \phi \leqslant \tan^{-1}\alpha\right) \\ 1 & (\phi \geqslant \tan^{-1}\alpha) \end{cases} \tag{33}$$

Figure 24 shows the three collapse mode regimes as a function of the ratio b/a of the braced elliptical tube. Dimensionless collapse load versus bracing angle responses for three ratios b/a are shown in Fig. 25.

6.3 Computer Simulation of Crash Events

Most currently available vehicle impact/crash simulation codes were initially developed many years ago, and are limited by their obsolete structures and restrictive simplifying assumptions and approximations. The fact is that their use does not permit a detailed, accurate modeling of most real-life vehicle impact scenarios. Specific limitations include the following.

- Modeling of both the vehicle and roadside safety structures is often two-dimensional.
- The number of modeling elements and the degrees of freedom are unnecessarily limited.
- Inelastic material properties of both vehicle and roadside structures and vehicle inertial properties are crudely modeled.
- Numerical instabilities often occur for large barrier displacements.
- Codes cannot predict vehicular instability.
- Vehicle/barrier contact and tire/road profile algorithm representations are limited.
- The terrain is not always modeled.
- Codes have not been adequately validated for most impact scenarios.

The state of the art in computer hardware and software development has evolved to the point where a powerful, versatile, user-friendly vehicle impact/handling simulation code can and should be produced. The availability of such a computer simulation tool would enable design engineers and researchers to develop safer roadways and more effective roadside safety features. Current full-scale crash testing of safety appurtenances such as longitudinal barriers, crash cushions, terminals, and luminaire supports are almost entirely limited to a few impact scenarios involving tracking vehicles. Most actual accidents bear little resemblance to these idealized crash test conditions. However, the significant expense associated with full-scale testing (some of which will always be necessary), coupled with the practical limitations of crash testing technology, combine to limit the number and variety of impact scenarios that can be crash-tested. An improved capability to accurately simulate vehicular dynamic responses and impacts with roadside features would result in most cost-effective roadway designs and roadside safety features. It would also permit a reduction in the number and expense of full-scale crash tests needed to develop new hardware. Most importantly, lives would be saved, since a better understanding of hardware performance would improve hardware designs.

6.4 Side Impact Crashes

In the United States, 225 000 people are involved in side impact collisions with fixed roadside objects every year. One in 3 is injured and 1 in 100 is killed.[40,41] This level of injury represents a societal loss of more than $3 000 000 000. These figures are based on the 1986 accident costs recommended by the Federal Highway Administration for cost-effectiveness analysis.[42] Approximately 910 000 vehicle occupants are involved in collisions with fixed objects each year, and almost 9000 are fatally injured. Collisions with the sides of vehicles account for one-quarter of the cases in both the National Accident Sampling System (NASS) and Fatal Accident Reporting System (FARS) databases.

Both the current testing procedures contained in Report 230 and the proposed new guidelines recommend impact conditions only with tracking, unbraked vehicles. Side impact collisions have little to do with these highly idealized test conditions. The recognition of the importance of addressing this ongoing side impact issue resulted in a research effort by the Federal Highway Administration that included a full-scale side impact crash testing program. It included the impact response of light automobiles during low-speed broadside collisions with slipbase luminaire supports and an energy dissipating pole developed in Sweden. The slipbase pole is designed so that an errant vehicle can pass by with little change in velocity after activating the slip mechanism.

The Swedish (ESV) pole uses a completely different principle to achieve the desired result; minimizing risk to the occupant. This device is designed to stop a vehicle rather than let it pass. The collapse mechanism has been designed so that the occupant response in a frontal collision is well below the threshold of serious injury. In over a decade of use throughout Scandinavia, there have been no fatalities reported in accidents involving ESV-like poles. By changing the dimensions of pole, the energy dissipation can be optimized for different vehicle masses and speeds. The outside of the pole is a hexagon formed from 1·35 mm thick steel skin. Four 9·5 mm diameter round steel rods are tack-welded to the inside of the sheet. The four rods are butt-welded to 25·4 mm diameter anchor rods. The pole can be installed using either a soil-mounted anchor or by directly attaching it to a rigid base. In a collision, the tack welds fracture and the cross-section flattens. The flattened section is then pulled around the bumper or sill, continuously flattening the cross-section. Energy is absorbed by the flattening and wrapping of the pole around the vehicle.

The complete results of this crash test program are reported elsewhere,[43] but a typical crash test is illustrated in Fig. 26. An 8·0 kN vehicle impacted the Swedish pole at 66·8 km/h. The pole responded like a crash cushion, and an extensive amount of kinetic energy was dissipated as approximately two-thirds of this pole length collapsed during the impact event.

Side impacts involving vehicles and fixed roadside objects represent an extremely dangerous and costly type of accident. Side impacts with utility poles, trees, and other narrow objects account for 75% of all fatal fixed-roadside side impact accidents. Recent studies have focused on vehicle-to-vehicle side impacts and the thoracic trauma that usually occurs in such collisions. In side impacts with narrow objects, however, the important injury mechanisms can be quite different. The full-scale crash test program referred to here shows that the occupant's head is often the first portion of the body to contact the vehicle interior. If the occupant and struck object are aligned, the occupant effectively strikes the pole directly, causing extremely large impulses very early in the impact event. Even breakaway luminaire supports have a high probability of fatally injuring the occupant, who strikes the pole long before the latter breaks away. An understanding of this important impact scenario is vital to making effective design changes to the pole and automobile. Installing interior padding, for example, without strength-

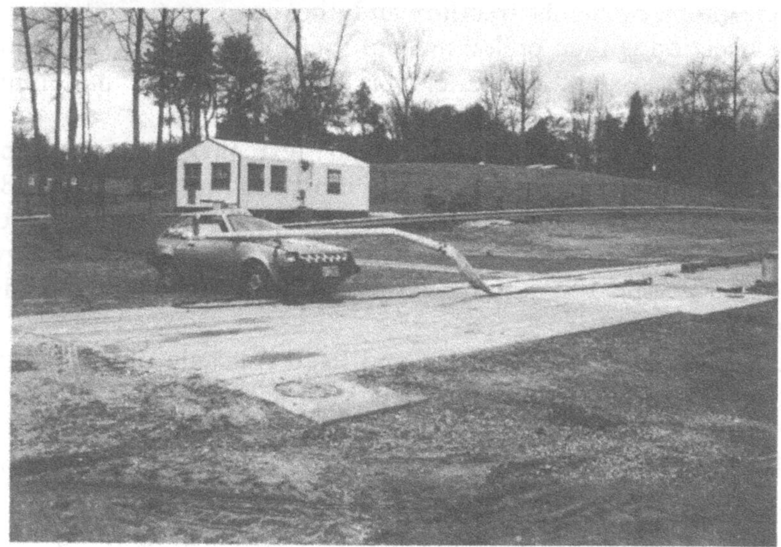

Fig. 26. Side impact crash test.

ening the door structure may have little beneficial effect for occupants of vehicles involved in side impacts with narrow objects.

One fact that is clearly illustrated by recent crash test results is that the Report 230 flail space occupant risk procedures yield little of value for this type of crash scenario. In all tests, the occupant impact velocities and ridedown accelerations calculated were well below the recommended design values of 9 m/s and 15*g*, respectively. However, unacceptably high HIC and TTI values were obtained, indicating that a severe or fatal injury would have resulted in most of the test cases.

Although non-tracking side impacts with roadside appurtenances are very hazardous crash events, this type of crash test is not currently conducted when hardware is being developed. There is a need to establish non-tracking, side impact test and evaluation criteria that could have a significant effect on the highway safety community's ability to design safe and effective roadside hardware.

7 CONCLUSIONS

This chapter has provided an overview of the advances that have been made in developing effective impact attenuation devices used in motorway safety applications. The intelligent use of these energy dissipation systems has saved many lives and reduced injuries and human suffering for over 25 years. Specific devices discussed have included innovative guardrail end treatments and both narrow and wide crash cushions. Energy is dissipated in these devices by plastically deforming metal structural shapes, through friction between moving parts, by metal tearing, by crushing polyurethane foam, through a transfer of momentum, by forcing fluid through orifices, and by deforming 'smart' materials that remember to return to their original structural configurations. The systems chosen for inclusion in this chapter are not all-inclusive. Several very effective impact attenuation devices have not been discussed. Special emphasis has been directed at considering those devices that are both highly effective (or have the potential of being highly effective) at dissipating energy and lend themselves to efficient mathematical modeling of their energy dissipation mechanisms. The look into the future dealt with several potentially useful design concepts and other safety issues involving hardware, computer software, and improved occupant risk prediction procedures.

Motorway related deaths and injuries are a major, worldwide, health problem. In both absolute terms and in relation to the occurrence of other injuries and diseases, the opportunities for motorway injury reductions are numerous. Given the size of the problem, it is clear that inadequate research funding is being directed towards its alleviation. Investment in the development of impact attenuation devices is an investment in cost-effective injury prevention. These systems will protect all elements of the driving population—the high-risk groups (intoxicated, elderly, and teenage drivers) and everyone else.

REFERENCES

1. Baker, S. P., O'Neill, B. & Karpf, R., *The Injury Fact Book.* Lexington Press, MA, 1984.
2. *National Transportation Safety Board,* Data, 1986.
3. *Injury in America, A Continuing Public Health Problem.* National Academy Press, Washington, DC, 1985.
4. Full-scale testing procedures for guardrails and guide posts. Highway Research Board Circular No. 482, Highway Research Board Committee on Guardrails and Guide Posts, 1962.
5. Michie, J. D. & Bronstad, M. E., Recommended procedures for vehicle crash testing of highway appurtenances. NCHRP Report 153, National Cooperative Highway Research Program, 1974.
6. Recommended procedures for vehicle crash testing of highway appurtenances. Transportation Research Circular No. 191, Committee A2A04, 1978.
7. Michie, J. D., Recommended procedures for safety performance evaluation of highway appurtenances. NCHRP Report, 230, National Cooperative Highway Research Program, 1981.
8. Hackney, J. R., Monk, M. W., Hollowell, W. T., Sullivan, L. K. & Willke, D. T., Results of the national highway safety administration's thoracic side impact protection research program. SAE Technical Paper Series 840886 1984.
9. Gennarelli, T. A., Champion, H. R., Sacco, W. J., Copes, W. S. & Alves, W. M., Mortality of patients with head injury and extracranial injury treated in trauma centers. *J. Trauma,* **29** (1989) 9.
10. Chou, C. C., Howell, R. J. & Chang, B. Y., A review and evaluation of various HIC algorithms. SAE Technical Paper 880656, 1988.
11. Michie, J. D. & Bronstad, M. E., Guardrail performance: end treatments. Report of Southwest Research Institute, San Antonio, TX, 1969.
12. Nordlin, E. F., Field, R. N. & Folsom, J. J., Dynamic tests of short sections of corrugated metal beam guardrail. Highway Research Record 259, Highway Research Board, 1969.
13. Michie, J. D. & Bronstad, M. E., Guardrail crash test evaluation—new concepts and end designs. NCHRP Report 129, Highway Research Board, 1972.
14. Quresky, A. B., Development of a new end treatment for W-beam guardrail. Ph.D. Thesis, Texas A & M University, 1992.
15. Sicking, D. L., Bleigh, R. P., Ross, H. E., Jr & Buth, C. E., Development of new guardrail end treatments. Research Report 404-1F, Texas Transportation Institute, Texas A & M University, 1988.
16. Sicking, D. L., Quresky, A. B. & Ross, H. E., Jr, Development of guardrail extruder terminal. TRB Transportation Research Record 1233, Design and Testing of Roadside Safety Devices, 1989.
17. Brakemaster. Energy Absorption Systems, Inc., 1 East Wacker Drive, Chicago, IL 60601-2076.
18. Bronstad, M. E., Hancock, K. L., Meczkowski, L. C. & Humble, W. P., Crash test evaluation of the vehicle-attenuating terminal. TRB Transportation Research Record 1133, Roadside Safety Features, 1987.

19. Guardrail energy absorbing terminal. Energy Absorption Systems, Inc., 1 East Wacker Drive, Chicago, IL 60601–2076.
20. Reid, S. R., Laterally compressed metal tubes as impact energy absorbers. In *Structural Crashworthiness,* ed. N. Jones & T. Wierzbicki. Butterworth, London, 1983, p. 1.
21. Shim, V. P.-W. & Stronge, W. J., Lateral crushing of thin-walled tubes between cylindrical indenters. *Int. J. Mech. Sci.,* **28** (1986) 683–707.
22. Shim, V. P.-W. & Stronge, W. J., Lateral crushing in tightly-packed arrays of thin-walled metal tubes. *Int. J. Mech. Sci.,* **28** (1986) 709–29.
23. Stronge, W. J. & Shim, V. P.-W., Microdynamics of crushing in cellular solids. *J. Engng Mater. Technol.,* **110** (1988) 185–90.
24. Reid, S. R., Carney, J. F., III & Drew, S. L. K., Energy absorbing capacities of braced metal tubes. *Int. J. Mech. Sci.,* **25** (1983) 649–67.
25. Reddy, T. Y., Reid, S. R., Carney, J. F., III & Veillette, J. R., Crushing analysis of braced metal rings using the equivalent structures technique. *Int. J. Mech. Sci.,* **29** (1987) 655–68.
26. Carney, J. F., III & Pothen, S., Energy dissipation in braced cylindrical shells. *Int. J. Mech. Sci.,* **30** (1988) 203–16.
27. Veillette, J. R. & Carney, J. F., III, Collapse of braced tubes under impact loads. *Int. J. Impact Engng,* **7** (1988) 125–38.
28. Carney, J. F., III, Dougan, C. E. & Hargrave, M. W., The Connecticut impact-attenuation system. TRB Transportation Research Record 1024, Application of Safety Appurtenances, 1985.
29. Merchant, W., On equivalent structures. *Int. J. Mech. Sci.,* **7** (1965) 613–19.
30. Gill, S. S., Large deflection rigid–plastic analysis of a built-in semi-circular arch. *Int. J. Mech. Engng Educ.,* **4** (1976) 339–55.
31. Reddy, T. Y. & Reid, S. R., Lateral compression of tubes and tube systems with side constraints. *Int. J. Mech. Sci.,* **21** (1979) 187–99.
32. Redwood, R. G., Discussion on crushing of a tube between rigid plates, by J. A. DeRuntz and P. G. Hodge, Jr. *J. Appl. Mech.,* **31** (1964) 357–58.
33. Carney, J. F., III & Sazinski, R. J., A portable energy-absorbing system for highway service vehicles. *Transportation Engng J. ASCE,* **104** (1978) 407–21.
34. Carney, J. F., III, Crash testing of a portable energy absorbing system for highway service vehicles. TRB Transportation Research Record 833, 1981.
35. Carney, J. F., III, The Connecticut narrow hazard crash cushion. TRB Transportation Research Record 1233, Design and Testing of Roadside Safety Devices, 1989.
36. Logie, D. S., Carney, J. F., III & Ray, M. H., An expert system for highway crash cushion design. TRB Transportation Research Record 1233, Design and Testing of Roadside Safety Devices, 1989.
37. Hayes, G. G., Ivey, D. L. & Hirsch, T. J., Performance of the hi-dro cushion cell barrier vehicle-impact attenuator. Highway Research Record 343, 1971.
38. Nordlin, E. F., Stoker, J. R. & Doty, R. N., Dynamic tests of an energy-absorbing barrier employing sand-filled plastic barrels. Highway Research Record 386, 1972.
39. Ray, M. H. and Carney, J. F., III, An improved method for reducing

vehicle transducer data and calculating occupant impact velocities. TRB Transportation Research Record 1233, Design and Testing of Roadside Safety Devices, 1989.
40. Ray, M. H., Troxel, L. A. & Carney, J. F., III, Characteristics of fixed-roadside-object side-impact accidents. *J. Transportation Engng ASCE,* **117** (1991) 281–87.
41. Troxel, L. A., Ray, M. H. & Carney, J. F., III, Side impact collisions with road-side obstacles. TRB Transportation Research Record 1302, Roadside Safety Features, 1991.
42. Motor vehicle accident costs. FHWA Technical Advisory T 7570.1, Federal Highway Administration, 1988.
43. Carney, J. F., III & Ray, M. H., Side impact crash testing of highway safety hardware. In *Proceedings of the International Conference—Strategic Highway Research Program and Traffic Safety on Two Continents, 18–20, September 1991,* Research Institute, 1991, pp. 25–42.

Chapter 12

GROUNDING DAMAGE OF SHIPS

TOMASZ WIERZBICKI & PAUL THOMAS

Department of Ocean Engineering, Massachusetts Institute of Technology Cambridge, Massachusetts 02139, USA

ABSTRACT

Damage prediction models for oil tankers are developed through the application of plasticity and fracture mechanics to the primary failure modes of the major structural members involved in grounding. Four primary energy absorbing mechanisms are identified: plate cutting, plate tearing, girder tearing and web crushing. The plate cutting force is derived from a recently published closed-form solution (due to Wierzbicki and Thomas) for the wedge cutting force in the initiation phase, and from a series of steady-state cutting experiments conducted at MIT. The plate tearing and girder tearing forces are derived from simplified models of their respective failure modes. The solution for crushing of the web girder is derived from the concept of a superfolding element. This solution is validated on small-scale tests and then applied to assess the damage of a unidirectionally stiffened double hull in the vertical indentation mode. Critical spreading angles of a reef are determined for which fracture of outer and inner hull will occur. The predictions of the longitudinal damage based on the tearing model are compared with earlier results from Vaughan's method for a typical 39 000 DWT double-hull vessel. Plots are presented showing how the energy is partitioned between the primary failure mechanisms over a range of parameters. The present models offer significant improvements over the existing damage prediction methods because they are not empirical, they distinguish the contributions of individual members to the overall energy absorption of the structure, and they are general enough to be broadly applied.

1 INTRODUCTION

1.1 General

Since the groundings of the tanker *Exxon Valdez* in Alaska's Prince William Sound in 1989, and the tanker *American Trader* off the coast of

Southern California in 1990, the public in the United States and worldwide has been awakened to the dangers of transporting large amounts of oil and other hazardous bulk cargos on the sea. For the first time the commercial shipbuilding industry, and the agencies that regulate it, are being asked to provide protection to the environment through improved tank vessel design. Traditionally, naval architects have designed vessels that are intended to be safe under 'normal' operating conditions. Now, however, the industry is forced to address the problem of vessel performance in collision or grounding.

A great deal of the recent literature on ship grounding has consisted of comparative studies,[1–3] which are intended to evaluate the relative effectiveness of alternative tank vessel designs in limiting the extent of damage and loss of cargo due to high-energy grounding. To a large extent, these studies have been undertaken to meet the immediate need of US and international regulatory agencies and political bodies for information on which to base long-term policy decisions regarding the next generation of tanker vessels. Unfortunately, the state of the art in structural analysis of ships has not kept pace with this need for information, and, as a result, these decisions have been based on technology that is over a decade old. The recent studies have not served to advance the state of the art in damage prediction, but rather they have relied on methods that are outdated and/or lack generality.

The objective of this chapter is to present the state of the art in damage prediction for steady-state grounding of tanker vessels. Two distinct types of groundings are discussed here: grounding on a sharp, narrow reef, and grounding on blunt or wide obstacles. The narrow reef grounding primarily involves local damage due to plate cutting and tearing, while the wide reef grounding involves less localized bending and crushing of the underframe, and shearing and fracture of hull plating. These two type of groundings require separate analysis.

In the case of narrow reef groundings, the existing damage prediction methods fall into two broad categories: (1) those based on small-scale experiments involving a specific hull configuration, and (2) those based on Vaughan's[4–7] empirical analysis of plate cutting experiments. Both of these general types of approach have limitations that restrict their application.

The formulation of a damage prediction method from small-scale experiments is fundamentally correct, but is severely restricted in its application. The RR701 group of the Shipbuilding Research Association of Japan presented such a formulation to the International Maritime Organization (IMO), based on a series of tests that involved cutting a scale section of a double hull.[3] They proposed a formula for the cutting force in the form

$$F = \eta \sigma_0 A \tag{1}$$

where η is an empirically determined proportionality constant, σ_0 is the material flow stress, and A is the cross-sectional area of damage. The constant η contains many important parameters of the problem, including critical strain and fracture toughness of the material, friction coefficient, reef geometry, arrangement of stiffeners and girders, and double-bottom separation. As a result, eqn (1) can only be applied to structures that closely resemble the test section, and cannot be used for comparative studies. Additional validation of eqn (1) was recently performed by Kuroiwa and co-workers,[8] who proposed a modified expression for F involving two proportionality constants: one for the plating and one for the stiffeners.

On the other hand, not only is Vaughan's method of damage prediction limited in application owing to its empirical nature, but it is also fundamentally incorrect because it is based on plate cutting experiments that involved the initiation portion of the cutting process only. When a plate is cut by a wedge of finite width, the cutting force increases with the length of the cut until the process reaches steady state, and then the force levels out. For the most part, the ground event must be considered as a *steady-state* penetration of a wedge into a stiffened plate, which involves a constant cutting force. It is incorrect to formulate a steady-state damage prediction model based on initiation experiments alone. The Vaughan method of damage prediction will be discussed further in this chapter because it has been the basis of many of the recent comparative studies.

In this chapter a damage prediction model for narrow reef groundings is presented that is both fundamentally based and general enough for broad practical application. The approach used here follows that developed by Wierzbicki and others[9–13] in applying the concepts of 'crashworthiness' to the analysis of ship structures. Three primary energy absorbing mechanisms of the hull girder are separately analyzed using the principles of plasticity and fracture mechanics, and then combined into an overall damage prediction model. These mechanisms are (1) plate cutting, (2) plate tearing, and (3) girder tearing. Although the plate cutting is the primary focus of the present work on the narrow reef problem, plate tearing and girder tearing are also included, but in a less rigorous fashion.

This chapter comprises seven sections, starting with a detailed analysis of the cutting of a ship's hull by a wedge. Then, as more failure mechanisms and structural members are considered, the description loses some of its original rigor in favor of a more global viewpoint and practical applicability. Section 2 reviews the mechanics of plate cutting by a sharp and narrow reef during the initiation phase of the process, as developed by Wierzbicki and Thomas.[14] Vaughan[7] and Lu and Calladine[15] have previously analyzed the initiation problem empirically. The closed-form solution developed by the present authors matches the empirical results for the given specific configurations, but it also allows direct scaling for application to ships.

In Section 3, the steady-state plate cutting problem is addressed. Experimental data is presented showing that the closed-form solution for the initiation phase can be used to estimate the steady-state plate cutting force. An expression for the force to tear the plate transversely is also derived in this section.

In Section 4, a steady-state damage prediction model for double-hull tank vessels is developed using the analysis of Sections 2 and 3, together with some additional analysis of the girder tearing. This model is intended to eliminate the major limitations of the past narrow reef damage prediction methods, through application of the steady-state plate cutting solution and the use of a non-empirical analysis for the primary failure modes of the major structural members involved in grounding. The model is applied to damage assessment for a typical uni-directionally stiffened double-hull (USDH) tank vessel. A comparison between Vaughan's method and this new model is presented for the absolute and comparative damage assessment cases.

Grounding on wider and less sharp obstacles is likely to result in failure modes other than plate cutting, plate tearing, and girder tearing. Typically, these modes will include bending of longitudinal stiffeners, crushing of transverse frames, and shearing of the hull plating. In the wide reef scenario, failure of the hull may be caused by tensile fracture of the plating and/or weld failure at the stiffeners due to excessive shear forces. These modes have been the subject of several recent studies,[10,16,17] the results of which are summarized in Section 5.

All of the above predictive methods have a common denominator in that none of them consider the vertical component of the reaction force between the ship and the grounding surface. McKenney[11] addressed the vertical indentation of a wedge into a USDH vessel in the case of stranding on a sharp, narrow reef. A similar problem was solved numerically by Ueda,[18] while Ito *et al.*[19] and Amdahl[20] reported on an comprehensive theoretical–experimental program. The work of McKenney is briefly described in Section 6.

Finally, Section 7 summarizes the progress made to date in this important and very complex area of structural crashworthiness, and indicates direction for further research.

1.2 Brief Literature Review

The problem of penetration of a wedge into thin metal plates has been the subject of intensive research over the past decade.[4–6,15,21–31] An excellent summary of current work in this area was recently presented by Lu and Calladine.[15] It is generally agreed among investigators that the plate resistance to wedge indentation involves two direct mechanisms: 'far-field' plastic deformation, and 'near-tip' fracture. A third, indirect mechanism addressed in some studies is friction. Attempts have been made in the literature to isolate the above mechanisms and to quantify the individual work or force contributions of each from experimental data. These attempts have not been fully successful, as reviewed in Ref. 15.

Lu and Calladine performed a series of well-controlled experiments and showed that a single-term empirical formula for the work of cutting W adequately described the test results. The proposed the following formula for the work of cutting plates:

$$W = C\sigma_0 l^n t^{3-n} \tag{2}$$

where σ_0 is the material flow stress, l is the length of cut, and t is the plate thickness. In this form, certain parameters of the process (e.g. the semi-angle of the wedge and the friction coefficient) are lumped together into a single numerical constant C. In 39 tests on mild steel and other metals, Lu and Calladine found this numerical constant to vary in the range $C = 0{\cdot}9\text{–}3{\cdot}5$, while the exponent was in the range $n = 1{\cdot}2\text{–}1{\cdot}4$. The corresponding power to which the thickness is raised in eqn (2) is $3 - n = 1{\cdot}8\text{–}1{\cdot}6$. It is interesting to note that in a steady-state 'trousers test' type of experiment involving a shear fracture mode, Yu *et al.*[32] found the tearing energy to be proportional to $t^{1{\cdot}61}$.

While Lu and Calladine recognized the existence of two essentially distinct processes of plastic deformation involved in the near-tip and far-field regions of the plate, they also observed that the sum of the two effects may be adequately approximated by the single-term eqn (2), and that apparently no fracture parameter is necessary for correlation of the experimental data.

This point of view has been opposed by Atkins,[33,34] who suggested that the far-field and near-tip events should give rise to separate, independent terms in the expressions for the energy absorption and hence the indentation force. An attempt to resolve the above controversy was made by Wierzbicki and Thomas,[14] who showed that the global and local deformation (far-field and near-tip) mechanisms *are not independent, as previously believed, but are related through a single* geometric parameter. This parameter is the instantaneous bending or rolling radius R of cylindrical flaps formed in the wake of the cut. Specifically, the farfield bending work is inversely proportional to R, while the membrane work in the near-tip fracture zone increases in a non-linear way with the rolling radius. Optimization of the instantaneous indentation force with respect to R yields a simple single-term relationship identical in form with that of eqn (2). The calculation of membrane work near the crack tip in this analysis was based on the crack opening displacement (COD) parameter δ_t.

Except for the already mentioned Refs 3 and 8 and the continuing work at Det norske Veritas, reported by Astrup,[35] the literature on steady-state plate cutting is virtually non-existent, but the research is vital to the development of a practical damage prediction model for ships in high-energy groundings. The empirical formulations for plate cutting force of both Vaughan[7] and Lu and Calladine[15] are based on experiments involving the initiation phase only. As a result, these formulations, as well as the closed-form solution of Ref. 14, have the cutting force increasing indefinitely with the cut length. Previous experiments did not drive the wedge far enough into the plate to observe the transition to steady state. It is incorrect to base a damage prediction model for groundings on initiation experiments, because, for the most part, the grounding process must be thought of as a steady-state event in which the resisting force of the ship structure is constant. Despite its limitation, Vaughan's empirical analysis was used by Det norske Veritas (DnV) as part of their comparative study of various very large crude carrier (VLCC) designs[1] and by the David Taylor Research Center (DTRC) for their collision and grounding assessment for double-hull tankers prepared for the US Coast Guard.[2]

2 FORMULATION OF THE PROBLEM OF PLATE CUTTING INITIATION

Consider a sharp rigid wedge pressing into a thin plate, as shown in Fig. 1. The equilibrium of the wedge–plate system is expressed via the principle of virtual work

$$F_p V = \dot{E}_b + \dot{E}_m \tag{3}$$

where F_p is the plastic resistance force, V is the wedge velocity, and $\dot{E}_b$ and $\dot{E}_m$ are the rates of bending and membrane work respectively. In a plane condition these quantities are defined by

$$\dot{E}_b = \int_S M_{\alpha\beta}\dot{K}_{\alpha\beta}\, dS + \sum_i M_0^{(i)}\dot{\phi}^{(i)}l^{(i)} \tag{4}$$

$$E_m = \int_S N_{\alpha\beta}\dot{\varepsilon}_{\alpha\beta}\, dS \quad (\alpha, \beta = 1, 2) \tag{5}$$

where $\dot{K}_{\alpha\beta}$ and $\dot{\varepsilon}_{\alpha\beta}$ are the curvature-rate and the strain-rate tensors respectively, and $M_{\alpha\beta}$ and $N_{\alpha\beta}$ are the corresponding bending-moment and membrane-force tensors. The second term on the right-hand side of eqn (4) is the rate of work on a discontinuous velocity field (rotation rate $\dot{\phi}$), where M_0 is a fully plastic bending moment, σ_0 is the flow stress of the material, and t is the plate thickness. The integration in the continuous deformation field is performed over the area S of the plastically deforming regions, while the contribution of a discontinuous field is summed over a finite number of straight line segments of length $l^{(i)}$ each.

For rigid–plastic, non-hardening material, eqn (3) gives an approximate expression for the instantaneous force F_p if a kinematically admissible velocity field $\dot{u}_\alpha(x, y)$ can be constructed compatible with the kinematic boundary condition (wedge motion V) and the strain-rate field ($\dot{K}_{\alpha\beta}$, $\dot{\varepsilon}_{\alpha\beta}$, $\dot{\phi}$). *It should be noted that eqn* (2) *is also valid for a cracked body, where the fracture process enters the solution through continuously changing geometry of the wedge–plate system.*

The problem consists in choosing a suitable, kinematically admissible flow field, and optimizing it with respect to one or more free parameters in order to arrive at the lowest cutting force. In addition to plastic resistance, friction has an influence on the total cutting force. The contribution of friction to the indentation force F_f is added later.

2.1 Kinematics of the Cutting Process

The kinematic model used to formulate the present theory is shown in Fig. 1. It is based in part on a visual analysis of the experimental results presented in Ref. 15, and those conducted by the present authors[36], Fig. 2.

Consider a wedge of a semi-angle θ advancing into a plate with a constant velocity V. A careful examination of the geometry of the flaps formed in the wake of the cut reveals that the plate material on both sides of the wedge is bent and curled into two variable curvature cylindrical surfaces as in Fig. 1(a). The plastic energy absorption in the process in concentrated in two primary areas: (i) the transverse membrane stretching and tearing of the plate material in the near-tip zone ahead of the wedge, and (ii) the bending of the plate into flaps in the far-field zone.

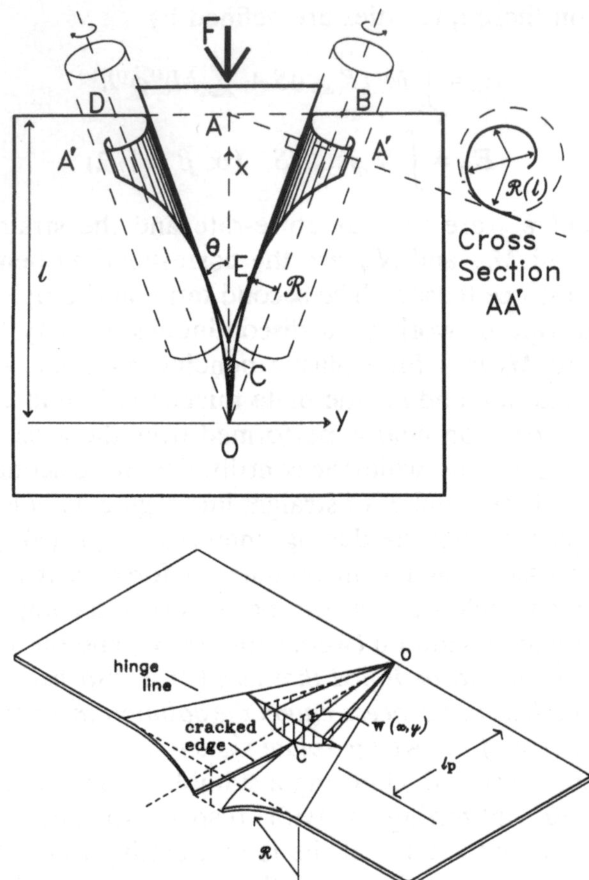

Fig. 1. Overall kinematics of the wedge indentation with curled cylindrical flaps. A close-up view from beneath shows lift in the near-tip region.

The total resistance to the wedge in plate cutting is the sum of the plastic resistance force F_p and the frictional force f_f, given by

$$F_c = F_p + F_p = 1{\cdot}67\sigma_0(\bar{\delta}_t)^{0\cdot2}l^{0\cdot4}t^{1\cdot6}g(\theta) \tag{6}$$

where

$$g(\theta) = \frac{1}{(\cos\theta)^{0\cdot8}}\left[(\tan\theta)^{0\cdot4} + \frac{\mu}{(\tan\theta)^{0\cdot6}}\right] \tag{7}$$

Details of the derivation of eqns (6) and (7) can be found in Ref. 14.

A plot of the function $g(\theta)$ (Fig. 3) reveals that the cutting force depends very weakly on the magnitude of the wedge semi-angle, particularly in a certain range. For example, with $\mu = 0{\cdot}3$, the difference in $g(\theta)$ between $\theta = 10°$ and $\theta = 20°$ is 5·88%. This is in

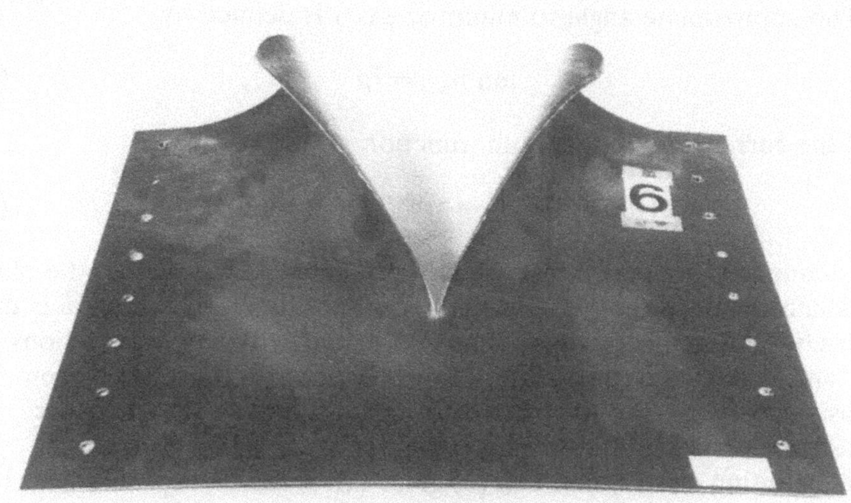

Fig. 2. Deformation mode of a cut plate in an initiation mode.

perfect agreement with the experimental observations of Lu and Calladine, who reported this difference to be about 5%.

Because of this weak dependence on θ, it is reasonable to simplify the expression for the cutting force, eqn (6), by evaluating it at the minimum value of the function $g(\theta)$ for each value of μ plotted in Fig.

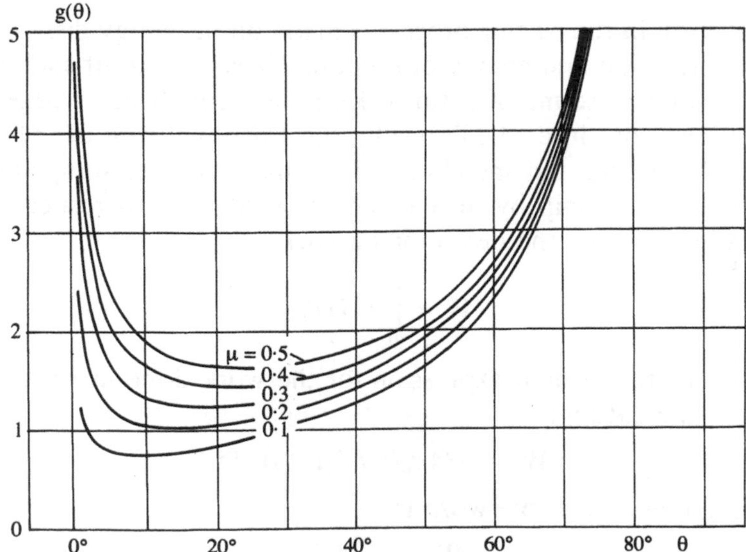

Fig. 3. Plot of the function $g(\theta)$ showing a weak dependence of the solution on the wedge semi-angle.

3. The approximate angle to minimize $g(\theta)$ is defined by

$$\tan \theta_{min} = \tfrac{3}{2}\mu \tag{8}$$

and the corresponding minimum function g is

$$g(\theta)_{min} = [(\tfrac{3}{2})^{0.4} + (\tfrac{2}{3})^{0.6}]\mu^{0.4} = 1.96\mu^{0.4}. \tag{9}$$

This equation implies that, at the optimum angle θ_{min}, 60% of the plate resistance comes from plastic dissipation and the remaining 40% is due to friction. Again, this result is in agreement with the observations of Lu and Calladine, who estimated the contribution of friction by measuring the ratio of the force of withdrawal to the force of penetration and found it to be $\eta = 0.4$.

Substituting eqn (9) into eqn (6) gives the practical expression for the approximate penetration force

$$F_c = 3.28\sigma_0(\bar{\delta}_t)^{0.2}\mu^{0.4}l^{0.4}t^{1.6} \tag{10a}$$

or the non-dimensional penetration force

$$\frac{F_c}{\sigma_0 t^2} = C\left(\frac{l}{t}\right)^{0.4} \tag{10b}$$

where

$$C = 3.28\mu^{0.4}(\bar{\delta}_t)^{0.2} \tag{10c}$$

2.2 Energy Absorption

The total work in the cutting process is made up of energy dissipated by plastic deformation and energy of friction. These two components have been expressed in terms of a single formula through the wedge angle minimization procedure. Additionally, the fractional rate of the plastic energies of bending and membrane deformations were lumped into a single-term relationship by means of minimization with respect to the rolling radius.[14] From the definition of work

$$W = \int_0^l F(l)\,\mathrm{d}l \tag{11}$$

we can calculate the final expression for the work done in resisting the wedge in plate cutting:

$$W_c = 2.34\sigma_0 l^{1.4}t^{1.6}\mu^{0.4}(\bar{\delta}_t)^{0.2} \tag{12a}$$

In dimensionless form, the work is

$$\frac{W_c}{\sigma_0 t^3} = C_1\left(\frac{l}{t}\right)^{1.4} \tag{12b}$$

These equations are identical in form with those proposed by Lu and Calladine, *and they give an explicit expression for the numerical constant* C_1:

$$C_1 = 2{\cdot}34\mu^{0{\cdot}4}(\bar{\delta}_t)^{0{\cdot}2} \tag{13}$$

For example, taking $\mu = 0{\cdot}3$ (which is reasonable for copper–steel sliding contact[37] and $\bar{\delta}_t = 1$,[38] the constant given by (13) is $C_1 = 1{\cdot}44$. This is in remarkably close agreement with the results of tests in copper sheets reported in Ref. 15, where the constant was found to be $C = 1{\cdot}4$ when the exponent was $n = 1{\cdot}4$. In general, Lu and Calladine found that for a variety of materials tested, the exponent n was always in the range 1·2–1·4.

According to the present theory, the work of plate cutting is dissipated through the three primary mechanisms as follows:

bending	24%,
membrane	36%,
friction	40%

This result is independent of the values of μ and θ over a wide range of wedge angles in the vicinity of the optimum angle θ_{min} given by eqn (8). For larger or smaller angles, the distribution may be different.

3 PLATE CUTTING EXPERIMENTS AND FORMULATION OF THE STEADY-STATE PROBLEM

When a plate is cut by a finite-width wedge, the process involves two distinct stages: initiation, and steady state. During the initiation stage, the cutting force increases with the length of the cut, as was described in Section 2 and observed experimentally by Lu and Calladine.[15] At some point, however, the cutting process involving a finite-width wedge must transfer to a steady state, and the cutting force will become constant. The length of the cut at the point that the transfer to a steady state occurs is defined here as the critical cut length, or critical length l_{cr}. Plate cutting experiments were conducted by Thomas[36] to verify the closed-form solution presented in Section 2 for the force during the initiation stage, and to determine the critical length at which the process transfers to the steady-state stage so that a method of estimating the steady-state cutting force can be developed.

3.1 Theory

During the initiation phase of the plate cutting process, the radius of the flaps in the wake of the wedge grows with the cut length, and the flaps form at an angle equal to the wedge angle. When the process transfers to a steady state, however, the flaps become of constant radius and bend parallel to the axis of the wedge. This hypothetical steady-state cutting mode is shown in Figures 4 and 8. When the transition to a steady state occurs, the cutting force becomes constant.

Commonsense dictates that the process must transfer to a steady state, because for a finite-width wedge the cutting force cannot become unbounded. At some point, it takes more energy to continue to increase the radius of the flaps than to stretch and bend them parallel at a constant radius. The length of the cut at this point is the critical length.

Once the critical length l_{cr} has been determined, the closed-form solution presented in Section 2 can be used to estimate the constant steady-state cutting force F_c by evaluating eqn (10a) at $l = l_{cr}$.

The closed-form solution for the plate cutting force accounts for the transverse stretching of the plate in the near-tip region. In steady-state cutting, however, the plate also undergoes longitudinal stretching in the far-field region due to the angle at which the flaps are formed. If the cutting wedge is narrow enough, the plate flaps will stretch lon-

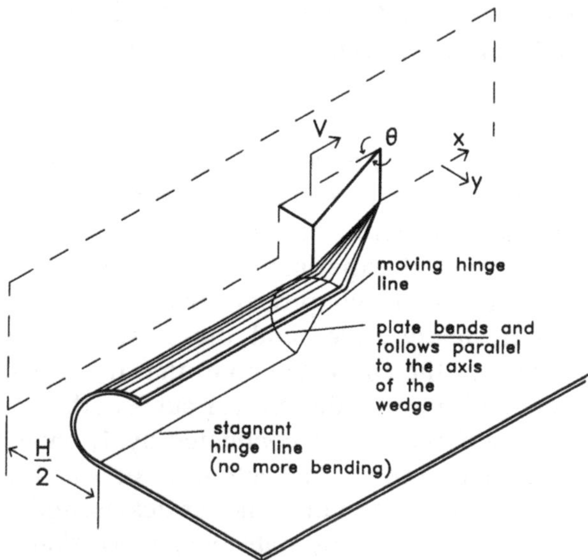

Fig. 4. Deformation mode in a steady-state cutting process.

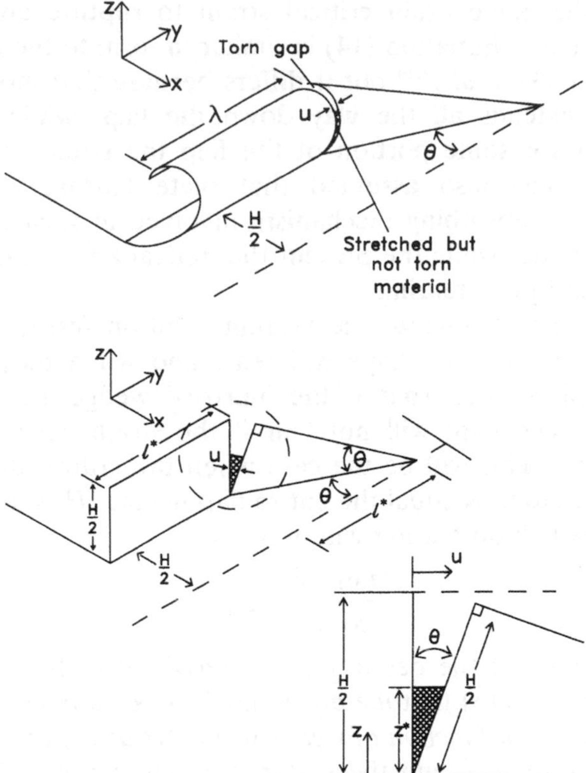

Fig. 5. Transverse tearing of the curled and flat flaps in the wake of a wide wedge. The insert shows details of the torn gap.

gitudinally and then bend parallel to the wedge axis as described above, and shown in Fig. 4. In the narrow wedge case, the membrane energy involved in the longitudinal stretching is small. For a wider wedge, however, the longitudinal stretching exceeds the critical strain to rupture ε_{cr}, and the plate flaps will tear transversely at some characteristic interval λ. The wide wedge mode is shown in Fig. 5, where λ is the wavelength at which the transverse tears occur along the length of the flaps.

The energy involved in tearing the flaps is significant, and the associated force must be accounted for. The force required to tear the flaps transversely was calculated by Thomas,[36] using a maximum-strain fracture criterion, and the interested reader is referred to this work for details of the derivation.

The expression for the 'tearing' force is

$$F_T = t\sigma_0\varepsilon_{cr}H\left[1 - \frac{\varepsilon_{cr}\cos\theta}{(\tan\theta)^2}\right] \tag{14}$$

where ε_{cr} is the plane-strain critical strain to rupture and $2H$ is the width of the wedge. Equation (14) is similar in form to the tearing force used by Wierzbicki *et al.*,[9,10] but it differs because that model assumed that the tear extends all the way down the flap, while the present analysis allows for some portion of the flap the stretch but not tear. Wierzbicki's model also assumed that plate tearing was the only significant energy absorbing mechanism involved in steady-state deformation of the plate, while the present theoretical analysis includes both plate cutting and plate tearing.

It is of interest to analyze the tearing solution further in order to determine when the plate flaps will tear, and when they will simply bend in one piece; i.e. what is the 'narrow' wedge and what is the 'wide' wedge? The flaps will not tear if the strain never reaches the critical value ε_{cr}. This will be the case when the critical distance z^* of the fracture initiation is equal the entire flap height $\frac{1}{2}H$ (see Fig. 5). The condition for which no tearing will occur is

$$\frac{(\tan\theta)^2}{\cos\theta} = \varepsilon_{cr} \tag{15}$$

Again, the details of the derivation are provided in Ref. 36, but it is apparent from eqn (15) that *the difference between a narrow wedge and a wide wedge is not a function of wedge width* $2H$ *at all, but a function of the wedge semi-angle* θ and the material parameter ε_{cr}. Dropping the cosine term in the denominator of eqn (15), an approximate expression for the critical wedge angle can be found:

$$\theta_{cr} \approx \arctan\sqrt{\varepsilon_{cr}} \tag{16}$$

This equation gives a simple way to predict the occurrence of transverse tearing of the plate flaps in the wake of a wedge. Wedges with semi-angle $\theta < \theta_{cr}$ will not cause tearing, and therefore the tearing force given by eqn (16) need not be added. For example, taking $\varepsilon_{cr} = 0{\cdot}2$, the critical angle is approximately 24°.

3.2 Experimental Apparatus and Test Specimens

The experiments were designed to test the plate cutting and plate tearing theories presented in the previous section. Test specimens were mounted in a rigid frame and placed in an INSTRON model 8500 dynamic testing system. A wedge mounted on the crosshead of the machine was then driven through the plate at a constant speed, and crosshead displacement and cutting force were automatically recorded. The apparatus is shown in Figure 6. Details of the frame and wedges are given in Ref. 36.

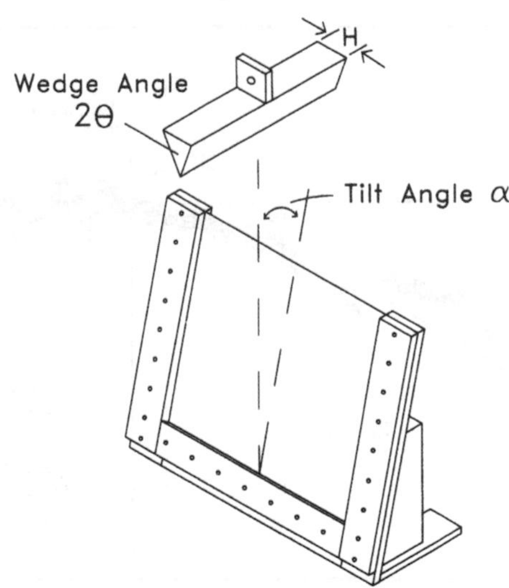

Fig. 6. Apparatus used for steady-state plate cutting experiments.

The test specimens were aluminum and mild steel plate cut to size from larger sheets. The aluminum specimens were 3003 alloy, $\frac{1}{16}$ in (0·0159 mm) and $\frac{1}{32}$ in (0·007 94 mm) nominal thickness, with an approximate yield stress of 114 MPa. The steel specimens were ASTM A366 Grade C cold rolled structural carbon steel, 16-gauge and 14-gauge nominal thickness, with an approximate yield strength of 227 MPa.

3.3 Representative Test Results

Eight experiments were conducted, involving two different wedge semi-angles, three wedge lengths, four plate thicknesses, and two materials. Only one tilt angle, $\alpha = 20°$, was tested. The width of the wedge shoulder is denoted by b.

The results of a typical test are shown in Figs 7 and 8. This is the test condition ($\theta = 20°$, $\alpha = 20°$) for which the kinematic model of Ref. 14 was developed, and which most closely matches the previous plate cutting work of Vaughan[6] and Lu and Calladine.[15] The first part of the plot in Fig. 7, labeled OA, is the initiation portion of the cutting process, during which the force increases with length. This looks very similar to the results presented in Ref. 15 for a mild steel plate under the same conditions. During this phase, the flaps in the wake of the wedge continued to form at an angle equal to the wedge angle, and the radius of the flaps increased with cut length as expected. Equation (10a) is also shown in Fig. 7. It generally follows the trace of the

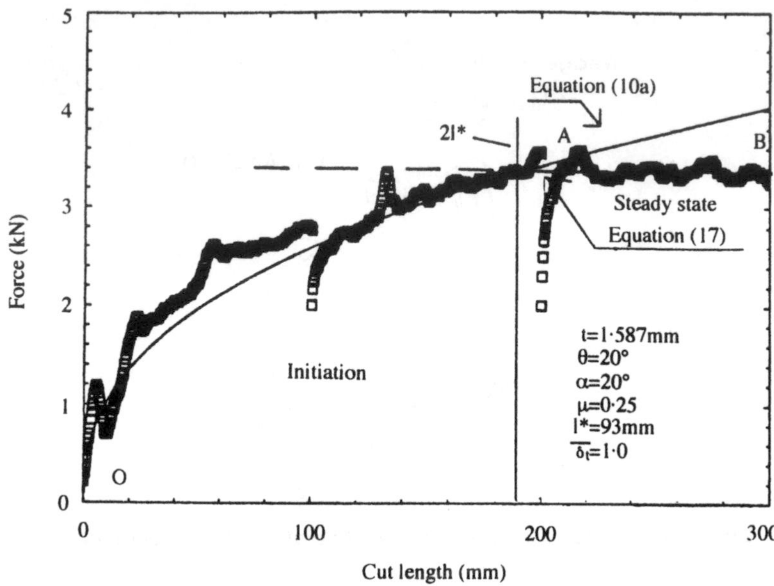

Fig. 7. Measured and calculated force–displacement characteristics of the aluminum plate.

experimental data throughout the initiation portion of the process. The values of μ and $\bar{\delta}_t$ used in plotting eqn (10a) were chosen to be 0·2 and 1·0 respectively. These values are used consistently throughout this chapter.

The second part of the trace, AB, is the steady-state portion of the

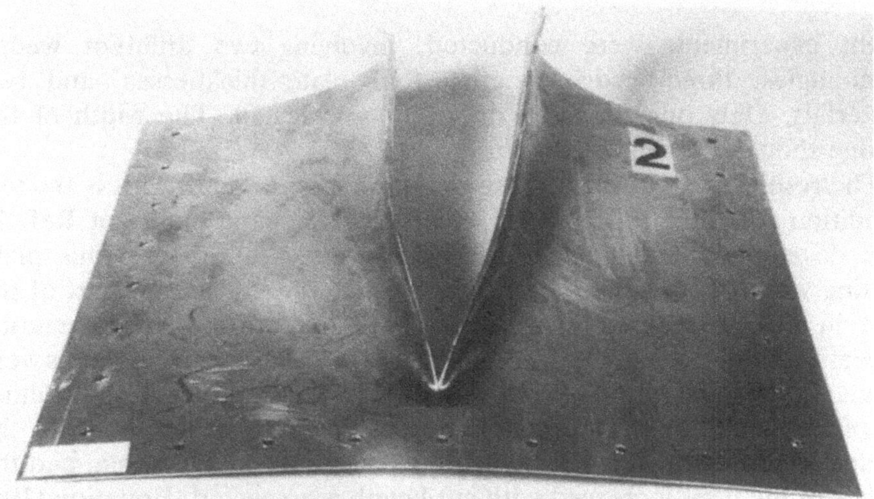

Fig. 8. Photograph of the damage specimen reaching a steady-state condition.

process. The cutting force no longer increases with length, but remains essentially constant as the wedge continues to penetrate the plate. Note that the process appears to transfer to a steady state shortly after the cut length is equal to twice the wedge length l^*. Note too that eqn (10a) gives a good estimate of the steady-state cutting force when evaluated at $l = 2l^*$.

Some of the test specimens showed the transition to a steady state in terms of physical kinematics. When the cutting wedge was about 200 mm (or twice the wedge length) into the plate, the flaps began to bend noticeably, to be parallel to the axis of the wedge rather than remain at 40°. The parallel flaps can be seen in the photograph in Fig. 8. The transition to a steady state was gradual, and continued throughout the remainder of the experiment.

As the wedge approached the bottom of the test specimen, and the lower edge of the support frame, the cutting force increased sharply, and the plate appeared to begin buckling ahead of the wedge tip. This was caused by the rigid boundary condition at the bottom edge of the plate, which interfered with the cutting process. The boundary effects caused by the support frame are observed much sooner in the process when $\theta = 30°$ than when $\theta = 20°$. In the 30° case, it was the sides of the frame, not the bottom, that interfered with the cutting process. As the wedge travels through the plate, the moving hinge lines in the center of the flaps migrate toward the edges of the plate. At some point, the ends of the hinge furthest from the origin of the cut reach the sides of the frame, and can go no further. When this happened, the flaps of the plate near the tip of the wedge began to knuckle, and the force increased sharply. These results may be significant for the application of these experiments to ship grounding, because it indicates that when a stiffened plate is cut by a wedge, the deformation mode can be quite different.

The test results show that eqn (10a) yields a very good approximation of the actual cutting force during initiation, throughout the range of plate thicknesses, wedge angles, and materials tested. Nearly every major feature of the kinematic model proposed in Section 2 was observed during the plate cutting tests. The flaps of the cut plate in the wake of the wedge were seen to be inclined cylinders with helical cross-sections, not variable-radius cones. Accordingly, the cut edge of the plate formed a modified helical line, and touched the wedge at only one location, as theorized in Fig. 1.

The results of the preliminary experiments point out the need for further plate cutting tests designed to ensure that the near-tip zone is torn, not cut (see Fig. 9). In a ship grounding, it is likely that the reef

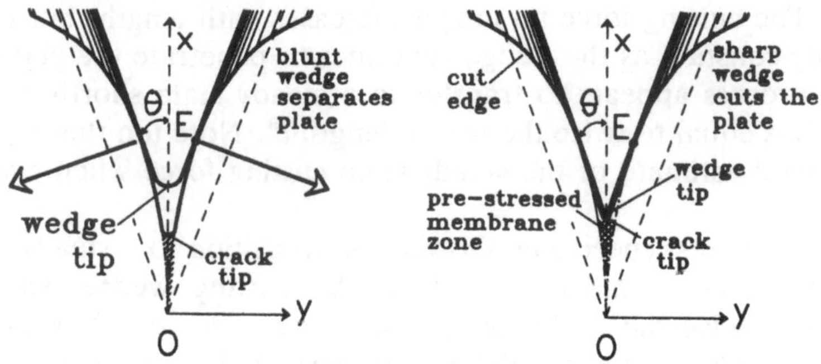

Fig. 9. Comparison of crack tip zones in blunt wedge and sharp wedge cutting experiments.

will not be as sharp as the wedges used here, and as a result the plate will crack owing to separation, not cutting. This is the mechanism that the present experiments were intended to investigate, but it appears likely that the sharp wedges used here actually resulted in some combined action of transverse tearing and longitudinal cutting in the near-tip zone. The same is true of the experiments of Ref. 15, since the tip of the wedge always coincided with the tip of the crack in those tests as well.[39] Future experiments should be conducted with blunt or rounded wedges to ensure that the near-tip zone is torn, not cut.

All of the test results clearly showed that the plate cutting process does leave the initiation phase and undergo a transition to a steady state; that is, the cutting force became essentially constant at some point after the shoulder of the wedge has penetrated the plate. The data also showed that the transition consistently occurs when the cut length is approximately equal to twice the wedge length. Thus, within the range of parameters tested, it is reasonable to approximate the critical length $l_{cr} = 2l^* = H \cot \theta$. On substituting this result into eqn (10a), an approximate expression for the steady-state cutting force is found:

$$F_{steady} = 1{\cdot}67\sigma_0(\bar{\delta}_t)^{0{\cdot}2}t^{1{\cdot}6}H^{0{\cdot}4}(\cos\theta)^{-0{\cdot}8}(1 + \mu \cot\theta) \qquad (17)$$

Further experimental work is required to specifically address the following aspects of plate cutting:

(1) Cut the plate with a blunt or rounded wedge so that the near-tip zone is torn transversely, not sheared longitudinally.

(2) Cut larger specimens so that the process has a chance to truly enter a steady state. Conduct a parametric study to determine which factors control the transition length, and how much the force increases during transition.

(3) Design an experiment specifically to detect transverse tearing. This may require the use of less ductile material and/or wider wedges. The experiments should be designed so that the same material is cut in the 'narrow wedge' mode and then in the 'wide wedge' mode so that eqn (15) can be verified.

4 DAMAGE PREDICTION MODEL FOR UNIDIRECTIONALLY STIFFENED DOUBLE-HULL (USDH) VESSELS

The analysis and results of Sections 2 and 3 have been employed to formulate a method for calculating the work done by the horizontal resisting force of the bottom plate during a steady-state, high-energy grounding of a unidirectionally stiffened double-hull (USDH) vessel. The work is then set equal to the kinetic energy of the vessel to determine the longitudinal extent of damage in the steady-state grounding mode. The steady-state problem is an important part of the overall grounding process, and it has not been thoroughly addressed in previously published models.

4.1 Formulation of the Problem

Consider an USDH tank vessel of mass M running onto a narrow rigid reef at velocity V_0, as shown in Fig. 10(a). The reef is originally a height Δ_r above the baseline of the vessel, and first contacts the ship bottom near the bow. The reef geometry is as shown in Fig. 10(b), where α is the sloping angle, β is the spreading angle and θ is the wedge angle at the front of the rock. The distance l^* is the wedge length of the idealized reef, and has the same meaning as the l^* defined in Section 3.

When the bow of the vessel hits the reef, the reaction force has both vertical and horizontal components. As a result of the horizontal force, the ship will actually be lifted a distance Δ_L. The amount of lift may be quite substantial near the bow of the vessel, and the damage to the bottom structure is limited to crushing without penetration.

As the vessel travels over the reef, however, the local reaction force between the ship and the reef will increase to the point that one or both of the hulls is penetrated. When the bottom plate ruptures, the penetration depth Δ differs from the reef height Δ_r by the amount of lift Δ_L, so $\Delta_r = \Delta + \Delta_L$. The transition between 'ride over' with no penetration, which occurs near the bow, and steady-state damage with full penetration, which occurs later, is the initiation and development of

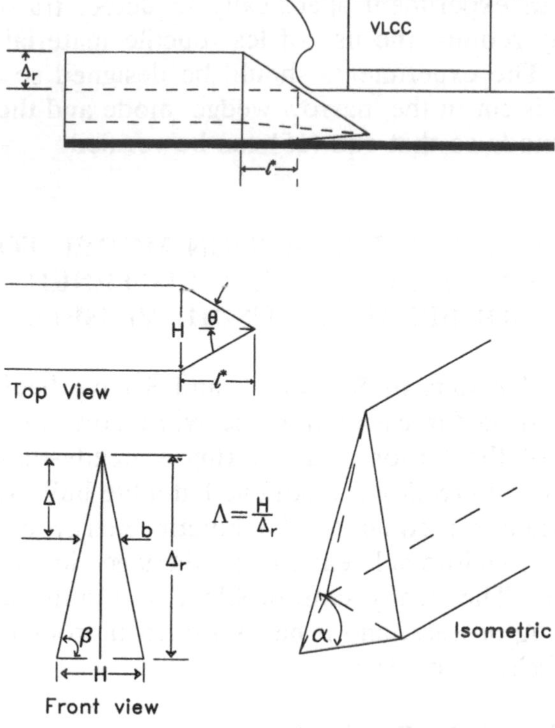

Fig. 10. Typical grounding scenario and reef geometry.

localized damage stage of the grounding event. It is a complex problem because it involves changing geometries and varying ratios of horizontal, vertical, and frictional forces. The initiation problem is discussed in Section 8.

4.2 Total Resisting Force and Longitudinal Extent of Damage

In general, the total resisting force F for the USDH is found by summing the contributions of cutting of the inner and outer bottom plates, F_{c_1} and F_{c_2}, tearing of the inner and outer bottom plate, F_{T_1} and F_{T_2}, and tearing of the longitudinal girders, F_g (see Fig. 11):

$$F = F_{c_1} + F_{c_2} + F_{T_1} + F_{T_2} + F_g$$

This force is constant for a given steady-state scenario, but the magnitude of the constant force depends on the geometry of the hull and the reef, determined by the parameters Λ and θ (Fig. 10). For a given hull and reef, the magnitude of the steady-state force depends primarily on the penetration depth Δ. As a result, the total force of resistance is quite cumbersome to list explicitly for the general case, but

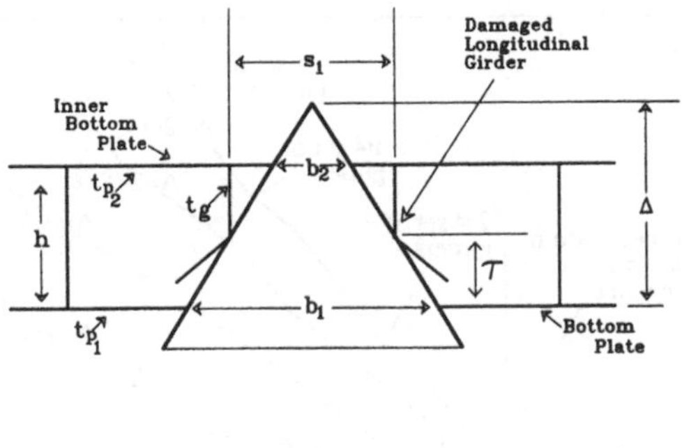

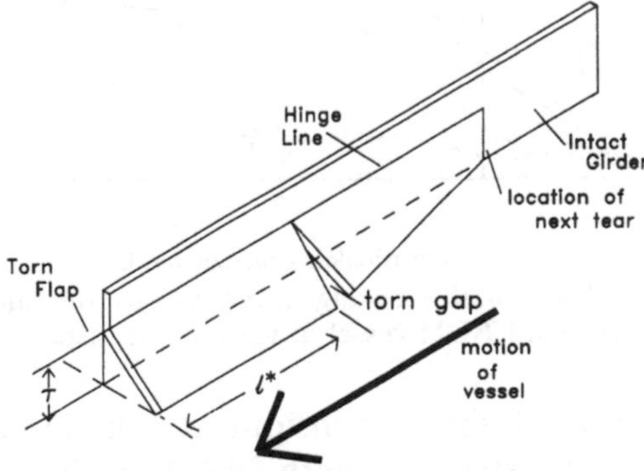

Fig. 11. Steady-state deformation mode of the longitudinal girders of the unidirectionally stiffened double hull (USDH).

it lends itself readily to computerization. Details of all the calculations are given by Thomas.[36]

The longitudinal extent of damage is found by equating the kinetic energy of the vessel when it entered steady-state grounding to the work done by the plate and girders. Since the resisting force F is assumed constant, the longitudinal extent of damage S_{max} is readily evaluated:

$$S_{max} = \frac{MV^2}{2F} \tag{18}$$

where M and V are respectively the mass (including added mass) and velocity of the vessel and F is the total force from eqn (17) in consistent units.

4.3 Damage Prediction Model Results

Results of the model are presented here in dimensionless plots for a typical USDH tank vessel. The exact equations used to generate these plots are presented in Ref. 36. For all the plots, the free parameters in the model were chosen as

$$\bar{\delta}_t = 1{\cdot}0, \qquad \mu = 0{\cdot}3, \qquad l_{cr} = 2l^*, \qquad \varepsilon_{cr} = 0{\cdot}2$$

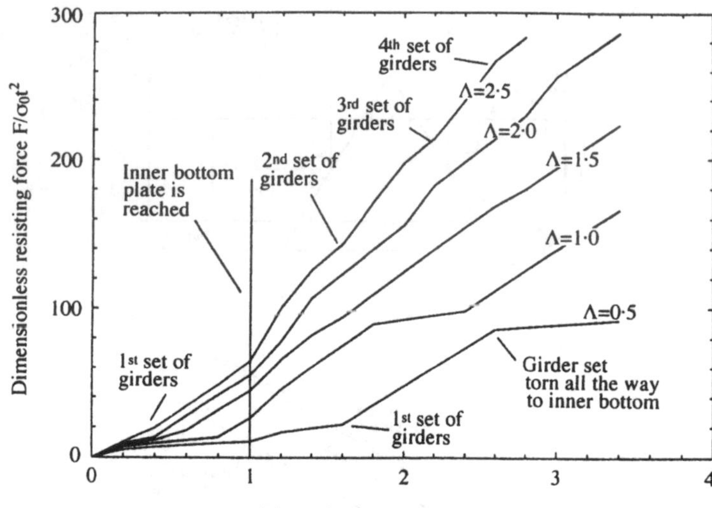

Fig. 12. Dimensionless horizontal resisting force versus dimensionless penetration depth for a typical USDH vessel at various reef aspect ratios Λ.

Figure 12 is a plot of the dimensionless horizontal resisting force versus dimensionless penetration depth (Δ/h) for five values of the reef aspect ratio Λ. All of these curves show step increases in the force when the dimensionless penetration depth is equal to 1. This reflects the additional contribution of the inner bottom plate, which begins when the penetration depth Δ is equal to the double-bottom spacing h. The other step increases on these curves occur each time the rock reaches a set of longitudinal girders and additional resistance is added. The plot shows that it is the aspect ratio of the reef compared with that of the hull (Λ versus s_1/h) that determines whether the reef will reach the inner bottom plate or the girder first when the penetration depth is increased.

Figure 13 is a dimensionless plot of resisting force versus penetration depth for three different wedge angles at $\Lambda = 2{\cdot}0$. Holding Λ constant means that the breadth of damage is constant for a given penetration depth in all three cases, but the wedge length l^* increases with θ. As expected, the total force increases with the reef wedge angle not because more girder sets are damaged (the number of girders set is constant for all curves in Fig. 13) but rather because both the cutting force and the tearing force of the plate are increased. In Fig. 13, the value of the critical strain to rupture $\varepsilon_{cr} = 0{\cdot}2$, so that $\theta_{cr} \approx 24°$. This means that the force to tear the plate transversely is not included in the $\theta = 20°$ curve, while it is in the other two, which accounts for the large jump in force between $\theta = 20°$ and $\theta = 30°$ curves.

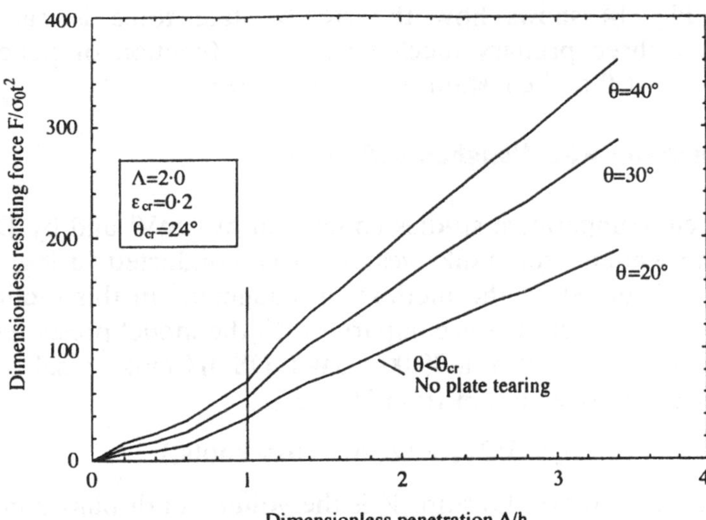

Fig. 13. Dimensionless horizontal resisting force of a typical USDH vessel versus dimensionless penetration length at three different reef wedge angles θ.

One distinct advantage that the present model has over past models is that it accounts for each energy absorbing mechanism separately, rather than lumping them into one single empirical term, as in eqn (1). As a result, it is possible to determine which mechanism, if any, dominates the energy absorption under a give set of circumstances. For

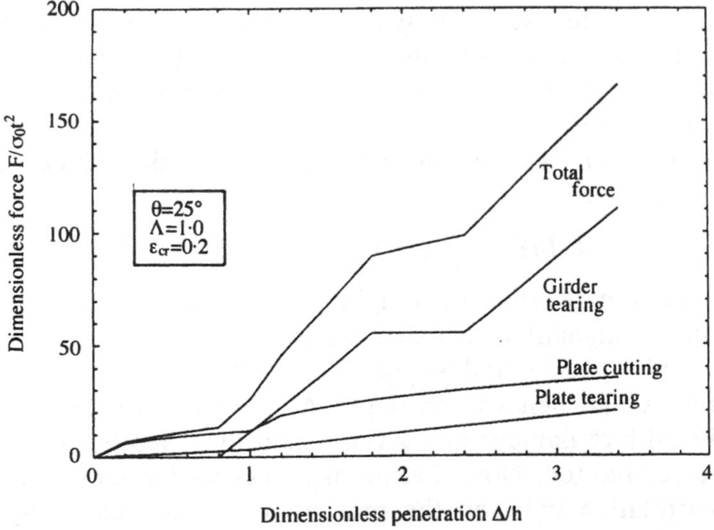

Fig. 14. Contribution of the three primary energy absorbing mechanisms to the total resistance force.

example, Fig. 14 shows how the steady state force is partitioned between the three primary mechanisms as a function of penetration depth, for $\Lambda = 1{\cdot}0$ and constant wedge angle $\theta = 25°$.

4.4 Comparison with Vaughan's Method

In two recent comparative studies conducted by DnV[1] and by DTRC,[2] damage assessment for tank vessels was conducted using similar methods, both based on the method of Vaughan.[7] In this section, the model used in those studies is compared with the model presented here, via a damage assessment of a 39 000 DWT USDH tank vessel.

Vaughan's energy equation from Ref. 7 is

$$W_s = 352V_s + 126A_s \quad \text{(ton-knots}^2\text{)} \tag{19}$$

where W_s is the work of the ship, V_s is the volume of displaced material, A_s is the area of torn material, and the coefficients 352 and 126 include all of the other parameters of the problem such as wedge angle, friction coefficient, and material constants. In eqn (19), the values of V_s and A_s are calculated using equivalent plate thickness in order to account for the longitudinal stiffeners that are damaged by the reef. This method has been widely used in the past, and is generally accepted as valid.[1,2,7,9,10] For the problem under consideration here, however, the equivalent thickness concept does not apply, since the USDH vessel does not have any longitudinal stiffeners. One could argue that the longitudinal girders of the USDH vessel should be incorporated into eqn (19) through the use of equivalent thickness as well, since they are not explicitly accounted for in the equation. In previous practical applications of Vaughan's method,[1,2] however, no mention has been made of the girders.

This comparison was conducted using particular values for the material parameters:

$$\sigma_0 = 300\ \text{MPa}, \qquad \varepsilon_{cr} = 0{\cdot}2, \qquad \mu = 0{\cdot}3, \qquad \bar{\delta}_t = 1$$

Figure 15 compares the total longitudinal extent of damage predicted by eqn (19) (Vaughan) with that predicted by the present model as a function of ship speed V and wedge angle θ, for a constant value of reef aspect ratio Λ and penetration depth Δ. Holding Λ constant ensures that the breadth of damage at a given penetration depth is the same for all four curves plotted. Note that at large reef wedge semi-angles there is good correlation between the present model and eqn (19), but at smaller wedge angles there is considerable difference between the two. Vaughan's method is insensitive to wedge angle, while the present

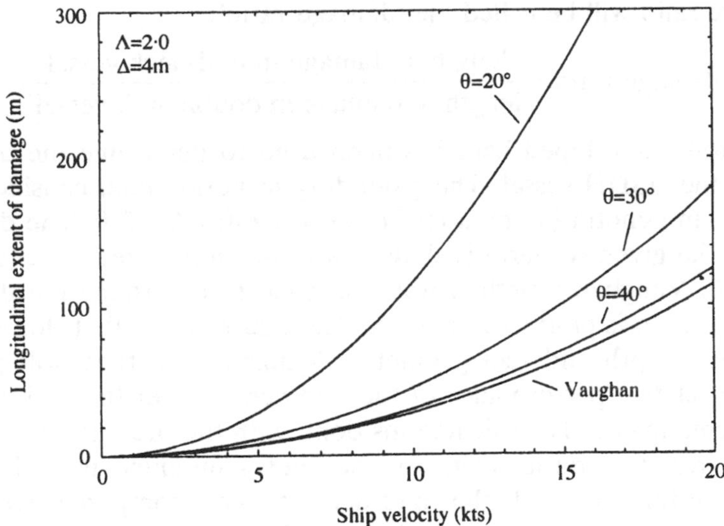

Fig. 15. Length of damage vs. ship speed for various reef wedge semi-angles.

model has dependence on θ in both the plate tearing and plate cutting forces explained in Section 3.2. As a result, when smaller wedge semi-angles are considered ($\theta < 30°$), Vaughan's method substantially underpredicts the damage.

4.5 Application to Comparative Studies

Although the model presented here has been formulated for a USDH tank vessel, the same principles and basic equations for plate cutting force, plate tearing force, and girder tearing force can be applied to any hull configuration. For a conventional double-hull tanker, for example, the longitudinal stiffeners can be 'smeared' to equivalent thicknesses and the model applied as described above. For a single-hull vessel, the formulation is identical but somewhat less complicated since the hull cell aspect ratio s_1/h does not enter. The damage prediction model presented here can, then, be used for comparative studies similar to those reported in Refs 1 and 2. When this is done, the shortcomings of Vaughan's method for the purpose of conducting comparative studies becomes more apparent.

Consider, for example, a comparison of a mid-deck tanker with a double-hull tanker for the purpose of determining relative performance in grounding. It is the ratio of the length of damage in the mid-deck tanker to the length of damage in the double-hull tanker (for a given grounding scenario) that is critical to the comparative study. For clarity, the above ratio will be called the 'damage ratio':

$$\text{damage ratio} = \frac{\text{length of damage in mid-deck vessel}}{\text{length of damage in double hull vessel}}$$

The model developed here has been used to determine the damage ratio for the USDH vessel. The grounding scenario under consideration involves 4 m penetration by a reef of aspect ratio $\Lambda = 2{\cdot}0$. The damage ratio for the given scenario is shown as a function of reef wedge angle in Fig. 16. Vaughan's method was also used to determine the damage ratio, and it too is plotted on Fig. 16. The figure shows that, for a given penetration depth and reef geometry, Vaughan's method will predict the same relative performance of the two vessels over the entire range of reef semi-angles. The calculations done with the present model show that this result is misleading. As the reef semi-angle is varied, the relative performance of the mid-deck tanker compared with the double-hull tanker (in terms on length of damage) changes by up to 25%, depending on the specific reef geometry and penetration depth examined.

It should be mentioned here that the recent comparative studies conducted for the IMO employ Vaughan's method, and that the results of those studies should be considered valid only for a small range of reef wedge semi-angles, $\theta \approx 45°$.

The general trend holds over the entire range of penetration depths, reef aspect ratios, and hull cell ratios. When smaller reef wedge

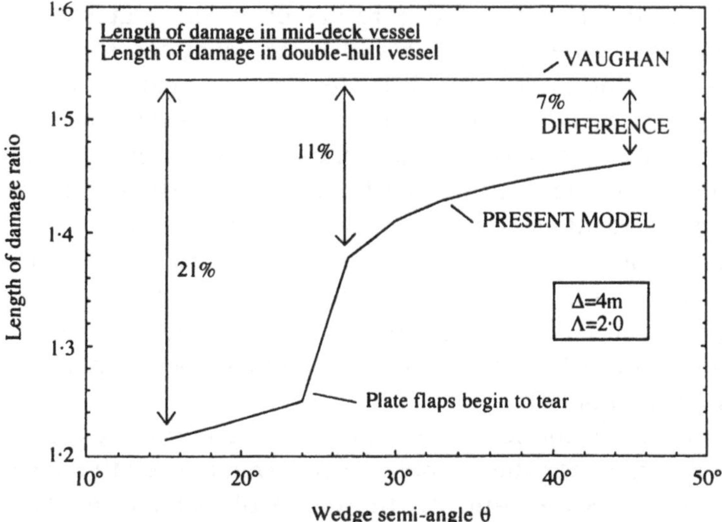

Fig. 16. Relative performance of a mid-deck tanker compared with a double-hull vessel as predicted by Vaughan and by the present model.

semi-angles of less than 40° are considered, Vaughan's method not only overpredicts absolute performance, it also incorrectly predicts relative performance. For grounding scenarios in which the plate cutting and tearing modes are significant relative to the total energy absorption (for example, low-aspect-ratio reefs and small penetration depths) the present model will yield a substantially more realistic estimate of the damage ratio.

The damage prediction model developed in this chapter offers significant improvement over previous damage prediction methods for the analysis of groundings on narrow, low-aspect-ratio reefs that involve relatively low dimensionless penetration depths. Analysis of groundings on wide (high reef wedge angle) or blunt (high reef aspect ratios) reefs requires a different type of model, since the primary energy dissipation mode in these scenarios does not include plate cutting or tearing, but is dominated by girder crushing, plate buckling, and plate shearing. This problem is addressed in subsequent sections.

5 CRUSHING DAMAGE OF WEB GIRDERS

In previous sections, the cutting and tearing failure of hull plating, and the attached longitudinal stiffeners were considered. In grounding and collision events, a major part of the hull girder resistance comes from local crushing of bulkheads, transverse frames, or longitudinal web girders. The problem of ultimate strength of web girders has been extensively studied for civil engineering and naval architecture applications. However, the post-failure behavior of structural members under localized in-plane load is not yet well understood. Recently Culbertson-Driscoll[17] and Wierzbicki and Culbertson–Driscoll[40] developed a simple computational model model for the crushing resistance of a web girder, and correlated the theory with some limited test results.

A typical structural unit considered in the abovementioned studies consisted of a web girder of depth H and a portion of an attached plating of breadth B (Fig. 17). The applied 'knife' load P causes the upper edge of the web to indent a distance Δ.

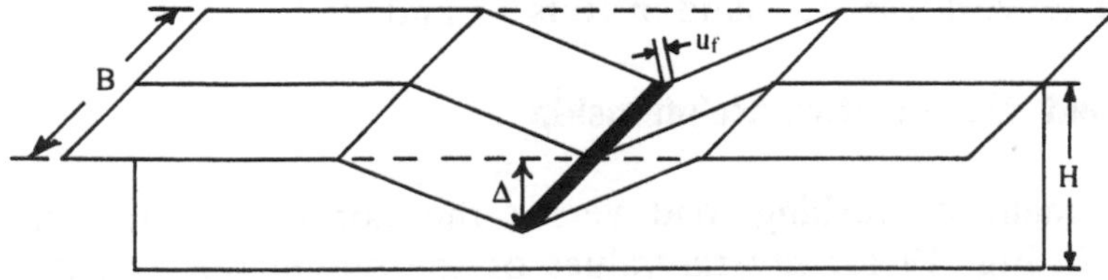

Fig. 17. Geometry of the deformed flange of a web girder.

5.1 Equilibrium

The solution of the stated problem is obtained by choosing a kinematically admissible field of displacements Δ, displacement rates $\dot{\Delta}$, strain rates $\dot{\varepsilon}_{\alpha\beta}$ and rotation rates $\dot{\theta}$ and applying the principle of virtual velocities. Note that the velocity on the left-hand side of eqn (3) should now be replaced by $\dot{\Delta}$.

Equations (3)–(5) incorporate the assumption that the plastic bending resistance is decoupled from the membrane resistance. Furthermore, the bending deformations are concentrated only in the plastic hinge line, so that the continuous deformation term on the right-hand side of eqn (4) vanishes. To further simplify the calculations, only the transverse component of the membrane strain rate (perpendicular to the load direction) is considered.

5.2 A Simplified Kinematic Model

Based on the observation of failure modes in the scale-model experiments reported in Ref. 17, a stationary hinge collapse model for the structural unit was developed. The model is shown in Fig. 18 in the undeformed and deformed stages. Six stationary hinge lines define the boundaries of triangular plates that rotate with respect to each other and deform as the indentation depth Δ increases. The first hinge line forms at the junction between the web and the upper flange. The second hinge line forms at an angle α below the first. The third line forms at an angle β below the second. The angle between the first and third hinge lines is taken as a process parameter ϕ. Experiments showed the length of the deformed region in the longitudinal direction to be approximately twice the length in the vertical direction. This characteristic was included in the model by fixing the angle $\alpha + \beta = \frac{1}{4}\pi$. The horizontal length of the deformed zone is taken as a free parameter ξ. As the indentation distance Δ increases, the second hinge line moves out of the plane, causing membrane deformations in the web. The shaded area in Fig. 18 illustrates the 'amount' of stretch in the deforming web. Specific expressions for the rate of bending and membrane work can be found in Refs 17 and 40.

5.3 Load–Deformation Relationship

The normalized crushing load versus the normalized indentation is plotted in Fig. 19 for several values of the parameter ξ in the range $\xi^* = \xi/H = 0{\cdot}1$–$1{\cdot}0$. The dimensions of the structural unit were $H/t_w =$

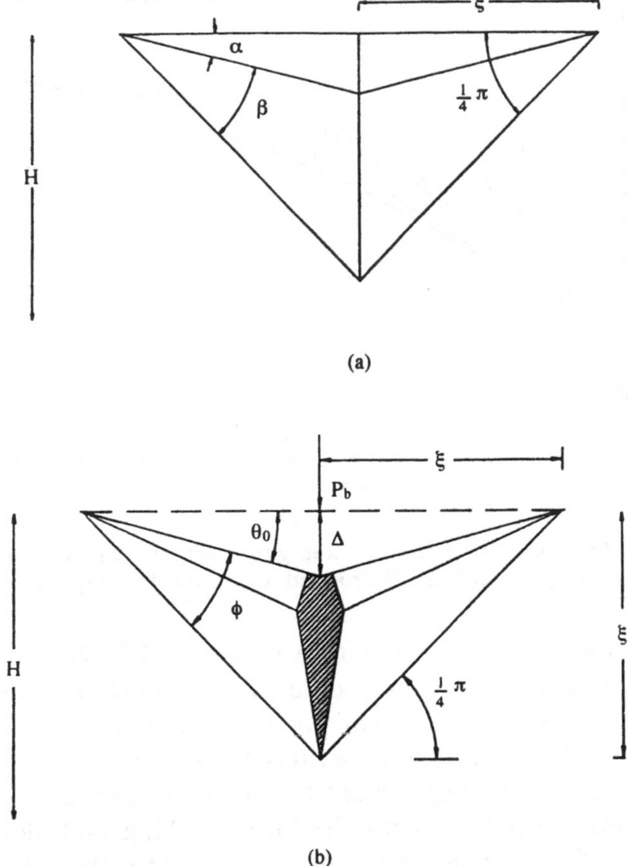

Fig. 18. Initial (a) and deformed (b) geometries of the web girder.

100, $B/t_p = 75$, $t_p/t_w = 1$, and $\alpha = \frac{1}{8}\pi$. It can be seen that the actual solution is an envelope of the family of curves, each corresponding to a different ξ^*. The extent of plastic deformation in the lateral deflection is thus an increasing function of Δ. In this respect, the response of the web girder is qualitatively similar to the behavior of tubes under lateral concentrated force, where a similar effect was observed.[41]

5.4 Comparison with Tests

When the vertical extent of damage matches the depth of the web girder, $\xi = H$, a modified mechanism is found that allows for overall rotation of the two areas of the girder without additional extension of the flange. This type of deformation is referred to as a generalized plastic 'hinge'. Three such hinges form a one-degree-of-freedom beam collapse mechanism. The reader is referred to Ref. 17 for details.

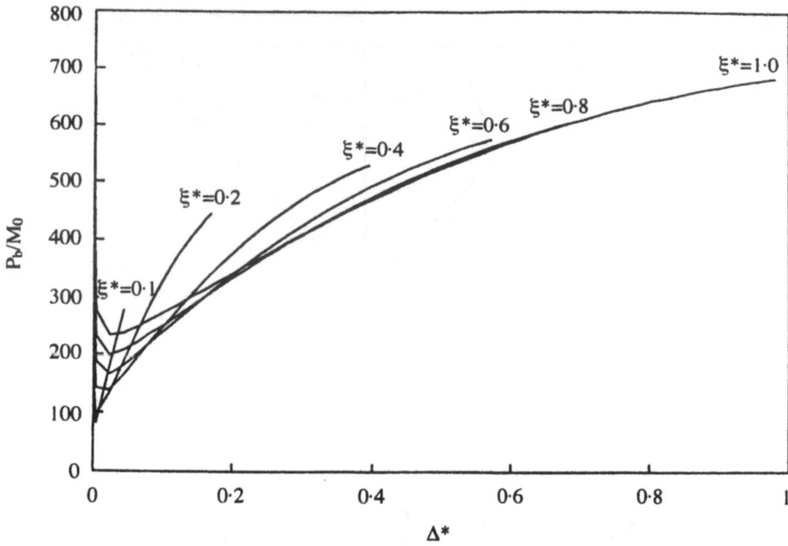

Fig. 19. Load–displacement curve for a web girder with growing deforming region as an envelope of curves constructred for fixed deforming regions.

A limited experimental program was reported in Ref. 17, in which double-cell extruded aluminum profiles were subjected to local crushing or three-point bending. The load was introduced through a long rigid cylindrical punch. The two experimental configurations along with photographs of damaged specimens are shown in Fig. 20.

The theoretical prediction for the local crushing and global bending collapse models are compared in Fig. 21 with the experimentally measured load–deflection curve. The present solution initially overestimates the crushing resistance of the web girder, but later follows the experimental trend very closely.

The application of the present simplified kinematic model to the determination of grounding resistance of hull girders is given in Section 6.

6 STRANDING DAMAGE OF USDH SHIPS

The resistance of a hull girder to a purely vertical load was analyzed by Ueda,[18] Ito *et al.*[19] and Amdahl.[20] The objective of such an analysis is twofold. First, it provides an estimation of the amount of damage to the hull in a stranding situation, when a reef is pushed up into the bottom of a ship. Secondly, the continuous damage model usually involves horizontal and vertical components of resisting forces. The solution to

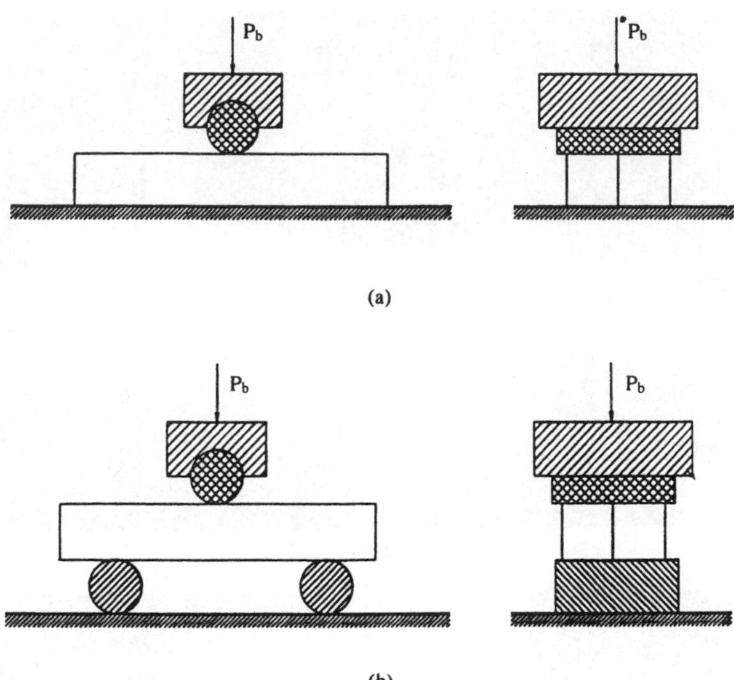

Fig. 20. Set-up for denting (a) and bending (b) experiments on a web girder. Photographs of dented (c) and bent (d) specimens.

the problem of vertical penetration gives important insight into the mechanics of an actual grounding event.

McKenney[11] developed a simplified computational model for stranding damage of USDH ships and studied the effects of penetration depth, slope, and spread angles of a reef on the total resisting force. This model and some representative results are briefly discussed here.

6.1 Failure Geometry

A sequence of failure patterns of the hull girder is shown in Fig. 22. The global geometry of the USDH is characterized by the hull separation and the cell width c. The reef is assumed to be a diamond shape with a slope angle α and spread angle β.

The reef comes first in contact with the outer plate. A diamond shape deformation zone is assumed, as shown in Fig. 22(a). The cell width c is the observed width for the initial plate bending. The length L is determined through minimization of membrane and bending energy.

The initial phase continues until the outer plate ruptures or the first set of girders is reached. After the diamond section of the plate

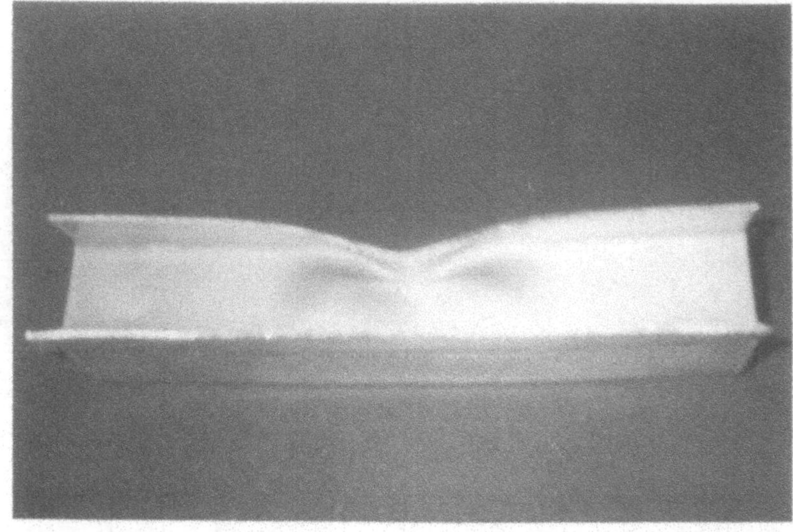

(c)

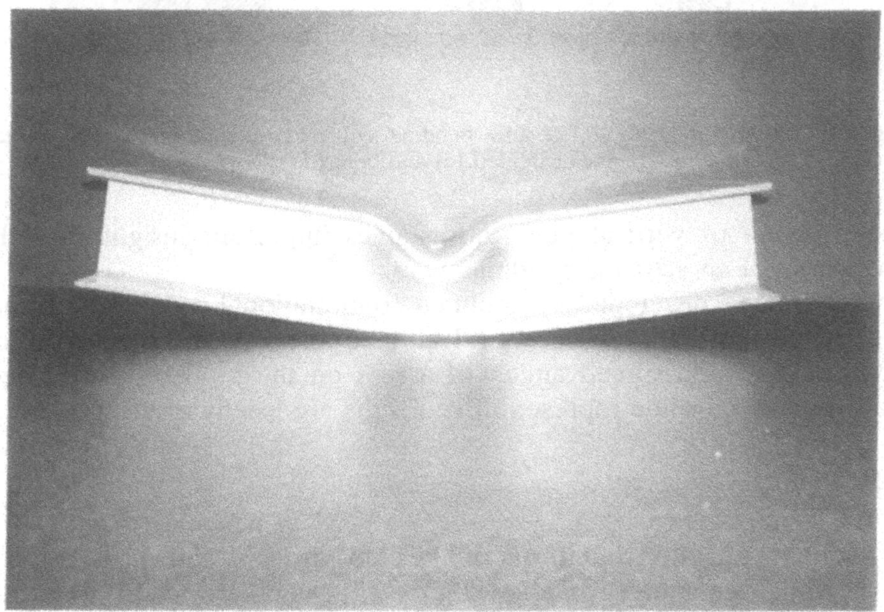

(d)

Fig. 20 (*continued*).

ruptures, it no longer contributes any membrane energy. The ruptured outer plate is shown in Fig. 22(b).

Once the first set of girders has been reached by the rock, a second phase begins. A second set of hinges in the outer plate are established, extending out to the next set of girders as shown in Fig. 22(c). The

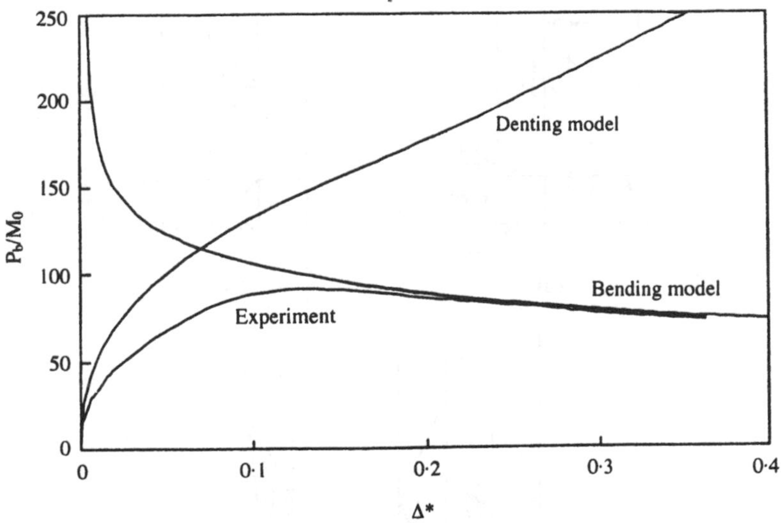

Fig. 21. Correlation of the denting and bending solutions with experiments.

girders are deformed as shown in Fig. 22(d). This phase continues until a second rupture occurs, as shown in Fig. 22(e), the next set of girders are reached by the sides of the reef, or the inner plate is reached by the apex of the reef, as shown in Fig. 22(f), depending on β, c, and h.

The inner plate deforms and ruptures in a manner similar to the outer plate, with one exception. For moderate angles β, the rock will already be pressing against the first set of girders when the top of the rock reaches the inner plate. It is observed that this results in the first set of hinges in the outer plate forming a diamond limited by the second girder position. This pattern of the hinge extending out to the next set of undeformed girders continues as the damage extends laterally over n cells. Figure 22(g) shows the initial inner plate rupture over three cells.

6.2 Rupture of Plating

The above kinematics of the indentation process can be combined with a simple fracture criterion to predict the onset of rupture in the outer and inner plating. The tensile membrane strain in the theory of the moderately large deflection of plates is defined as

$$\varepsilon = \frac{1}{2}\left(\frac{dw}{dy}\right)^2 \tag{20}$$

where w is the out-of-plane deflection and y is the plate coordinate in the transverse direction. It is assumed that fracture occurs when the

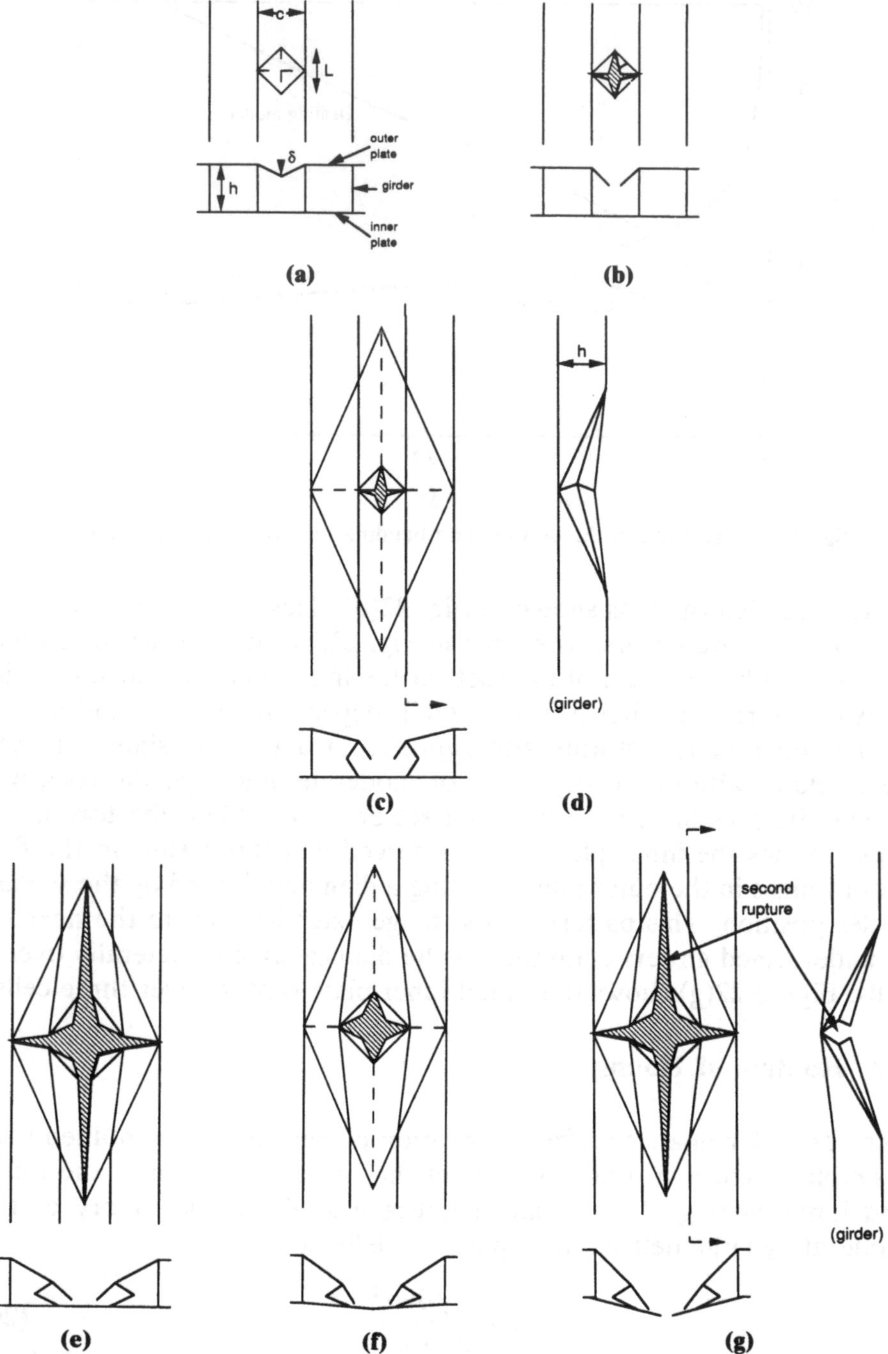

Fig. 22. Progression of a damaged zone in a USDH: (a) local denting; (b) first rupture; (c,d) spread of denting deformation to the neighboring cells; (e) second rupture of outer plate; (f) inner plate deformation; (g) inner plate rupture.

maximum strain reaches the critical value, $\varepsilon = \varepsilon_{cr}$. The critical spread angle of the reef causing fracture can be shown to be

$$\beta_{cr} = \arctan \sqrt{2\varepsilon_{cr}} \tag{21}$$

For example, for $\varepsilon_{cr} = 0{\cdot}1$,[42] $\beta_{cr} = 24°$. If β is less than this value, rupture of the outer plate due to membrane straining will never occur, since the hinge position in the USDH ships will continue to move to each successive set of girders before β_{cr} is reached. In reality, failure of the outer plating will occur for smaller critical angles when the bending component of strains are added as shown by Jones.[43]

Performing a similar analysis for the inner plate, it was found[11] that for a square-cell hull, $c = h$,

- a very steep reef $\beta > 68°$ (top spread angle of reef less than 44°) will rupture over once cell;
- a moderately steep reef $44° < \beta < 68°$ ($44° <$ top angle $< 84°$) will rupture over three cells;
- a gradually sloped reef $\beta < 44°$ (top angle $> 84°$) will not rupture the inner plate.

The above results, although obtained through elementary calculations, are of great practical importance because they define the envelope of reef angles that distinguishes between the purely crushing deformation model of hull (no fracture), the fracture of outer hull only, and the fracture of both inner and outer hulls. It is important to note that the critical failure angle depends on the critical strain to rupture ε_{cr} and the cell aspect ratio h/c rather than on hull separation h or cell width c alone.

6.3 Hull Resisting Force

The force–deflection relationship for the USDH can be obtained using the principle of virtual velocity, where the horizontal wedge velocity V in eqn (3) is replaced by the vertical rate of deflection $\dot{w}$. The rate of bending and the membrane energies are calculated using the failure model of a web girder, explained in Section 5. The kinematics of the indentation process involves one unknown parameter, the longitudinal extent of deformation L. Following the procedure developed in Ref. 41, the parameter L is determined from the minimization of the resistance force, for a given lateral extent of damage nc, where n is an integer. Representative force–deflection curves for three different reef spread angles are shown in Figs 23–26.

Figure 23 shows the force-deformation relationship for a USDH and

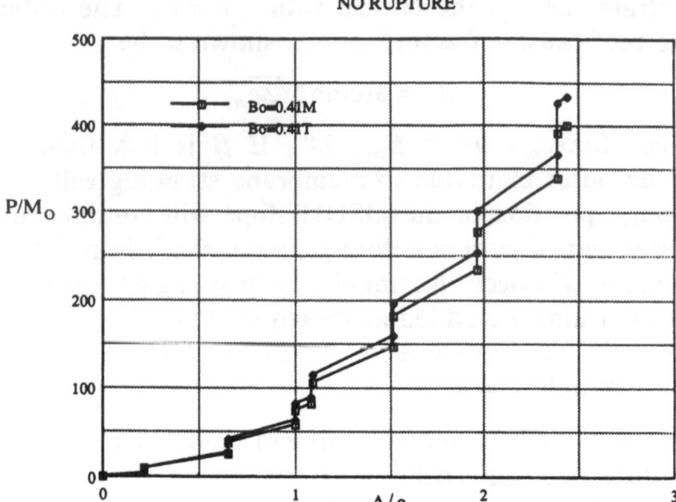

Fig. 23. Predicted force–indentation curves for a reef spread angle $\beta_0 = 0{\cdot}41$ rad.

a reef with a small (gradual) spreading angle. There is a steady step-like increase in force with increasing penetration as each additional set of cells is reached. There is an additional step increase at $\Delta/c = 1$ when the inner plate is reached. No rupture occurs.

Figure 24 shows the force–deformation relationship for a USDH and a reef with a moderate spreading angle. A sudden loss of strength is noted when the outer plate ruptures, and again when a second rupture

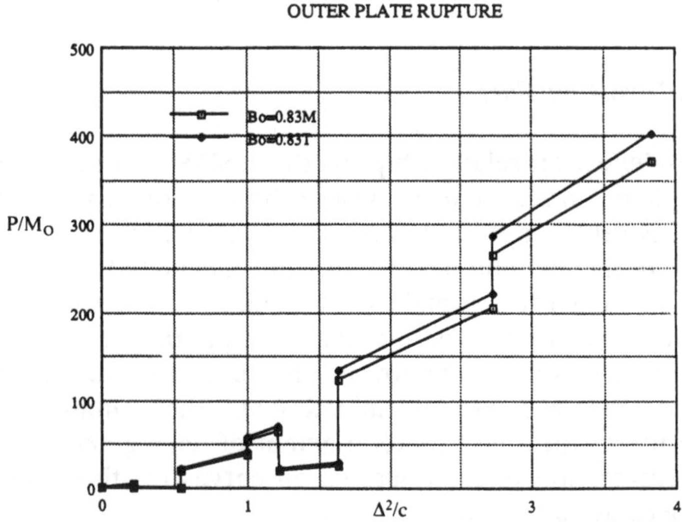

Fig. 24. Predicted force–indentation curves for a reef spread angle $\beta_0 = 0{\cdot}83$ rad.

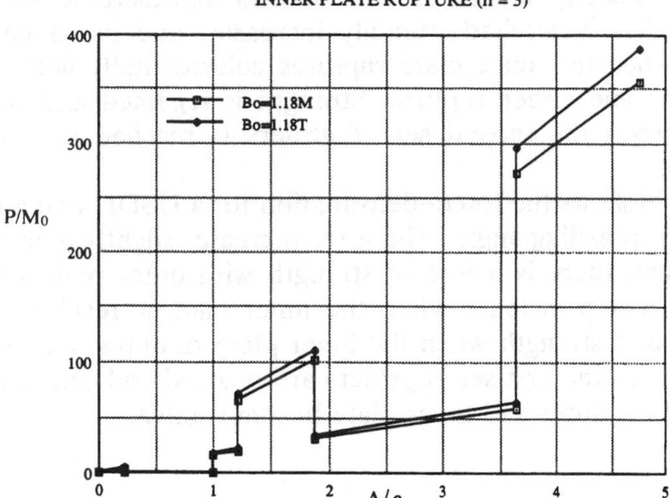

Fig. 25. Predicted force–indentation curves for a reef spread angle $\beta_0 = 1{\cdot}18$ rad.

of the outer plate and the first set of girders occurs. This pattern of step increases and sudden losses of strength will continue as each set of girders is reached and ruptures.

Figure 25 shows the force–deformation relationship for a USDH and a reef with a steep spreading angle. The force increases slightly with outer plate deformation, there is a loss of strength with outer plate rupture, the force has a step increase when the inner plate is reached

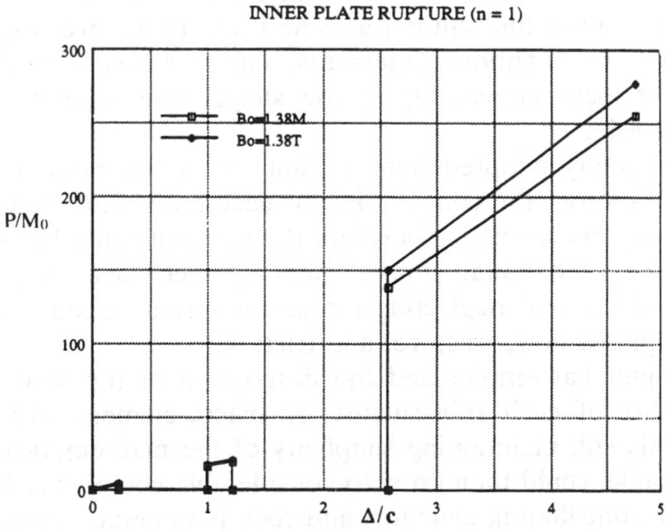

Fig. 26. Predicted force–indentation curves for a reef spread angle $\beta_0 = 1{\cdot}38$ rad.

($\Delta/c = 1$), steadily increases, has a second step increase when the first set of girders is reached, steadily increases, and has a large loss of strength when the inner plate ruptures coincidentally with the second outer plate and girder rupture. Strength is regained and continues to increase when the second set of girders is reached and the pattern continues.

Figure 26 shows the force–deformation for a USDH and a reef with a very steep spreading angle. The force increases slightly with outer plate deformation, there is a loss of strength with outer plate rupture, the force has a step increase when the inner plate is reached, there is a second loss of strength when the inner plate ruptures, and a large step increase when the first set of girders are reached and consequently new regions of the outer and inner plate become active.

7 CONCLUSIONS

Two damage prediction models have been developed for ships undergoing a grounding accident. The first describes the longitudinal cutting of a single or double hull by a sharp rock. Also, a wider and blunter rock shape is considered, in which case there could be considerable crushing of the supporting structure. The second model applies to the vertical indentation of the hull by a rigid obstacle.

A ship's hull is a complex structural system consisting of plating attached through weldments to longitudinal stiffeners, transverse frames, bulkheads, and web girders (in the case of double-hull ships). Each component of the hull deforms and fails in its own way, and also interacts with its neighboring elements. The ship is subjected to a wide spectrum of loads depending on the shape, size, and strength of a ground obstacle.

The philosophy adopted here is that, from the point of a global structural response, the contribution of each individual member to the energy absorption is small. Therefore the solution could be constructed by assembling individual simple models, each describing only the first-order effects and neglecting the second-order effects. These main effects are plastic flow, fracture, and friction.

This chapter has emphasized the description of the load–deflection characteristics of each substructure or macro-element with sufficient realism while still maintaining simplicity of the mathematical solution. A global model could then grow to considerable complexity if sufficient numbers of contributing elements and rock parameters were used. The present methodology has been proven useful for predicting the longitudinal, lateral, and vertical extent of damage to oil tankers and other ships.

An important limitation of the presented method is that coupling between the longitudinal and vertical damage of the hull has not been considered. This problem will be addressed in future research, both at the level of an individual member and the overall lifting and ride-over kinematics of the ship.

ACKNOWLEDGEMENT

This work was supported in part by the Joint MIT–Industry Consortium on Tanker Safety. Helpful discussions on various aspects of the grounding problem with Frank McClintock, Chris Calladine, Tony Atkins, Norman Jones, Jennifer Culbertson-Driscoll, and Steve Price are gratefully acknowledged.

REFERENCES

1. Kohler, P. E. *et al.*, Potential oil spill from tankers in case of collision and/or grounding—a comparative study of different VLCC designs. DnVC Report 90-0074, May 1990.
2. Snyder, B. & Mcafee, D., Collision assessment of double hull tankers. DTRC Report SSPD-92-173, January 1992.
3. Report of the RR701 Committee to the Maritime Environmental Protection Committee of the IMO, Shipbuilding Research Association of Japan, 1978.
4. Vaughan, H., Bending and tearing of plate with application to ship-bottom damage. *The Naval Architect* **97** (May 1978).
5. Vaughan, H., The tearing and cutting of mild steel plate with application to ship grounding damage. In *Proceedings of 3rd International Conference on Mechanical Behavior of Materials,* Vol. 3., ed. K. F. Miller & R. F. Smith. Pergamon Press, Oxford, 1979, pp. 479–87.
6. Vaughan, H., The tearing of mild steel plates. *J. Ship Res.*, **24** (1980) 96.
7. Vaughan, H., Damage to ships due to collision and grounding. Det norske Veritas Report DnV 77–345.
8. Kuroiwa, T., Private Communication, April 1992.
9. Wierzbicki, T., Rady, E., Peer, D. & Shin, J. G., Damage estimates in high energy groundings of ships. Joint MIT–Industry Program on Safe Tankers, Report 1, June 1990.
10. Wierzbicki, T., Peer, D. & Rady, E. The anatomy of tanker grounding. Joint MIT–Industry Program on Safe Tankers, Report 2, May 1991.
11. McKenney, T. L., Grounding resistance of unidirectionally stiffened double hulls. Joint MIT–Industry Program on Safe Tankers, Report 3, May 1991.

12. Peer, D. B., Coupling of global motion and local deformation in tanker grounding. Joint MIT–Industry Program on Safe Tankers, Report 4, May 1991.
13. Thomas, P. & Wierzbicki, T., Performance criteria vs design standards for commercial tank vessels. Joint MIT–Industry Program on Safe Tankers, Report 5, October 1991. Presented at 5th International Symposium on Practical Design of Ships and Mobile Units, May 1992.
14. Wierzbicki, T. & Thomas, P., Closed-form solution for wedge cutting force through thin metal sheets. Joint MIT–Industry Program on Tanker Safety, Report 6, March 1992.
15. Lu, G. & Calladine, C. R., On the cutting of a plate by a wedge. *Int. J. Mech. Sci.*, **32** (1990) 295–313.
16. Price, S. R., Plastic shear buckling of ship hull plating induced by grounding. Joint MIT–Industry Program on Tanker Safety, Report 9, May 1992.
17. Culbertson–Driscoll, J., Crushing characteristics of web girders in unidirectionally stiffened double hull structures. Joint MIT–Industry Program on Tanker Safety, Report 10, May 1992.
18. Ueda, K. *et al.*, Ultimate strength of the double bottom stranding on a rock.
19. Ito, H. *et al.*, A simplified method to analyze the strength of double hulled structures in collisions. *J. Soc. Naval Architects Japan.*
20. Amdahl, J., Analysis and design of ship structures for grounding collision. In *Practical Design of Ships and Mobile Units*, Vol. 2. Elsevier Applied Science Publishers, London.
21. Jones, N. & Jouri, W. S., A study of plate tearing for ship collision and grounding damage. *J. Ship Res.*, **31** (1987) 253.
22. Akita, Y., Ando, N., Fujita, Y. & Kitamura, K., Studies on collision-protective structures in nuclear powered ships. *Nucl. Engng Des.*, **19** (1972) 368.
23. Akita, Y. & Kitamura, K., A study on collision of elastic stem to a side structure of ships. *J. Soc. Naval Architects Japan,* **131** (1972) 307.
24. Minorsky, V. U., An analysis of ship collisions with reference to protection of nuclear power plants. *J. Ship Res.* **3** (1959) 1.
25. Wosin, G., Comments on Vaughan: The tearing of mild steel plates. *J. Ship Res.*, **26** (1982) 50.
26. Jones, N., Scaling of inelastic structures loaded dynamically. In *Structural Impact and Crashworthiness,* Vol. 1, ed. G. A. O. Davies. Elsevier Applied Science Publishers, Barking, Essex, 1984, pp. 45–74.
27. Jones, N., Jouri, W. S. & Birch, R. S., On the scaling of ship collision damage. In *Proceedings of 3rd International Congress on Marine Technology, Athens International Maritime Association of East Mediterranean.* Phivos Publishing, Greece, 1984, pp. 287–94.
28. Goldfinch, A. C., Plate tearing energies. Part II Project, Engineering Department, Cambridge University, 1986.
29. Prentice, J., Wedge drop test to investigate plate tearing characteristics. Part II Project, Engineering Department, Cambridge University, 1986.
30. Shu, D., Energy dissipation in cutting plates and splitting tubes. M.Sc. Thesis, Peking University, 1986 (in Chinese).

31. Atkins, A. G., Scaling in combined plastic flow and fracture. *Int. J. Mech. Sci.*, **30** (1988)173.
32. Yu, T. X., Zhang, D. J., Zhang, Y. & Zhou, Q., A study of the quasi-static tearing of thin metal sheets. *Int. J. Mech. Sci.*, **30** (1988) 193–202.
33. Atkins, A. G., Note on scaling in rigid–plastic fracture mechanics. *Int. J. Mech. Sci.*, **32** (1990) 547–8.
34. Atkins, A. G., Letter to the Editor. *Int. J. Mech. Sci.*, **33** (1991) 69–71.
35. Astrup, O., An experimental study on cutting of thick plates by a wedge. DnV–MIT Workshop on the Mechanics of Ship Collision and Grounding, Oslo, 16–17 September, 1992.
36. Thomas P. F., The mechanics of plate cutting with application to ship grounding. Masters Thesis, MIT, May 1992.
37. Baumeister, T., ed., *Mark's Standard Handbook for Mechanical Engineers*, 8th edn. McGraw-Hill, New York, 1978.
38. Parks, D. M., Freund, L. B. & Rice, J. R., Running ductile fracture in a pressurized line pipe. In *Mechanics of Crack Growth*. American Society for Testing and Materials, ASTM STP 590, Philadelphia, 1976, pp. 254–62.
39. Calladine, C. R., Private Communication, February 1992.
40. Wierzbicki, T. & Culbertson-Driscoll, J., Plastic crushing of web girders. *J. Construction Steel Res.* (1993).
41. Wierzbicki, T. & Suh, M. S., Indentation of tubes under combined loading. *Int. J. Mech. Sci.*, **30** (1988) 229–46.
42. McDermott, J. F. *et al.*, Tanker structural analysis for minor collisions. *Trans. SNAME*, **82** (1974) 382–414.
43. Jones, N., Plastic failure of ductile beams loaded dynamically. *Trans. ASME, J. Engng Ind.*, **98** (1976) 131–6.

INDEX

Milton Keynes UK
Ingram Content Group UK Ltd.
UKHW021920071024
449327UK00022B/1688

9 780367 864743